Characterization of Metal and Polymer Surfaces

VOLUME 1   Metal Surfaces

ACADEMIC PRESS RAPID MANUSCRIPT REPRODUCTION

Characterization of Metal and Polymer Surfaces

Volume 1 Metal Surfaces

EDITED BY

LIENG-HUANG LEE

Xerox Corporation
Rochester, New York

ACADEMIC PRESS, INC. New York San Francisco London 1977
A Subsidiary of Harcourt Brace Jovanovich, Publishers

COPYRIGHT © 1977, BY ACADEMIC PRESS, INC.
ALL RIGHTS RESERVED.
NO PART OF THIS PUBLICATION MAY BE REPRODUCED OR
TRANSMITTED IN ANY FORM OR BY ANY MEANS, ELECTRONIC
OR MECHANICAL, INCLUDING PHOTOCOPY, RECORDING, OR ANY
INFORMATION STORAGE AND RETRIEVAL SYSTEM, WITHOUT
PERMISSION IN WRITING FROM THE PUBLISHER.

ACADEMIC PRESS, INC.
111 Fifth Avenue, New York, New York 10003

United Kingdom Edition published by
ACADEMIC PRESS, INC. (LONDON) LTD.
24/28 Oval Road, London NW1

Library of Congress Cataloging in Publication Data

Symposium on Advances in Characterization of Metal
    and Polymer Surfaces, New York, 1976.
    Characterization of metal and polymer surfaces.

    Sponsored by the Division of Organic Coatings
and Plastics Chemistry, the Division of Analytical
Chemistry, and Cellulose, Paper, and Textile
Division of the American Chemical Society.
    Includes bibliographical references and indexes.
    CONTENTS: v. 1.   Metal surfaces.
    1.   Metallic surfaces—Congresses.   2.   Polymer
surfaces—Congresses.   3.   Spectrum analysis—
Congresses.   I.   Lee, Lieng-Huang,   Date
II.   American Chemical Society.   Division of
Organic Coatings and Plastics Chemistry.
III.   American Chemical Society.   Division of
Analytical Chemistry.   IV.   American Chemical
Society.   Cellulose, Paper, and Textile Division.
V.   Title.
QD506.A1S95   1976   541'3453   77-2255
ISBN 0–12–442101–6 (v. 1)

PRINTED IN THE UNITED STATES OF AMERICA

Contents

Contributors	ix
Preface	xi
Contents of Volume 2	xiii

Centennial Tribute: Surface Science and Polymer Technology LIENG-HUANG LEE	1

PART I: Atom-Probe and Mössbauer Spectroscopy

Plenary Lecture: Surface Analysis at Atomic Level Using the Atom-Probe ERWIN W. MÜLLER AND S. V. KRISHNASWAMY	21
Applications of Mössbauer Spectroscopy to the Study of Corrosion G. W. SIMMONS AND H. LEIDHEISER, Jr.	49
Characterization of Bulk and Surface Properties of Heterogeneous Ruthenium Catalysts by Mössbauer and ESCA Techniques C. A. CLAUSEN, III, AND M. L. GOOD	65
Discussion	99

PART II: Auger Electron Spectroscopy and Electron Microprobe

Introductory Remarks M. L. GOOD	103
Plenary Lecture: Low Energy Electrons as a Probe of Solid Surfaces ROBERT L. PARK, MARTEN DEN BOER, AND YASUO FUKUDA	105
Surface Characterization by Electron Microprobe IAN M. STEWART	127

Auger Electron Spectroscopy of Solid Surfaces   133
    J. T. GRANT

A Study of Passive Film Using Auger Electron Spectroscopy   155
    C. E. LOCKE, J. H. PEAVEY, O. RINCON, AND M. AFZAL

Discussion   181

PART III: Low Energy Electron Diffraction

Introductory Remarks   185
    D. L. ALLARA

Plenary Lecture: LEED Studies of Surface Layers   187
    PEDER J. ESTRUP

The Use of Direct Methods in the Analysis of LEED   211
    DAVID L. ADAMS AND UZI LANDMAN

Surface Structure by Analysis of 'LEED' Intensity Measurements   271
    P. M. MARCUS

Computation Methods of LEED Intensity Spectra   299
    N. STONER, M. A. VAN HOVE, AND S. Y. TONG

Discussion   347

PART IV: Secondary Ion Mass Spectrometry

Plenary Lecture: Ion Microscopy and Surface Analysis   351
    G. H. MORRISON

Surface Characterization by Ion Microprobe Analyzer   367
    IAN M. STEWART

Study of Adhesive Bonding and Bond Failure Surface Using ISS-SIMS   375
    W. L. BAUN

Discussion   391

PART V: Photoelectron and Electron Tunneling Spectroscopy

Introductory Remarks 397
  RUTH ROGAN BENERITO

Plenary Lecture: Surface Characterization Using Electron Spectroscopy (ESCA) 399
  DAVID M. HERCULES

Photoemission Study of Chemisorption on Metals 431
  THOR RHODIN AND CHARLES BRUCKER

The Study of Organic Reactions on the Surface of Magnetic Pigments by X-Ray Photoelectron Spectroscopy (ESCA) 467
  ROBERT S. HAINES

Molecular Spectroscopy by Inelastic Electron Tunneling 477
  KENNETH P. ROENKER AND WILLIAM L. BAUN

Discussion 495

About Authors *499*
Author Index *501*
Subject Index *509*

Contributors

DAVID L. ADAMS   University of Aarhus, Aarhus, Denmark
M. AFZAL   Electrocast Steel Company
D. L. ALLARA   Bell Laboratories, Inc.
W. L. BAUN   Air Force Materials Laboratory, Wright-Patterson AFB
RUTH ROGAN BENERITO   Southern Regional Research, Center, USDA
CHARLES BRUCKER   Cornell University
C. A. CLAUSEN, III   Florida Technological University
MARTEN den BOER   University of Maryland
PEDER J. ESTRUP   Brown University
YASUO FUKUDA   University of Utah
M. L. GOOD   University of New Orleans
J. T. GRANT   Universal Energy Systems, Inc.
ROBERT S. HAINES   International Business Machines Corporation
DAVID M. HERCULES   University of Georgia
S. V. KRISHNASWAMY   The Pennsylvania State University
UZI LANDMAN   University of Rochester
LIENG-HUANG LEE   Xerox Corporation
H. LIEDHEISER, JR.   Lehigh University
C. E. LOCKE   University of Oklahoma
P. M. MARCUS   International Business Machines Corporation
G. H. MORRISON   Cornell University
ERWIN W. MÜLLER   The Pennsylvania State University
ROBERT L. PARK   University of Maryland
J. H. PEAVY   Motorola Semiconductor Product Division
THOR RHODIN   Cornell University
O. RINCON   University of Zulia, Venezuela
KENNETH P. ROENKER   Thomas More College
G. W. SIMMONS   Lehigh University
IAN M. STEWART   Walter C. McCrone Associates, Inc.
N. STONER   University of Wisconsin, Milwaukee
S. Y. TONG   University of Wisconsin, Milwaukee
M. A. VAN HOVE   University of Wisconsin, Milwaukee

Preface

1976 was the Centennial year of the American Chemical Society. During this Centennial Meeting, the Symposium on Advances in Characterization of Metal and Polymer Surfaces was held to mark the achievements of surface science. Because of its broad appeal, the Symposium was jointly sponsored by the Division of Organic Coatings and Plastics Chemistry, the Division of Analytical Chemistry, the Division of Colloid and Surface Chemistry, and Cellulose, Paper, and Textile Division. A total of eight sessions took place between April 5 and 8 in the Statler Hilton Hotel, New York.

We are grateful to have had as many physicists as chemists speak on various subjects of surface science. As a result, some sessions had lively discussions, and it was not easy to recapture all those words in writing. Nonetheless, we have attempted, to our best ability, to assemble the reviewed papers and most of the discussions in these two volumes. It is probable that one hundred years from now in some corner of the world, these two volumes may still be preserved in book form or on microtapes to mark the historic event of this conference.

Not all papers in these two volumes follow the original order of presentation owing to regrouping under the subtitles *Metal Surfaces* for Volume 1 and *Polymer Surfaces* for Volume 2. The volumes are further subdivided into parts as follows.

Volume 1: METAL SURFACES

    I. Atom-Probe and Mössbauer Spectroscopy
    II. Auger Electron Spectroscopy and Electron Microprobe
    III. Low Energy Electron Diffraction
    IV. Secondary Ion Mass Spectroscopy
    V. Photoelectron and Electron Tunneling Spectroscopy

Volume 2. POLYMER SURFACES

    I. Electron Spectroscopy for Chemical Analysis
    II. Infrared and Laser Raman Spectroscopy
    III. Microscopy for Polymers
    IV. Surface-Chemical and Radiation Analyses

For each part, there is at least one paper to provide an authoritative survey of the subject matter. Other contributed papers then present recent research results related to the same theme. Two postconference contributions have been included, while several original papers were published elsewhere.

We would like to thank Session Chairpersons and all contributors to these two volumes. We would like to acknowledge the Petroleum Research Fund, administered by the American Chemical Society, for assisting our speakers from overseas. We sincerely appreciate the editorial assistance of Robert M. S. Lee of Princeton University.

Lieng-Huang Lee

Contents of Volume 2

PART I: ELECTRON SPECTROSCOPY FOR CHEMICAL ANALYSIS

Introductory Remarks
   D. M. Hercules

Plenary Lecture: Application of ESCA to Structure and Bonding in Polymers
   D. T. Clark

Sputter-Induced Compositional Change during ESCA/Sputtering of Polymers
   Dwight E. Williams and Lawrence E. Davis

Characterization of Chemically Modified Cottons by ESCA
   Donald M. Soignet

Surface Analysis of Plasma Treated Wool Fibers by X-Ray Photoelectron Spectroscopy
   Merle M. Millard

Plasma Modification of Polymers Studied by Means of ESCA
   D. T. Clark and A. Dilks

XPS Studies of Polymer Surfaces for Biomedical Applications
   Joseph D. Andrade, Gary K. Iwamoto, and Bonnie McNeill

Discussion

PART II: INFRARED AND LASER RAMAN SPECTROSCOPY

Introductory Remarks: Surface Characterization of Polymers by Infrared and Raman Spectroscopy
   Lieng-Huang Lee

Plenary Lecture: Transmission and Reflection Spectroscopy, Nature of the Spectra
   N. J. Harrick

The Study of Thin Polymer Films on Metal Surfaces Using Reflection-Absorption Spectroscopy—Oxidation of Poly(1-Butene) on Gold and Copper
  D. L. Allara

Use of Laser Raman Techniques in the Study of Polymers
  R. D. Andrews and T. R. Hart

Discussion

PART III: MICROSCOPY FOR POLYMERS

Introductory Remarks
  L. H. Princen

Microscopical Analysis of Chemically Modified Textile Fibers
  Wilton R. Goynes and Jarrell H. Carra

Laboratory Study of Fiber Fracture Using the Scanning Electron Microscope
  Alfredo G. Causa

The Investigation of Poly(tetrafluoroethylene) Wetting Behavior by Scanning Electron Microscopy
  N. E. Weeks, G. M. Kohlmayr, and E. P. Otocka

Fluoropolymer Surface Studies, II
  David W. Dwight

Structural Characterization of Poly(N-vinyl-carbazole)
  C. H. Griffiths

Discussion

PART IV: SURFACE-CHEMICAL AND RADIATION ANALYSES

Characterization of Latexes by Ion Exchange and Conductometric Titration
  J. W. Vanderhoff

Surface Area of Polymer Latexes by Angular Light Scattering
  Robert L. Rowell and Raymond S. Farinato

Evaporative Rate Analysis: Its First Decade
  John Lynde Anderson

Radiation Absorption for Polymers
  John R. Hallman, J. Reed Welker, and C. M. Sliepcevich

About Authors, Author Index, Subject Index

Centennial Tribute: Surface Science and Polymer Technology

Lieng-Huang Lee

Wilson Center for Technology
Xerox Corporation
Webster, New York 14580

The historical background of the emerging surface science is briefly reviewed. The applications of surface analyses to polymer technology are presented for further discussion during the Centennial symposium.

Tributes are paid to those who discovered or developed surface analytical techniques during the last 100 years.

INTRODUCTION

During this Centennial meeting, several symposia are dedicated to marking the progress of chemistry. In looking back 100 years, there was neither surface science nor polymer technology. Though there were early innovations at the turn of the century, surface science[1] did not grow exponentially until the 1960's, presumably as a byproduct of space research. So far, only one or two surface techniques have been applied to polymer technology. We believe that new applications will be developed in the future which will enable polymer technologists to solve complex materials problems.

In this paper, I shall try first to describe the background of each important innovation of surface analyses. Then I shall discuss possible applications of each technique to polymer technology. I am hoping that this brief introduction can provide our readers with an overview of the emerging surface science presented to this Symposium. Furthermore, we can pay tribute to those early pioneers who contributed significantly to the progress of surface science during the last 100 years.

HISTORICAL BACKGROUND

Early Innovations

In the Middle Ages, the most important analytical tool alchemists relied upon was wet chemistry.[2] Presumably, the first physical instrument was the early spectrometer developed in 1860 by Gustav R. Kirchoff and Robert W. Bunsen.[3] Since the discovery of X-rays by Wilhelm C. Röntgen[4] in 1895, chemists inherited another spectroscopic method to study the structure of matter. In 1912, Max. V. Laue[5] discovered the diffraction of Röntgen rays by crystals. One year later, William H. Bragg[6] built the first X-ray spectrometer for the study of crystal structure. In the same year, H.G.J. Moseley[7] established the basis for qualitative and quantitative X-ray spectrochemical analysis. But it was X-ray diffraction that became the first important analytical tool for surface analysis of polycrystalline materials. Today, the use of X-ray beams for surface analysis has surpassed the diffraction method, and the scope of these new techniques will be discussed throughout this Symposium.

Besides X-rays, ion beams were also explored for surface analysis. In 1905, J.J. Thompson,[8] the discoverer of the electron, built the first mass spectrometer for the study of the electrical conduction of gases at reduced pressure. Later, in the 1920's, F.W. Aston[9] constructed a new mass spectrometer to give high sensitivity. The simplified spectra were formed by bringing all ions of a given m/e to a common focus on a photographic plate. However, not until 1945 was mass spectrometry widely recognized to be a versatile analytical tool for both surface and bulk of materials.

Field Ion Microscopy

One of the important methods involving the use of ions is field ion microscopy (FIM). The story of FIM is also the story of a devoted scientist, Dr. Edwin W. Müller,[10] of the Pennsylvania State University. Interestingly, the original innovation, field emission microscopy, did not involve ions. In 1936, Müller[11] applied a large electrostatic field, approximately 30 million V/cm to a cold cathode so that electrons could tunnel from the solid through the forbidden barrier in the vacuum. The emitter was a sharp point (about 1000Å in radius) and the enlarged image (10^6 times) of the spatial distribution of the tunneling electrons could be projected onto a fluorescent screen. This work formed the foundation of field emission microscopy.

In 1951, Müller[12] used hydrogen ions as the imaging particles. Field ion microscopy achieved a resolution approaching the atomic scale, about 3-6Å instead of 20Å by the field emission technique. Later, the discovery of field evaporation[13] in addition to field ionization led to the development of field ion microscopy (FIM). By 1960, most of experimental techniques were well developed to give high-quality images of lattice defects. Only then did FIM start to be pursued by other laboratories as a research tool.

In 1968, Müller et al.[14] combined FIM and a time-of-flight mass spectrometer to form the atom-probe. Recently, the atom-probe incorporated a channel-plate converter together with improved electronics to become atom-probe FIM. This interesting story will be told to this Symposium by Dr. Müller and his co-worker.[15]

Mössbauer Spectroscopy

A remarkable post World War II development was the flourishing research in nuclear phenomena. In 1958, Rudolf L. Mössbauer[16] during his graduate work at Heidelberg, discovered the gamma-resonance effect, i.e., the recoil-free emission and resonant absorption of nuclear gamma rays in solids. Shortly after his discovery, chemists and physicists around the world recognized the importance of the Mössbauer effect in the characterization of surface and bulk of materials. Since then, Mössbauer spectroscopy has been used to study the oxidation and corrosion of metals. In this book, a paper by Simmons and Leidheiser[17] reviews the background of this specific application. The mechanisms of catalytic reactions[18] have also been investigated with the aid of the Mössbauer effect. In view of the significance of Mössbauer's contribution to chemistry and physics, he was awarded a Nobel Prize in Physics in 1961, only three years after his discovery.

Auger Electron Spectroscopy

The 1960's marked the beginning of the space age. As a by-product of space research, ultrahigh vacuum was developed to simulate the outer space environments. Clean surfaces of materials were produced to equal those which would have been found on the moon or other planets. As a direct beneficiary of space research, surface science staged a spectacular growth. In fact, before 1967, there was not a single technique which could be considered to be suitable for the study of clean surfaces.

Since then, several surface techniques [19-22] were revived or perfected ready for practical applications. Among them were Auger electron spectroscopy (AES), low-energy electron diffraction (LEED), scanning electron microscopy (SEM), electron microprobe (EMPA), ion microprobe (IMMA), secondary ion mass spectro-

TABLE I. CHARACTERISTICS OF SURFACE ANALYTICAL TECHNIQUES

SURFACE ANALYTICAL METHOD	ABB'N.	INCIDENCE BEAM BEAM	INCIDENCE BEAM ENERGY (eV)	INCIDENCE BEAM DIAMETER (mm)	EMITTED BEAM OR PARTICLE BEAM OR PARTICLE	EMITTED BEAM OR PARTICLE ENERGY (eV)	TYPICAL BACKGROUND PRESSURE (TORR)
Auger Electron Spectroscopy	AES	Electron	100–5,000	0.1–1.0	Secondary Electron	20–2,000	10^{-9}–10^{-10}
Electron Microprobe Analyzer	EMPA	Electron	500–40,000	10^{-3}–0.25	X-ray, Photon	200–10,000	10^{-5}–10^{-10}
Low Energy Electron Diffraction	LEED	Electron	15–500	1.0	Same Electron as Incident	Within 0.5 ev of Incident Energy	10^{-10}
Scanning Electron Microscope	SEM	Electron	1,000–60,000	0.5×10^{-5}–2.0×10^{-5}	Secondary Electron	0–10	10^{-4}–10^{-9}
Secondary Ion Mass Spectrometry	SIMS	Ar^+ or other ions	100–100,000	10^{-3}–3.0	Secondary Ion From Surface	0–20	10^{-10}
Ion Microprobe Analyzer	IMMA	Ar^+ or other ions	10,000	2×10^{-3}–3.0	Secondary Ion From Surface	0–20	10^{-7}
Electron Spectroscopy for Chemical Analysis	ESCA (XPS)	X-rays	1,000–10,000	1.0–3.0	Electron	0–10,000 (width ~1 ev)	10^{-7}–10^{-10}
Ultraviolet Photoelectron Spectroscopy	UPS	UV Photon	4–40	0.06–0.2	Electron	0–30	10^{-10}–5×10^{-10}

TABLE I. (Cont'd)

TYPE	TARGET ELEMENTAL COVERAGE	SURFACE DESTRUCTION	DISTANCE BELOW SURFACE (Å)	SENSITIVITY (MONOLAYERS)	TYPE OF INFORMATION OBTAINED
Single crystal or polycrystal metal or semiconductor (with or without adsorbate)	Li → U	For some adsorbed layers and insulators	0-10	3×10^{-3}	Elemental composition
Single crystal or polycrystal metal or oxide	B → U	Partly, at low energy in particular	200-20,000	$\sim 1 \times 10^{-2}$	Elemental composition of surface region
Single crystal metal, semiconductor or insulator (with or without adsorbate)	Li → U	For some adsorbed layers & insulators, but not for metals & semiconductors	0-10	2×10^{-3} (if ordered)	Symmetry & lateral atomic spacings of ordered structures in surface region & in adsorbed layers
Almost any material, including polymers		Features sought are not usually destroyed	50-100		Surface topography
Polycrystal or single crystal metal or insulator	H → U	When depth profile needed	1st. exposed layer only at a given moment	10^{-5}	Approximate elemental composition of surface and underlying layers
Polycrystal metal, semiconductor, or insulator	Li → U	Yes	1st. exposed layer only at a given moment	Impurities to parts per billion	Appox. elemental composition as a function of depth(0-1000μ); sputtering rates from < 1 monolayer sec^{-1} to 500Å
Almost all solid materials including polymers & many frozen liquids & gases	Li → U	No	5-20	$\sim 1 \times 10^{-2}$	Elemental composition. Electronic states. Chemical binding. Core level widths.
Polycrystal or single crystal (with or without adsorbate)	Li → U	No	0-30	$\sim 2 \times 10^{-2}$	Electronic states of adsorbed layer & substrate. Work function. Vibration levels.

metry (SIMS), electron spectroscopy for chemical analysis (ESCA), and ultraviolet photoelectron spectroscopy (UPS). AES, LEED, SEM and EMPA all use electrons as the primary beam.

The innovation of Auger electron spectroscopy can be traced back to 1923 when Pierre Auger[23] first studied the photoelectric effect in gases in a cloud chamber. To his excitement, he observed that additional electrons were generated after the initial production of photoelectrons. Those additional electrons were later referred to as Auger electrons. Approximately thirty years later, James Lander[24] identified Auger electrons in secondary electron distributions of bombarded surfaces. However, not unitl 1967 did Lawrence Harris[25] demonstrate that the electron energy distribution could be electronically differentiated to improve and simplify the detectivity of Auger peaks. With this simplification, Auger spectroscopy became a practical method for surface analysis (Table 1). Since then, there have been many publications [26,27] about AES. In this volume, readers may find the papers by Park[28] and by Grant[29] about the principles and applications of AES to be interesting and informative.

Low Energy Electron Diffraction

The development of LEED almost follows a similar pattern - innovation in the Twenties and revival in the Sixties. In 1925, Clinton Davisson and Lester Germer[30] of Bell Telephone Laboratories studied the back-scattering of slow electrons from a polycrystalline nickel target. They accidentally observed that the angular distribution of the electrons changed after the surface was oxidized at high temperatures. Their subsequent studies on electron diffraction led to the confirmation of de Broglie's hypothesis on the wave nature of electrons and to the development of LEED. Harry Farnsworth[31] in 1930 confirmed Davisson and Germer's observations during his study on Cu (100) and Ag (100). However, from then on LEED nearly hibernated for over thirty years.

In the meantime, high energy electron diffraction (HEED) had been widely used since the early work of Thomson.[32] The reasons are simple. Experimentally, HEED requires lower vacuum (10^{-4} to 10^{-5} torr) and simpler electron optics than LEED (Table 1). HEED is generally carried out at an electron energy of > 10 kv, in contrast to LEED at 0.02 to 0.5 kv (Table 1). With these advantages, HEED was developed earlier than LEED for the study of polycrystalline materials in place of X-ray diffraction. HEED has also been useful for the examination of epitaxy and the growth of thin layers of metal or oxides on various substrates.

Since 1959, Germer[33] has improved experimental techniques for LEED, and much tedious work can now be carried out. With ultrahigh vacuum facilities, LEED has become a powerful tool for the study of the surface monolayer of a single crystal, the corrosion of metal and the formation of epitaxial layers. The most important application so far has been the research on the catalyst surface. A review on LEED by Estrup[34] in this book provides us with principles and applications of this technique.

During the last several years, much attention has been given to actual calculations of intensity profiles or spectra. Theoretical difficulties related to the strong scattering of elastic and inelastic electrons of surface atoms had to be overcome. Only recently has surface structure analysis[35] been carried out to some satisfaction. Thus, readers will find in this volume three important papers by Adams,[36] Marcus[37] and Tong et al.[38] specifically dealing with theoretical calculations of LEED.

Scanning Electron Microscopy and Electron Microprobe

The uses of electron beams to probe solid surfaces were established earlier than LEED. The first SEM was built by M.V. Ardenne[39] in 1938. Since 1948, Oatley and his co-workers[40] at Cambridge University continued the studies with SEM; their work finally brought forth the first commercial unit of SEM in 1965. Today, SEM (Table 1) is being used extensively for the study of metal as well as polymer surfaces.

Another important tool using the electron beam is the electron microprobe analyzer (EMPA), which was built first by R. Castaing[41] in 1949. EMPA[42,43] can perform both qualitative and quantitative surface analyses provided a standard is available. EMPA has also been attached to a SEM. It is often used in conjunction with other techniques for thin film studies.

EMPA can determine the chemical composition of a solid specimen weighing as little as 10^{-11} gram and having a volume as small as one cubic micron (Table 1). Thus, the unique characteristic of EMPA is the possibility of obtaining a quantitative analysis of a very small sample, e.g., a pollutant or a dust particle. A selected area of the sample is bombarded with a beam of electrons, and the emitted X-ray spectrum can then be analyzed for the presence of elements and the composition by measuring the intensities of spectral lines. EMPA is therefore suitable for examining single crystals or polycrystals of metals and oxides.

Ion Microprobe and Secondary Ion Mass Spectrometry

Unlike the electron microprobe, there have been different early versions of the ion microprobe analyzer (IMMA), for example, with or without a secondary ion mass spectrometer. With SIMS, R.H. Sloane and R. Press[44] studied in 1938 the sputtering of negative ions upon positive ion bombardment. The microprobe without SIMS was innovated by v. Ardene[45] in 1939 when he focused an ion beam from a canal-ray tube with a single lens to a probe 30 microns in diameter. In 1949, R.F.K. Herzog and F.P. Viehböck[46] of the University of Vienna developed the ion-optical system used for spark-source mass spectrometers and investigated SIMS as a surface characterization technique. In the development of instruments for systematic study of the field, Liebl and Herzog[47] started to build elaborate apparatus in the U.S.A. and Germany. A chronological study of various ion microprobes by Liebl[48] should be consulted for details about the developments.

In principle, sputtering of surface layers by ion beam results in the production of secondary ions.[49-51] Thus, by monitoring the appropriate mass versus time, one or more elements can be depth profiled. The secondary ion images from the lateral elemental distributions can be obtained either directly by the use of stigmatic secondary ion optics as in the ion microscope or by rastering a finely focused ion beam (1-3um) as in the ion microprobe (Table 1).

The ion microscope was developed in 1962 by R. Castaing and G. Slodzian[52] based on the concept of having ion beams to carry the image information. The microscope can display an image of the surface of a substrate with intensities directly related to the concentrations of a given isotope. The image actually viewed or photographed is that produced by tertiary electrons on a fluorescent screen. The details of the ion microscope are described by Morrison[53] in this book.

Photoelectron Spectroscopy

Two of the more recent surface analytical techniques involving photoelectrons are electron spectroscopy for chemical analysis (XPS or ESCA) and ultraviolet photoelectron spectroscopy (UPS). Like AES and LEED, ESCA began in the 1900's. In 1914, H. Robinson and W.F. Rawlinson[54] observed that the primary X-radiation of a sufficiently high frequency gave rise to what might be called a "line spectrum" composed of groups of electrons from different electronic levels. The main thrust of the development since 1946 has come from Kai Siegbahn[55] and his co-workers at the Uppsala University in Sweden. In 1964, Siegbahn[56] finally established ESCA as an important surface analytical method with new instrumentation and ultrahigh vacuum. Today, ESCA is widely used for the characterization of inorganic and organic surfaces including polymers (Table 1). Several reviews[57-59] on ESCA should be interesting reading for those who want to be familiarized with the subject.

Unlike ESCA, ultraviolet photoelectron spectroscopy has not yet grown into an analytical technique. In the past ten years, the advances in preparing clean surfaces in an ultrahigh vacuum system and the attainment of intense monochromatic UV radiation sources prompted the development of UPS (Table 1). It has been primarily used for the study of gas adsorption on metal surfaces. Electronic energy levels of adsorbed gases[60] and the density of states near the surface by the adsorbed levels[61] can also be determined with UPS. In this volume, Rhodin[62] reviews the chemisorption phenomena studied with UPS.

Because of space limitations we are unable to describe historical backgrounds for other surface analytical techniques, e.g., infrared reflectance spectroscopy, laser Raman spectroscopy, radiation analyses. Details of these techniques will be described in the second volume of this book.

APPLICATION OF SURFACE ANALYSES TO POLYMER TECHNOLOGY

Most of the physical methods mentioned in the above section have been adopted for analytical characterization of solid surfaces. However, there are a few exceptions, e.g., LEED and UPS, which are primarily employed for the research of clean surfaces under ultrahigh vacuum. New applications need to be developed.

It is comforting to think that there are still many new methods which can help solve complex problems. Examples of those techniques that we have not discussed are inelastic electron tunneling (IET),[63] Rutherford backscattering (RBS),[64] reflection high energy electron diffraction (RHEED),[65] scanning Auger electron spectroscopy (SAES),[66] soft X-ray appearance potential spectroscopy (SXAPS),[67] ion scattering spectroscopy (ISS),[68] ion neutralization spectroscopy (INS),[69] glow discharge mass spectrometry (GDMS),[70] etc. Furthermore, any combination of those methods can be more powerful than a single technique.

Polymer technology[71] also had a humble beginning. At the turn of the century, scientists were already interested in the search for "man-made" fibers. For studying the structure of fibers, X-ray diffraction was first used in 1913 by S. Nishikawa and S. Ono.[71] In 1928 H. Mark and K.H. Meyer[71] obtained sufficient diffraction data to support their findings of the partial crystalline structure of cellulose.

Besides X-ray diffraction,[72] conventional physical techniques,[73] e.g., infrared reflection spectroscopy, laser Raman spectroscopy,[74] light-scattering, nuclear magnetic resonance,[75] radioisotope, scanning electron microscopy,[76] have been extensively applied to determine polymer structures. Today, polymer technology has produced many diversified materials, and we are constantly dealing with complex surface problems, such as metal-polymer adhesion,[77,78] friction and wear between metal and polymer,[79,80] coating of metal with polymer or electroplating of polymer with metal. The structure and composition at the interface are of utmost importance in deciding a proper mechanism for a specific surface interaction. Therefore, the application of new surface analytical techniques in conjunction with the conventional methods in solving interfacial problem is always welcomed.

The scope and the depth of research work presented to this Symposium can reflect current interests in the application of surface techniques to solving polymer problems. The first volume of the proceedings deals predominately with metal surfaces. The first volume contains atom-probe, Mössbauer spectroscopy, Auger electron spectroscopy, low energy electron diffraction, ion microprobe, secondary ion mass analysis and photoelectron spectroscopy. Most of these methods have been employed for the study of metal, oxide, insulator or semiconductor surfaces.

The second volume consists of topics on electron spectroscopy for chemical analysis, infrared spectroscopy, laser Raman spectroscopy, scanning electron microscopy and surface-chemical and radiation analyses. The emphasis is on polymer surfaces. Several review papers by Clark,[81] Harrick,[82] Andrews and Hart,[83] Vanderhoff,[84] and Anderson[85] should be helpful in providing background information for various surface techniques.

Since most of the methods discussed here have not been applied to solve polymer-related problems, I am listing some potential applications to stimulate discussions throughout the Symposium:

1. Field ion microscopy (FIM)[86]:

   a. Adhesion and diffusion at interface.

   b. Friction and wear.

   c. Gas-surface reactions: e.g. carburising, nitriding and hydrogenation.

d. Macromolecules, e.g., copper phthalocyanine, nylon, etc.

e. Structure of electrodeposited layer.

2. Mössbauer spectroscopy[87]:

a. Adsorption and chemisorption

b. Catalyst surface.

c. Contamination of surface.

d. Corrosion of metals.

e. Microcrystalline and thin film properties of substrates.

f. Surface composition and structure.

3. Auger spectroscopy (AES):

a. Adhesion of coatings.

b. Contamination of surface.

c. Diffusion at interface.

d. Friction and wear.

e. Passivation of metal surfaces.

f. Profile and composition

4. Scanning electron microscopy (SEM)[76]:

a. Adhesive bonding and fracture.

b. Coating integrity.

c. Friction and wear.

d. Surface contamination.

e. Surface composition by X-ray distribution microscopy.

f. Surface topography.

g. Structure of biological materials.

h. Structure of fiber.

i. Structure at the interface of composite.

5. Electron microprobe (EMPA):

    a. Concentration gradient in thin films (1-100u)

    b. Electroplating of surface.

    c. Metal impurities at interface.

    d. Quantitative analysis of very thin surface layer (0.1 - 1.0u).

    e. Surface contamination.

    f. Wear debris.

6. Ion microprobe (IMMA) and SIMS:

    a. Adhesion and diffusion.

    b. Coating imperfection.

    c. Corrosion and oxidation of metal.

    d. Elemental distribution mapping.

    e. Fracture surface.

    f. Partial identification or organic materials by spectral matching.

7. X-ray photoelectron spectroscopy (ESCA):

    a. Adhesion and diffusion.

    b. Catalyst surfaces.[88]

    c. Corrosion and oxidation.

    d. Friction and wear.

    e. Identification of biomedical materials.

    f. Polymer composition and structure.

    g. Polymer modification.

h. Surface contamination.

i. Structure of pigments.

8. Ultraviolet photoelectron spectroscopy (UPS)[89]

    a. Adsorption and chemisorption.

    b. Catalyst surface.

    c. Corrosion and oxidation.

    d. Electronic state of adsorbate.

    e. Solid state device surface.

9. Low energy electron diffraction (LEED)

    a. Adsorption and chemisorption.

    b. Catalyst surface.

    c. Corrosion and oxidation.

    d. Epitaxial layer.

    e. Metal-adsorbate interaction.

    f. Surface reactions.[90]

    g. Surface topography.

    h. Tribological interactions.

The fulfilment of some of the above applications of surface science will make the coming decade an exciting era of rewarding research.

REFERENCES

1. C.B. Duke and R.L. Park, Physics Today 25, No. 8, 23 August, 1972.
2. F. Greenaway, Anal. Chem. 48, No. 2, 148A (1976).
3. G. Kirchhoff, Pogg - Ann 109, 275 (1859), quoted in G.W. Ewing "Analytical Chemistry: The Past 100 Years", Chem. & Eng. News 128, April 6, 1976.
4. W.C. Röntgen, Sitzungsber Phys. -Med. Ges. Würzburg, (137) (1895), Ann. Phys. Chem. 64, 1-11 (1898).
5. W. Friedrich, P. Knipping and M. von Laue, Ber Bayer Akad. Wiss. Muchen 303 (1912); Ann. Phys. 41, 971 (1913).
6. W.H. Bragg and W.L. Bragg, Proc. Phys. Soc. (London) A88, 428 (1913).
7. H.G.J. Moseley, Phil. Mag. 26, 1024 (1913), and 27, 703 (1914).
8. J.J. Thompson, *Rays of Positive Electricity and Their Application to Chemical Analysis,* Longmans, Green & Co., London, 1913.
9. J.B. Farmer, in *Mass Spectrometry,* Ed. C.A. McDowell, Chapt. 2, McGraw Hill, New York (1963).
10. E.W. Müller in *Field-ion Microscopy,* Ed. J.J. Aren and S. Ranganathan, Chapt. 1, Plenum Press, New York (1968).
11. E.W. Müller, Phys. Z. 37, 838-841 (1936).
12. E.W. Müller, Z. Physik 131, 136-142 (1951).
13. D.G. Brandon, in *Field Ion Microscopy,* Ed. J.J. Hren and S. Ranganathan, p. 28-52, Plenum Press, New York (1968).
14. E.W. Müller, J.A. Panitz and S.B. McLane, Rev. Sci. Instr. 39, 83 (1968).
15. E.W. Müller and S.V. Krishnaswamy, The Proceedings of this Symposium, Vol. 1.
16. R.L. Mössbauer, Z. Phys. 151, 124 (1958).
17. G.W. Simmons and H. Leidheiser, Jr., The Proceedings of this Symposium, Vol. 1.
18. C.A. Clausen, III and M.L. Good, The Proceedings of this Symposium, Vol. 1.
19. P.F. Kane and G.B. Larrabee, Ed. *Characterization of Solid Surfaces,* Plenum Press, New York 1974.
20. J.P. Hobson, Japan J. Appl. Phys. Suppl. 2, Pt. 1, 317 (1974).
21. R. Heckingbottom, Phys. in Technol. 47, Mar. 1975.
22. C.A. Evans, Jr., Anal. Chem. 47, No. 9, 818A, Aug. 1975.
23. P. Auger, J. Phys. Radium 6, 205 (1925).
24. J.J. Lander, Phys. Rev. 91, 1382 (1953).
25. L.A. Harris, J. Appl. Phys. 39, 1419 (1968).
26. C.C. Chang, Surface Sci. 25, 53 (1971).
27. C.C. Chang, in *Characterization of Solid Surfaces,* Ed. P.F. Kane and G.B. Larrabee, Chapt. 20, Plenum Press, New York (1974).

28. R.L. Park, The Proceedings of this Symposium, Vol. 1.
29. J.T. Grant, The Proceedings of this Symposium, Vol. 1.
30. C.J. Davisson and L.H. Germer, Phys. Rev. 30, 705 (1927); L.H. Germer, Z. Physik 54, 408 (1929), A translation is given in Bell Syst. Tech. J. 8, 591 (1929).
31. H. Farnsworth, Phys. Rev. 34, 679 (1930); 40, 682 (1932).
32. G.P. Thomson and W. Cochran, *Theory and Practice of Electron Diffraction,* Macmillan (1939).
33. L.H. Germer and C.D. Hartman, Rev. Sci. Instr. 31, 784 (1960).
34. P. Estrup, The Proceedings of this Symposium, Vol. 1.
35. T.N. Rhodin and D.S.Y. Tong, Physics Today 26, No. 10, 23 (1975).
36. D.L. Adams and U. Landman, The Proceedings of this Symposium.
37. P.M. Marcus, The Proceedings of this Symposium, Vol. 1.
38. N. Stoner, M.A. Van Hove and S.Y. Tong, The Proceedings of this Symposium, Vol. 1.
39. M.V. Ardenne, Z. Phys. 109, 553 (1938).
40. C.W. Oatley, W.C. Nixon and R.F.W. Pease, Advances in Electronics and Electron Physics, 21, 181 (1965).
41. R. Castaing, Thesis, University of Paris, Paris, France, 1951; Publ. ONERA, No. 55.
42. L.S. Birkes, *Electron Probe Microanalysis,* Interscience Publishers, 1963.
43. G.A. Hutchins, in *Characterization of Solid Surfaces,* Ed. P.F. Kane and G.B. Larrabee, Chapt. 18, Plenum Press, New York (1974).
44. R.H. Sloane and R. Press, Proc. Roy. Soc. Ser. A, 168, 284 (1938).
45. M.V. Ardenne, Z. Tech. Phys., 20, 344 (1939).
46. R.F.K. Herzog and F.P. Viehböck, Phys. Rev. 76, 855 (1949).
47. R.F.K. Herzog, W.P. Poschenrieder, H.J. Liebl and A.E. Barrington, Solid Mass Spectrometer, NASA Contract NSA W-839, GCA Technical Report No. 65-7N47 (1967).
48. H.J. Liebl, Anal. Chem. 46, No. 1,22A, (1974).
49. W.K. Huber, H. Selhofer and A. Benninghoven, J. Vac. Sci. Tech. 9, 482 (1972).
50. C.A. Evans, Jr., Anal. Chem. 44, No. 13, 67A, (1972).
51. H.W. Werner, Surface Sci. 47, 301 (1975).
52. R. Castaing and G. Slodzian, J. Microscopie 1, 395 (1962).
53. G.H. Morrison, The Proceedings of this Symposium, Vol. 1; G.H. Morrison and G. Slodzian, Anal. Chem. 47, (11), 932A (1975).
54. H. Robinson and W.F. Rawlingson, Phil. Mag. 28, 277 (1944).
55. N. Svartholm and K. Siegbahn, Arkiv. f. Mat. Astr. Fys. 33A, 21 (1946).
56. K. Siegbahn, C. Nordling, A. Fahlman, R. Nordberg, K. Hamrin, J. Hedman, G. Johansson, T. Bergmark, S.-E. Karlsson, I. Lindgren and B. Lindberg: *ESCA: Atomic, Molecular and Solid State Structure Studied by Means of Electron Spectroscopy.* Nova Acta Regiae Soc. Sci. Upsaliensis, Ser. IV. Vol. 20 (1967).

57. K. Siegbahn, C. Nordling, A. Fahlman et al., *Electron Spectroscopy for Chemical Analysis,* Technical Report AFML-TR-68-189, Oct. 1968.
58. S.H. Hercules and D.M. Hercules, in *Characterization of Solid Surfaces,* Ed. P.F. Kane and G.B. Larrabee, Chapt. 13. Plenum Press, N.Y. (1974). Also, D.M. Hercules, the Proceedings of this Symposium, Vol. 1.
59. D.T. Clark, In *Advances in Polymer Friction and Wear,* Ed. L.H. Lee, Vol. 5A, 241, Plenum Press, New York (1974).
60. D.E. Eastman and J.K. Cashion, Phys. Rev. Lett. $\underline{27}$, 1520 (1971).
61. G.E. Becker and H.D. Hagstrum, J. Vac. Sci. Technol. $\underline{10}$, No. 1, 31 (1973).
62. T.N. Rhodin and C. Brucker, The Proceedings of this Symposium, Vol. 1.
63. N.M. Brown and D.G. Walmsely, Chem. in Brit. $\underline{12}$, No. 3, 92 (1976).
64. W.D. Mackintosh, in *Characterization of Solid Surfaces,* Ed. P.F. Kane and G.B. Larrabee, Chapt. 16, Plenum Press, New York (1974).
65. H.M. Kennett and A.E. Lee, Surf. Sci. $\underline{33}$, 377 (1972).
66. N.C. MacDonald and J.R. Waldrop, Appl. Phys. Lett. $\underline{19}$, 315 (1971).
67. R.L. Park, J.E. Houston and D.G. Schreiner, Rev. Sci. Instr. $\underline{41}$, 1810 (1970).
68. D.P. Smith, Surf. Sci. $\underline{25}$, 25 (1971).
69. H.D. Hagstrum, Science $\underline{178}$, 275 (1972).
70. J.W. Coburn and E. Kay, Appl. Phys. Lett. $\underline{18}$, 435 (1971); $\underline{19}$, 350 (1971).
71. H.F. Mark, Chem. & Eng. News 176, April 6, 1976.
72. B.K. Vainshtein, *Diffraction of X-Rays by Chain Molecules,* Elsevier, Amsterdam (1966).
73. R.H. Grieser, Prog. in Org. Coatings, $\underline{3}$, 1 (1975).
74. P.J. Hendra and P.M. Stratton, Chem. Revs. $\underline{69}$, 325 (1969).
75. F.A. Bovey, Proc. Inter. Sym. on Macromolecules, Rio de Janeiro, Ed. E.B. Mano, Elsevier, Amsterdam (1975). p. 169.
76. P.R. Thornton, *Scanning Electron Microscopy,* Chapman and Hall, London (1968).
77. L.H. Lee, Ed. *Adhesion Science and Technology,* Part A p. 1, Plenum Press, New York (1975).
78. W.A. Brainard and D.H. Buckley, Wear $\underline{26}$, 75 (1973).
79. L.H. Lee, Ed. *Advances in Polymer Friction and Wear,* Plenum Press, New York (1974).
80. T.F.J. Quinn, *The Application of Modern Physical Techniques to Tribology,* Von Nostrand Reinhold (1971).
81. D.T. Clark, The Proceedings of this Symposium, Vol. 2.
82. N.J. Harrick, The Proceedings of this Symposium, Vol. 2.
83. R.D. Andrews and T.R. Hart, The Proceedings of this Symposium, Vol. 2.
84. J.W. Vanderhoff, The Proceedings of this Symposium, Vol. 2.

85. J.L. Anderson, The Proceedings of this Symposium, Vol. 2.
86. K.M. Bowkett and D.A. Smith, *Field Ion Microscopy,* Elsevier, New York (1970).
87. G.K. Wertheim, *Mössbauer Effect: Principles and Applications*, Academic Press, New York (1964).
88. J.S. Briner, Accounts. Chem. Res. $\underline{9}$, 86 (1976).
89. B. Feuerbacher, Surf. Sci. $\underline{47}$, 115 (1975).
90. G.A. Somorjai, Accounts. Chem. Res. $\underline{9}$, 248 (1976).

PART I

Atom-Probe and Mössbauer Spectroscopy

Plenary Lecture: Surface Analysis at the Atomic Level Using the Atom-Probe

Erwin W. Müller and S.V. Krishnaswamy

Department of Physics
Pennsylvania State University
University Park, Pennsylvania 16802

The atom-probe FIM as the most sensitive microanalytical tool combines the single atom resolution of a field ion microscope with mass spectrometric single ion identification. The early, straight time-of-flight instrument, however, suffers from insufficient mass resolution as ion energy deficits cause a spill-over to apparently heavier masses, and from random artifact signals caused by residual gas or by afterpulses of the detector. A curved energy-focusing drift tube and a double channel plate detector eliminate these problems. By aiming at individual impurity atoms seen as interstitials, at dislocations or at grain boundaries, their nature can be identified. Atom-layer by atom-layer depth profiling of alloys with adjacent isotopes and the analysis of small atomic clusters is unambiguous. The occurrence of metal-hydride, metal-helide, or metal-neide ions formed with some metals in the presence of the imaging gas is now well established. Gas-metal surface reactions with nitrogen and carbon monoxide are found to be affected by dissociation due to the special conditions in the FIM, such as the high field and possible excitation of the adsorbate by electron impact.

INTRODUCTION

In heterogeneous reactions the role of the solid-gas interface is of paramount importance. The situation promises to be most easily interpreted when the surface can be well characterized. What we must primarily know for basic advances are three features, the topography of the surface structure, the chemical identity of its constituents, and its electronic states. There are now more than 50 different methods available to give us pieces of such information, but the puzzle remains difficult as each technique is heavily weighted for one of the features. Atom-probe field ion microscopy has the advantage of providing quite directly the first two answers, while the third one still remains evasive.

Surface characterization owes its advances to a variety of intricate instrumental developments based on simple physical phenomena. Often specimen preparation is a major undertaking, such as the preparation of a "good" crystal surface for seeing its reciprocal lattice in a LEED apparatus. The LEED diffraction pattern is made up of the contribution of some 10^{13} atoms. Perhaps 10^{10} of them may not sit at the "proper" lattice sites, thereby remaining undetectable, and yet may just cause the decisive step in a surface reaction such as catalysis. Auger electron spectroscopy (AES), electron spectroscopy for chemical analysis (ESCA) and secondary ion mass spectroscopy (SIMS) do give us a chemical "surface" analysis, which, however, is averaged over a depth of 5 to 20Å, an area encompassing millions of possibly undefined surface sites, and is subject to errors by estimates of cross sections and matrix contributions.

The atom-probe is a logical development of the now 25 year old field ion microscope.[1] The FIM has remained the only device capable of routinely showing the structure of the specimen in atomic resolution, with the restrictions to metals and the need of shaping the specimen to a fine needle point, the cap of which is radially projected onto a screen by the ions of an imaging gas, usually He, Ne, Ar or H_2. The specimen is simultaneously the "lens" which obtains its perfectly rounded shape by the process of field evaporation, in which an applied high voltage removes the irregular, more protruding surface areas of locally enhanced field strength until an evenly curved surface is achieved. All essential applications of field ion microscopy[2] were established before 1960, such as viewing individual lattice defects, vacancies, interstitials, dislocations, grain boundaries, and performing in-situ experiments such as radiation damage, thermally activated surface migration, bulk diffusion of interstitials to the surface, and fatigue studies. The general acceptance and proliferation of this technique in the sixties brought the quantitative aspects of field ion microscopy into the foreground,[3] one of which was the need of identifying the chemical nature of the constituents of the specimen surface. This atomic surface analysis has been achieved by combining a field ion microscope with a mass spectrometer of single ion sensitivity. The selection of a site is obtained by a probe hole in the screen relative to which the field ion image can be shifted by tilting the tip.[4]

THE STRAIGHT TOF ATOM-PROBE

Of the various types of mass spectrometers capable of single ion sensitivity, the time-of-flight version proves to be most useful. The necessary discrimination from noise is effectively achieved by time gating the electron multiplier detector for the possible arrival time of the ion species some one to twenty microseconds after the pulse initiated field evaporation event. The identifying mass-to-charge ratio m/n is obtained from the kinetic energy equation

$$1/2\ mv^2 = n\ e\ (V_{dc} + V_p) \tag{1}$$

when the velocity of the ion is known for the length ℓ of the drift path and the measured time-of-flight t of a particle whose energy is the sum of the image voltage V_{dc} and the superimposed evaporation pulse V_p. (Fig. 1) The reliability of the mass determination depends, with the integer multiplicity of charge varying from 1 to 5 and generally known from experience, on the precision by which the acceleration voltages V_{dc} and V_p are known and by which the time-of-flight can be measured.

$$m/n = 2\ e\ (V_{dc} + V_p)\ t^2/\ell^2. \tag{2}$$

The ToF atom-probe had been conceived by the author in 1967. An improved 1973 version[5,6] of the author's laboratory is shown in Fig. 2, the microscope section featuring an external gimbal system for the manipulation of the tip direction, a concentric pulse feeding line to the tip, cryogenic cooling of the sapphire insulated tip by a flexible connection to the cold finger, and a 75 mm

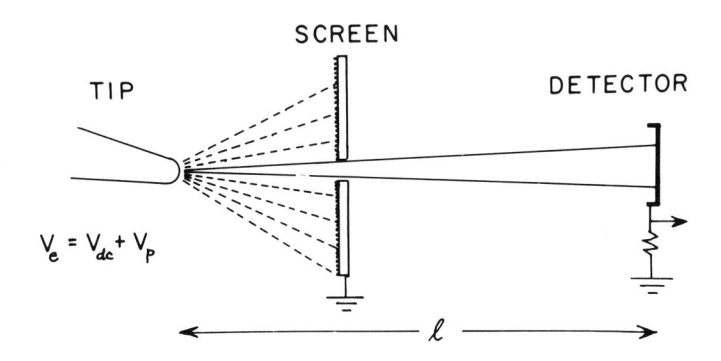

Fig. 1. Principle of the ToF atom-probe.

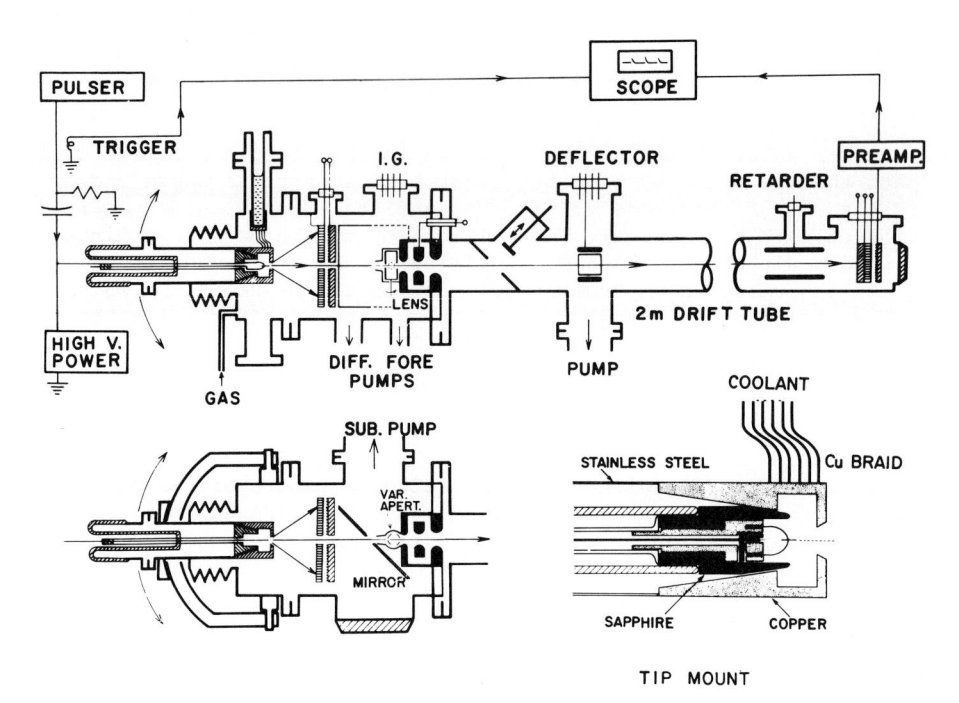

Fig. 2. A straight atom-probe of 1973 (Ref. 5).

diameter microchannel plate-screen assembly for viewing the intensified field ion image. The effective diameter of the probe hole may be varied by an aperture in front of an einzel lens which focuses the beam onto a 2 m distant detector. This is a double channel plate-screen assembly with a 10^7 fold intensification of the single ion impact. The time-of-flight is measured by an oscilloscope, with the 0.5 to 20 μsec sweep started by a trigger from the evaporation pulse. For highest resolution, the trigger may be delayed to permit a fast sweep rate within a narrow time gap.

A small number of very similar instruments have been constructed elsewhere.[7-9] For the handling of larger amounts of data, as are desirable in depth analysis, time-of-flight measurements with a 100 megacycles digital clock may be used advantageously.[10] For simple analysis of the tip bulk, without caring for identifying specific atomic surface sites, electronic data processing of large

numbers of signals from cycled evaporation pulses have also been
employed. However, for really high resolution as obtainable with
the new energy focused ToF atom-probe the desirable time readout
to a precision of 1 or 2 nsec still requires oscilloscopical time
determination.

OTHER TYPES OF ATOM-PROBES

Magnetic sector mass spectrometers in connection with a field
ion source have been used since Inghram and Gomer's pioneering
work,[11] and field ion mass spectrometry of externally supplied
gases has become a successful branch of research with the work of
Beckey.[12] Utilizing the ample supply of ionizable species obtained
by continuous or repeated adsorption or storage on large surface
whisker emitters, products of field desorption can also be readily
analyzed with magnetic sector or quadrupole mass spectrometers.
The narrow mass range that such instruments can receive at one time
while all the other emitted ions are wasted made the detection of
single ionic species from a selected site practically impossible.
A useful compromise is our magnetic sector atom-probe[13] in which a
good section of the entire mass spectrum, ranging from a selected
lower mass m_0 to about 1.8 m_0, is displayed in the form of "spectral lines" on a phosphor screen. A set of two microchannel plates
before the screen make every impinging single ion appear as a
bright spot that can easily be recorded photographically. A mass
line is fairly unambiguously distinguished from noise spots by
the lineup of 3 or 4 ion spots at the position of an integer mass
number. Spectral lines of the three neon isotopes at 20, 21 and
22 are about 10 mm apart and have a width of 0.1 mm, giving a mass
resolution of $\Delta M/M = 1/2000$. Metal tips can be slowly evaporated
by slightly raising the negative dc voltage of an intermediate
electrode, while the tip remains at a fixed potential to maintain
the ion energy constant. As shall be shown later, this instrument
with its slow evaporation rate is in some respects complementary
to the fast pulsed ToF atom-probe.

At the time of its introduction, 25 years ago,[1] the FIM was
conceived as a desorption microscope. The image gas hydrogen
was admitted with the intent to be pulled off the surface in ionic
form from a continuously replenished adsorbed state. However, the
attainment of self-imaging of the tip surface by field desorbing
or evaporating just one surface layer was foreseen to require a 10^6
time image intensification by some electron multiplier devices.
Thus, we had to wait 20 years until the field desorption microscope[14] could be made to work with the advent of the microchannel
plate. A field desorption image, Fig. 3, obtained by evaporating
a number of surface atoms equal to about a monolayer does not display the beautiful crystallographic symmetry and completely resolved net plane patterns we are familiar with from field ion
microscopy. There are at least three reasons for it: 1) The

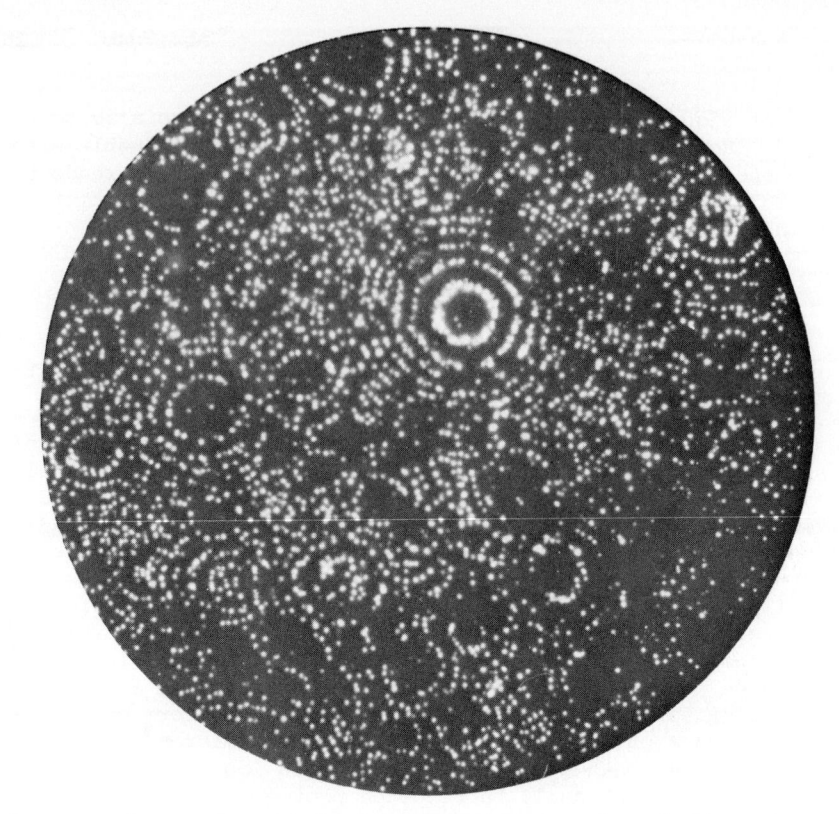

Fig. 3. Single shot field desorption image of a rhodium tip.

microchannel plate has a detection sensitivity of about 50%.[5] Thus, only one-half of the atoms evaporated are imaged in a statistically random way. This alone makes a regular net plane geometry hard to recognize. 2) The evaporation rate over the entire tip cap is not uniform. While one crystal plane may field evaporate completely, just a few kink site atoms may come off from other planes. Very near the onset of field evaporation, there seems to be a feedback of energy from the first evaporating ion, producing more ions to come off within patches of some 30Å diameter. 3) There is a possibility that at some areas field evaporation occurs as a second step after the atom has laterally moved away from its original site.[15]

Despite these limitations, the use of a field desorption microscope for a ToF mass identification is an interesting proposition as it promises an overall picture of the distribution of various atom species over the crystal hemisphere of the specimen. This was first considered by Müller et al.[16] who gated the channel plate-screen assembly with a pulse delayed from the evaporation pulse by the time-of-flight of a selected ion species. However, the very limited tip-to-detector distance makes all the drift times shorter than a microsecond, which in connection with the difficulties encountered with fast precise pulsing of the large capacity channel plate assembly poses so serious limitations to the resolution, that we did not continue this approach. Subsequently, Panitz[17] with his "10 cm Atom-Probe" introduced some improvements by employing a curved chevron-microchannel plate concentric to the tip and by retarding the ions to slightly increase the times of flight. Still, his channel plate gating would not allow a separation of the 50 ns time difference (at 5 kv) between Ir^{+++} and $IrNe^{+++}$ which would require a mass resolution of 1/10. Without gating and by picking up the total ion current from the entire screen[18] and by evaporating 60 consecutive monolayers a mass spectrum of the molybdemum isotopes was obtained with a resolution (at FWHM) of $\Delta M/M \sim 1/50$. Forsaking the idea of localizing the ionic species, this, of course, is no more a single-atom mass spectroscopy. To return to the atom-probe concept, Panitz also used a limited area pick-up at the output screen, replacing the probe hole by a flexible, small aperture photomultiplier and subsequent oscilloscopic recording. If in such an arrangement a constant retarding potential is used to at most double the time-of-flight, the FIM magnification of the pulsed image differs from that of the dc helium gas image used for selecting a crystallographic site, causing an aiming error.

Although any field ion microscope with a single channel plate may be converted into a ToF analytical device, caution is necessary to eliminate or at least recognize artifacts. The marginal sensitivity of a single channel plate requires intense pulsing, to the limit of what the channel plate can stand, as shown by Waugh.[19] He concluded from gated desorption images of iridium tips in neon the ample concurrence of $IrNe^{++}$ and $IrNe^{+++}$ species. However, these are most likely artifacts, as in our high resolution atom-probe under the same tip pulsing conditions and the same crystallographic areas these species are found to have an abundance of less than 1/1000.

ENERGY DEFICIENCIES AND THE ENERGY FOCUSED ATOM-PROBE

Artifact signals have plagued atom-probe field ion microscopy since its beginning. Molecules of the image gas or of residual gas are ionized in free space, particularly when in the early instruments before the introduction of channel plate image intensification the image gas pressure had to be as high as 10^{-3} Torr.[20,21] In a large sequence of evaporation events, when needed for statistical weight of the data, the entire mass spectrum is covered with scattered signals (Fig. 4) colloquially referred to as "grass", making it very difficult to find real ion species when their abundance is less than a few percent of the main peaks,[22] and to reliably identify molecular ions as products of field induced reactions

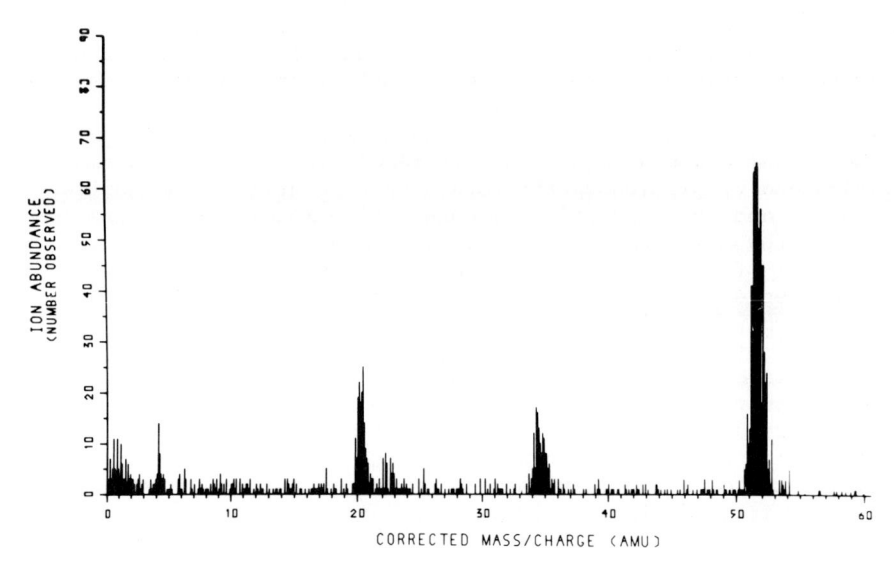

Fig. 4. Histogram of a mass spectrum of 2089 ion signals obtained by evaporating Rh in the presence of He-Ne as an imaging gas, with many artifact signals (from J.A. Panitz, Ph.D. thesis, The Pennsylvania State University 1969).

between the tip metal and gaseous adsorbates. Another annoying artifact is delayed signals from afterpulses in the detector, first investigated by Brenner and McKinney.[23] Many electron multiplier detectors produce a second signal following by some 50 to 600 nsec the primary output pulse due to a real ion impact. Indistinguishable from a real ion signal, afterpulses may be misinterpreted as a second ion species of slightly higher mass, particularly a metal hydride ion, but this problem is considerably reduced by increasing the resolution of the atom-probe with a longer drift path.[24] In addition, a double microchannel plate with the phosphor screen as a collector turned out to be free of afterpulses to at least one in 1000 primary events, and this detector is now used in the more successful atom-probes.

The most serious limitation of atom-probe resolution is due to an unavoidable predicament placed by the shape of the evaporation pulse. Since some time we realized that in a histogram of repeated pulse evaporations the "mass lines" exhibit a tail extending over one to three atomic mass units toward heavier mass. Energy analysis of the pulse evaporated ions with the retarder electrode[25] in the straight atom-probe of Fig. 2 and more precisely with a 90° electrostatic deflection analyzer[6,26] revealed an energy spread of up to several hundred eV in repeated evaporations. Clearly, the energy of the ions was not exactly the sum of the applied dc voltage and the superimposed pulse voltage, and the energy deficits were not even the same for ions of different masses coming off during the same pulse. Although the rise time is made as short as possible, of the order of 5×10^{-10} sec, an evaporating ion typically acquires one-half of its final energy within 10^{-11} sec over the first 10^{-4} cm of its path. The cause of the energy deficit trouble is that the evaporation event takes place before the evaporation pulse has matured to its nominal plateau.[25,27] If more than one ion is coming off in the area defined by the probe hole, these ions may leave the surface sequentially from a net plane step, each experiencing a different time dependent acceleration voltage during the initial path near the tip. The true pulse shape at the tip, at an unterminated pulse line, is difficult to determine, but most likely there is an overshoot as shown in Fig. 5. Evaporation of the first and possibly subsequent ions may occur at any time between points A and C, and their final energy will also depend on the mass as the lighter species will get most of its energy nearer its respective starting potential than a slow moving heavier ion. If highest resolution is aspired, the pulse voltage should be so low that evaporation takes place only at B near the top of the overshoot, so that at least the starting time is defined to within a

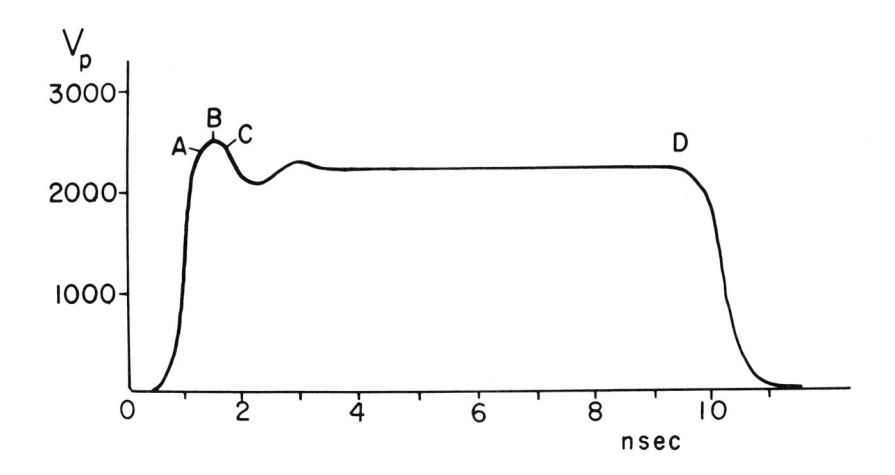

Fig. 5. Shape of a nominal 10 nanosec, 2200 V evaporation pulse, which is superimposed to the dc voltage of the tip.

nanosec or less. This consideration also shows that even an "ideal" rectangular pulse shape would not be desirable, as an evaporation event could occur anytime on the flat top several nanosec after the pulse rise from which the time-of-flight count begins. In practice, the exact evaporation event may occur even before point A, when for instance an adsorbed gas atom, possibly during its surface migration, causes a temporary reduction of the evaporation field of a metal substrate atom. In short, there seems to be no way to assure the basic requirement of a precise time-of-flight atom-probe, which is to have all particles entering the drift path with a well defined energy.

The solution of this problem is the compensation of energy deficits by ion optical means. We first used the energy discrimination of a 90° cylindrical or spherical electrostatic sector field and inclined the detector front plate with respect to the optical axis in such a way that the more deflected slower ions traveled a shorter path than the more energetic species.[5,6,28,29] However, this system works well only for very small acceptance angles, that is small probe holes. Fortunately, various other energy deficit compensating systems for a ToF mass spectrometer with a conventional electron impact ion source had already been conceived by Poschenrieder,[30] and we were able to adapt one of his energy focusing ion optical configurations, a 163° toroidal electrostatic deflector, to our existing atom-probe.

In an electrostatic sector field, the path length of an ion of lower energy is always shorter (dashed line in Fig. 6) than of an ion having the nominal energy which would move at the (curved) optical axis. Moreover, the energy deficient ion traveling nearer the negative deflector electrode, is accelerated all the time to a velocity above the one of the nominal ion. By properly balancing the length of the over-compensating sector with the straight path sections, the low energy particle gains just enough time inside the sector field that it arrives at the detector isochronously with the nominal energy ion. This system also works perfectly for a fairly large acceptance angle, as the crossing over of the wide angle trajectories (full lines in Fig. 6) exactly half way through the sector field compensates for the time gain and loss in the respective sector halves, again assuring isochronous arrival at the stigmatic focus at the detector. In our system (Fig. 7), the toroidal deflector plates have principal radii of R_1 = 31.1 cm and R_2 = 35.2 cm, while the axial radii are 139 and 143 cm. The voltages are slightly asymmetric with respect to ground in order to provide zero potential for an ion of nominal total evaporation energy V_e which travels along the optical axis. The range of energy acceptance is defined by the width between the plates half way through the deflector where an intermediate astigmatic focus forms an energy spectrum. The time- and energy-focused stigmatic ion image of the tip appears on the screen of the chevron channel plate detector as a spot of 1 mm diameter, after a drift path of

Atom-Probe 31

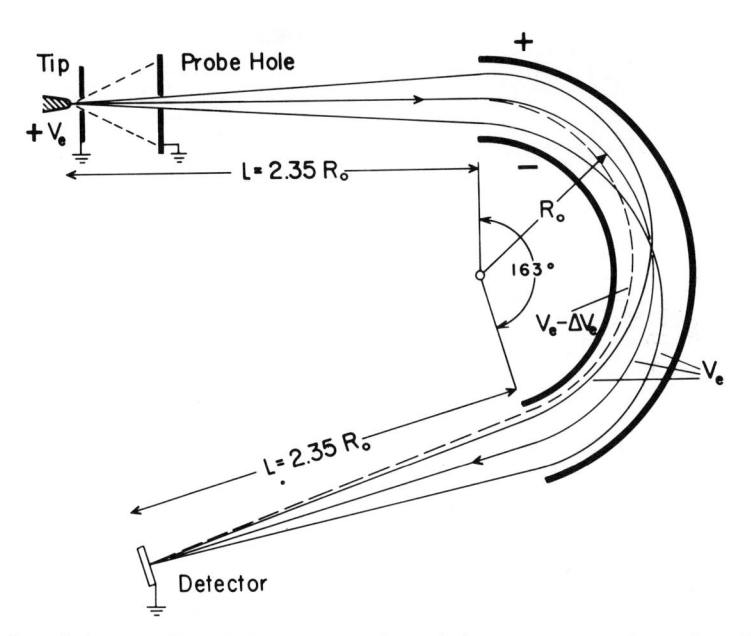

Fig. 6. Scheme of a ToF atom-probe with an energy focusing 163°
electrostatic sector field (from ref. 26).

2.5 m. The acceptance angle or effective probe hole diameter is
determined by an externally adjustable aperture of 20 or 12 mm
diameter just in front of the deflector. The evaporation voltage
within which the actual experimental time focusing is better than
±2 nanosec ranges to 4% on both sides of the nominal V_e to which
the deflector voltages are set. In practice, these voltages are
taken from two ganged precision power supplies which are set by
reading a digital voltmeter, connected over a once preadjusted
potentiometer. Thus, the operator has only to determine by ob-
serving the FIM screen, the approximate evaporation voltage
$V_e = (V_{dc} + V_p)$ of a given tip and dial a deflector voltage to
$V_{defl} \approx 0.1\ V_e$ in order to exactly time-focus all ion species
with energies from 0.96 to 1.04 V_e. After obtaining the precise
oscilloscopic time readout of the ion signals the exact identifying
m/n values are calculated from Eq. 2 by using the $(V_{dc} + V_p) =$
10 V_{defl}, without having to know the actual precise values of V_{dc}
or V_p. This way of operation facilitates the data taking because
only V_{defl} needs to be kept constant within 4 digit accuracy. To
demonstrate this with a sequence of 40 oscilloscope tracings of

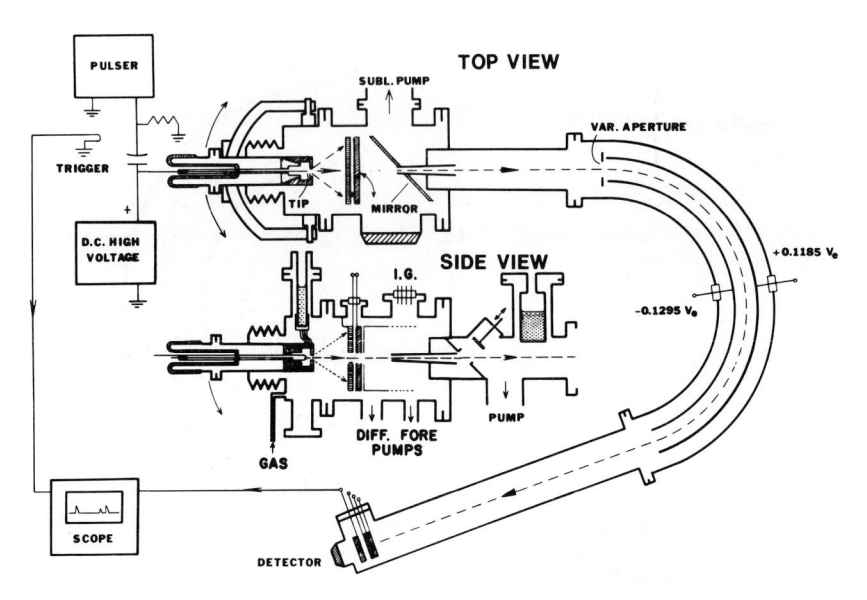

Fig. 7. The 1975 high resolution, energy focused ToF atom-probe. Top view shows the microscope chamber equipped with a swing-away channel plate-screen assembly, while the side view shows a previously used channel plate-screen with a probe hole.

CO desorbed from tungsten, V_{dc} was gradually raised from 19,400 to 21,000 V to maintain an about constant evaporation rate while the tip radius increased. (Fig. 8) The arrival time of 6.75 μsec remained constant within the dt = 2 nsec readout accuracy of the 50 nsec/division sweep rate of the oscilloscope, corresponding to a mass resolution dM = 2 M dt/t = 2 x 28 x 0.002/6.75 = 0.0166 amu, or dM/M = 1/1700. Using a less extreme range of the evaporation voltage and the small probe hole aperture a resolution up to 1/5000 has been verified. This contrasts with the performance of the straight atom-probe in which the width of a mass peak at one-half of peak height in a histogram of many repeated shots at one precisely measured evaporation voltage is about 0.5 amu, while the base of the distribution is three times wider,[22] thereby making impossible an unambiguous separation of adjacent integer mass numbers.

The unique capability of the energy focused ToF atom-probe to definitely identify a single atom with one shot, that is not relying

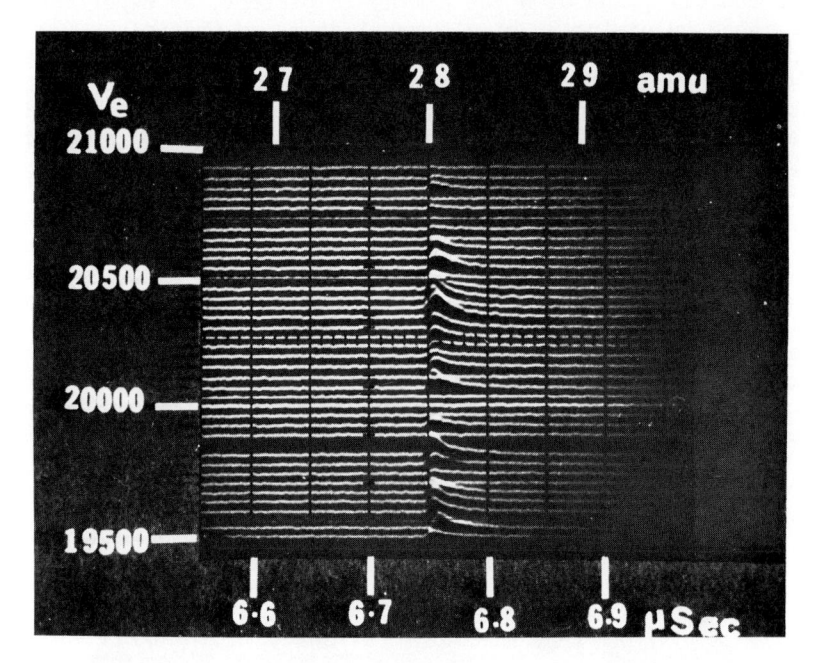

Fig. 8. Oscilloscope readout of CO at 28 amu desorbing from a tungsten tip. In a sequence from bottom to top 38 desorption events were recorded at a sweep rate of 50 nanosec/div., while the dc voltage was raised from 19,500 to 21,000 V without changing the deflector voltage of 2000 V.

on a statistical repetition, requires also the assurance of an efficient aiming by the probe hole.[31] It has been realized since the early atom-probe experiments[7,32,33] that the impact of the evaporated ion on the screen may be slightly displaced with respect to the helium image spot of the original surface atom. This deviation may be due to the difference in origin of an image gas ion, about 4.5 Å above the surface, and the surface atom itself, and there is also the possibility of a lateral displacement of a kink site atom if a suggested two-step evaporation process is taking place.[15] With a double channel plate-screen, it is easy to image a surface with helium and then superimpose on the same photograph the ion spots from a subsequent evaporation pulse.[17] The correlation is not too good, frequently showing aberrations of the order of one lattice spacing. In most tip areas the displacement is away from a net plane edge towards the ledge, except for the rim of the densely packed net plane 011 on W and 001 and 111 of the fcc lattices, where the evaporated ions arrive inside the ring which marks the net plane edge in the helium ion image.[7] More towards the center of the 011 plane of W the aiming is perfect, as experimentally established by aiming at single metal atoms on that plane.[31,34] Aiming experiments also give a satisfactory score over the entire tip surface when the probe hole is large enough to cover an area of about 10 Å diameter, as can be determined by identifying single bright atom spots typical for the minor constituent of some dilute alloys.[31]

A quantitative small area analysis by counting atom-probe signals has to include the crystallographic variation of ion density over the tip hemisphere. When a field desorption image[14] is recorded by photographically integrating some 5 to 50 atomic layers, the image surprisingly is not random but shows very distinct crystallographic features with regions of reduced ion density, particularly along certain zone lines which form an intricate lace-like pattern. (Fig. 9) Such an image was first published by Müller and Tsong[35] and explained as an ion optical effect caused by the diversion of

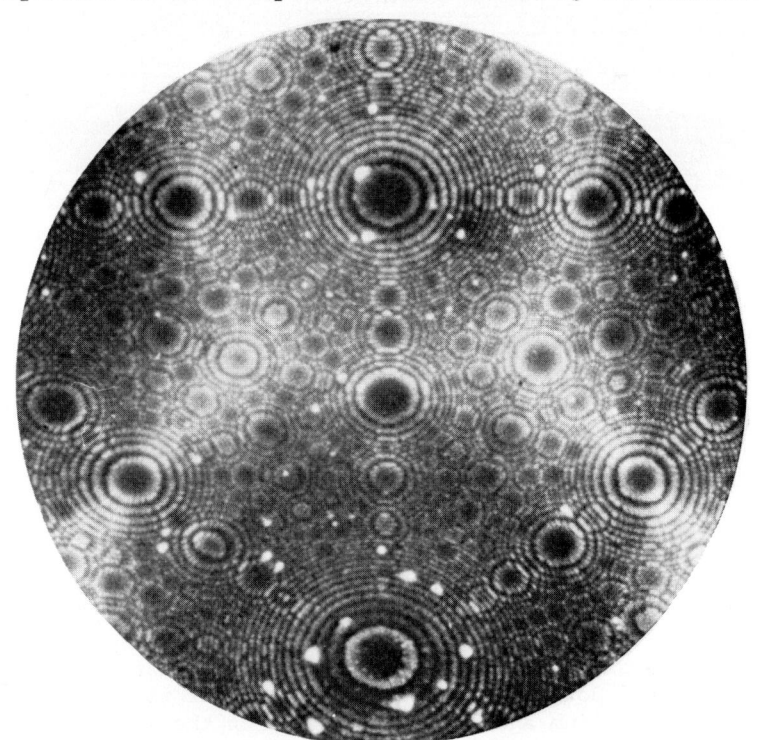

Fig. 9a. Field ion microscope image of a rhodium tip, showing the (001) plane near the top and the (111) plane near the bottom. The random extra bright spots are due to Pt atoms in a concentration of 0.1%.

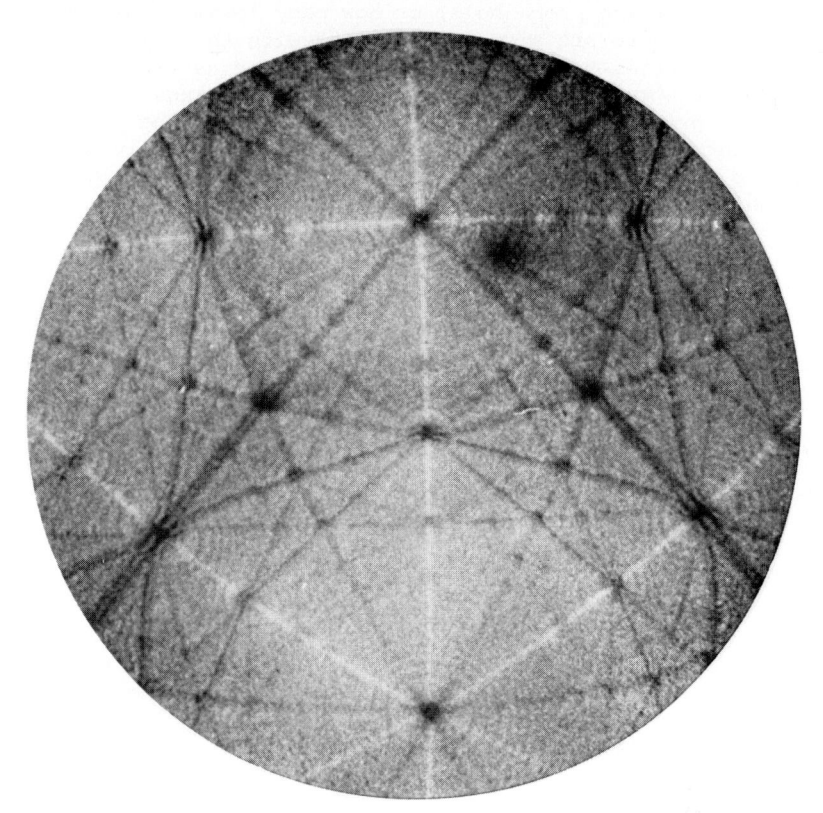

Fig. 9b. Integrated multilayer field desorption image of the same rhodium tip.

ion tragectories from the exactly radial projection one could expect if the emitter were an ideal sphere. Actually, it is a polyhedron from which the vicinals on both sides of a zone project their ions in a nearly normal direction, leaving the zone line itself dark. Similar integrated multilayer patterns were later obtained by other authors,[15,36] who invoked the idea of a short path surface migration of an ion immediately preceding its evaporation. While such a mechanism may apply for the displacement of ions from the edge of the 011 plane of W, we believe that our original explanation of the dark zone lines as an ion optical effect is sustained by the fact that time gated desorption images of He$^+$ and Ne$^+$ by Panitz[37] show exactly the same pattern feature.

One practical disadvantage of aiming with a probe hole is that one cannot see the atom spot at the moment of aiming, whether it is exactly centered, whether two adjacent target spots are in the field of view of the probe hole, or how many imaged substrate atoms are covered. There are also difficulties with the ion optical, defocusing properties of the channel plate-screen assembly, which either requires turning off the channel plate voltages before pulsing, or providing a ground potential lining of the probe hole which further increases the invisible area. We found it more convenient to use a channel plate-screen assembly without a probe hole, which can be manually swung out of the beam in a fraction of a second before the evaporation pulse is applied. The effective probe hole area is marked on the unperturbed screen by an optically projected circle of light. The accurate position of this marker can easily be established when the helium ion image is viewed on the detector screen by defocusing the deflector voltage slightly from the voltage of 0.1 V_{dc}, so that not the tip itself is focused but rather the entrance pupil of the system. This contains the projection image of the true effective probe hole area with all the atomic details as seen by the system.

APPLICATIONS OF THE ENERGY FOCUSED ATOM-PROBE

Surface Analysis

The practical use of the atom-probe has been along the basic two lines already foreseen with the introduction of the new instrument.[4] One is surface characterization such as the identification of individual bright spots appearing in field ion images following exposure of a clean surface to the ambient residual gas or to intentionally introduced gases, or of the chemical complexes formed with mostly invisible adsorbates by their possibly field induced reactions with the substrate. The second is the analysis of metallic bulk structures obtainable through controlled layer-by-layer field evaporation. Because of the unsurpassed discrete visibility of atomic species and ultra fine agglomerate structures, well below the range of conventional transmission or scanning electron microscopy, the atom-probe FIM offers the identification of impurity atoms, of interstitials, of atomic species in ordered alloys, or diffusion dependent compositional changes with increasing depth below the surface. Controlled field evaporation in the atom-probe does not affect the composition and structure of the underlying layers, in contrast to the deep reaching lattice damage by atomic displacement and implantation encountered in the more conventional microanalytic methods of depth profiling using ion impact sputtering for layer removal. Bulk analysis on an atomically fine scale is particularly successful in determining segregations at grain boundaries,[38] which can be followed into the depth of the specimen, as well as in revealing the composition of nuclei of precipitates in special alloys,[39] or depth profiling of diffusion depleted surface layers.[40] Most of the applications of the atom-probe have so far been in the study of bulk properties of the specimen, but in the context of the present symposium we will emphasize the aspects of surface analysis by the ToF atom-probe, pointing to the particular advantages of the high resolution version of the instrument. Inasmuch as atom-probe bulk analysis also depends upon the proper interpretation of the imaging process of the FIM and of the intricacies of field evaporation, the understanding of the surface effects involved here is again of paramount importance.

The first analysis with a magnetic sector mass spectrometer of the products of field evaporation under the operational conditions of the FIM by Barofsky and Müller[41] showed that several metals field evaporate in the form of doubly charged ions, as had been predicted by the image force theory.[42,43] So it came as a surprise when the prototype ToF atom-probe revealed the occurrence of triply and even quadruply charged ions.[4] Further work with improved versions of the instrument[44,45,46] showed the evaporation of W^{4+} and W^{3+} from tungsten tips, of Ta^{4+} and Ta^{3+} from tantalum, or Ir^{3+} and Ir^{2+} from iridium, and of Rh^{3+} and Rh^{2+} from rhodium.

Field Absorption of Helium and Neon

The next unexpected observation was the occurrence of ion signals at the respective masses of the noble imaging gases helium, neon or argon.[47,48] Since at 10^{-5} Torr image gas pressure the arrival of an imaging gas molecule and ionization at an atomic surface site occurs at a typical rate of only 10 to 10^3 per second, the presence of a transient noble gas atom at almost any nanosecond instant of the atom-probe pulse is very unlikely. Thus, a noble gas atom must be waiting at the surface for the evaporation pulse, that is, it must be adsorbed up to a surprisingly high temperature of 200 K. More detailed experimental[48,49] and theoretical[50,51] considerations show the adsorption being due to a field induced dipole-dipole interaction bond of up to 0.2 eV, with the adsorbate located at the apex site[52] of the imaged kink site atoms of the substrate. As the binding energy increases with the square of the field, the noble gas remains adsorbed until the evaporation field of the substrate atom is reached. Then, surprisingly, field desorption of helium often occurs in the form of a molecular ion compound with the metal. While earlier observations with the straight ToF atom-probe were plagued by the uncertainties of a limited mass resolution resulting from premature field evaporation and by artifacts from afterpulses of the detector, our high resolution energy deficit compensated atom-probe has assured us that helium is field adsorbed at all imaged surface sites of the refractory metals (Fig. 10). From the brightly imaged crystallographic areas such as (111) of a tungsten tip, helium desorbs as a separate ion together with W^{3+} metal ions, while in the dimly imaged regions in the vicinity of (011) no He^+ signals at 4 amu are observed. Yet helium must be adsorbed at all kink sites as the metal atoms are

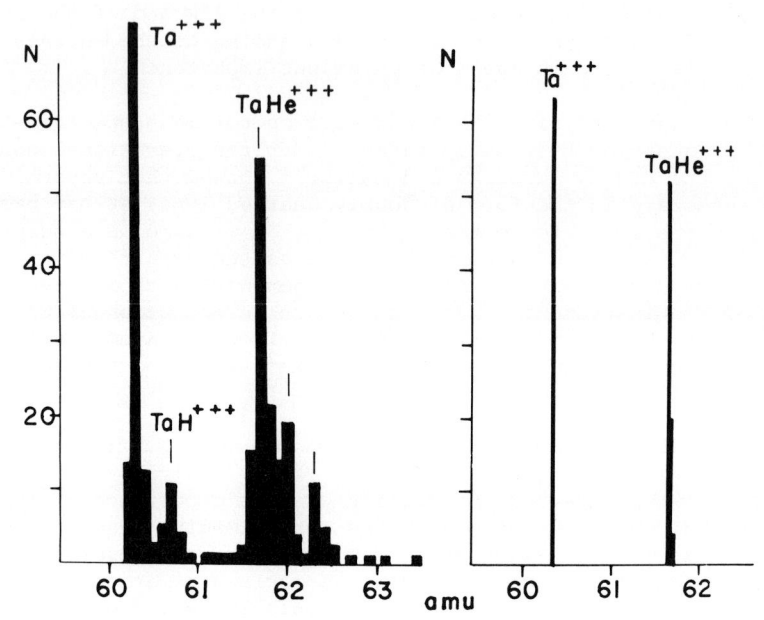

Fig. 10. Histograms of mass spectra of tantalum field evaporated in the presence of helium. At left, best performance of a straight atom-probe with tails due to afterpulses and energy deficits (from E.W. Müller, Naturwiss. 57, 222 (1970)). At right, a high resolution spectrum from the new energy focused atom-probe.

coming off as WHe^{3+}.[53] Subsequent to our work, Panitz[37] confirmed the regional distribution of He$^+$ desorbing from tungsten, using a time-gated field desorption microscope. Because of the poor mass resolution of this device it is impossible to identify the molecular ion WHe^{3+}. By generalizing our results with tungsten, Panitz also concluded the existence of IrHe^{3+} and IrNe^{2+} as being highly probable by assuming that "dark regions of the inert gas micrograph which correspond to bright regions in the substrate micrograph indicate areas of stable molecular ion formation". This conclusion is unjustified as long as it is not proved that helium is indeed adsorbed in these dark regions of the time-gated field desorption image. We were unable to confirm the existence of IrHe^{3+} and IrNe^{2+} speculated by Panitz to be highly probable. These species were also "imaged" by Waugh[19] but are most likely afterpulse artifacts from an excessively pulsed, time-gated single channel plate.

Although the determination of the areal distribution of ion species through the narrow confines of the probe hole is tedious compared to the more appealing one-shot imaging of a large section of the crystal hemisphere by the time-gated field desorption microscope, the results with the energy focused ToF atom-probe are more reliable.[54]

The occurence of metal helides (Table 1) varies considerably with crystallographic orientation and substrate temperature, peaking for the metals studied in detail around 80 K with 40% to near 100% of all metal ions coming off.[54] The highest stages of ionization of the metal ions do not form helide compounds.

The experience with the high resolution ToF atom-probe convinced us that some of the metal-neon ion compounds as well as many other metal-residual gas compound ions reported from work with the early atom-probe[49] were artifacts from free-space ionization. Before the availability of channel plate image intensification, the neon gas pressure had to be as high as 10^{-3} Torr for sufficient image brightness. Now, operating with several orders of magnitude lower image gas pressure, and rejecting free-space gas ions which have an energy deficit beyond the acceptance range of the 163° focusing deflector, the background of the ToF mass spectra remains free of artifacts to a level of one signal per 1000 specimen ions, mostly from small amplitude afterpulses at the double channel plate detector.

Table 1

Multiply Charged Metal and Metal-Helide Compounds Ions.
(low abundances)

W	$W^{5+}(2 \times 10^{-4})$	W^{4+}	W^{3+}, WHe^{3+}	W^{2+}
Mo		Mo^{4+}	Mo^{3+}, $MoHe^{3+}$	Mo^{2+}, $MoHe^{2+}(10^{-2})$
Ta		Ta^{4+}	Ta^{3+}, $TaHe^{3+}$	Ta^{2+}, $TaHe^{2+}(2 \times 10^{-2})$
Re		Re^{4+}	Re^{3+}, $ReHe^{3+}$	Re^{2+}
Ir			Ir^{3+}	Ir^{2+}, $IrHe^{2+}$
Pt			Pt^{3+}	Pt^{2+}, $PtHe^{2+}$
Rh			Rh^{3+}	Rh^{2+}, $RhHe^{2+}$

The search for metal-neon molecular ions became interesting again when with a magnetic sector atom-probe[13] palladium was found to form PdNe$^+$ ions in great abundance when hydrogen was added. With the very good mass resolution of this instrument all combinations of the six Pd isotopes with both ^{20}Ne and ^{22}Ne were clearly seen. Similar neides are found with Mo, Nb, Ti and Zr.[55] In the magnetic sector atom-probe field evaporation occurs at a rate 10^7 times slower than in the pulsed ToF instrument. Each metal-neon adsorbate surface complex is subject to excitation by electrons impinging with up to several hundred eV energy from free-space ionized hydrogen. It may be possible that an additional hydrogen atom at the surface is needed for the formation of the metal neïde ion, as these seem to occur only with metals that form hydrides. As an electronic excitation of the surface complex during the subnanosecond pulsing of the ToF atom-probe is unlikely, a large abundance of metal-neïdes in the ToF atom-probe or the time-gated field desorption microscope is not to be expected. Indeed, a thorough search for metal-neon molecular ions in the high resolution ToF atom-probe showed neïdes of W, Mo, Ta, Re, Ir and Pt to appear with a probability of less than 10^{-3} of all evaporated metal atoms.[54] Because of the limited viewing area of the probe hole we cannot claim to have looked at all crystal regions, but we did aim at the zones for which Waugh[19] reported high abundances. We did not find neïdes above the detection limit of gated field desorption microscopes.

Field Desorption of Hydrogen

The adsorption of hydrogen on metal surfaces is of fundamental interest, and although hydrogen atoms cannot be seen in the FIM, their presence as well as the formation of metal hydrides is readily detected by the atom-probe. The large mass difference makes the areal distribution of desorbing hydrogen easily accessible in the gated desorption microscope, while the detection of metal hydrides as desorbing species requires the high resolution of the energy focused ToF atom-probe or the magnetic sector atom-probe. Even at a low ambient pressure, hydrogen is seen to come off abundantly from all metals as H$^+$, and occasionally as H$_2^+$. As expected from the small size of the adsorbate, there seems not to be a distinct crystallographic preference. Panitz[56] published gated field desorption micrographs of H$^+$ from iridium, in which some preferential adsorption in the (321) region is suggested. However, the images also show random absorption all over the tip cap, with some clustering perhaps located at the many crystal defects seen in the helium FIM image of this particular tip.

In early work with the ToF atom-probe, it seemed that field desorbed metal hydride ions were obtained from all metals studied. However, when the tail of the mass lines of the metal ions was recognized as being due to energy deficits or detector afterpulses, and disappeared with the advent of the energy focused high resolution instrument, the occurrence of definitely established metal hydrides also declined. For instance with the isotope spectrum of tungsten fully resolved and no overspill of ion signals between the individual mass lines, we were unable to find any WH^{3+} or WH^{2+} species. At the same time, the presence of hydrogen at the surface was evidenced by abundant H^+ signals, some rare H_2^+ from the (111) plane, and the general appearance of W^{++}, which only shows up when adsorbed hydrogen reduces the evaporation field. Similarly, tantalum showed no TaH^{3+} (Fig. 10) while at room temperature Ta^{2+}, TaH^{2+} and TaH_3^{2+} are clearly seen. In earlier work, we had concluded a dissociation of field evaporating TaH^{3+} in the high field region near the tip, resulting in Ta^{2+} ions with an excess of kinetic energy, and in H^+ ions with an energy deficit.[21] Thus, in the straight ToF atom-probe these species were recognizeable by an apparent mass range from 85 to 90, peaking at 88 amu for Ta^{2+} and at 1.01 to 1.15 amu for the hydrogen. In the energy focused ToF atom-probe, these species are indistinguishable from Ta^{2+} at 90.5 amu and H^+ at 1.00 amu. This is one case in which the energy deficit compensation of the new atom-probe suppresses information. Although we may not have looked into all crystallographic regions, it appears that there are no stable hydrides formed with the field evaporation of rhodium, iridium, and platinum. On the other hand, hydrides are definitely present as field desorption products of beryllium (BeH^+ and BeH_3^+), of copper (CuH^+, most abundantly CuH_2^+, some CuH_3^+, and some CuH_4^+), and probably also with the non-refractory transition metals, of which ZrH^{2+} and PdH^+ are well established. In the magnetic atom-probe in which the desorption field needed for the seven orders of magnitude lower desorption rate is much reduced, several more metals are seen to come off as hydrides.[55] This suggests that a dissociation process like that of TaH^{3+} in the high field near the tip[21] may also be occurring with other metals when the fast pulsing of the ToF atom-probe is used.

Field Desorption of Other Gases

Field desorption of the other common molecular gases as well as of their chemical and field induced reaction products with the substrate metal have all been studied in a very preliminary and exploratory fashion only, and mostly with the simple straight atom-probe. Next to the field desorption of barium from tungsten,[57] the removal of an adsorbed oxygen layer is the oldest example of field desorption.[58] The atom-probe shows that from tungsten the adsorbate comes off as O^+, as well as triply and doubly charged WO, WO_2 and WO_3. The early mass spectrometric work by Vanselow and Schmidt[59] at high temperatures also produced various oxides of platinum. While nitrogen is assumed to be molecularly adsorbed on rhodium, the atom-probe spectra[25] reveal dissociation by the species N^{2+}, N^+ besides the prevalent N_2^+. The nitride RhN^{2+} is also seen. Similarly, the adsorption of CO on rhodium gave predominantly CO^+, but also many C^{2+}, some C^+, and the corresponding number of O^+ ions, as well as some RhC^{2+}. Corresponding ion species are also found with CO adsorbed on iridium. On tungsten field desorption of CO produces again C^{2+}, C^+, O^+, CO^+, W^{3+} and WC^{3+}. When neon was introduced as an imaging gas in these experiments with CO on W, all the above species appeared again, together with Ne^+, but the signal at 28 amu of undissociated CO^+ was absent. Most likely, all adsorbed CO was dissociated or desorbed, at a field much lower than the pulsed desorption field, by the electron shower that fell onto the surface from free-space ionized neon. Not surprising, the relative abundance of the dissociation products and of CO^+ in the absence of neon was strongly dependent on the pressure of CO in the 10^{-8} to 10^{-6} Torr range, due to electron bombardment from free-space ionization of CO. An interesting observation was made with the adsorption of hydrogen disulfide on rhodium, interacting at 78 K. The field desorption products were H^+, S^{3+}, S^{2+}, Rh^{3+} and Rh^{2+}. Neither singly charged sulfur nor sulfur molecule ions were detected at the high fields at which the metal substrate was evaporating. This is in contrast to the conditions at an about 10 times lower field where Block[60] found the desorption of ion species S_2^+ to S_8^+ from a multilayer covered sulfurized tungsten emitter, using a quadrupole mass filter and a temperature range from 150 to 500 K.

In all of the exploratory work listed before, very few attempts have been made of defining the crystallographic specificity of the various desorption products, although this information will eventually become a major objective of atom-probe work in the direction of catalysis and corrosion research. Also, very little has been done to identify the nature of scattered individual bright spots that appear in the field-ion image upon exposure to various reactive gases. Many of these adsorbates come off at relatively low fields, so that in a straight ToF atom-probe premature evaporation during the pulse rise produces ions with energy deficits. Thus, in their early work Brenner and McKinney[61] could identify only one-half of the mass signals obtained. Employment of the energy focused atom-probe should provide a more efficient identification of weakly bound surface constituents.

Aiming at Metallic Impurities

Aiming the atom-probe at one specific atomic site poses no particular problem, if the effective probe hole is not too small. With the largest 4.7×10^{-4} sr angular aperture of the energy focused atom-probe the aiming efficiency was experimentally determined[31] to be 42% over the entire tip surface, and 60% on the flat (011) plane of tungsten. On the latter plane initial ion trajectory deviations due to the local crystallographic surface structure are minimal, and the aiming yield equals the measured detection efficiency of the channel plate detector,[62] which is

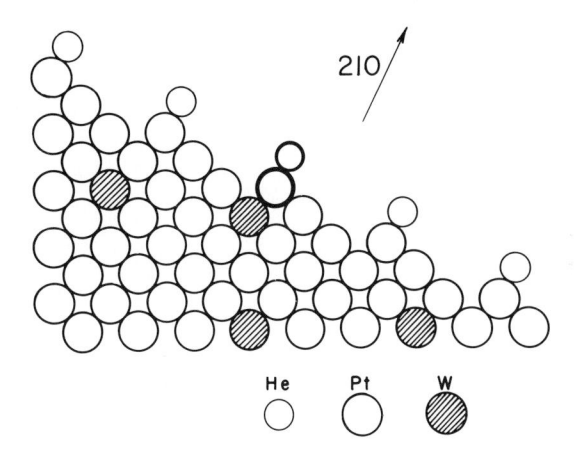

Fig. 11. Schematic cross section through a (210) plane of platinum with 8% tungsten. The protruding Pt atoms of the (210) surface are crowned by apex-adsorbed gas atoms.

essentially the ratio of capillary openings to total plane surface. For the aiming experiments, the extra bright spots seen on the surface of a Pt-8% W alloy (Fig. 11) were determined to actually represent a disturbance of the Pt lattice by a W atom in the next lower layer. A similar effect was seen with the aiming at the individual bright spots which remain on the (011) plane of a W-3% Re alloy. These are single atoms or clusters of 2 to 6 atoms. The 60% detected with the evaporation of these spots were 36% Re^{3+}, 10% $ReHe^{3+}$, and 14% W^{3+} or WHe^{3+}. Thus, about one in four of all the spots were in fact tungsten atoms, which owed their extra brightness to the electronic effects on the field ionization probability of the helium image gas by the strong bond of a Re atom in the next lower layer (Fig. 12). The latter was indeed detected subsequently with the field evaporation of the next tungsten layer using the same probe hole position. Such an analysis at the atomic level would have been very unreliable with the insufficient resolution of the straight atom-probe. The isotope spectrum of tungsten with masses at 180, 182, 183, 184 and 186 amu overlaps with that of rhenium at 185 and 187 amu, and a further complication arises from the formation of compound ions with helium as the imaging gas (Fig. 13).

While the signal at 185 is definitely due to ^{185}Re, the signal at 186 may be either ^{186}W or ^{182}W^{4}He, and at 187 may be ^{187}Re or ^{183}W^{4}He ions. Signals at 189 and 191 are definitely due to rhenium helides. Thus, a full investigation of the behavior of the Re spots on the (011) plane of W requires a good number of data for sufficient statistical weight. The detection limit of low percentage constituents of a bulk specimen again depends upon the statistical weight one wishes to achieve. Experiments with a rhodium tip and aiming at some extra bright random impurity spots (Fig. 9a) yielded 9 He$^+$, 78 Rh^{2+}, 37 RhHe^{2+}, 13 Pt^{2+}, 10 PtHe^{2+} and 1 Pt^{3+} ions. Subsequent fixed probe hole evaporation through a total of 3000 Rh atoms yielded 3 Pt^{2+} signals, affirming a concentration of a 0.1% Pt impurity as the cause of the extra bright spots in this particular specimen of a rhodium wire material that was first imaged in an FIM some 17 years ago.[63]

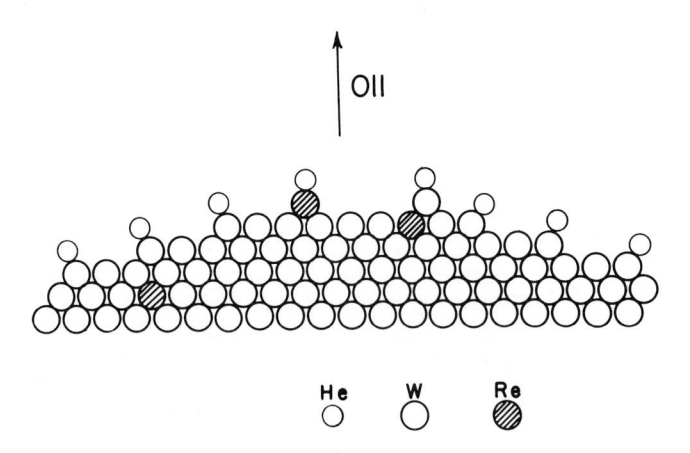

Fig. 12.   Schematic cross section through the (011) pole of a tungsten tip, containing 3% Re. A rhenium and a tungsten adatom remaining on the (011) plane are holding helium atoms at their apex.

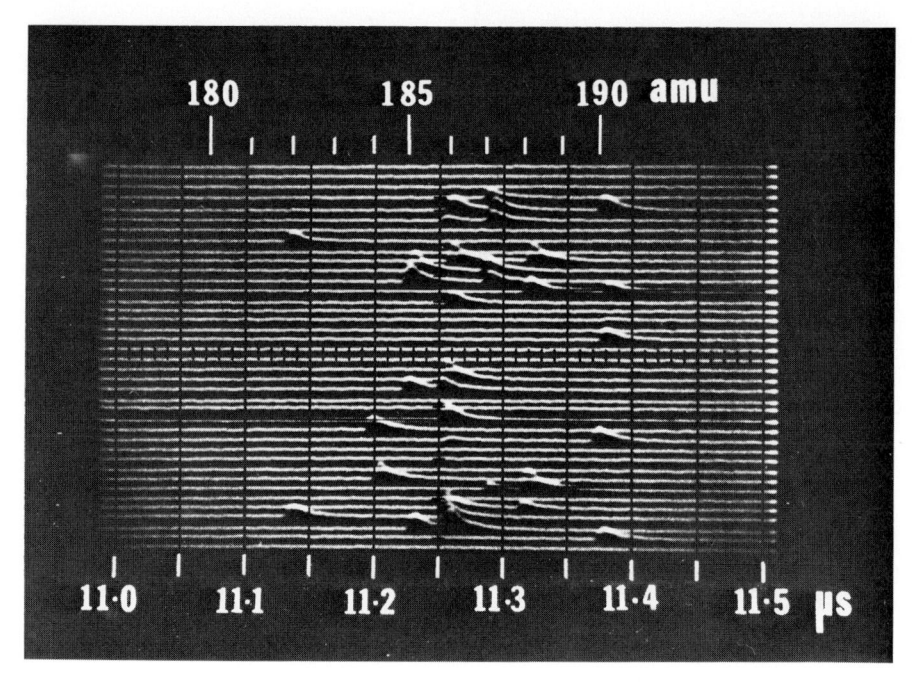

Fig. 13. Oscilloscope traces from a W-26% Re tip imaged in 5 x 10^{-6} Torr He, taken at a sweep rate of 50 nanosec/div. The mass signals of the triply charged ions line up within ±3 nanosec, representing a mass resolution ΔM/M better than 1/1900 (ref. 31).

CONCLUSION

The atom-probe is a unique microanalytical tool of ultimate sensitivity. In its most advanced form of the energy focused device, it can unambiguously determine the nature of a single surface particle as seen in the atomically resolved image of a field ion microscope. The atom-probe has greatly extended our understanding of the imaging mechanism[64] of the FIM and of the physical processes going on at the surface of the specimen. As its limitations due to the peculiar necessity of preparing a tip from the metallic specimen and exposing it to an extremely high field are recognized, the atom-probe will become increasingly useful as a very effective tool of surface analysis[65] in fundamental as well as applied research in areas such as physical metallurgy, chemisorption, catalysis and corrosion.

REFERENCES

1. E. W. Müller, Z. Physik, 131, 136 (1951).
2. E. W. Müller, in *Advances in Electronics and Electron Physics*, Vol. XIII, 83, (1960), L. Marton, Ed., Academic Press, New York.
3. E. W. Müller and T. T. Tsong, *Field Ion Microscopy, Principles and Applications*, Elsevier, New York (1969).
4. E. W. Muller, J. A. Panitz and S. B. McLane, Rev. Sci. Instrum., 39, 83 (1968).
5. E. W. Müller and T. T. Tsong, *Progress in Surface Science*, Vol. 4, Part 1, (1973), S. G. Davison, Ed., Pergamon Press, Oxford.
6. E. W. Müller, LABEX Lecture, London (1973), Laboratory Practice, 22, 408 (1973).
7. S. S. Brenner and J. T. McKinney, Surface Sci., 23, 88 (1970).
8. P. J. Turner, B. G. Regan and M. J. Southon, Vacuum, 22, 447 (1972).
9. T. M. Hall, A. Wagner, A. S. Berger and D. S. Seidman, Cornell University Materials Science Center Report #2357 (1975).
10. A. S. Berger, Rev. Sci. Instrum. 44, 592 (1973).
11. M. G. Inghram and R. Gomer, J. Chem. Phys., 22, 1274 (1954).
12. H. D. Beckey, *Field Ionization Mass Spectrometry*, Pergamon Press, Oxford (1971).
13. E. W. Müller and T. Sakurai, J. Vac. Sci. & Technol., 11, 878 (1974).
14. R. J. Walko and E. W. Müller, Phys. Stat. Sol.,(a) 9, K 9 (1972).
15. A. J. W. Moore and J. A. Spink, Abstracts, p. 29, 21st Field Emission Symposium, Marseille (1974).
16. E. W. Muller, S. V. Krishnaswamy, S. B. McLane, T. Sakurai and R. Walko, Abstracts, p. 61, 19th Field Emission Symposium, Urbana, Illinois (1972).
17. J. A. Panitz, J. Vac. Sci. & Technol., 11, 206 (1974).
18. J. A. Panitz, Critical Reviews in Solid State Sciences, 5, 153 (1975).
19. A. R. Waugh, Abstracts, p. 69, 22nd Field Emission Symposium, Atlanta, Georgia (1975).
20. E. W. Müller, S. B. McLane and J. A. Panitz, Surface Sci., 17, 430 (1969).
21. E. W. Müller, S. V. Krishnaswamy and S. B. McLane, Surface Sci., 23, 112 (1970).
22. P. J. Turner, B. G. Regan and M. J. Southon, Surface Sci., 35, 336 (1973).
23. S. S. Brenner and J. T. McKinney, Rev. Sci. Instrum., 43, 1264 (1972).

24. E. W. Müller, S. V. Krishnaswamy and S. B. McLane, Rev. Sci. Instrum., 44, 84 (1973).
25. E. W. Müller, Ber. d. Bunsenges., 75, 979 (1971).
26. E. W. Müller and S. V. Krishnaswamy, Rev. Sci. Instrum., 45, 1053 (1974).
27. S. V. Krishnaswamy and E. W. Müller, Rev. Sci. Instrum., 45, 1049 (1974).
28. E. W. Müller, J. Microscopy (Oxford), 100, 121 (1974).
29. E. W. Müller, Proc. Second Intern. Conf. on Solid Surfaces, Kyoto, Japan, Japan J. Appl. Phys., Suppl. 2, part 2, 1 (1974).
30. W. P. Poschenrieder, Int. J. Mass Spectrom. Ion Phys., 9, 357 (1972).
31. S. V. Krishnaswamy, S. B. McLane and E. W. Müller, Rev. Sci. Instrum., 46, 1237 (1975).
32. E. W. Müller, Abstracts, p. 108, 15th Field Emission Symposium, Bonn (1968).
33. J. A. Panitz, Ph.D. Thesis, The Pennsylvania State University, 1969.
34. R. S. Chambers and G. Ehrlich, Abstracts, p. 74, 22nd Field Emission Symposium, Atlanta, Georgia (1975).
35. E. W. Müller and T. T. Tsong, Reference 5, page 48.
36. A. R. Waugh, E. D. Boyes and M. J. Southon, Nature, 253, 342 (1975).
37. J. A. Panitz, J. Vac. Sci. Technol. 12, 210 (1975).
38. M. J. Southon, E. D. Boyes, P. J. Turner and A. R. Waugh, Surface Sci., 53, 554 (1975).
39. S. R. Goodman, S. S. Brenner and J. R. Low, Jr., Metall. Trans., 4, 2371 (1973).
40. S. V. Krishnaswamy, S. B. McLane and E. W. Müller, J. Vac. Sci. Technol., 11, 899 (1974).
41. D. F. Barofsky and E. W. Müller, Surface Sci., 10, 177 (1968).
42. E. W. Müller, Phys. Rev., 102, 618 (1956).
43. D. G. Brandon, Surface Sci., 3, 1 (1965).
44. E. W. Müller, in *Applications of Field-Ion Microscopy*, p. 59, R. F. Hochman, E. W. Müller and B. Ralph, Eds., Georgia Inst. Technol. (1969).
45. E. W. Müller, S. B. McLane and J. A. Panitz, 4th European Reg. Conf. Electron Microscopy, p. 135, Tipografia Poliglotta Vaticana, Rome (1968).
46. S. S. Brenner and J. T. McKinney, Appl. Phys. Letts., 13, 29 (1968).
47. E. W. Müller, Centenary Lecture, Chem. Soc., London, Quart. Revs., 23, 177 (1969).
48. E. W. Müller, S. B. McLane and J. A. Panitz, Surface Sci., 17, 430 (1969).
49. E. W. Müller, S. V. Krishnaswamy and S. B. McLane, Surface Sci., 23, 112 (1970).
50. T. T. Tsong and E. W. Müller, Phys. Rev. Letts., 25, 911 (1970).
51. T. T. Tsong and E. W. Müller, J. Chem. Phys., 55, 2284 (1971).

52. E. W. Müller and S. V. Krishnaswamy, Surface Sci., 36, 29 (1973).
53. E. W. Müller, S. V. Krishnaswamy and S. B. McLane, Phys. Rev. Letts., 31, 1282 (1973).
54. S. V. Krishnaswamy and E. W. Müller, J. Vac. Sci. Technol. 13, 665 (1976).
55. S. Kapur and E. W. Müller, to be published.
56. J. A. Panitz, in *Hydrogen Energy*, Part B, T. N. Veziroghu, Ed., Plenum Publishing Company, New York, 1975, p. 1079.
57. E. W. Müller, Naturwiss., 29, 533 (1941).
58. E. W. Müller, Z. Elektrochem., Ber. d. Bunsenges., 59, 372 (1955).
59. A. Vanselow and W. R. Schmidt, Z. Naturforsch., 21 a, 1690 (1966).
60. J. H. Block, in *Methods of Surface Analysis*, Chapter 9, A. W. Czanderna, Ed., Elsevier, New York 1975.
61. S. S. Brenner and J. T. McKinney, Surface Sci., 20, 411 (1970).
62. E. W. Müller and T. T. Tsong, Reference 5, page 24.
63. E. W. Müller, Reference 2, page 163.
64. E. W. Müller, Critical Reviews in Solid State Sciences, 6, issue 2, 85-109 (1976).
65. E. W. Müller, in *Methods of Surface Analysis*, Chapter 8, pp. 329-378, A. W. Czanderna, Ed., Elsevier, New York 1975.

Applications of Mössbauer Spectroscopy to the Study of Corrosion

G.W. Simmons and H. Leidheiser, Jr.

*Center for Surface and Coatings Research
Lehigh University
Bethlehem, Pennsylvania 18015*

    Each of the transmission, reflection and emission Mössbauer spectroscopic techniques provide a unique physical method for studying corrosion. An introduction to each of these methods is presented and examples are given to demonstrate how these methods have been used to study corrosion phenomena.

INTRODUCTION

    The purpose of this presentation is to provide an introduction to the application of Mössbauer spectroscopy as a method for studying corrosion phenomena. This subject has already been reviewed recently in considerable detail.[1,2] The present effort consequently consists of some of the highlights from these reviews. Emphasis is placed on concepts of Mössbauer spectroscopy, on the various experimental methods and on a few typical examples of the application of each of these methods for studying corrosion. Although Mössbauer spectroscopy is applicable to only a limited number of metals, the technique nevertheless has provided and will continue to provide a tool for practical as well as fundamental corrosion studies primarily of iron, tin, cobalt, and alloys containing these elements. The considerable information available in the literature on the characterization by Mössbauer spectroscopy of iron and tin compounds provides a strong basis for applying γ-ray resonance spectroscopy to corrosion studies. The isomer shift, quadrupole splitting, and magnetic hyperfine interaction have been measured for a large number of organic and inorganic compounds formed during corrosion of these metals. In particular, the oxides and hydroxides of iron have been studied extensively. Mössbauer spectroscopy is, therefore, readily applicable to qualitative analysis of corrosion products. In many cases, direct quantitative measurements can be made of the corrosion product(s) that consist either of a single phase or of a complex mixture of corrosion species. A particular advantage of the Mössbauer technique is that analysis can be made of either amorphous or crystalline corrosion products. In addition to the analytical applications, it is possible to determine some of the chemical and physical properties of oxides that are basic to the understanding of corrosion phenomena. Experimental techniques have been developed that permit studies of corrosion films with thickness ranging from tenths of a nanometer to several micrometers. Furthermore, it is possible in many cases to conduct studies <u>in situ</u>. Mössbauer spectroscopy has been used for qualitative and quantitative analyses of corrosion products, for studies of defect structure of corrosion products, kinetic studies, for studies of passivity and corrosion inhibition, for determining extent of corrosion beneath a coating, in studies related to stress corrosion cracking and hydrogen embrittlement, and has been used in studies of diffusion in oxides. This presentation, however, does not include all of these subjects. Most of the examples of applications are taken from studies made in our laboratory. A more complete treatment of Mössbauer studies of corrosion related phenomena has been given in the reviews referenced above.

PRINCIPLES OF MÖSSBAUER SPECTROSCOPY

The Mössbauer effect is simply the resonant emission and absorption of gamma radiation that takes place in the nuclei of certain isotopes. This resonance is possible only if the recoil energy associated with the emission and absorption processes is much less than the natural line width of the gamma radiation. Furthermore, these essentially recoilless emission and absorption events are possible only if the nuclei are rigidly bound and if the recoil energy does not greatly exceed the lattice excitation energy of the solid in which the nuclei are bound. Since active isotopes do not exist for all elements and since the energy and lifetimes of the gamma radiation from isotopes that are active must be within certain limits to meet the criteria required for resonance, Mössbauer spectroscopy is not possible for many of the elements. The Fe^{57} isotope of iron, however, is Mössbauer active which fortunately provides the opportunity for the application of Mössbauer spectroscopy to studies of corrosion of iron and steel as well as other alloys that contain iron. Most of the material presented in this paper, therefore, is about the Mössbauer spectroscopy of Fe^{57}.

The narrow line width of the 14.4 keV gamma radiation from Fe^{57} makes it possible to resolve the small perturbations induced on the nuclear energy levels by the electron environment at the nucleus. The possible types of interaction and their effect on the nuclear levels of Fe^{57} are shown in Fig. 1. The isomer shift is produced by coulombic interaction of the s electrons with the nucleus. The magnitude of this interaction is a function of a nuclear and an electron contribution. The nuclear contribution

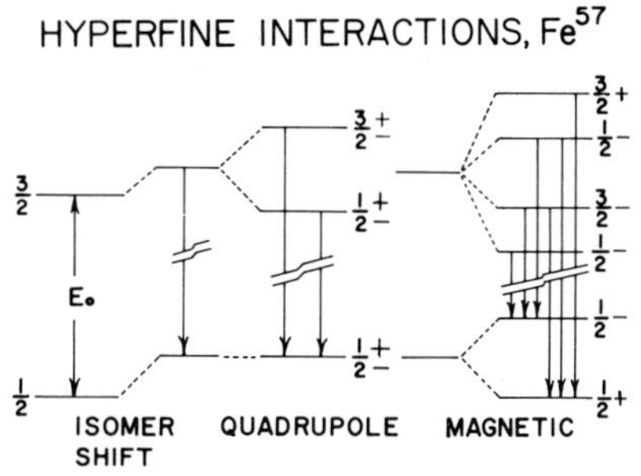

HYPERFINE INTERACTIONS, Fe^{57}

Fig. 1. Schematic energy level diagram of the possible hyperfine interactions of the Fe^{57} nucleus.

is associated with the difference in the size of the nucleus between the excited and ground states, and the electron contribution is related to the s electron density at the nucleus. The non-spherical shape of the nucleus in the I = 3/2 excited state gives rise to a quadrupole splitting of this level when the symmetry of the electron environment is other than cubic. Spin polarization of the s electrons is the major contribution to the internal magnetic field at the nucleus which produces the Zeeman splitting of the nuclear levels as depicted in Figure 1.

The measured values of these hyperfine interactions are used to characterize the chemical and physical properties of compounds containing Fe^{57}. The gamma ray resonance spectrum of a specimen is generated by adsorption of radiation emitted from a source (for Fe^{57} the parent source isotope is Co^{57}) which has a single emission line. The gamma radiation from the source is given a range of energies by moving the source at various velocities relative to the absorber. Mössbauer spectra are obtained by measuring the amount of gamma ray absorption as a function of velocity of the source. A more complete treatment of Mössbauer spectroscopy can be found elsewhere.[3]

APPLICATIONS TO STUDIES OF CORROSION

Mössbauer spectra can be obtained by the transmission, scattering and emission techniques shown schematically in Fig. 2. Each of these spectroscopic methods has features that make it unique to particular types of corrosion studies. Each of these

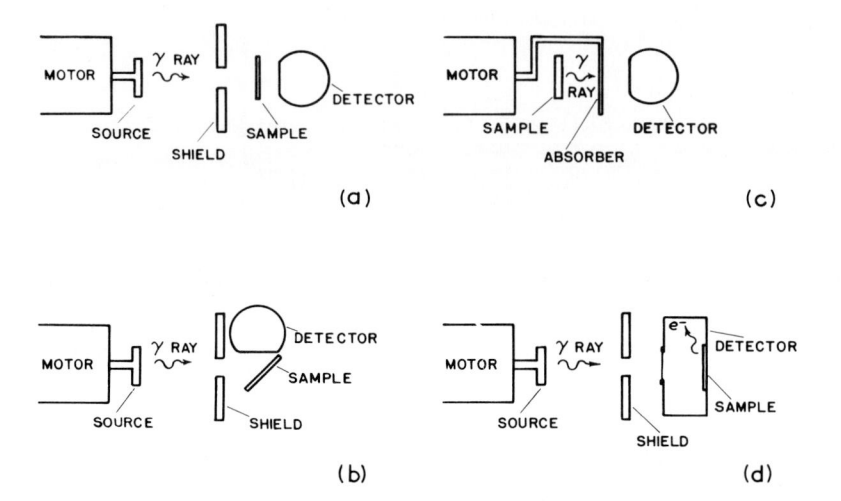

Fig. 2. Experimental arrangements used in Mössbauer experiments (a) transmission, (b) x-ray and γ-ray scattering, (c) emission and (d) electron scattering.

techniques is described separately along with typical application(s).

Transmission Techniques

Transmission Mössbauer spectroscopy is directly applicable to corrosion studies of thin metal foils, and the major application has been in studies of relatively thick corrosion products (10^3 to 10^5 nm). In situ studies of the corrosion of thin films and foils are possible with suitably designed reaction cells with windows that are essentially transparent to the Mössbauer γ-rays. The corrosion products on thick materials can be studied by transmission after removal from the substrate. The relative concentrations for any number of phases 1, 2, 3 ... n can be determined from the following relationship:

$$\frac{N_n}{\sum_{j=1}^{n} N_j} = \frac{A_n/f_n}{\sum_{j=1}^{n}(A_j/f_j)} \tag{1}$$

where for each phase, N is the number of atoms/cm^3 of the Mössbauer element, A is the area under the resonance peak(s) and f is the recoil-free fraction. Absolute quantitative analysis is possible, but is more difficult since it is necessary to determine accurately the background counts and the fundamental parameters such as recoil-free fraction, linewidth of the source, and the resonance cross sections.

Channing and Graham[4] have extensively studied by Mössbauer spectroscopy the growth behavior of Fe_3O_4 and α-Fe_2O_3 formed on iron at 450° and 550°C. The transmission technique was employed, and the specimens were metal foils (∼10μm) that were oxidized in one atmosphere of oxygen. Changes in the relative amounts of the phases Fe, Fe_2O_3 and Fe_3O_4 as a function of time for different temperatures are summarized in Figure 3. At all temperatures, it can be seen that Fe_3O_4 is formed initially and after the iron foil is nearly completely oxidized the α-Fe_2O_3 phase begins to form at the expense of the Fe_3O_4 phase. The rate of oxidation of the magnetite to hematite followed a parabolic rate law. The oxidation was, therefore, controlled by ionic diffusion through the α-Fe_2O_3 layer. It was not possible, however, to identify whether the differing species were cations or anions.

Reflection Techniques

Gamma rays, x-rays or electrons produced after resonance absorption may be used for reflection spectroscopy. The energies and yields for the resonantly scattered radiation in the case of iron will now be given. After resonance absorption of the 14.4 keV γ-rays by Fe^{57}, approximately 90% of the excited nuclei revert to the ground state by conversion electron emission, of which 80% are K-electrons with energies of 7.3 keV. The other 10% of the nuclei decay to the ground state by re-emission of 14.4 keV γ-rays. Approximately 60% of the atoms that are ionized as a result of K-conversion subsequently release energy by the emission of KLL Auger electrons with 5.4 keV energy, and the remaining 40% undergo de-excitation by emission of 6.3 keV K_α x-rays. The resonantly scattered γ-rays and x-rays are applicable for studying advanced stages of corrosion, and the conversion and Auger electrons are more suitable for investigating thin corrosion layers. The major advantage of reflection spectroscopy is that bulk specimens can be studied while the corrosion films are still attached to the substrate. Another significant advantage of scattering techniques

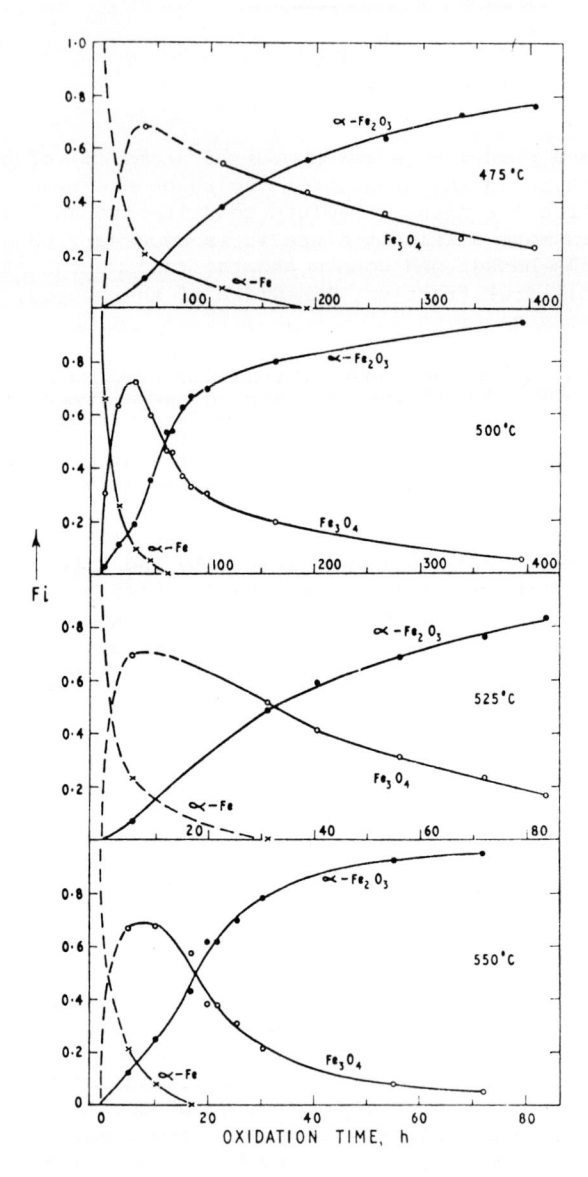

Fig. 3. The fraction of total Fe[57] contained within each component of an iron foil as a function of heating time in one atmosphere of oxygen at the temperature specified. Starting foil thickness 9.6 μm (Channing and Graham[4]).

is the relatively high signal-to-noise ratio. In transmission experiments, the background count rate (or nonresonance count rate) is largely due to the γ-rays that have not been absorbed by the specimen, consequently, the total background counts do not differ appreciably from the number of incident γ-rays. On the other hand, in scattering geometry the detector is shielded from direct irradiation by the source, and most of the background, therefore, originates from non-resonant scattering of radiation in the absorber. The cross sections for the non-resonant phenomena in scattering experiments are a few orders of magnitude lower than for the resonance effects, consequently, higher signal-to-noise is possible for reflection than for transmission spectroscopy. Because of the high internal conversion coefficients, most reflection Mössbauer spectroscopy studies have utilized the x-rays and electrons associated with internal conversion rather than the resonantly scattered γ-rays. The main features of the x-ray and electron scattering techniques, in terms of applicability to corrosion studies, will be discussed separately.

There is a significant difference in surface sensitivity between conversion x-ray and conversion electron reflection Mössbauer spectroscopy owing to differences in depths from which the two types of measured radiation originate within a specimen. The applicability of the x-ray reflection method to corrosion studies will be discussed first. Swanson and Spijkerman[5] have determined experimentally that 78% of back-scattered x-ray signal comes from the first 5×10^{-4} cm of surface depth in iron and that 93% of the signal originates in the first 1.2×10^{-3} cm. Quantitative analysis in x-ray scattering experiments requires a detailed description of both the attenuation of resonant γ-rays as they enter the absorber and the attenuation of resonantly produced x-rays that are emitted from specific depths within the scatterer. Terrell and Spijkerman[6] have derived theoretical expressions for the intensity of the resonantly scattered x-ray signal as a function of resonator thickness. The determination of thickness for single phase on iron is possible with these expressions by using the appropriate resonant absorption cross section for the 14.4 keV γ-ray and mass absorption coefficient for the 6.3 keV x-ray. Thickness determinations are, however, difficult for multiphase corrosion layers, since the attenuation of resonant γ-rays and scattered x-rays would have to be considered separately for each phase.

Resonant scattering of electrons is a Mössbauer spectroscopic technique that offers the opportunity for obtaining qualitative and quantitative information about the chemical and physical properties of thin corrosion layers. The relatively high yield of resonantly produced electrons of moderately low energies makes possible Mössbauer spectroscopy with a high surface sensitivity, since the escape depth of electrons originating within an absorber is limited by a high probability for inelastic scattering. Swanson and Spijkerman[5] have determined experimentally the escape depth of resonantly produced electrons in natural iron by taking Mössbauer spectra of known thickness of iron films deposited onto a stainless steel substrate. From the areas under the resonance peaks for iron and stainless steel, they calculated that 65% of the signal originates within the first 60 nm of the surface. Simmons et al.[7] have demonstrated that oxide films of approximately 5 nm in thickness can be detected on Fe^{57}-enriched iron surfaces. For back-scattered K-conversion electrons, Krakowski and Miller[8] have derived expressions for the area resonance peaks and for the electron signal intensity at maximum resonance as functions of resonator thickness. The integrals describing the electron signal are expressed in terms of the reduced resonator thickness $\mu_K t$ and the ratio μ_R/μ_K, where μ_K and μR are the linear absorption coefficients for conversion electrons and for resonant γ-rays, respectively. Quantitative analysis, in this case, is possible by using the appropriate values of μ_K and μ_R for the specimen under study. Single phase corrosion films are amenable to quantitative analysis, but unfortunately such analysis is not easily accomplished for multiphase and/or multilayered films. The major shortcoming of the resonant electron scattering technique is that the specimens must be placed inside a flow counter. Since changes in the composition of the flow gas strongly influence the counting efficiency, it is not possible to conduct experiments either under in situ conditions or at low temperatures. Furthermore, some corrosion products may change composition by dehydration in the relatively dry environment of the flow counter.

Mössbauer spectra of corrosion films as a function of depth can be obtained by energy analysis of K-conversion electrons, since the electron energy losses are a function of escape depth. Krakowski and Miller[8] have determined theoretically the factors that limit the depth resolution of electron-scattering Mössbauer experiments. They concluded that for scattered electrons with energies greater than approximately one-half of the initial conversion electron energy, the energy resolution versus depth is sufficient to permit Mössbauer spectra to be obtained from selected regions near a specimen surface. Furthermore, for good depth resolution sufficient energy separation between the conversion and Auger electrons is required. The latter criterion is satisfied in the case of tin since there is a 16.6 keV difference between the energies of the L-conversion and LMM Auger electrons. Prospects for the application of this spectroscopic technique with good depth resolution are not as favorable in the case of iron since energy separation of the K-conversion and KLL Auger electrons is only 1.9 keV.

Simmons et al.[7] used the K-conversion electron reflection method to study oxide films formed by the thermal oxidation of Fe^{57} enriched iron specimens at 225°, 350°, and 450°C. Mössbauer spectra were obtained for oxide thicknesses ranging from approximately 5 to several tens of nm. Figure 4 shows the spectra of iron after oxidation at 225°C for various times. The Fe_3O_4 phase shown in Figure 4 was estimated to be approximately 12 nm thick. The equal intensities of the A site and B site resonance lines indicate that the oxide formed at 225°C is either non-stoichiometric F_3O_4 or is a mixture of the stoichiometric oxides Fe_3O_4 and γ-Fe_2O_3. In either case, it can be concluded that oxidation at 225°C produces a cation-deficient $Fe_{3-v}O_4$ oxide phase. Oxidation at 350°C for short times produced a duplex film consisting of Fe_3O_4 and α-Fe_2O_3 as shown in Figure 5. The spectrum obtained after oxidation at 450°C indicated that a single oxide, nearly stoichiometric Fe_3O_4, was formed. The absence of α-Fe_2O_3 after the short-time oxidation at 450°C was attributed to an increased cation flux at the higher temperature (see Figure 6).

Bonchev et al,[9] have obtained Mössbauer spectra of corrosion films on tin as a function of depth by energy analysis of Sn^{119} L-conversion electrons. An electron spectrometer was used to focus conversion electrons of a definite energy onto a scintillation detector. A Sn^{119}-enriched tin specimen was exposed to bromine vapor for approximately 10 seconds and the Mössbauer spectra were obtained as a function of the magnetic focusing field of the spectrometer. For electrons with lowest measured energy, a Mössbauer spectrum of the β-Sn substrate was produced and for electron of higher energies, corrosion layers of $SnBr_2$ and $SnBr_4$ were detected. A similar experiment was performed with tin which was exposed to fuming nitric acid vapors. The relative intensities of the β-Sn and SnO_2 Mössbauer resonance lines observed in this case were measured as a function of electron energy. The relatively small dependence that was found for the β-Sn/SnO_2 intensity ratio as a function of electron energy was attributed to a highly inhomogeneous oxide film.

Emission Techniques

Emission spectra of Mössbauer nuclei are obtained by doping a specimen with a source isotope, and performing a conventional transmission experiment with a single line resonant absorber. In the case of Co^{57}-doped specimens, for example, the emission Mössbauer spectra arises from the Fe^{57} "probe" atoms, and the observed isomer shift, quadrupole splitting, and magnetic hyperfine splitting provide chemical and physical information about the host matrix. For corrosion studies, this technique has the important advantage that investigations can be readily carried out with possible surface sensitivity of less than one atomic layer. Surface sensitivity is obtained simply by controlling the thickness of the doped layer on the specimen surface. The low scattering cross-section of γ-rays allows in situ studies of corrosion in aqueous environments as well as nondestructive determination of the extent of corrosion beneath thin organic coatings.

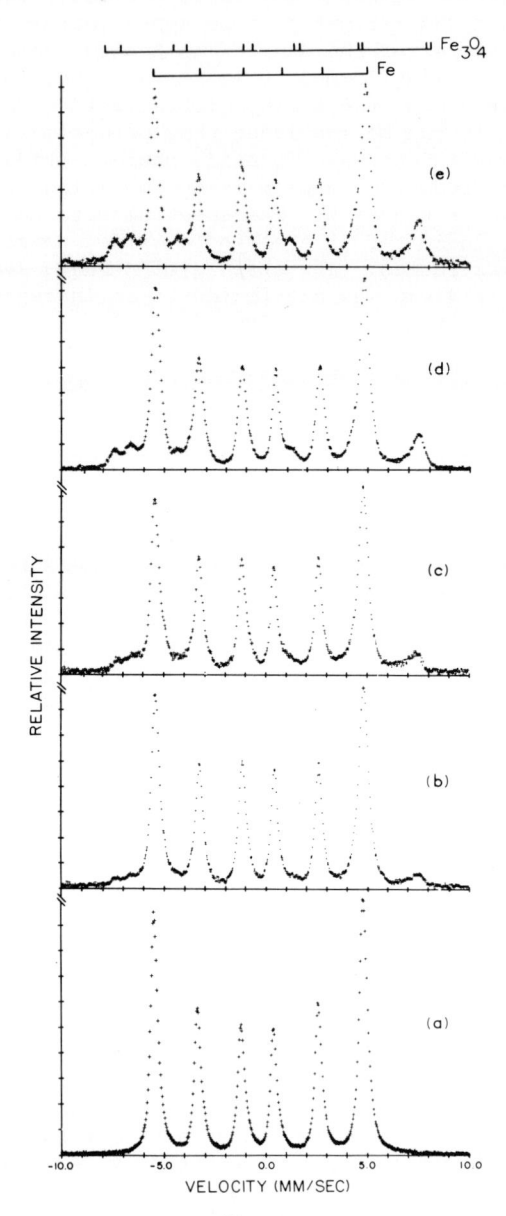

Fig. 4. Conversion electron Mössbauer spectra of iron oxidized at 225°C for specific times (a) before oxidation, (b) 5 minutes, (c) 15 minutes, (d) 120 minutes and (e) 1,000 minutes (Simmons, et al.[7]).

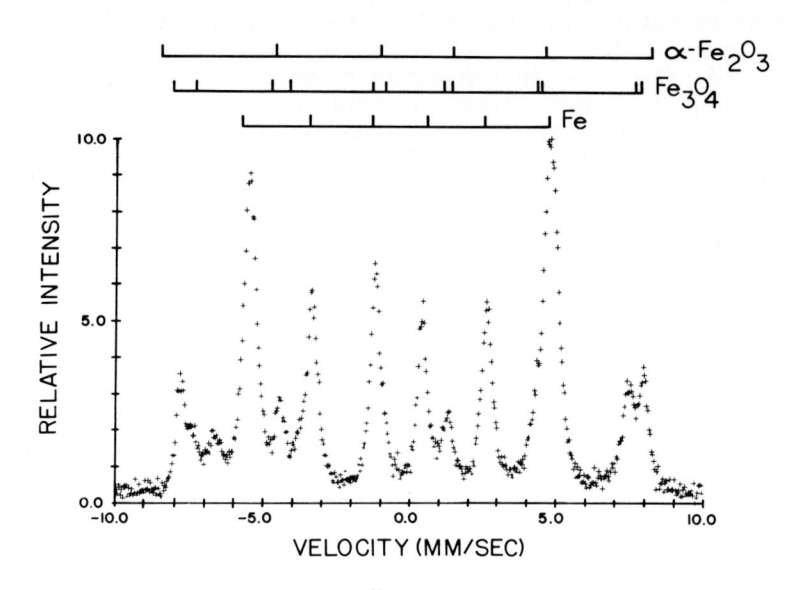

Fig. 5. Conversion electron Mössbauer spectra of iron after 5 minutes oxidation at 350°C (Simmons, et al.[7]).

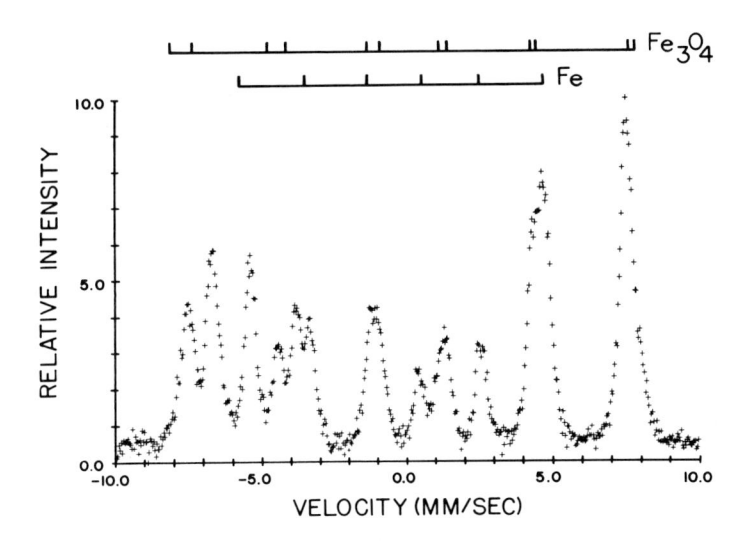

Fig. 6. Conversion electron Mössbauer spectra of iron after 10 minutes oxidation at 450°C (Simmons et al.[7]).

Some potential problems arise with the application of the emission technique when the source isotope decays by electron capture. In the case of Co^{57}, for example, electron capture produces an ionized K-level in Fe^{57} daughter atoms, and the de-excitation that follows by Auger cascade can produce charge states as high as +7 in the valence levels of Fe^{57} (Pollak[10]). If these nonequilibrium charges on the daughter Fe^{57} atoms have a lifetime on the order of the nuclear excited state ($\sim 10^{-7}$s), then the emission spectrum will show resonance lines that do not represent the intrinsic properties of the parent Co^{57} atoms. The stability of charges produced by Auger aftereffects depends upon the chemical and physical environment of the Fe^{57} nucleus in the host. Normal charge states are observed with emission spectroscopy for metals and alloys that are doped with Co^{57}, since charges on the Fe^{57} atoms are rapidly equilibrated by highly mobile conduction electrons. The charge equilibration is slower for insulating materials, and nonequilibrium charges as high as +3 have been observed (Wickman and Wertheim[11]). Stability of these charge states in insulators has been attributed, in some cases (CoO for example), to localized lattice effects and nonstoichiometry. Lattice energy, crystal field effects, and electron exchange with neighboring cations, however, may also contribute in some compounds to the formation or stabilization of ionic charges different from the parent ionic state. The possible formation of charged Fe^{57} species different from the original charge on Co^{57} imposes a complication in the interpretation of spectra, nevertheless emission spectroscopy can be useful for obtaining important information of chemical and physical interest to corrosion studies.

Leidheiser, et al.[12] have demonstrated that emission Mössbauer spectroscopy may be applied to nondestructive studies of the rate of corrosion at the metal-coating interface of a polymer-coated cobalt surface. Cobalt doped with Co^{57} was electrodeposited onto a cobalt substrate. The mass deposited was equivalent to 5 nm assuming uniform deposition. The specimen was then coated to a thickness of 0.001 cm with polybutadiene, and cured at 200°C in air for 30 min. Comparisons were made of the emission spectrum of freshly prepared specimen with the spectrum after the specimen had been exposed to 3% NaCl solution for 84 hours. In addition to the six lines from the unreacted metal, a central line(s) originating from ionic cobalt was observed. In the case of freshly prepared sample, the ionic cobalt contribution is from the thin oxide that was present on the surface prior to the application of the coating. The increase in the intensity of the center line after exposure to the salt solution was attributed to corrosion that had taken place beneath the polymer film. Changes in the ratio of resonance areas from the ionic and metallic cobalt as a function of exposure time to the salt solution provided a means of evaluating the protective properties of the coating. No attempt, however, was made to identify the corrosion product(s) in this study.

Emission Mössbauer spectroscopy technique was successfully demonstrated to be an effective method for in situ studies of changes in cobalt surfaces as a function of polarization (Simmons and Leidheiser[13,14]). Polarization of cobalt in buffered borate solution, pH 8.5, produced a classical potential versus current dependence. An active-to-passive transition occurs between -500 mV and -300 mV (vs. SCE), and cobalt remains passive at higher anodic potentials up to +500 mV. Above +500 mV, thick anodic film formation and oxygen evolution occurred commensurate with an increase in anodic current. Specific polarization potentials were chosen for study which represented the different characteristic regions of the polarization curve. The surface sensitivity required to study thin films was obtained by electrodepositing from 2 to 20 nm thick Co[57] active layers on cobalt surfaces. Emission Mössbauer spectra from Fe[57] daughter ("probe") atoms were obtained during polarization of these specimens.

Despite the possible ambiguities introduced by effects associated with the emission technique, characteristic spectra were found for cobalt surfaces as a function of applied potentials. These results are shown in Fig. 7 and are summarized as follows: (a) The cobalt was shown to be essentially free of a corrosion film during cathodic polarization (-1100 mV). (b) Resonance lines from both +2 and +3 oxidation states were found in the emission Mössbauer spectra of anodic films formed at low passivating potentials (-100 mV). (c) At potentials in the passive region of the polarization curve (+200 mV and +500 mV) the spectra indicated that the passive film contained primarily +3 oxidation state. (d) The anodic film formed at transpassive potentials (+800 mV) was found to consist of +3 and +4 oxidation states, and the +3 component of this film was shown to be likely the same as that formed at the passive potentials. Auger aftereffects and chemical effects may give rise to a charge state on the Fe[57] daughter ("probe") that is different from the original charge on the parent Co[57]. Further interpretation of the spectra obtained in this study, however, is possible with reference emission spectra for the oxides, hydroxides and oxyhydroxides of cobalt. Progress on this phase of research has already been made, and a description of the composition and structure of the anodic films formed during anodic polarization of cobalt in buffered borate solution, pH 8.5, will be presented in the near future.

CONCLUSIONS

The experimental methods and results that were presented herein indicate that Mössbauer spectroscopy provides a unique means for studying corrosion behavior. The variety of techniques available allows for the study of corrosion from the formation of the first atomic layer to the development of corrosion layers many

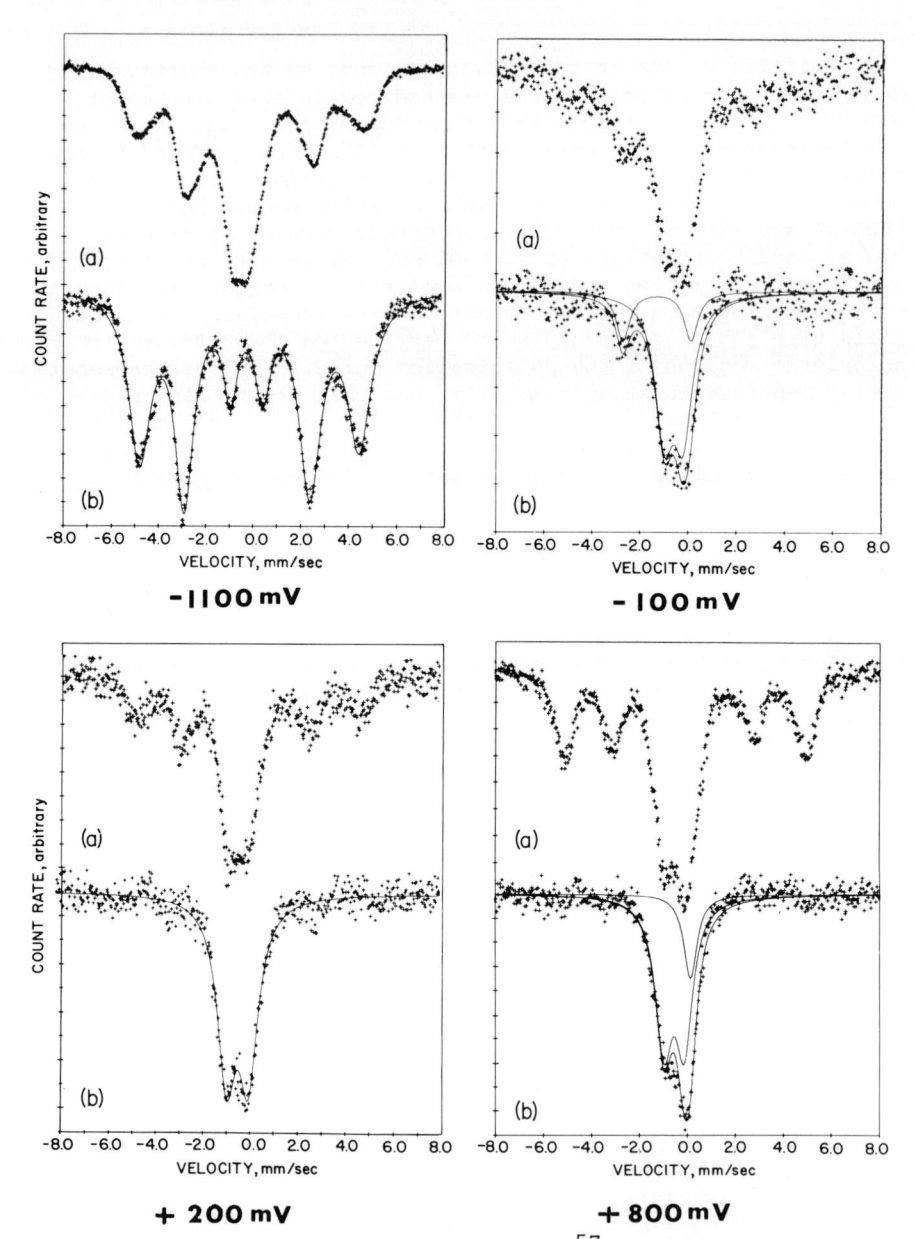

Fig. 7. Emission Mössbauer spectra of Fe^{57} in cobalt polarized at different potentials (vs. SCE). The -1100 mV potential is cathodic where (a) was taken in air and (b) was taken during polarization. The -100 mV and +200 mV are passive potentials and +800 mV is transpassive. In each of the latter three spectra (a) indicates before and (b) indicates after the unreacted metal background has been subtracted from the spectra (Simmons[13,14]).

microns thick. Experiments can be designed for investigations in situ so that corrosion processes may be studied while the metal is immersed in a solution or while it is covered with a protective coating. The limitation of corrosion studies to iron, tin, and cobalt and alloys of these elements is a serious shortcoming of the Mössbauer method, but many important fundamental and applied problems can be examined.

REFERENCES

1. H. Leidheiser, G.W. Simmons and E. Kellerman, Croatica Chemica Acta 45, 257 (1973).
2. G.W. Simmons, and H. Leidheiser, in *Applications of Mössbauer Spectroscopy*, ed. R.L. Cohen, Academic Press (1976).
3. V.I. Goldanskii and R.H. Herber, Eds. *Chemical Applications of Mössbauer Spectroscopy*, Academic Press, New York (1969).
4. D.A. Channing and M.J. Graham, Corrosion Science 12, 271 (1972).
5. K.R. Swanson and J.J. Spijkerman, J. Appl. Phys. 41, 3155 (1970).
6. J.H. Terrel and J.J. Spijkerman, Appl. Phys. Lett. 13, 11 (1968).
7. G.W. Simmons, E. Kellerman and H. Leidheiser, Corrosion 29, 227 (1973).
8. R.A. Krakowski and R.B. Miller, Nucl. Instrum. Methods 100, 93 (1972).
9. Zw. Bonchev, A. Jordanov and A. Ninkova, Nucl. Instrum. Methods 70, 36 (1969).
10. H. Pollack, Phys. Status Solidi 2, 270 (1962).
11. H.H. Wickman and G.K. Wertheim, in *Chemical Applications of Mössbauer Spectroscopy*, Eds. V.I. Goldanskii and R.H. Herber, pp. 604-614, Academic Press, New York (1968).
12. H. Leidheiser, G.W. Simmons, and E. Kellerman, J. Electrochem. Soc., 120, 1516 (1973).
13. G.W. Simmons, E. Kellerman and H. Leidheiser, J. Electrochem. Soc., 123, 1276 (1976).
14. G.W. Simmons and H. Leidheiser, *10th Mössbauer Methodology Symposium*, Ed. I.J. Gruverman, Plenum Press, New York (1976).

Characterization of Bulk and Surface Properties of Heterogeneous Ruthenium Catalysts by Mössbauer and ESCA Techniques

C. A. Clausen, III

Department of Chemistry
Florida Technological University
Orlando, Florida 32816

and

M. L. Good

Department of Chemistry
University of New Orleans
New Orleans, Louisiana   70122

The wide-spread use of metallic and supported metal heterogeneous catalysts in a variety of chemical processes is well known. However, the level of understanding of the fundamental chemistry of these catalytic processes has remained at a relatively primitive level. The recent development of instrumental methods for probing the surface of these materials promises significant progress in our understanding of these systems. This report contains the first efforts to simultaneously probe the bulk and surface properties of a series of supported ruthenium catalyst models by the use of ESCA and Mössbauer spectroscopy. Extensive Mössbauer data and preliminary ESCA results are presented for a variety of ruthenium systems. Highly dispersed ruthenium metal on silica and alumina supports has been investigated. The stabilization of ruthenium by BaO has been evaluated and the effects of exchanging a ruthenium complex into a Y-type zeolite have been observed. The results indicate that the two techniques, Mössbauer and ESCA, are complementary and that the correlation of ESCA and Mossbauer data can provide new insights into the properties of these important materials.

INTRODUCTION

The utilization of metallic and supported metal heterogeneous catalysts in a variety of chemical processes is the backbone of a large segment of the chemical industry. However, the level of understanding of the fundamental chemistry of these materials, particularly that occurring on the surface, has remained relatively primitive. Although the area has enjoyed a high rate of success in the empirical improvement of catalytic materials, the goal of "tailor-made" heterogeneous catalysts has been elusive. However, recent developments in instrumentation now promise detailed information about the chemical processes taking place on the surface of these materials. Presently, extensive work is going on in the area of atomically clean single crystal surfaces and on supported, high surface area, dispersed metal catalysts. The ability to monitor the chemical species on the surface of such systems at the same time that the chemical composition of the bulk material is determined, should provide the ultimate tool in heterogeneous catalyst evaluation. The combination of ESCA (electron spectroscopy for chemical analysis) and Mössbauer spectroscopy should provide such a tool. Both of these two techniques were developed primarily during the 1960's and have recently been applied to the characterization of pure metals and their surfaces and to supported metal systems. Although Mössbauer parameters can be related to certain characteristics of surface species as outlined below, the major contribution from Mössbauer spectroscopy is the determination of bonding properties and the identification of specific chemical species in the bulk sample. ESCA on the other hand, provides similar information about the surface species although some information about bulk properties can be obtained by successive "ion-etching" of the sample. Thus, the simultaneous application of these two techniques to metal catalysts, both those utilizing pure metals or alloys and those consisting of metallic particles supported on various activating and inert supports, should provide definitive information about the chemical properties of these important materials. Although the complete elucidation of the catalytic processes will require extensive kinetic data gathered _in situ_, the comparison of the chemical properties of useful catalytic materials with those of catalytically inactive materials should provide significant insight into the production of "made-to-order" catalyst systems.

The Mössbauer effect is produced by the recoilless resonant absorption of nuclear gamma rays. The recoil-free fraction, or signal amplitude, will be enhanced for low transition energies, for solids in which the nuclei are strongly bound and for experiments performed at low temperatures. The application of Mössbauer spectroscopy to problems of chemical interest typically involves two distinct steps: the extraction of the basic Mössbauer effect parameters (isomer shift, electric field gradient tensor at the nucleus and the magnetic hyperfine splittings) from the experimental data and the interpretation of these parameters in terms of structure and bonding. The isomer shift is generally the most useful parameter for chemical purposes since it can be correlated directly with the electron density at the nucleus. For a relatively simple model of the nucleus, the absolute isomer shift is given by the expression:

$$\text{I.S.} = \frac{2\pi}{3} Ze^2 [<r_{ex}^2> - <r_{gd}^2>]\{|\psi_{abs}(0)|^2 - |\psi_{source}(0)|^2\}$$

where Z is the charge on the nucleus, $<r_{ex}^2>$ and $<r_{gd}^2>$ are the expectation values for the nuclear radius squared for the excited and ground state, and $|\psi_{abs}(0)|^2$ and $|\psi_{source}(0)|^2$ are the electron densities at the nucleus for the absorber and source respectively. Generally, the I.S. is measured relative to some reproducible standard material and can thus be expressed:

$$\text{I.S.} = k \cdot \pm \text{ nuclear factor} \cdot |\psi_{abs}(0)|^2$$

Thus, the isomer shift is highly sensitive to the effective oxidation state of the metal: the removal of s-electrons decreases $|\psi(0)|^2$, whereas the removal of electrons of higher ℓ decreases the s-electron shielding of the nucleus and leads to an increase in $|\psi(0)|^2$. Correlations of I.S. with oxidation state, ligand strength, and π-bonding contributions have been achieved for a number of Mössbauer species. Note that the I.S. will either increase or decrease with increasing $|\psi(0)|^2$, depending on the sign of the nuclear factor for the nuclide of interest; for example, the nuclear factor is negative for ^{57}Fe and positive for ^{99}Ru. The origin and the factors which affect the I.S. are shown diagrammatically in Figure 1.

The second Mössbauer parameter, the quadrupole splitting, arises from the interaction of the electric field gradient with the nuclear quadrupole moment. This splitting of the nuclear energy levels is exhibited as multiple line spectra with the degree of spectral complication a function of the nuclear spin states involved in the transition. The magnitude of this splitting provides information about the chemical environment of the nucleus, particularly its symmetry. The absence of a quadrupole splitting indicates that the electronic environment about the nucleus is essentially cubic. A splitting, indicative of a non-zero EFG (electric field gradient) can occur from both lattice effects (determined by nearest neighbors) or valence effects (arising from asymmetric electronic configurations about the nucleus). In any case, the quadrupole splitting parameter can be interpreted in terms of the electronic structure surrounding the Mössbauer nuclide.

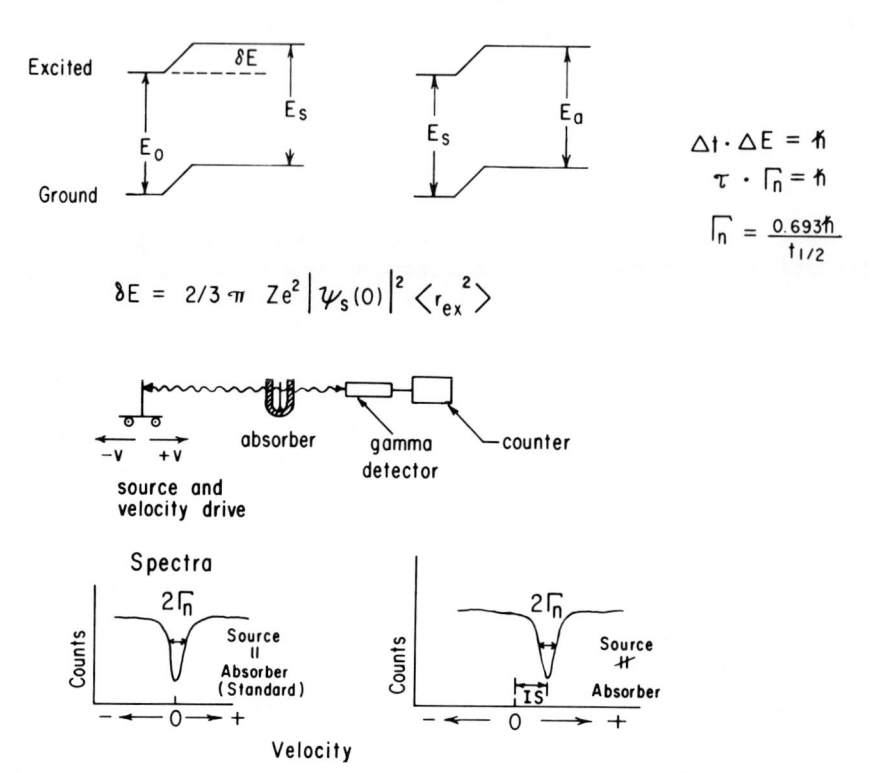

Fig. 1. Origin of the Isomer Shift (I.S.) in the Mössbauer Effect (τ, lifetime of the nuclear excited state; Γ_n, linewidth of the excited state; $t_{1/2}$, the half-life of the excited state).

$$\delta E = 2/3 \pi\, Ze^2 |\psi_s(0)|^2 \langle r_{ex}^2 \rangle$$

$$\Delta t \cdot \Delta E = \hbar$$
$$\tau \cdot \Gamma_n = \hbar$$
$$\Gamma_n = \frac{0.693\hbar}{t_{1/2}}$$

Magnetic hyperfine splitting is observed in the spectra of a Mössbauer nuclide which is influenced by a magnetic field, either a field intrinsic to the sample or an externally applied field. For ferromagnets, the intensity and direction of the magnetic field at the nucleus can be determined. Similar information can be obtained for paramagnetic species with long relaxation times, in addition to effective rate constants for the relaxation process. The details of the quadrupole and magnetic hyperfine effects for Mössbauer spectroscopy are shown diagrammatically in Figure 2.

To observe a chemically significant Mössbauer spectrum, the following criteria for the Mössbauer nuclide must be met: a source of excited nuclei; a spectral linewidth which is small compared to the I.S. but large enough to be easily seen; a reasonable isotopic

Ruthenium Catalysts   69

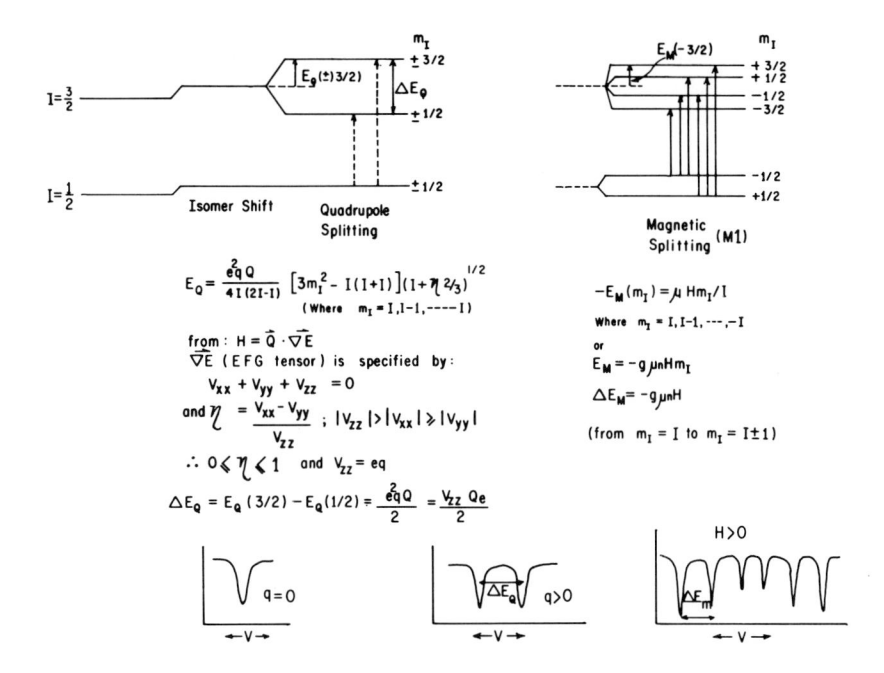

Fig. 2. Origin of the Hyperfine Splitting in Mössbauer Spectra
(I, nuclear spin state; E_Q, quadrupole interaction;
ΔE_Q, the quadrupole splitting; M_I, magnetic quantum
numbers; Q, nuclear quadrupole moment; η, asymmetry
parameter; EFG, electric field gradient, q, electronic
charge distribution; E_M, magnitude of the magnetic
interaction; ΔE_M, magnetic splitting; g, nuclear g-
factor; μ, nuclear magnetic moment; $μ_n$, nuclear Bohr
magneton; H, magnetic field strength).

abundance; a relatively high recoil-free fraction (a relatively small gamma energy); and a matrix which produces an unbroadened, single line source. These criteria are met for several nuclei, the most favorable being ^{57}Fe and ^{119}Sn. Significant Mössbauer data have been obtained for many other nuclei, although in some cases with great experimental difficulty. For example, the 90 Kev gamma transition in ^{99}Ru (as compared to the 14 Kev transition in ^{57}Fe) reduces the recoil-free fraction to such an extent that data must be taken at very low temperatures (usually at 4.2°K) to compensate. The essential properties of the ^{57}Fe and ^{99}Ru Mössbauer transitions are shown in Figure 3.

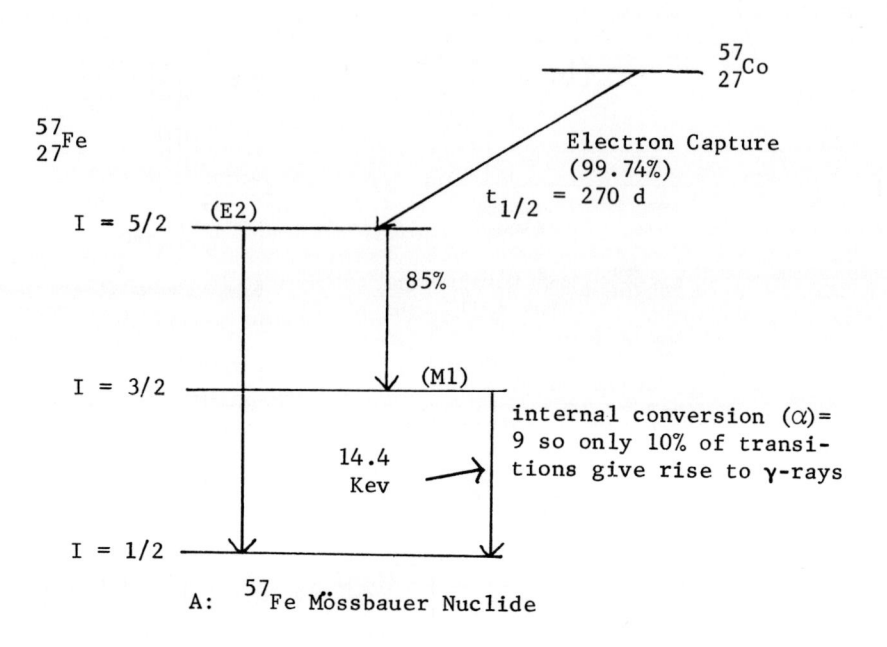

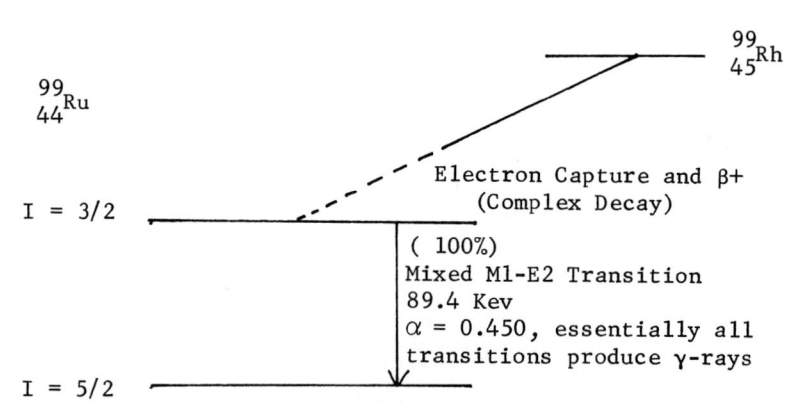

Fig. 3. Nuclear Characteristics and Radioactive Precursors for the Mössbauer Effect in ^{57}Fe and ^{99}Ru.

The Mössbauer experiment is generally carried out in the transmission mode where the sample is placed between the gamma source and the detector. The resonance absorption thus occurs throughout the sample and the measured parameters are characteristic of the bulk properties of the material. For example, Hightower and coworkers[1] have recently determined the bulk structural changes in ferrite catalysts used in the oxidative dehydrogenation of butene to butadiene. The Mössbauer experiment can also be designed to detect the conversion electrons or the resulting x-ray where the measured parameters will be characteristic of the outer layers of the sample, i.e., for conversion electrons the escape depth is of the order of 100 Å for most materials[2]. To obtain parameters characteristic of surface species, investigators have devised methods for dispersing the Mössbauer nuclide on various supports[3]. The success of these methods depends on obtaining a high surface to volume ratio for the material of interest. Thus if suitable samples can be prepared, the Mössbauer Effect can be used to probe the chemical and structural properties of both surface and bulk species. Specific surface effects in iron and tin systems which have been evaluated by Mössbauer techniques include: size of dispersed particles on the surface of a support and the dispersed particle-support bonding as a function of the recoil-free fraction[4]; asymmetry of surface sites and the size of dispersed particles as evaluated by the magnitude of the quadrupole splitting[5]; determination of dispersed particle size by the interpretation of their magnetic hyperfine parameters[6]; and the chemical state of surface species as a function of their I.S. values[7,8]. Thus, Mössbauer spectroscopy has been utilized to evaluate both the bulk properties and the properties of highly dispersed iron and tin catalyst systems and should be suitable for analogous ruthenium systems. However, if there are questions about the surface area versus volume ratio for dispersed species or if low surface area materials are of interest, other techniques which provide unequivocal surface information would provide appropriate complementary data.

ESCA spectroscopy is the determination of the photoelectron spectra created by irradiation of the sample with mono-energetic x-rays. The effect is confined to the outer layers of the surface of samples since the escape depth of the photoelectrons will be limited to a few angstroms (of the order of 10-20 Å for heavy metals). The energies of the photoelectrons are related to the energies of the incident x-rays by the following expression:

$$E_{h\nu} = E_B + E_K + \phi_{sp}$$

where E_B is the binding energy of the ejected electron, E_K is the kinetic energy of the ejected electron and ϕ_{sp} is the work function of the spectrometer material. If $E_{h\nu}$ is known for the x-ray generator, E_K is measured in the spectrometer and ϕ_{sp} is known or assumed constant for the given system, then E_B, the electron binding energy can be calculated. For most work, ϕ_{sp} is determined by calibration with a substance of known binding energy although care must be taken to be sure that the sample and the standard material have the same electrostatic properties. Generally, this is done by incorporating the standard material into the sample matrix; for example, a thin layer of gold is vacuum sputtered onto the sample[9]. The core electron binding energies measured in this way will be a function of the chemical environment of the atom involved although the relationship is not necessarily simple. For example, the binding energy represents the work required to remove an electron from a charged atom and that required to remove the electron from the field of the surrounding charged atoms. Jolly[9] presents these effects in the form of the potential model equation

$$E_B = kQ_i + V + \ell$$

where k and ℓ are empirical constants, Q_i is the charge on the ionizing atom and V is the Madelung potential energy created by the surrounding charged atoms. This model has been further exploited by Kim and Winograd[10] who have calculated the binding energy <u>shifts</u> to be expected for a common ion in two different ionic matrices. They represent these shifts as:

$$\Delta E_B = \Delta q/r - \Delta V - \Delta E_R$$

where Δq is the difference in valence electronic charge, r is the radius of the valence shell, ΔV is the difference in crystal field potential (the Madelung potential) and ΔE_R is the difference in extra- and intra-atomic relaxation energies. The first term, $\Delta q/r$ (or kQ_i in the Jolly equation above) is the "chemical shift" caused by the electron density in the valence shells of the ionizing ion and the second and third terms are dependent on the host matrix. Thus, correlations between the core binding energies and other atomic properties which depend on valence electron density, such as the Mössbauer I.S., must be made carefully, being fully cognizant of the possible large contributions to E_B from the surrounding lattice, particularly in ionic solids[11]. Linear correlations have been reported for the core binding energies and Mössbauer I.S. values in several iron and tin systems where the materials were molecular solids.[12] Also, the presence of two different oxidation states for an element in a given compound have been confirmed by both Mössbauer and ESCA studies on K_xFeF_3[13] and $[(NH_3)_5Ru-N\underline{O}N-Ru(NH_3)_5]^{+5}$.[14] Comparisons of the E_B and I.S. values for some gold[15] and iridium[16] systems have been carried out. A unique study on iron-phthalocyanine polymers used a bulk Mössbauer measurement and surface ESCA data to show that the loss of electrocatalytic activity in the sample was not a function of the iron complex but was related to the partial oxidation of the active carbon surface[17].

Efforts in our laboratory in recent years have been devoted to the development of Mössbauer spectroscopy as a diagnostic chemical tool, particularly for ruthenium systems. The multiplicity of oxidation states exhibited by ruthenium and the common occurance of multi-nuclear systems makes this an ideal target for a technique which can distinguish oxidation states even when they are found in the same compound. A recent review of ruthenium Mössbauer spectroscopy indicates the scope of the studies which are possible via this technique.[18] An evaluation of these Mössbauer studies and a review of the applications of ESCA to catalytic research[19] indicate that the evaluation of heterogeneous ruthenium catalyst systems should be feasible.

This paper is a review of our efforts in the application of Mössbauer spectroscopy to the solution of structure and bonding problems in several supported ruthenium systems and our initial evaluation of the value of ESCA spectroscopy as a corollary technique. To provide the reader with an overview of the scope of the studies carried out, all of the various systems studied so far are described, although the detailed Mössbauer studies of the alumina and silica supported systems have been published[20], and the detailed Mössbauer data for the zeolites and the automotive emission control catalysts have been submitted for publication[21,22]. The ESCA data and the material on the alkali-metal promoted catalysts have not been reported elsewhere.

EXPERIMENTAL METHODS

Mössbauer Data

The Mössbauer spectra were obtained with the spectrometer and cryogenic system described previously[20,23]. The use of a germanium-lithium drifted detector (Elscint Ltd., Model GP/GC) resulted in improved resolution over that previously reported. All spectra were obtained at 4.2°K by use of a Kontes/Martin glass dewar system where both the source and absorber were immersed directly in the liquid helium well. The source used to study the alumina and silica supported catalyst samples consisted of approximately 7 mCi of 16 day ^{99}Rh contained in a host lattice of ruthenium metal. This source exhibited linewidths of 0.28-0.32 mm/sec for a natural ruthenium metal absorber. The source used to study the zeolite, automotive emission control and promoted catalysts consisted of approximately 7 mCi of ^{99}Rh contained in a host lattice of rhodium metal. This source exhibited linewidths of 0.45 ± 0.30 mm/sec for a natural ruthenium metal absorber. Both sources were prepared by New England Nuclear Corporation, Boston, MA.

In general, the base line for each spectrum contains between 1 and 2 million counts per channel and the relative percent absorption of the Mössbauer peaks are in the range of 0.1-0.5 percent. Data reduction was carried out on a PDP-10 computer system. The spectra were subjected to a least-squares fit to a Lorentzian line shape with both the experimental points and the calculated least-squares curve plotted out directly by a Calcomp Model 563 plotter. The Mössbauer hyperfine parameters were calculated from the least-squares fit. Error analyses for the isomer shift, quadrupole splitting, and peak full width at half maximum values are given along with the data.

ESCA Data

All of the ESCA spectra were obtained with a DuPont 650B Electron Spectrometer equipped with a magnesium x-ray source. Actual data were taken by Dr. C. R. Ginnard of the DuPont ESCA Applications Laboratory in Monrovia, California. Survey spectra (1000-0 eV) were obtained with a multi-channel analyzer accessory. High resolution spectra of significant energy levels were obtained in the analog operative mode. Samples were prepared by compacting the dry powders into recessed 650B probe tips. Those supported catalyst materials studied were shipped in their sealed cell from the vacuum treatment line and were handled under nitrogen in a glove-bag attached directly to the 650B spectrometer. Most of the binding energies reported have been adjusted to a carbon 1s binding energy of 285.0 eV to compensate for sample charging.

Crystallite Size Measurements

The average crystallite size of the supported ruthenium metal was determined for some samples. X-ray line broadening using Warren's correction as described by Klug and Alexander[24] and hydrogen absorption measurements were used. The hydrogen absorption isotherms were obtained with a conventional Pyrex glass, constant volume adsorption system using the method of Dalla-Betta[25].

Purification of Materials

Purified tank air was used in the calcination of catalyst samples. Hydrogen for the reduction steps was purified by passing it successively through a heated palladium catalyst, a 13X molecular sieve, a liquid nitrogen trap and finally through the cell. Anhydrous ammonia was purified by refluxing over sodium before distilling into storage bulks. Matheson carbon monoxide, 99.5% pure, was passed through a trap at 195°K before use.

RESULTS AND DISCUSSION

A. Silica and Alumina Supported Systems

Davison silica gel Grade 923 (100-200 mesh, surface area approximately 285 m^2/g) and Davison η-alumina Grade 992-F (100-200 mesh, surface area approximately 210 m^2/g) were used as support materials. The model catalysts were prepared by impregnating the support materials with aqueous solutions of ruthenium trichloride (RuCl$_3$·1-3H$_2$O, A. D. Mackay, Inc.) by the incipient wetnesss method, followed by oven drying at 110°C for 24 hours. Catalyst samples were impregnated with 10 wt. % ruthenium. After drying, each sample was placed in a quartz cell and all further treatments were carried out on the sample in the cell attached to a vacuum line.

The supported particle sizes for the various samples investigated are shown in Table 1. Note that the average particle size increases for those samples that are calcined before being reduced and increases as the temperature and length of the calcination step increases.

Mössbauer Results

Mössbauer spectral data obtained for ruthenium on a silica support during various stages of treatment are given in Table 2 and Figure 4. Mössbauer data for a variety of known ruthenium compounds are given in Table 3 for comparison.

The Mössbauer spectrum for sample 1-A (RuCl$_3$·3H$_2$O impregnated on silica and then dried for 24 hours at 110°C) shows that the impregnated ruthenium complex is absorbed on the surface of the silica support without undergoing a chemical change. The Mössbauer parameters for this sample are the same within experimental error as that observed for unsupported RuCl$_3$·3H$_2$O.

After obtaining the Mössbauer spectrum for sample 1-A, it was reduced according to the previously described procedure. This reduced sample is called 1-B. After the accumulation of approximately 2 million counts in each channel, no absorption peaks could be detected in the Mössbauer spectrum for this sample. This was somewhat surprising since this sample gave a well-resolved spectrum prior to the reduction step. Chemical analysis showed that there was no loss of ruthenium from the catalyst sample during the reduction procedure. The absence of an observable spectrum for this sample must be the result of a decrease in the nuclear recoil-free fraction following the reduction of the complex to the metallic state.

It has been observed by Suzdalev, et al.[26] that in highly dispersed tin the probability of the Mössbauer effect diminishes as the particle diameter decreases. It has also been shown by Van

TABLE 1

PARTICLE SIZE OF RUTHENIUM METAL SUPPORTED ON SILICA AND ALUMINA

Sample No.	Support	Treatment	Wt % Ru	Ru surface area (m²/g)[a]	Av diam (Å) X-ray	Av diam (Å) Ads
1-B	SiO$_2$	H$_2$ reduction[b]	10	57	---	85
2-C	SiO$_2$	Calcined @ 400°C then reduced in H$_2$[c]	10	22	240	230
3	η-Al$_2$O$_3$	H$_2$ reduction[b]	10	45	95	108
4	η-Al$_2$O$_3$	Calcined @ 300°C then reduced in H$_2$[d]	10	33	160	151
5-C	η-Al$_2$O$_3$	Calcined @ 400°C then reduced in H$_2$[c]	10	18	295	275

a) Calculated from hydrogen adsorption data.
b) Reduction was carried out in flowing hydrogen for 2 hrs at 150°C, 2 hrs at 300°C and finally 2 hrs at 400°C.
c) Treatment details were: Calcination in flowing air for 2 hrs at 150°C, 2 hrs at 300°C and finally 3 hrs at 400°C followed by reduction in flowing hydrogen for 2 hrs at 150°C, 2 hrs at 300°C and 3 hrs at 400°C.
d) Treatment details were: Calcination for 2 hrs at 150°C and 2 hrs at 300°C followed by reduction in flowing hydrogen for 2 hrs at 150°C, 2 hrs at 300°C and 2 hrs at 400°C.

TABLE 2

MÖSSBAUER PARAMETERS FOR RUTHENIUM SUPPORTED ON SILICA

Sample No.	Treatment	Absorber thickness (mg Ru/cm^2)	Isomer[a] shift (mm/sec)	Quadrupole Splitting (mm/sec)	Peak Width (Γ) @ Half-Height (mm/sec)
1-A	Before Reduction	175	-0.34 ± 0.02	---	0.53 ± 0.04
1-B	After Reduction	175	(No spectrum observed)		
2-A	Before Reduction	165	-0.35 ± 0.02	---	0.54 ± 0.04
2-B	After Calcination	165	-0.27 ± 0.02	0.46 ± 0.02	$\Gamma_1 = 0.37 \pm 0.04$ $\Gamma_2 = 0.36 \pm 0.04$
2-C	After Reduction	165	$+0.02 \pm 0.02$	---	0.34 ± 0.03

a) Zero velocity is taken to be the center of the spectrum of a standard ruthenium metal sample.

TABLE 3

MÖSSBAUER PARAMETERS OF SEVERAL WELL CHARACTERIZED RUTHENIUM COMPOUNDS

Ruthenium Species	Absorber Thickness (mg Ru/cm^2)	Isomer[a] Shift (mm/sec)	Quadrupole Splitting (mm/sec)	Peak width (Γ) @ Half-Height (mm/sec)
RuCl$_3 \cdot$1-3H$_2$O	525	-0.34 $\pm$ 0.02	---	0.52 $\pm$ 0.04
Ru Metal Powder	185	0.00 $\pm$ 0.02	---	0.32 $\pm$ 0.03
RuO$_2$	380	-0.23 $\pm$ 0.03	0.51 $\pm$ 0.05	0.57 $\pm$ 0.03
RuO$_4$	340	+1.06 $\pm$ 0.01	---	0.28 $\pm$ 0.02
KRuO$_4$	520	+0.82 $\pm$ 0.02	0.37 $\pm$ 0.02	0.40 $\pm$ 0.04
BaRuO$_4 \cdot$H$_2$O	320	+0.38 $\pm$ 0.01	0.44 $\pm$ 0.02	0.30 $\pm$ 0.02
[Ru(NH$_3$)$_6$]Cl$_2$	367	-0.72 $\pm$ 0.02	---	0.33 $\pm$ 0.05
[Ru(NH$_3$)$_5$CO]Br	151	-0.54 $\pm$ 0.02	---	0.39 $\pm$ 0.05
[Ru(NH$_3$)$_5$NO]Cl$_3 \cdot$H$_2$O	142	-0.16 $\pm$ 0.02	0.34 $\pm$ 0.02	0.31 $\pm$ 0.05
[Ru(CO)$_3$Cl$_2$]$_2$	181	-0.31 $\pm$ 0.02	---	0.42 $\pm$ 0.04
Ru$_3$(CO)$_{12}$	735	-0.24 $\pm$ 0.02	---	0.51 $\pm$ 0.05

a) Zero velocity is taken to be the center of the spectrum of a standard ruthenium metal sample.

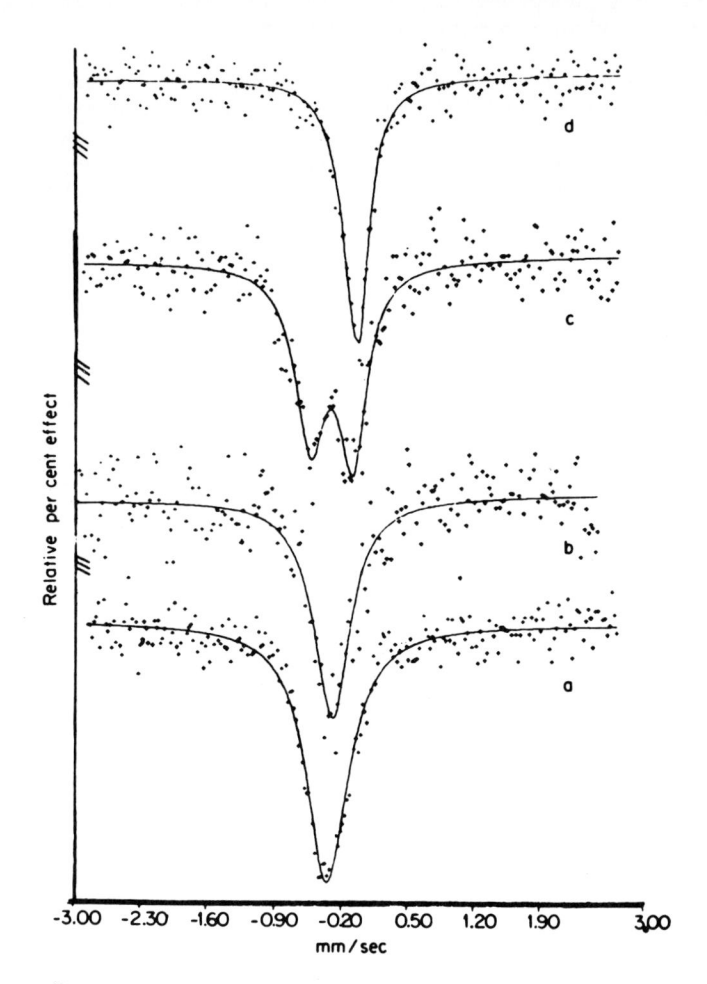

Fig. 4. Mössbauer spectra of: (a) $RuCl_3 \cdot 1-3H_2O$; (b) Sample 2-A ($RuCl_3 \cdot 1-3H_2O$ impregnated on a silica support); (c) Sample 2-B (ruthenium on a silica support after calcination); (d) Sample 2-C (ruthenium on a silica support after reduction).

Wilringen[4b] that in metal powders the particles may be so small that a single particle is unable to give a "recoilless" Mössbauer transition. Van Wilringen defined the critical size of a particle as being that mass which is just sufficiently large to absorb the recoil of the gamma quantum without observable exchange of energy. The critical size can be calculated if it is assumed that the recoil energy is unobservable when it gives rise to a line displacement less than the natural line width.

Using a value for the ruthenium-99 gamma recoil energy given by Stevens and Stevens[27] it follows that the mass absorbing the recoil energy should be at least 19.4×10^5 times the mass of a single ^{99}Ru nucleus. For spherical particles of ruthenium (density 12.3 g/cm^3), this leads to a critical particle diameter of 368 Å. Data in Table 1 show that the ruthenium particles (85 Å) in sample 1-B are much smaller than the critical size. Since no Mössbauer spectrum was observed for this sample, it must follow that either no recoil energy, or an insignificant amount, is transferred to the support. This suggests that the strength of the binding of the ruthenium to the silica support is very weak and that the binding forces between the atoms in the small catalyst particles are similar to those between ruthenium atoms in the powdered metal.

In order to increase the particle size of the supported metal, a sample was calcined before the reduction step. The new impregnated sample (2-A) before treatment exhibited a Mössbauer spectrum identical to that observed for sample 1-A. After the calcination step, a Mössbauer spectrum was obtained (sample 2-B). The data as given in Figure 4 and Table 2 show a well-resolved doublet corresponding to an isomer shift of -0.27 mm/sec and a quadrupole splitting of 0.46 mm/sec. These parameters agree very well with the isomer shift (-0.22 mm/sec) and quadrupole splitting (0.51 mm/sec) which have been observed for RuO_2.[28] The absence of any unidentified peaks in the spectrum indicates that essentially all of the ruthenium is present as small crystallites of RuO_2.

Sample 2-B was reduced according to the previously described procedure. Even though the average particle size of the metal (240 Å) in this sample is still less than the critical particle size, a Mossbauer spectrum was observed. This spectrum exhibited a single absorption peak with an isomer shift that agrees exactly within experimental error to that observed for powdered ruthenium metal. The absence of any other lines in the spectrum indicates that all of the ruthenium has been reduced to the zero valence, metal state. The fact that a Mössbauer spectrum was observed for this sample in spite of the subcritical particle size, indicates that the "effective" Mössbauer mass of the principles must be greater than the critical mass. This suggests that weak binding forces exist between the small metal particles and the silica support. Another possible explanation is that the observed Mössbauer effect may be due to a small fraction of metal particles that are larger than 368 Å. In either case, additional work is necessary in order to establish the absolute minimum ruthenium particle size on silica for which a Mössbauer effect can be observed.

Attempts were made to obtain Mössbauer spectra of chemisorbed CO, NH$_3$, and H$_2$S on the reduced ruthenium catalyst. In each case, the chemisorbed species were introduced to a total pressure of 50 Torr at 25°C in the sample cell. The Mössbauer spectra obtained for each of these samples exhibited a single line that was identical within experimental error to that observed for the reduced catalyst. This suggests that either the ratio of surface ruthenium atoms to bulk ruthenium atoms is not great enough to observe surface effects, (which is unlikely in view of the Ru surface areas reported in Table 1) or that the chemisorption of these molecules on a ruthenium atom does not perturb its electronic structure enough to bring about an observable change in the Mössbauer spectrum.

To investigate the nature of the ruthenium support on alumina, samples of η-alumina were impregnated with ruthenium trichloride and treated in a manner as previously described. Mössbauer spectra of catalyst samples are given in Table 4 and Figure 5.

A single narrow absorption peak was observed in the Mössbauer spectra of catalyst samples 3 and 4. It should be noted that each of these samples was evacuated to a pressure of 10^{-6} Torr at 400°C, therefore their spectra represent a surface free of chemisorbed hydrogen. Within experimental error, their Mössbauer parameters correspond exactly to those observed for ruthenium metal. As was observed in the case of the reduced silica catalysts, there is no evidence for the existence of any ruthenium species other than the reduced metal. Even though sample 3 was reduced directly, whereas sample 4 was calcined before reduction, the Mössbauer data show that other than for average particle size the state of the ruthenium is the same in both samples.

Both catalyst sample 3, with an average particle size of 95 Å, and sample 4, with a particle size of 160 Å, contain supported ruthenium crystallites that are much smaller than the critical particle size. The occurrence of a Mössbauer effect in these samples indicates that fairly strong binding forces exist between the metal particles and the alumina support. The data also show that the Mössbauer effect increases as the average particle size increases.

A Mössbauer spectrum was obtained for untreated impregnated ruthenium trichloride on an alumina support (5-A). The spectrum as shown in Figure 5 exhibits an asymmetric doublet with an isomer shift of -0.41 mm/sec and a quadrupole splitting of 0.45 mm/sec. This spectrum is significantly different from the spectrum obtained for unsupported RuCl$_3$·1-3H$_2$O and RuCl$_3$·1-3H$_2$O supported on silica. The isomer shift is slightly more negative than that observed for the unsupported ruthenium trichloride and falls in a region that borders on the upper end of isomer shifts observed for Ru(III) complexes and the lower end of isomer shifts for Ru(IV) complexes.

TABLE 4

MÖSSBAUER PARAMETERS FOR RUTHENIUM SUPPORTED ON ALUMINA

Sample No.	Treatment	Absorber Thickness (mg Ru/cm^2)	Isomer Shift (mm/sec)	Quadrupole Splitting (mm/sec)	Peak Width (Γ) @ Half-Height (mm/sec)	% Abs
3	After Reduction	185	+0.01 $\pm$ 0.02	---	0.41 $\pm$ 0.02	0.2
4	Reduced After Low temp calcination	180	+0.01 $\pm$ 0.02	---	0.43 $\pm$ 0.04	0.3
5-A	Before Reduction	190	-0.41 $\pm$ 0.03	0.45 $\pm$ 0.02	Γ_1 = 0.35 $\pm$ 0.04 Γ_2 = 0.49 $\pm$ 0.04	---
5-B	After Calcination	190	-0.27 $\pm$ 0.03	0.49 $\pm$ 0.02	Γ_1 = 0.33 $\pm$ 0.04 Γ_2 = 0.35 $\pm$ 0.04	---
5-C	After Reduction	190	-0.02 $\pm$ 0.02	---	0.38 $\pm$ 0.04	0.6

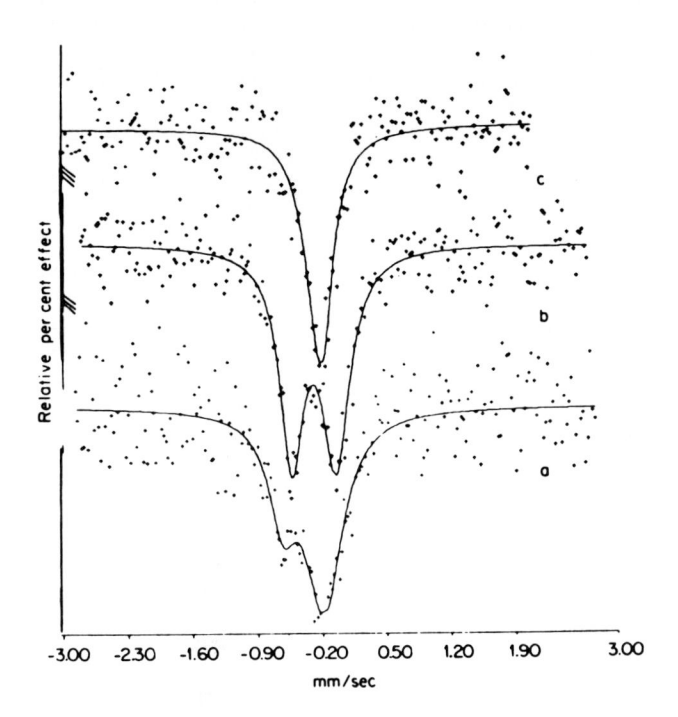

Fig. 5. Mössbauer spectra of: (a) Sample 5-A (RuCl$_3 \cdot$1-3H$_2$O impregnated on alumina); (b) Sample 5-B (ruthenium on an alumina support after calcination); (c) Sample 5-C (ruthenium on an alumina support after reduction).

Therefore, it is difficult to determine whether the ruthenium has undergone a change in oxidation state or has been coordinated to the support.

Sample 5-A was calcined to form catalyst sample 5-B. The Mössbauer spectrum for this sample shows that all of the ruthenium was converted to RuO$_2$. This sample was then reduced and evacuated to a pressure of 10^{-6} Torr at a temperature of 400°C. The spectrum for this sample (5-C) exhibited a single line with spectral parameters that agree with those observed for the other reduced alumina samples. The Mössbauer effect (0.6%) was greater than that observed for the other reduced alumina samples. This is expected, since the average particle size of sample 5-C (295 Å) is twice the average particle size of the other samples.

Again, attempts were made to obtain Mössbauer spectra for CO, NH$_3$, H$_2$O, O$_2$ and H$_2$S chemisorbed (at 25°C) on the reduced catalyst samples. The Mössbauer spectra obtained for each of these samples exhibited a single line that was identical within experimental error to that observed for reduced ruthenium on an alumina support. It is somewhat surprising that no chemisorption effects were observed with catalyst sample 3. The average metal particle size in this sample is only 95 Å, which should give a favorable, surface to bulk metal atom ratio.

ESCA Results

Several alumina supported catalyst systems prepared as discussed above were subjected to ESCA analysis. The results are shown in Figure 6 and Table 5. Also shown in the Figure and Table are values for ruthenium sponge before and after it has been etched. Several other ruthenium compounds are listed in the Table for comparison purposes. The results as shown in Figure 6 are most interesting. Note that the spectrum for ruthenium trichloride dispersed on alumina exhibits a higher value for the Ru $3d_{5/2}$ binding energy than any of the other samples which is not inconsistent with the observation given above that the Mössbauer I.S. for this material indicates an interaction with the support to give an apparent increase in the oxidation state of the metal to Ru(IV). Both techniques also indicate that mixed species are present in this sample. Spectrum B (the dispersed reduced metal) and spectra E and F of bulk ruthenium metal provide an interesting comparison. Although sample B exhibits an E_B characteristic of an "oxidation state" lower than that for sample A it is very different from normal ruthenium metal (samples E and F). These data, when compared to the Mössbauer results, indicate that the effect of the support material on the E_B values is very significant. A very similar effect has been recently reported by Kim and Winograd[10] for gold dispersed on silica. By an inert gas implantation technique, they were able to estimate the contribution to E_B from the matrix and to evaluate the "chemical shift". The very small E_B shifts exhibited by the supported materials investigated here implies that the matrix contribution is very large. This is consistent with the observation of large recoil-free fractions for the Mössbauer effect reported above for the small supported ruthenium particles which indicated strong metal-support binding.

The spectrum for the calcined sample (sample C) indicates that the ruthenium on the support surface (in sample B) has undergone partial oxidation. However, the rather complex spectrum of broad, overlapping peaks indicates that multiple chemical species are probably present. This would be consistent with the observation by Kim and Winograd[29] of RuO_3 on the surface of bulk RuO_2. It may be possible to resolve these peaks by convolution of spectra as that exhibited by sample C. However, more data will be required before definite results can be obtained. It is interesting that the exposure of sample B to atmospheric air (sample D) did not produce the same oxidation products as were formed by the

TABLE 5

BINDING ENERGIES (eV) FOR Ru OXYGEN SYSTEMS

	Ru 3d$_{5/2}$	O 1s
Ru Metal	280.0	------
	280.0[a]	------
RuO$_2$	280.9	529.4
	280.7[a]	529.4[a]
RuO$_3$	282.5[a]	530.7[a]
RuO$_4$	283.3[a]	------
Sample A	282.8	531.7
Sample B	280.7	531.0
Sample C	281.0	531.1
Sample D	280.6	531.5

a) Reference 29.

calcination of sample B at high temperatures. This implies that the mixed oxides are formed at high temperatures. This is an important result since the subsequent oxidation to the volatile RuO$_4$ results in the loss of ruthenium metal from the support.

These studies indicate the complimentary nature of Mössbauer and ESCA data and point out the problems involved in evaluating chemical species by ESCA alone. The combination of the two techniques may make it possible to more easily determine the role of the matrix in these supported systems.

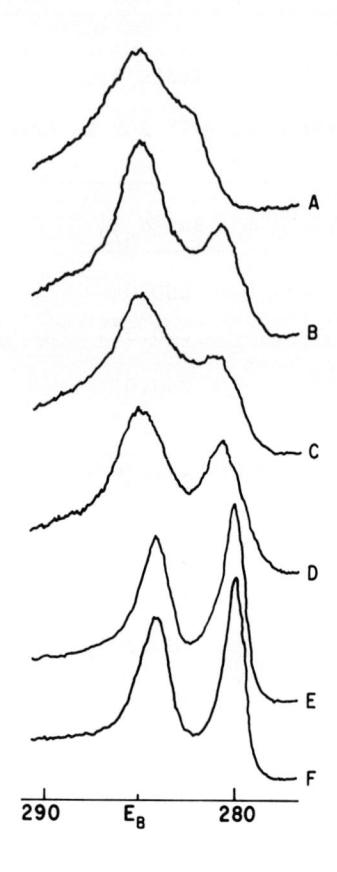

Fig. 6. ESCA Spectra for a Series of Alumina Supported Ruthenium Catalysts and Bulk Ruthenium Metal.

B. Alkali Metal-Promoted Ruthenium Systems

Catalyst samples were prepared by impregnation of activated coconut charcoal (Analabs 90-100 mesh) with solutions of aqueous ruthenium trichloride. The samples were then dried for 24 hours at 110°C. The ruthenium concentration on the support was 5 wt. percent. The dried catalyst was loaded into a quartz cell and evacuated to 10^{-3} Torr. The ruthenium was then reduced by circulating hydrogen at 300 Torr while heating the catalyst to 100°C for 2 hours, to 200°C for 2 hours and then slowly to 400°C where it was held for 3 hours.

The potassium (or sodium) was transferred to the catalyst layer by evaporation from a side tube in the vacuum line. The treated catalyst was then heated in circulating helium at 400°C for 6 hours to distribute the alkali metal over the entire catalyst. The cell was sealed and placed in the spectrometer.

The alkali metal content of the samples was determined by treating the catalyst with water and then titrating the filtrate with hydrochloric acid. The sodium treated sample contained 10.5% alkali metal and the potassium treated sample contained 12%.

Both the sodium and potassium promoted samples gave a Mössbauer spectrum that was identical to that of unpromoted ruthenium metal. This result is somewhat surprising in light of the reports that the catalytic activity of transition metals for ammonia synthesis is remarkably enhanced by the addition of alkali metals[30]. It has been proposed that the promoter action is likely to be provided by a charge transfer from alkali metal to transition metal which facilitates the formation of anionic intermediates over the transition metal. The Mössbauer results imply that the promoter action is not produced by a static charge transfer from alkali metal to ruthenium but it does not rule out the possibility of a dynamic charge transfer in the formation of the reaction transition state. ESCA results on these systems would be of value since the E_B values for both surface ruthenium and alkali metal could be determined.

C. Automotive Emission Control Catalysts

Ruthenium-containing catalysts have been found to have a pronounced selectivity for reduction of nitrogen oxides to molecular nitrogen, and attention has recently been focused on the development of these catalysts as a means of controlling nitrogen oxide emissions[31-32]. However, these studies have shown that ruthenium catalysts exhibit poor stability when the exhaust contains a net oxidizing composition. Analysis of spent catalysts revealed severe losses of the active component, which was readily explained by the formation and removal of the volatile ruthenium tetroxide[32].

One method which has been proposed to minimize the tendency of the ruthenium to volatilize is based on the formation of the nonvolatile barium ruthenate[31]. The barium ruthenate was prepared in situ on the alumina support by impregnation first with a solution of barium nitrate followed by calcination to convert the nitrate to the oxide. The support was then impregnated with a solution of ruthenium trichloride. The catalyst was dried and reduced in hydrogen and then "fixed" by rapid heating in air at 900°C. Catalyst samples prepared by this technique were found to exhibit considerable improvement in the prevention of ruthenium volatilization while maintaining the desirable selective catalytic reduction of nitric oxide to molecular nitrogen[31]. However, under vehicle operating conditions the loss of ruthenium from the stabilized catalyst was still found to be higher than acceptable.

88   C. A. Clausen, III and M. L. Good

The reason for this gradual loss in ruthenium can be accounted for by referring to the Mössbauer data in Table 6 and Figure 7 for some BaO "stabilized" catalyst systems.

The stabilized automotive emission control catalysts were prepared by the incipient wetness impregnation of the η-alumina support, first by a solution of barium nitrate followed by calcination at 900°C for 8 hours to convert the nitrate to the oxide, and secondly by a solution of ruthenium trichloride. The sample was then dried for 24 hours at 100°C. The dried samples were reduced in flowing hydrogen for 2 hours at 150°C, 2 hours at 300°C and finally 2 hours at 400°C. The very small ruthenium metal particles were then "fixed" by rapid heating in flowing air at 900°C for 1 hour, according to the procedure of Shelef and Gandhi[31].

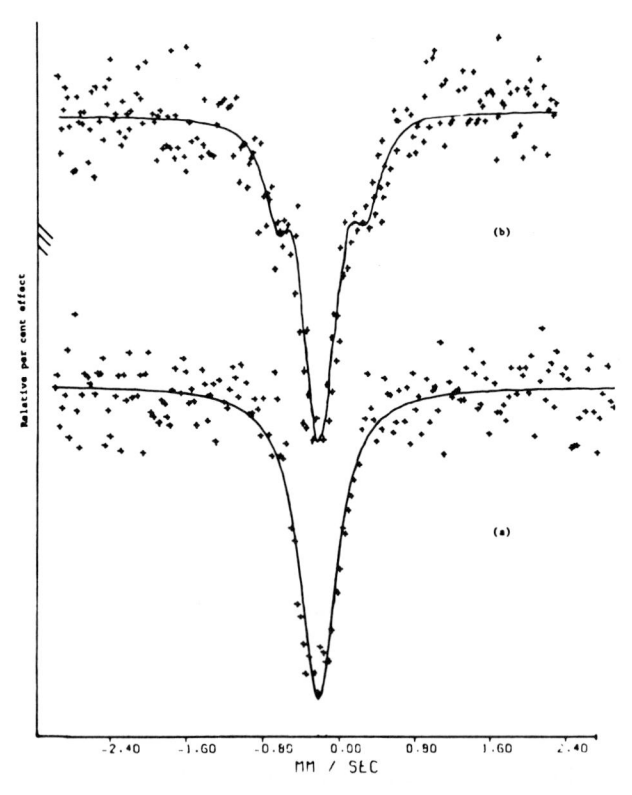

Fig. 7. Mossbauer spectra of: (a) Barium ruthenate; (b) Sample 4-A (12% barium and 4% ruthenium on an alumina support after initial "fixation" step).

TABLE 6

MÖSSBAUER DATA FOR STABILIZED RUTHENIUM AUTOMOTIVE EMISSION CONTROL CATALYSTS

Sample No.	Treatment @ 700°C	No. of Lines in Spectrum	Isomer Shift (mm/sec)	Quadrupole Splitting (mm/sec)	Peak Area[a] Ratio
20-A	Initial Fixed Sample	3	-0.30 ± 0.04 -0.24 ± 0.04	--- 0.53 ± 0.05	2.3:1.0
20-B	Sample 20-A heated for 10 hrs in SAE[b]	1	+0.02 ± 0.03	---	-------
20-C	Sample 20-B heated for 30 minutes in net oxidizing SAE atmosphere	3	-0.27 ± 0.04 -0.22 ± 0.03	--- 0.53 ± 0.03	2.0:1.0
20-D	Sample 20-B cycled between net reducing SAE and net oxidizing SAE for 150 hours	3	-0.28 ± 0.05 -0.23 ± 0.03	--- 0.52 ± 0.03	0.4:1.0
Barium Ruthenate	------------	1	-0.28 ± 0.03	---	-------
Ruthenium Dioxide	------------	2	-0.23 ± 0.03	0.51 ± 0.05	-------

a) This is the ratio of the area of the single peak to the area of the pair of quadrupole split peaks.

b) SAE = simulated auto exhaust

Sample 20-A is a sample containing 12 wt. % barium and 4 wt. % ruthenium prepared by the method described above. The Mössbauer data for this sample were taken after the 900°C "fixation" step. The three lines in the spectrum match those that would be found for a sample containing a mixture of barium ruthenate and ruthenium dioxide. The peak area ratio indicates that the barium ruthenate is present in a greater concentration. This piece of datum indicates that in this sample, every ruthenium atom has not been deposited in the vicinity of a stabilizing oxide so as to assure the formation of the ruthenate. Therefore, the loss of the non-stabilized ruthenium during operation would account for some of the ruthenium volatilization.

Sample 20-A was heated @ 700°C in a simulated auto exhaust (SAE) having the following composition:

Component	Content, mole %	Component	Content, mole %
H_2	0.33	CO_2	13.00
O_2	0.35	C_3H_8	0.10
H_2O	10.00	NO	0.10
CO	2.00	N_2	74.12

The treated sample was called 20-B and its Mössbauer data shown in Table 6 indicate that all of the ruthenium has been reduced to the metallic state by this treatment. The SAE was then made net oxidizing in composition by substituting 2% O_2 for the 2% CO. Sample 20-B was heated in the oxidizing SAE mixture for 30 minutes at 700°C. This sample after treatment is called 20-C. Its Mössbauer spectral parameters show that the ruthenium has been oxidized back to $BaRuO_3$ and RuO_2. However, the peak area ratio indicates a smaller composition of $BaRuO_3$ than was present in the initial sample. This sample was cycled at 700°C between the net reducing SAE for 50 minutes and 10 minutes in the net oxidizing SAE. The treatment was continued for 150 cycles over a period of 150 hours. The treated sample is referred to as sample 20-D. The Mössbauer spectrum for this sample indicates again that all of the ruthenium is in the form of barium ruthenate and ruthenium dioxide. However, the peak area ratio confirms that the $BaRuO_3$ concentration is now less than the RuO_2 concentration. This indicates that the cycling of these stabilized catalysts between a net reducing atmosphere and a net oxidizing atmosphere results in a significant separation between the ruthenium metal and the stabilizing agent. This explains why these catalysts do not have a satisfactory lifetime.

D. Zeolite Supported Ruthenium Systems

The zeolite materials provide an opportunity to disperse metals or metallic compounds almost mono-molecularly in the cage spaces of the crystals. Catalytic activity and chemisorption have been observed for iron zeolite systems[33,34]. Our initial study of the ruthenium systems supported on zeolites is outlined below.

Linde Na-Y zeolite (63.5% SiO_2, 23.5% Al_2O_3 and 13.0% Na_2O) was used to prepare the zeolite catalyst samples. Ruthenium was exchanged into the zeolite support by use of the $[Ru(NH_3)_5N_2]Cl_2$ complex which was prepared according to the method of Allen, et al.[35]. Cation exchange of this complex was performed in the following manner. The complex (1.8 - 2.2 g) was added to deoxygenated water (100 ml) and an appropriate amount of Na-Y zeolite (6-7 g) was then added and the exchange allowed to proceed for 12-16 hours with shaking under a nitrogen atmosphere. The zeolite was filtered, washed several times with water and then dried over P_2O_5 under vacuum in a desiccator for 48 hours. Based on the percent of sodium ions displaced, the exchange of $[Ru(NH_3)_5N_2]^{2+}$ was in the range of 60-70% for all samples prepared by this method.

Listed in Table 7 are the Mössbauer parameters obtained for the $[Ru(NH_3)_5N_2]^{2+}$-Y zeolite samples. Representative Mössbauer spectra for two of the zeolite samples are shown in Figure 8. The synthetic faujasite Y type zeolite was chosen for this study because a great deal of information about its structure, catalytic activity and the chemical nature of the cation exchange sites has been published[36]. For the purpose of introducing ruthenium atoms into the zeolite framework, we chose to use the dinitrogen complex cation $[Ru(NH_3)_5N_2]^{2+}$ because we felt that it offered the possibility for stripping the NH_3's and N_2 from the coordination sphere, leaving the bare ruthenium ion in the zeolite. Secondly, we felt that the N_2 group might serve as a pathway for reversibly introducing such groups as CO, NO, etc., into the ruthenium coordination sphere.

The sample referred to as 10-A in Table 7 corresponds to a portion of the $[Ru(NH_3)_5N_2]^{2+}$-Y zeolite after drying for 48 hours under vacuum. The sample was held under a vacuum of 10^{-5} Torr while the spectrum shown in Figure 8-(a) was obtained. The isomer shift and relative line intensities observed for this sample agree with those obtained for a crystalline sample of $[Ru(NH_3)_5N_2]Cl_2$. However, the quadrupole splitting for the zeolite sample (0.56 mm/sec) was greater than that observed for the $[Ru(NH_3)_5N_2]Cl_2$ sample (0.22 mm/sec). These data suggest that the $[Ru(NH_3)_5N_2]^{2+}$ group is exchanged without undergoing oxidation or ligand loss. However, the increase in quadrupole splitting indicates that some distortion in the coordination sphere has occurred upon exchange. The distortion may be produced by the rigid aluminosilicate back-

TABLE 7

MÖSSBAUER PARAMETERS FOR RUTHENIUM EXCHANGED ZEOLITES

Sample No.	Treatment	Absorber Thickness (mg Ru/cm^2)	Isomer Shift (mm/sec)	Quadrupole Splitting (mm/sec)	Peak Width (Γ) @ Half-Height (mm/sec)
10-A	[Ru(NH$_3$)$_5$N$_2$]$^{+2}$-Y dried and evacuated to 10^{-5} Torr @ 25°C	125	-0.80 ± 0.04	0.56 ± 0.04	0.61 ± 0.05
10-B	Sample 10-A exposed to air for 2 days @ 25°C	125	-0.37 ± 0.03	---	0.79 ± 0.05
10-C	Sample 10-B reduced in H$_2$ at 400°C for 4 hours	125	$+0.02 \pm 0.03$	---	0.61 ± 0.04
10-D	A portion of Sample 10-A reduced in H$_2$ at 400°C for 4 hours	110	$+0.01 \pm 0.02$	---	0.52 ± 0.03
10-E	Sample 10-D exposed to air for 24 hours @ 25°C	110	-0.10 ± 0.03	---	0.67 ± 0.04
Ru Metal Powder	-----------	225	0.00 ± 0.02	0	0.45 ± 0.03
[Ru(NH$_3$)$_5$N$_2$]Cl$_2$	-----------	175	-0.76 ± 0.04	0.22 ± 0.03	0.51 ± 0.03
[Ru(NH$_3$)$_5$OH]Cl$_2$	-----------	190	-0.39 ± 0.03	0	0.49 ± 0.03

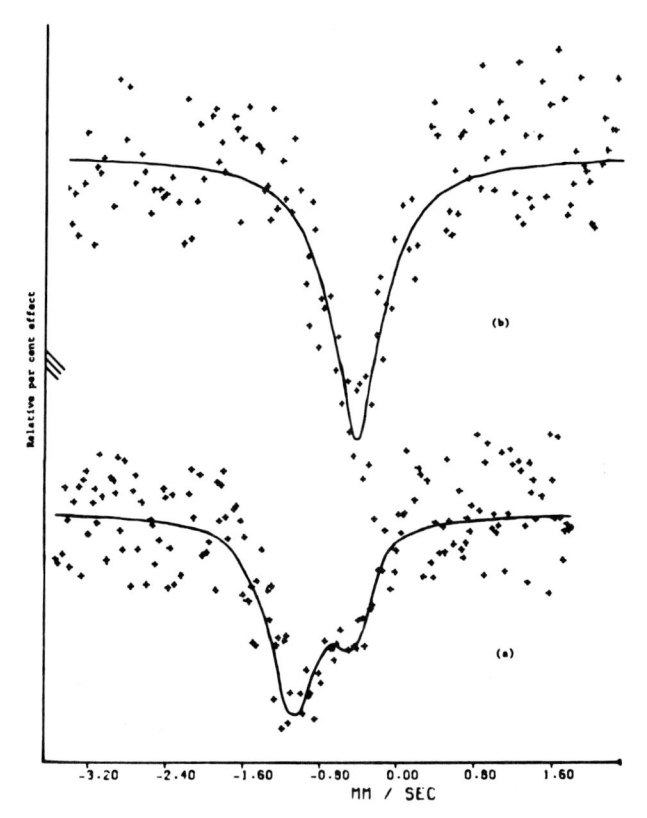

Fig. 8. Mössbauer spectra of: (a) Sample 10-A ([Ru(NH$_3$)$_5$N$_2$]$^{2+}$-Y zeolite after drying); (b) Sample 10-B (this is sample 10-A after exposure to air).

bone structure of the zeolite. For example, cations exchanged in a zeolite have been found to be capable of occupying several different sites within the zeolite framework[37,38]. Since these sites are located on the sides and at the distances of different size and shaped cavities, each site would impose its own characteristic structural and electronic requirements on the cation. Unfortunately, because of the limited number of studies dealing with this topic, it is not currently possible to identify the position of the [Ru(NH$_3$)$_5$N$_2$]$^{+2}$ cation in the Y-zeolite from its Mössbauer spectral parameters.

Upon exposure of sample 10-A to the atmosphere at 25°C, it slowly turned a wine color. After two days, the color appeared to stabilize and the spectrum shown in Figure 8-(b) was obtained. This sample is referred to as sample 10-B in Table 5. The Mössbauer parameters have changed significantly. The change in isomer shift from -0.80 to -0.37 mm/sec indicates that the ruthenium has undergone oxidation from the +2 state to the +3 state[23]. The only +3 wine colored ruthenium compound that could be found in the literature corresponds to the [Ru(NH$_3$)$_5$OH]Cl$_2$. The Mössbauer spectral parameters for this compound as shown in Table 7 agree with those observed for the wine colored compound in the zeolite. The broader linewidth for the zeolite sample may result from the presence of smaller concentrations of other ruthenium species or possibly [Ru(NH$_3$)$_5$OH]$^{+2}$ groups at different sites in the zeolite.

Laing et al.[39] have also observed that a [Ru(NH$_3$)$_5$N$_2$]$^{+2}$-Y zeolite sample decomposes in air to give a wine colored species. They proposed that the decomposition may occur by the following reaction:

$$[Ru^{II}(NH_3)_5N_2]^{+2}\text{-Y} + H_2O \rightarrow [Ru^{III}(NH_3)_5OH]^{+2}\text{-Y}$$

$$+ NH_4^+ + \text{other products}$$

Sample 10-C in Table 7 corresponds to sample 10-B after treatment in a stream of hydrogen for 4 hours at 400°C. The Mössbauer parameters for this sample indicate that all of the ruthenium has been reduced to the metallic state. However, x-ray analysis of this sample indicated that a significant amount of crystallinity in the zeolite framework was lost upon reduction. Therefore, a new portion of sample 10-A was reduced in a hydrogen stream and this sample is referred to as 10-D. Again, the Mössbauer data show that all of the ruthenium has been reduced to the metallic state. X-ray analysis of this sample indicated that most of the zeolite structure was maintained during the treatment and that all of the ruthenium metal particles were less than 80 Å in diameter. When this sample was exposed to the atmosphere, the Mössbauer data designated for sample 10-E in Table 7 was obtained. The observed change in isomer shift and linewidth upon exposure to the atmosphere indicates that some type of interaction has occurred between the small ruthenium metal particles and the gaseous components of air. The change in isomer shift by -0.10 mm/sec is greater than the experimental error in the measurement and indicates that the effective s-electron density has been reduced in a majority of the ruthenium atoms. This could be due to increased shielding brought about by chemisorbed groups occupying p and d ruthenium orbitals, or by direct s-electron withdrawal by chemisorbed groups. However, the important thing is that some form of interaction was observed. This indicates that a favorable surface atom to bulk atom ratio exists in the sample, since larger ruthenium metal particles supported on silica and alumina exhibited no interaction with air as observed by Mössbauer spectroscopy. Therefore, it appears that zeolite supported ruthenium can be used as a model system for studying chemisorption phenomena on ruthenium metal by Mössbauer spectroscopy.

CONCLUSIONS

It is evident that significant information about heterogeneous ruthenium catalyst systems can be obtained from solid-state Mössbauer data. The need for data related only to the surface species is also evident. Preliminary results from ESCA studies indicate that this technique may fulfill this need for ruthenium systems. Thus, the combination of Mössbauer and ESCA spectroscopies provide a powerful tool in the quest for a complete characterization of heterogeneous ruthenium catalysts.

ACKNOWLEDGEMENTS

The authors are grateful to the National Science Foundation (Grant No. GP-38054X) for financial support. Thanks are due to Dr. C. R. Ginnard of the DuPont ESCA Applications Laboratory in Monrovia, California for providing the initial ESCA measurements reported here.

REFERENCES

1. W.R. Cares and J.W. Hightower, J. Catal. $\underline{39}$, 36 (1975).
2. a) Za. Bonchev, A. Jordanov and A. Minkova, Proc. Conf. Applications Mössbauer Effect (Tihany, 1969), Akademai Kiado, Budapest, 1971, p. 333.
   b) R.L. Collins, Mössbauer Methodology, $\underline{4}$, (1968).
   c) R.A. Krakowski and R.B. Miller, Nucl. Inst. Methods, $\underline{100}$, 93 (1972).
3. a) V.I. Lisichenko, S.L. Korduk, O.L. Orlov and A.N. Smoilovskii, Proc. Conf. Applications Mössbauer Effect (Tihany, 1969), Akademai Kiado, Budapest, 1971, p. 339.
   b) H. Hobert and D. Arnold, ibid, p. 325.
4. a) M. Rich, Phys. Lett. $\underline{4}$, 153 (1963).
   b) J.S. Van Wilringen, Phys. Lett. A, $\underline{26}$, 370 (1968).
   c) H.M. Gager, M.C. Hobson and J.F. Lefelhocz, Chem. Phys. Lett., $\underline{15}$, 124 (1972).
   d) J.W. Burton and R.P. Godwin, Phys. Rev., $\underline{158}$, 218 (1967).
5. a) W. Kundig, K.J. Ando, R.H. Lingquist and G. Constabaris, Czech. J. Phys., $\underline{B17}$, 467 (1967).
   b) H.M. Gager, J.F. Lefelhocz and M.C. Hobson, Jr., Chem. Phys. Lett., $\underline{23}$, 386 (1973).
   c) A.Z. Hrynkiewicz, A.J. Pastowka, B.D. Sawicka and J.A. Sawicka, Phys-Status Solidi, A, $\underline{9}$, 607 (1972).

6. a) I.S. Jacobs and C.P. Bean, *Magnetism*, Vol. 3 (G.T. Rado and H. Sudl, eds.), Academic Press, N.Y., 1963, p. 271.
   b) W. Kundig, H. Bommel, G. Constabaris, and R.H. Lindquist, Phys. Rev., 142, 327 (1966).
7. V.I. Goldanskii and I.P. Suzdalev, Proc. Conf. Application of the Mössbauer Effect (Tihany, 1969), Akademai Kiado, Budapest, 171, p. 325.
8. H.M. Gager and M.C. Hobson, Jr., Catal. Rev.-Sci. Eng., 11, 117 (1975).
9. W.L. Jolly, Coord. Chem. Reviews, 13, 47 (1974).
10. K.S. Kim and N. Winograd, Chem. Phys. Letters, 30, 91 (1975).
11. G.K. Wertheim, *Perspectives in Mössbauer Spectroscopy*, (S.G. Cohen and M. Pasternak, Eds.), Plenum Press, New York, 1973, p. 41.
12. a) I. Adams, J.M. Thomas, G.M. Bancroft, K.D. Butler and M. Barber, J. Chem. Soc. Chem. Comm., 751 (1972).
    b) M. Barber, P. Swift, D. Cunningham and M.J. Frazer, Chem. Comm., 1338 (1970).
    c) W.E. Swartz, P.H. Watts, E.R. Lippincott, J.C. Watts and J.E. Huheey, Inorg. Chem., 11, 2632 (1972).
13. D.N.E. Buchanan, M. Robbins, H.J. Guggenheim, G.K. Wertheim and V.G. Lambrecht, Jr., Solid State Comm., 9, 583 (1971).
14. a) P.H. Citrin, Jour. Amer. Chem. Soc., 95, 6472 (1973).
    b) C. Creutz, M.L. Good and S. Chandra, Inorg. Nucl. Chem. Lett., 9, 171 (1973).
15. F. Holsboer and W. Beck, Z. Naturforsch, 27b, 884 (1972).
16. a) F. Holsboer, W. Beck and H.D. Bartunik, Chem. Phys. Letters, 18, 217 (1973).
    b) F. Holsboer, W. Beck and H.D. Bartunik, J.C.S. Dalton, 1828 (1973).
17. R. Larsson, J. Mrha and J. Blomquist, Acta. Chem. Scand., 26, 3386 (1972).
18. M.L. Good, Mössbauer Effect Data Index, p. 51 (1972).
19. W.N. Delgass, T.R. Hughes and C.S. Fadley, Catal. Rev., 4, 179 (1970).
20. C.A. Clausen, III, and M.L. Good, J. Catal., 38, 92 (1975).
21. C.A. Clausen and M.L. Good, Inorg. Chem., submitted. [A report on the zeolite supported materials.]
22. C.A. Clausen and M.L. Good, J. Catal., submitted. [A report on the automotive catalysts.]
23. C.A. Clausen, III, R.A. Prados and M.L. Good, Mössbauer Effect Methodology, 6, 31 (1971).
24. H.P. Klug and L.E. Alexander, *X-Ray Diffraction Procedures*, pp. 504-509, John Wiley and Sons, New York, N.Y. (1954).
25. R.A. Dalla-Betta, J. Catal., 34, 57 (1974).
26. I.P. Suzdalev, M.Y. Gen, V. I. Goldanskii and E.F. Markarov, Sov. Phys. JETP, 24, 79 (1967).
27. J.G. Stevens and V.E. Stevens, *Mössbauer Effect Data Index*, p. 226, Plenum Data Corp., New York, 1973.
28. C.A. Clausen, III, R.A. Prados and M.L. Good, Chem. Comm., 1188 (1969).

29. K.S. Kim and N. Winograd, J. Catal., 35, 66 (1974).
30. A. Ozaki, K. Aida and H. Hari, Bull. Chem. Soc. Japan, 44, 3216 (1971).
31. a) M. Shelef and H.S. Gandhi, Ind. Eng. Chem., Prod. Res. Dev., 11, 393 (1972).
    b) M. Shelef and H.S. Gandhi, Platinum Metals Rev., 18, 2 (1974).
32. R.L. Klimisch and K.C. Taylor, Envir. Sci. Tech., 7, 127 (1973).
33. W.N. Delgass, R. L. Garten and M. Boudart, J. Phys. Chem., 73, 2970 (1969).
34. R.L. Garten, W.N. Delgass and M. Boudart, J. Catal., 18, 90 (1970).
35. A.D. Allen, F. Bottomly, R.O. Hains, V.P. Reinsaln and C.V. Senoff, J. Amer. Chem. Soc., 89, 5595 (1967).
36. H.S. Sherry, Advan. Chem. Series, 101, 350, American Chemical Society (1971).
37. J.V. Smith, Advan. Chem. Series, 101, 171, American Chemical Society (1971).
38. D.H. Olson, J. Phys. Chem., 74, 2758 (1970).
39. K.R. Laing, R. Leubner and J. Lunsford, Inorg. Chem., 14, 1400 (1975).

Discussion

On the Paper by E.W. Müller and S.V. Krishnaswamy

H. Leidheiser (*Lehigh University*): There is great interest at present in the interaction of H2S with iron. Have you studied this reaction in the atom-probe?

E.W. Müller (*Pennsylvania State University*): No. Though we realize the interest in the Fe-H2S interaction so far we did not get to it. As I mentioned earlier we did study the interaction of H2S on Rh and Ni surfaces only using the straight atom-probe. We plan to study the Fe-H$_2$S system in the near future. We think the high resolution of our energy focused atom-probe should enable us to distinguish between the iron isotopes and the hyrides much more unequivocally than was possible with other straight atom-probes.

L.H. Lee (*Xerox Corp.*): This same day Dr. Müller will be admitted to the hospital for a throat surgery. He regretted that he could not be here in person to present his paper. However, we are fortunate to have Dr. Krishnaswamy present this paper to us. We all agree that Dr. Krishanswamy has done an excellent job in covering the entire subject. (This Symposium voted unanimously to send a telegram to Dr. Müller to wish his early recovery).

On the Paper by G.W. Simmons and H. Leidheiser

L.H. Lee: Dr. Leidheiser gave us a very useful survey about the application of Mössbauer spectroscopy for the study of corrosion. Many organic coatings have been used to protect metals from corrosion. It is thus important for surface chemists to understand mechanisms of corrosion in order to design new polymers for various end uses. Recent applications of numerous methods, e.g., atomic probe, Auger spectroscopy, photoemission, can all lead us to a better understanding of corrosion mechanisms.

On the Paper by C.A. Clausen and M.L. Good

L.H. Lee: New search in energy resources has stimulated heterogeneous catalysis studies in many universities. This paper by Dr. Good has demonstrated the applications of ESCA and Mössbauer spectroscopy for the investigations of catalyst surface. We all appreciate Professor Good for the excellent presentation.

PART II

Auger-Electron Spectroscopy and Electron Microprobe

Introductory Remarks:

M.L. Good

Department of Chemistry
University of New Orleans
New Orleans, Louisiana 70122

It is always a pleasure to participate in an outstanding symposium. However, it is particularly exciting to be a part of an interdisciplinary program such as this one where techniques developed in basic research laboratories are being discussed with the view of their direct application to problems of immediate interest in the polymer and metal surface fields. The papers presented yesterday were indicative of the scope and vision of researchers on the "cutting edge" of surface science and they indicate the impressive arsenal of physical techniques which are being brought to bear on the definition of surface chemistry. Our program this morning will continue this theme with an exhaustive look at the basic principles and applications of Auger Spectroscopy. The plenary lecture on the basic concepts of electron beams as surface probes will be followed by "state-of-the-art" reviews of electron spectroscopy as applied to surface studies. If symposia such as this one can be successful, our problem of "technology transfer" will be effectively eliminated. I would like to thank the Division of Organic Coatings and Plastics Chemistry and the organizers of this symposium for their invitation to participate. The Division and its members should be congratulated for their sponsorship of such a timely program.

Plenary Lecture: Low Energy Electrons as a Probe of Solid Surfaces

Robert L. Park, Marten den Boer and
Yasuo Fukuda

*Department of Physics and Center of Materials Research
University of Maryland
College Park, Maryland 20742*

Electron beams provide a convenient probe for the creation of excited states at the solid-vacuum interface. The energies of these states range from a few tens of milli-electron volts for vibrational states to several kilo-electron volts for excited core states. They are studied by three distinct classes of experiment: measurements of the characteristic energy losses suffered by electrons in the excitation process, measurements of the threshold energies for the creation of excited states, and analysis of the photons or electrons emitted to conserve energy in the decay back to the ground state. These approaches have in common a sensitivity to the surface region that is consequence of the short mean free path for inelastic scattering of low-energy electrons. It is a perverse fact, however, that the very inelastic damping that enables us to restrict our view to the surface region, renders that view highly distorted. These effects will be illustrated by spectra of aluminum and silicon surfaces.

INTRODUCTION

In a little over a decade the astonishing growth of solid state electronics has replaced technological interest in the solid-vacuum interface with concern over phenomena occurring at solid-solid interfaces. Meanwhile, in such areas as electrochemistry and catalysis, our limited understanding of solid-liquid and solid-gas interfaces tempers our response to the new economics of energy. Yet, it is the solid-vacuum interface that is the primary focus of a burgeoning basic research effort. The reason is simply that the exposed surface is accessible to the sort of poking and probing which enables us to better understand the force laws that govern all interfaces.

The surface region of a solid is best defined as the layer in which the atomic potentials differ from those of atoms still deeper within the solid.[1] This region includes those substrate atoms that sense the altered chemical environment imposed by the loss of translation symmetry, as well as adsorbed foreign atoms. For clean metals the surface region may include no more than the outermost two atomic layers.[2] The extent to which measurments are specific to this region is determined either by the attenuation of an incident probe beam, or by the attenuation of escaping particles excited by the probe beam. Perhaps the most fashionable probes of the surface are the intense beams of short wavelength photons that can be extracted from a synchroton storage ring.[3] For those who cherish the freedom of individual research, however, it should be noted that a twenty-five dollar electron gun can generally produce results comparable to those obtained with the best synchrotron sources. In this paper, we will review the use of low-energy electron beams to study the spectrum of excited states in the surface region of a solid.[4]

Secondary Electron Yield

One of the most fundamental quantities associated with a surface is its work function, which is the energy that would be expended in removing an electron to infinity from an infinite plane surface of the material. In practice, the quantity that is usually measured is the contact potential difference between the surface studied and a reference electrode. The most convenient reference electrode is the thermionic emitter. Measurement of contact potential differences by the retarding potential method is illustrated schematically by the energy level diagram in Figure 1. Electrons thermionically emitted from the source are accelerated to some arbitrary energy eV_o and focused into a parallel beam directed normally at a plane surface of the sample under study. When the external potential applied between the source and sample, called the retarding potential, just corresponds to the work function differences between them, normally incident electrons will have zero kinetic energy as they arrive at the sample surface. This condition, therefore represents a threshold, below which no current will be detected in the external circuit, and above which the current measures the secondary emission coefficient as a function of incident electron energy.

An important modification of this method, using a field emission source, was introduced nearly half a centry ago by Henderson.[5] The field emission retarding potential (FERP) method provides an absolute measure of the work function since source electrons tunnel directly from the Fermi sea of the emitter rather than escaping over a work function barrier. This method has been perfected by Strayer, Makie, and Swanson.[6]

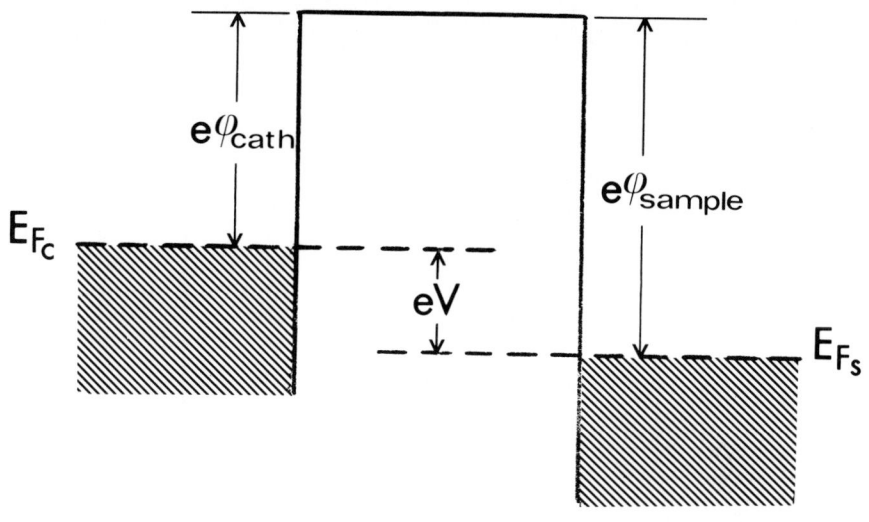

Fig. 1. Energy level diagram illustrating the measurement of contact potential differences by the retarding potential method. The condition shown represents the threshold, for electrons which have just surmounted the work function barrier of the cathode, to reach the sample, i.e., $V = \phi_{sample} - \phi_{cath}$. This condition corresponds to the knee of the retarding potential curve, since thermionically emitted electrons will all have some kinetic energy.

Above the retarding potential threshold, the sample current exhibits considerable structure. This is seen in Figure 2 which shows a retarding potential curve for a contaminated (111) Si surface. The structure is enhanced by examining the derivative of the regarding potential curve, which was obtained by the potential modulation method. In the simplest interpretation, this structure represents variations in the density of unfilled states lying more than the work function above the Fermi energy. Thus, the probability of an electron being transmitted into the bulk of the solid, rather than being reflected, depends on the availability of states at that energy.

Striking evidence of this interpretation is given by the retarding potential curves of Thomas for thin epitaxial films of gold on iridium single crystals.[7] He observed a variation in the retarding potential curve, the period of which was inversely proportional to the thickness of the gold film. This variation corresponds to the existence of quantum size states associated with standing electron waves in the gold film. A remarkable aspect of these measurements was that the incoming electron was able to "see" the gold-iridium interface even for film thicknesses

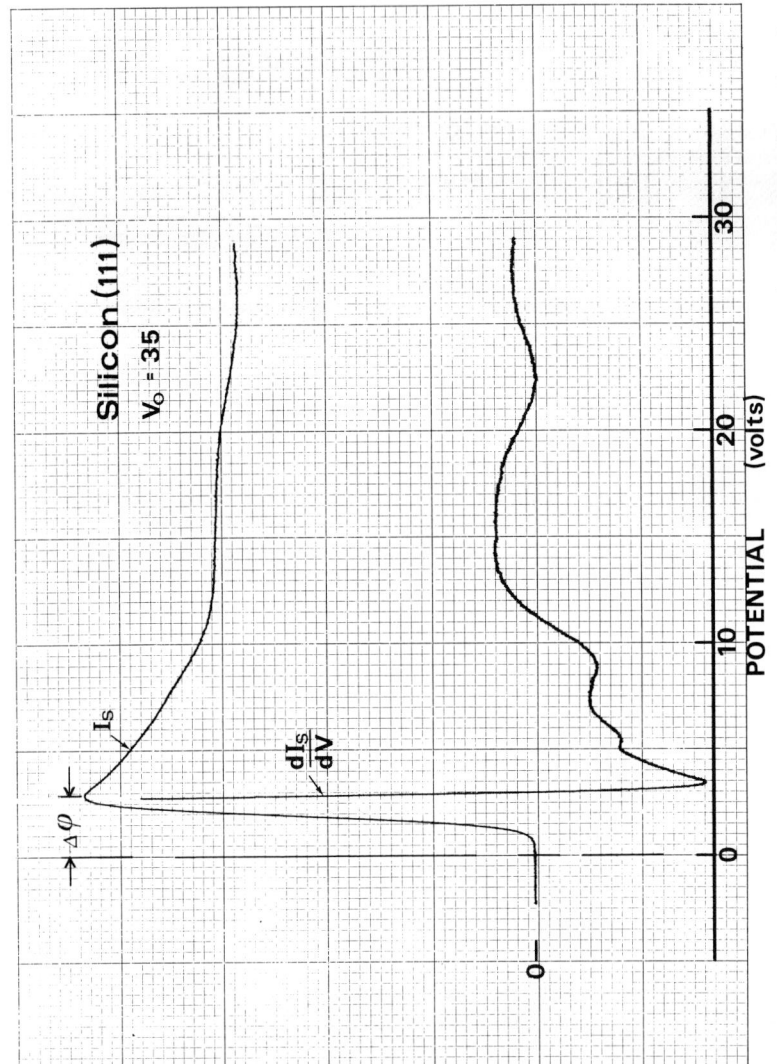

Fig. 2. Retarding potential curve for a silicon (111) surface, and its derivative obtained by the potential modulation technique. The derivative spectrum reveals structure that cannot be identified in the total current. At low energies the spectrum can be regarded as a measure of the electron reflectivity. The structure is a measure of the states available to electrons with the momentum of the incident beam.

of 65Å. As we will see, this is a consequence of the fact that there are few mechanisms for inelastic scattering of electrons in this very low energy.

We are justified in treating the retarding potential curve as a measure of the electron reflectivity of the surface, only if we can neglect inelastic scattering. It was shown by Farnsworth as early as 1925, two years before the wave nature of the electron was demonstrated, that at energies of a few electron volts most electrons are indeed reflected from a metal surface without measurable loss in energy.[8] Above the threshold for plasmon excitation, however, inelastic processes must be taken into account.

Plasmons are quantized collective oscillations of the valence electron fluid. For electrons in the energy range below a few keV collisions with nuclei are rare, and plasmon creation is the principal source of energy loss. The sensitivity of electron spectroscopies to the surface region is a consequence of the short mean free path for plasmon creation. The inelastic mean free path for plasmon excitation has been treated theoretically by Quinn,[9] and refined recently by Penn.[10] Since these theories consider only plasmon losses, the calculated mean free path becomes infinite below the plasmon energy, which is typically between 5 and 25 eV depending on the electron density. There are, of course, excitations other than plasmons by which electrons lose energy in the very low energy range, such as phonon creation and bremsstralung emission, but the cross sections for these events are relatively small, and, as the Thomas experiment[7] demonstrates, the mean free paths can be quite long. Coupling to plasmons also decreases at high energies, with the result that the minimum sampling depth for metals usually occurs for electrons between 50 and 100 eV. Experimental measurements of elastic escape depths have been collected by Powell.[11]

Plasmons typically have lifetimes of a few electron volts. They can decay either by the emission of phonons into the lattice,[12] by the emission of photons,[13] or by the ejection of single electrons in an Auger type process.[14] We would expect, if the latter process is very likely, to see changes in the secondary emission coefficient as we reach the threshold energy for creation of plasmons. The first such threshold should correspond to the creation of surface plasmons.[15] The threshold for creation of bulk plasmons should occur at an energy $\sqrt{2}$ greater than the surface plasmon energy. In fact, we can expect the spectrum to be more complicated than this would suggest, since any excited state of the solid may decay by electron emission. This includes interband single electron excitations and the excitation of shallow core states.

The complexity of the structure in the secondary electron yield discouraged early attempts to extend the Frank-Hertz experiment to the core levels of solids. Farnsworth[16] concluded that correlations of inflections in the secondary electron yield with the critical potentials for X-ray production were fortuitous. Richardson[17] reported, however, that although there were many unexplained inflections in the secondary emission yield, there was an increase in the yield at every potential for which an increase in soft X-ray emission could be detected. Our measurements show these thresholds to be extremely weak, and it seems unlikely that they could have been unambiguously identified in the total yield. Although weak, these thresholds are quite sharp, and above a few hundred electron volts, where band structure effects, plasmon thresholds, and interband transitions can all be neglected, thresholds for scattering from core electrons can be identified in the derivative of the secondary electron yield,[18,19] as shown in Figure 3.

The threshold for scattering from a core electron of binding energy E_B relative to the Fermi energy occurs at

$$eV = E_B - e\phi_{cath} - kT$$

and corresponds to the case in which both the incident and excited core electron are captured in states at the Fermi level. Above the threshold, the excitation probability depends on the possible final states of two electrons; the incident electron and the excited core electron as shown in Figure 4. Thus, assuming constant oscillator strengths, the two electron density of states, $N_{2c}(E)$, should vary as the self-convolution of the one electron density of conduction band states.[20] The actual transition density must, of course, include the initial state. The shape must, therefore, be broadened by the lifetime of the core hole. The actual measured shape will be further smeared by the energy spread of the incident electrons, $\sim kT$, and the amplitude of the potential modulations used to obtain the derivative plot. The combination of these experimental broadenings can be kept below .25 eV, which makes this one of the highest resolution techniques available for probing the core levels of atoms.

The inner shell electrons do not, of course, participate in chemical bonds, and might be thought to hold little interest for the chemist. Precisely because they are relatively unaffected by bonding, however, they provide a means of elemental identification. This is certainly an important piece of information, but for the purpose of this paper, we will be more concerned with what we can learn of the chemical environmental of the atoms from transitions involving the core levels. The core electron wave functions over-

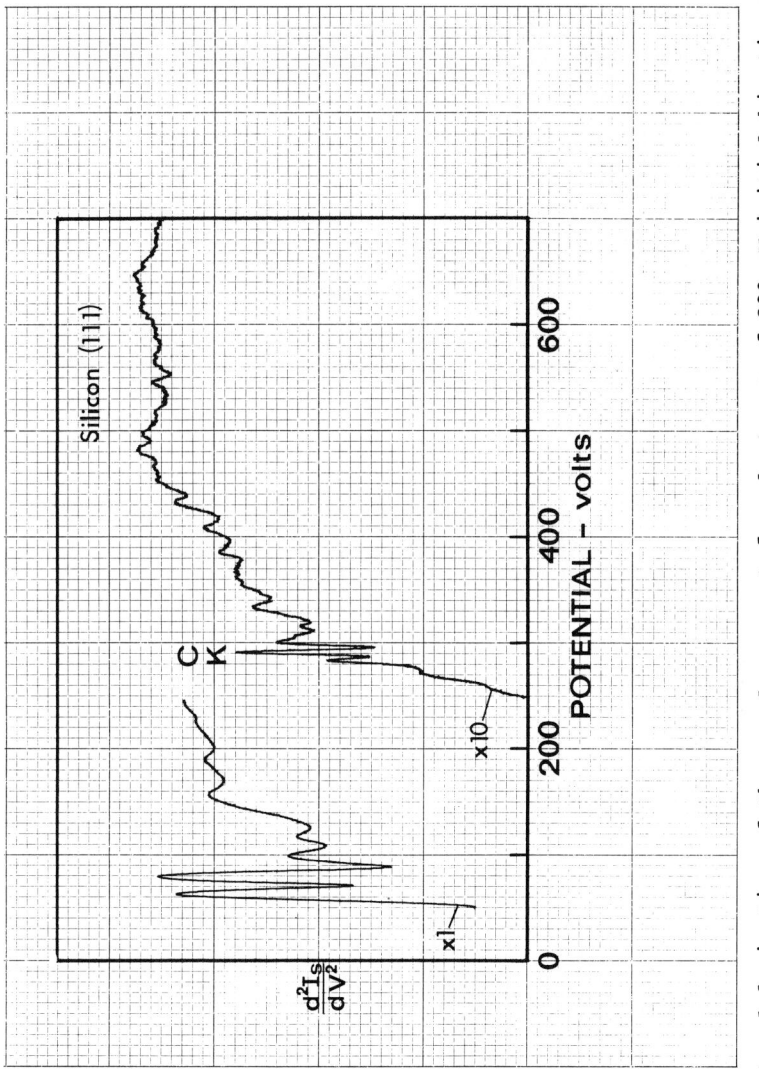

Fig. 3. Second derivative of the sample current for electrons of 800 eV initial kinetic energy on silicon (111). High energy band structure effects, plasmon thresholds, etc., produce strong variations, particularly at low energies. At higher energies, core level Auger electron appearance potentials, such as the carbon K shell spectrum from a surface impurity, can be identified. The shape of the C K spectrum is related to the density of unfilled states near the Fermi level.

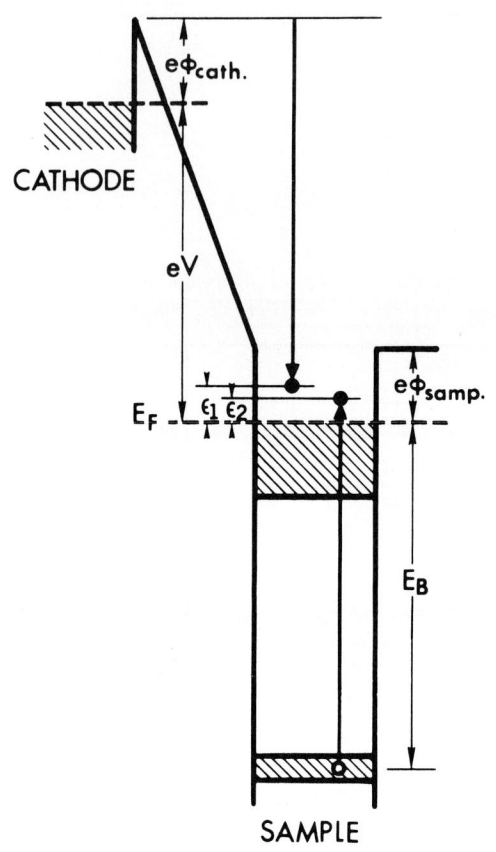

Fig. 4. Energy level diagram of the appearance potential experiment. An incident electron of energy $eV+e\phi_{cath}$ may be captured in a state ε_1, above the Fermi energy, exciting a core electron into a state $\varepsilon_2 = e\phi + e\phi_{cath} - \varepsilon_1 - E_B$ above the Fermi energy. The threshold for excitation of the core electron is at $eV + e\phi_{cath} = E_B$. The excitation probability above the threshold should vary as the self-convolution of the conduction band density of states, broadened by the lifetime of the core hole. Core electron excitations are signaled in soft X-ray appearance potential spectroscopy by an increased emission of X-rays, and in Auger electron appearance potential spectroscopy by an increase in secondary emission.

lap only a small region of the valence and conduction bands. By using the core levels as windows, a local density of states associated with a single element can be viewed separately.

Energy may be conserved in the decay of an excited state by photon emission rather than electron emission. For core levels with binding energies in the soft X-ray region, however, radiative recombination of a core hole is a relatively rare event,[21] with non-radiative Auger electron emission being favored by at least a factor of 10^2. Soft X-rays are, moreover, notoriously difficult to detect, the efficiency of photoelectric detectors in this range being no greater than 1%. It may seem remarkable, therefore, that threshold potentials for electron scattering from core levels of atoms in a solid were clearly observed in plots of the total X-ray yield as early as 1921[22] - fifty years before such thresholds were detected with certainty in secondary emission yields.

The reasons for this seemingly contradictory state of affairs is just that the background in the soft X-ray case is comparatively well behaved. The background, consisting mostly of bremsstrahlung, rises almost linearly. Plasmon decay by radiative emission, although it is known to occur,[13] seems to have little effect on the yield measured in this way. As a consequence of this well-behaved background, most appearance potential measurements are made by the soft X-ray technique.[20,23,24] The price that is paid is that much higher incident electron currents are required. Using photoelectron detectors, these currents are typically 1 to 10 milliamperes. Using solid state detectors, and aluminum filters to remove much of the very soft bremsstrahlung, Andersson et al., have succeeded in reducing the incident electron current requirements by two orders of magnitude.[25] These currents are nevertheless two orders of magnitude greater than those required for Auger electron appearance potential spectroscopy, and are capable of producing sample heating, and significant desorption of gases. Andersson and his colleagues[26] have, however, been able to obtain high resolution spectra of strongly adsorbed gases. As with Auger electron appearance potential spectroscopy, however, the region below a few hundred eV is largely inaccessible to this technique. Using the more conventional photoelectric detection technique, and taking the second derivative to suppress the diode characteristic, enables appearance potentials to be detected as low as 30 volts.[27] Thus, most soft X-ray appearance potential spectra have been obtained from surfaces of clean metals and comparatively stable compounds, where the high currents cause little difficulty.

Characteristic Losses

In the appearance potential techniques, the energies of excited states are determined from changes in the yield of either electrons or photons as the energy of the incident electrons is varied across the excitation threshold. Only the energy of the incident electrons is measured in such experiments. It is also possible to determine these characteristic energies from the energy lost by the incident electron in creating the excited state. This requires an analysis of the energies of backscattered electrons. At incident electron energies very much higher than the plasmon threshold, most of the secondary electrons are contained in the so-called "true secondary peak" which typically has its maximum at about 2 eV,[28] as shown in Figure 5. The distinction between "true" and "untrue" secondaries seems rather artificial. If, however, we define true secondary electrons as those resulting from the decay of excited states created by the incident electron beam, the distinction becomes very clear. We then can separate the secondary electron spectrum into a "loss spectrum," which includes only those secondaries whose energies are correlated with the energy of the incident electron beam, and an "emission spectrum," consisting of electrons emitted to conserve energy in the decay of excited states. Since decay times are generally very long compared to excitation times, the decay of an excited state is uncorrelated with the energy of the exciting particle.

Experimentally, the separation of emission and loss spectra is quite straightforward.[29] A small oscillation is superimposed on the incident electron energy. By detecting only those spectral features that exhibit the same modulation, the loss spectrum is obtained.

The separation of the secondary electron energy distribution of aluminum into emission and loss spectra is shown in Figure 6. The emission spectrum was obtained by modulating the sample potential. This has the effect of varying the analyzer window synchronously with the incident electron energy.

At the high energy limit of the loss spectrum, there is a sharp peak corresponding to electrons elastically or quasi-elastically scattered from the ion cores. The quasi-elastically scattered electrons are those which have lost energy to the creation of lattice phonons or of localized surface vibrations. These losses cannot be resolved in the spectra shown here since they amount to less than half of an eV. Using a very high resolution spectrometer, however, they were observed by Propst and Piper.[30] More recently, they have been examined by Ibach.[31] They correspond to infrared absorption energies, which have been studied for many years, but generally on poorly characterized surfaces with very large areas.[32] The electron energy loss measurements can be

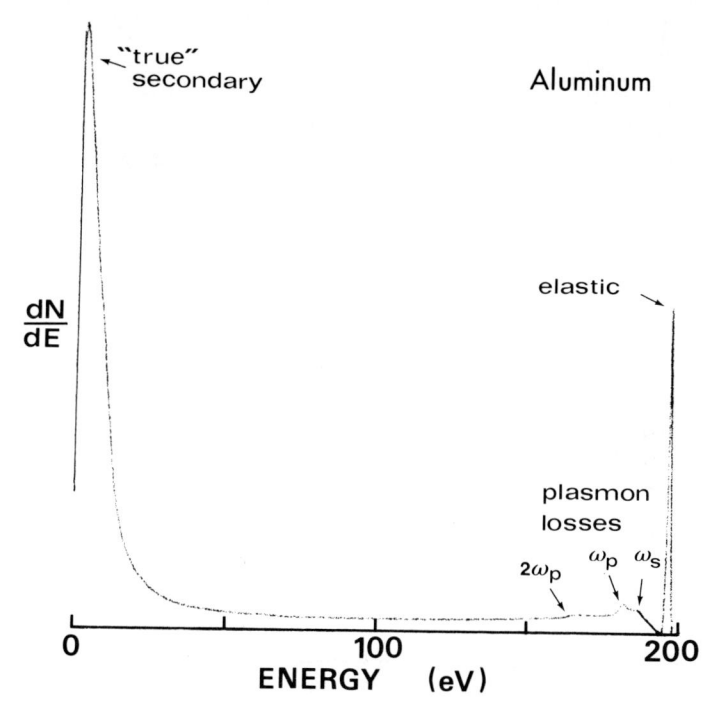

Fig. 5. Secondary electron energy distributed for a "clean" aluminum surface bombarded with 200 eV electrons. Most secondary electrons are contained in the large peak at low energies. Reflected electrons are contained in the elastic peak, along with quasi-elastic electrons which have excited vibrational modes. The principal energy loss mechanism is by plasmon creation including both surface, ω_s, and bulk, ω_p, plasmons.

carried out on extremely small surface areas, and may be sensitive to vibrations that are IR inactive.

Just below the peak in Figure 6 is a region of very distinctive loss structure corresponding to discrete energy losses from the excitation of plasmons and interband transitions. This structure can be seem in much greater detail by plotting the derivative of the spectrum as shown in Figure 7. In addition to structure corresponding to creation of single bulk and surface plasmons, additional features can be associated with multiple plasmon creation. The energy loss spectrum is determined by the same excitations that give rise to the optical properties of solids.[33] The principal differences lies in the extreme sensitivity of the electron energy loss measurements to the surface region.

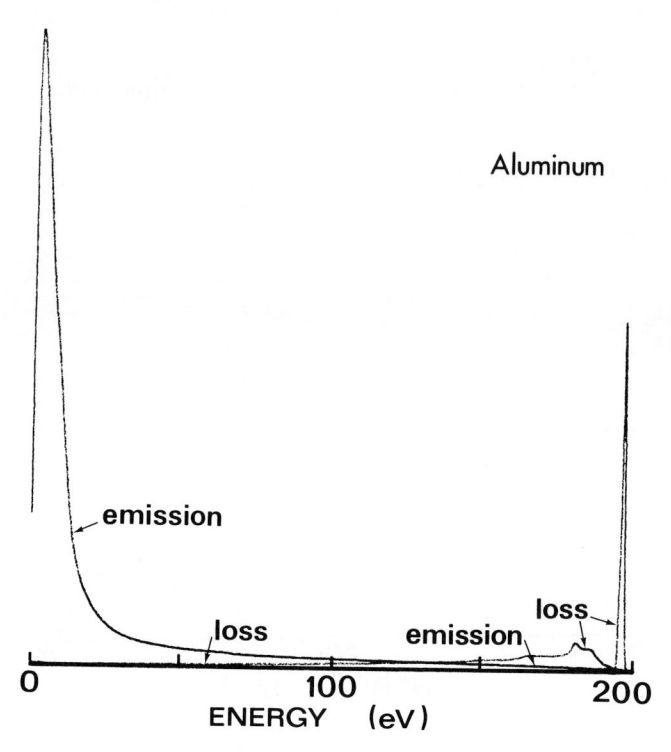

Fig. 6. Separation of the secondary electron energy distribution shown in Figure 5 into a loss spectrum, which is correlated with the incident electron energy, and an emission spectra which is uncorrelated with incident electron energy. Separation is accomplished by superimposing a small oscillation on the incident electron energy and detecting those features which do or do not exhibit a synchronous modulation.

Still further removed from the elastic scattering peak, we observe characteristic loss features resulting from inelastic scattering from the inner shells of surface atoms.[34] Unlike the plasmon and vibrational losses, however, the core level excitations do not result in discrete losses. Rather, they represent loss edges corresponding to two extreme cases: (1) the case in which the incident electron loses just enough energy to excite the core electron to the first unoccupied state, and (2) the case in which the incident electron is captured by the first unoccupied state above the Fermi level and yields all its energy to the core electron (Figure 8). The two cases are physically indistinguishable, and the amplitudes of the two paths, rather than their intensities, must be added.[35] Between these extremes, all combinations

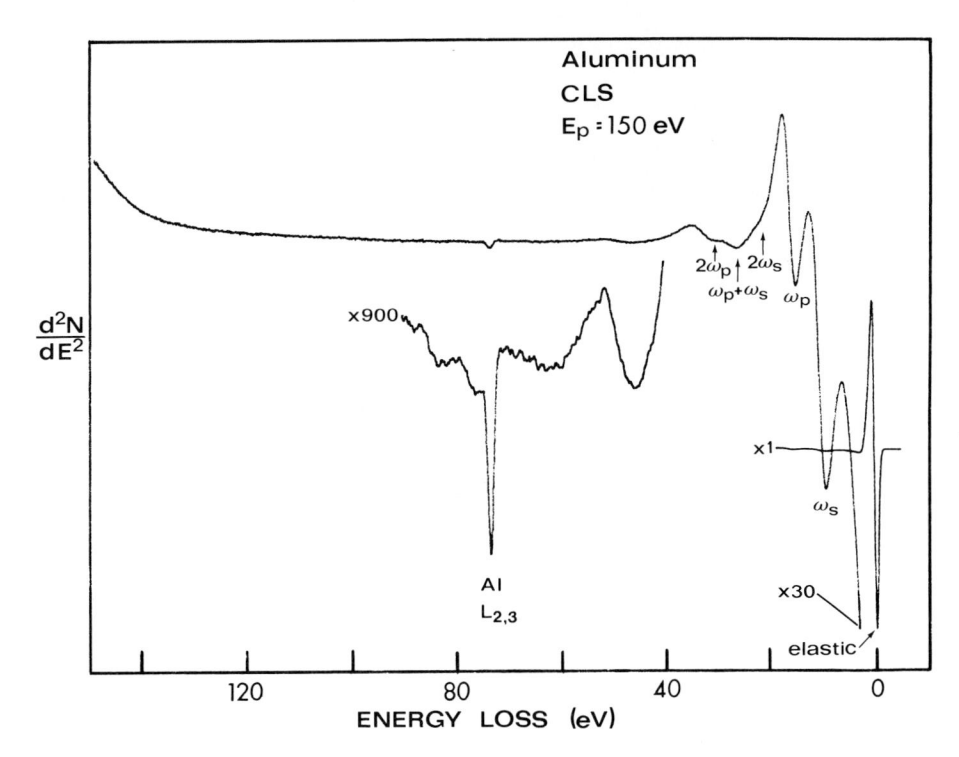

Fig. 7. Derivative of the energy loss spectrum of clean Al for 150 eV incident electrons. In addition to loss features associated with plasmon creation, the loss edge corresponding to excitation of the Al $L_{2,3}$ (2p) shell is distinctly observed. The shape of the loss edge is determined by the density of unfilled states above the Fermi level.

of final state energies of the incident and excited core electrons that satisfy the conservation of energy are possible.

The probability of a given combination of final state energies depends on the availability of states at the two energies. The density of states far above the Fermi level is a slowly varying function of energy. Hence, near the edge, the spectrum should reflect the density of unfilled states near the Fermi energy. To a first approximation, therefore, the core level characteristic loss edges resemble soft X-ray absorption edges.[36] Unlike X-ray absorption edges, however, the backscattered-loss spectrum probes only a shallow region near the surface.

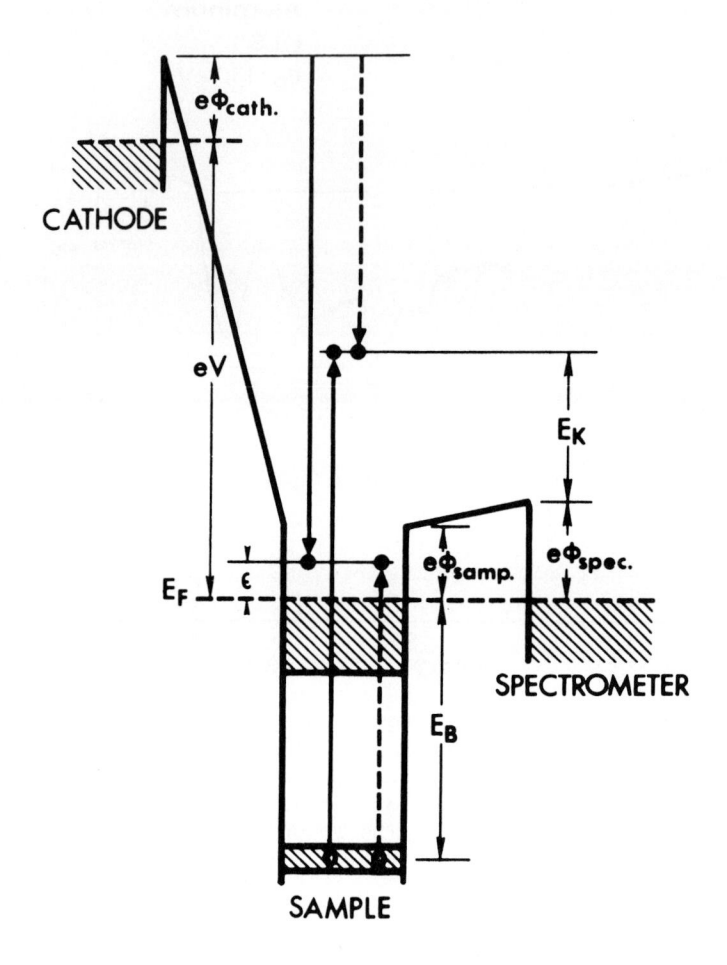

Fig. 8. Energy level diagram of the core level characteristic loss experiment. The maximum kinetic energy of the ejected core electron is $E_{K_{max}} = eV + e\phi_{cath} - E_B - e\phi_{spec}$ and corresponds to the case where $\varepsilon = 0$. The situation indicated by the dashed arrows is physically indistinguishable from that represented by solids arrows. The shape of the excitation edge reflects the density of unfilled states.

Emission Spectra

In comparison with the characteristic loss spectrum, the emission spectrum in Figure 6 seems relatively featureless, consisting of a single, very large, low-energy peak, and a broad tail extending up to the incident electron energy. As we will see, this apparent simplicity is deceptive. The conventional explanation of the low energy maximum is in terms of a "cascade" process in which excited electrons produced by the scattered primary beam, diffuse to the surface, and are emitted if they possess sufficient energy to overcome the work function barrier.[37] This sort of model can be made to yield a spectrum qualitatively resembling the observed shape,[38] but many of the parameters are poorly characterized and no attempt has been made at actual comparison with experiment.

This theoretical approach contains the explicit assumption that the energy loss process is dominated by the excitation of single electrons.[28] It is now generally accepted, however, that in the energy range of greatest interest in surface science, the principal loss mechanism is by the excitation of plasmons.[39] It seems important, therefore, that we reconsider the secondary emission process from this standpoint. Structure in the emission spectrum of aluminum was in fact associated with plasmon decay by von Koch,[14] but he failed to take account of the work function. An alternative explanation was advanced by Henrich[40] who concluded that the structure is the result of incident electrons losing energy by the creation of plasmons. Such an explanation is clearly contradicted by the correlation measurements in Figure 6 which clearly demonstrate that loss features in the low energy region account for only a small portion of the "true" secondary maximum.

We must conclude, therefore, that the shape of the secondary emission spectrum at low energies has not been satisfactorily explained. In view of the fact that plasmon creation is the principal energy loss mechanism for energetic electrons, however, it seems likely that the correct explanation will be found in the decay of plasmons by electron emission.

As we hinted at the beginning of this section, the emission spectrum is not as smooth as it might appear. The presence of subsidiary maxima was first reported by Haworth.[41] Lander[42] pointed out that some of these maxima could be associated with Auger transitions during the decay of excited core states. This was not exploited, however, until Harris[43] demonstrated the great increase in sensitivity that could be achieved by plotting the derivative of the secondary electron energy distribution. This increase is clear from Figure 9 showing the derivative of the emission spectrum of aluminum. The value of this technique for

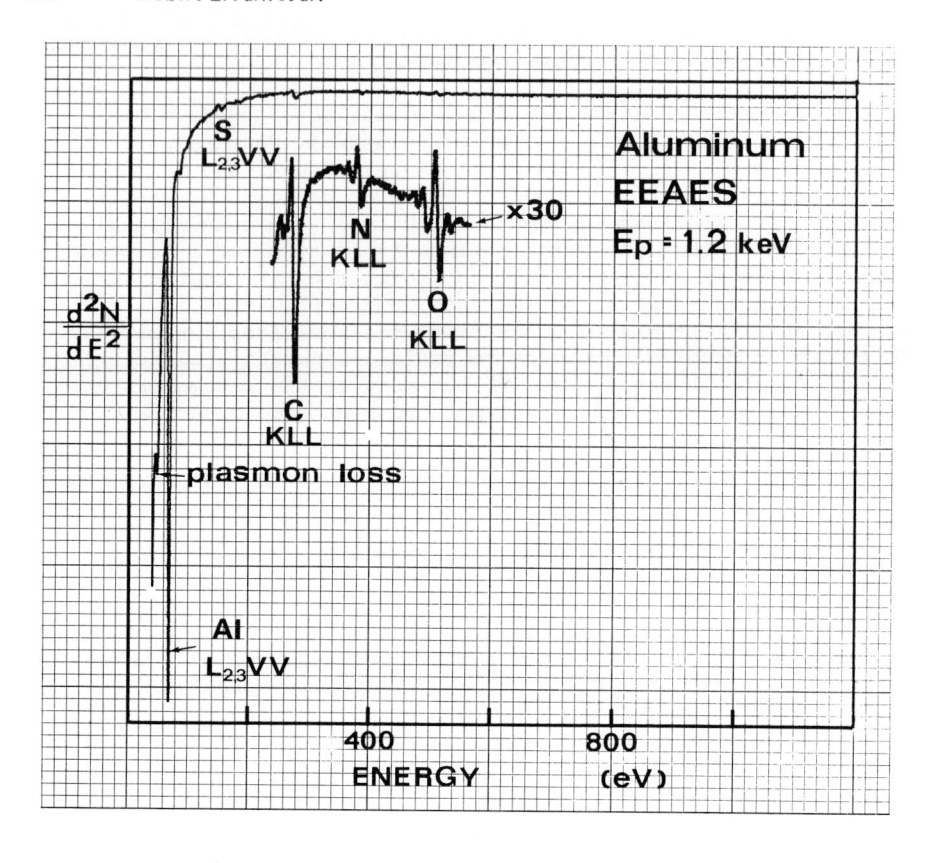

Fig. 9. Derivative of the emission spectrum, showing the remarkable enhancement of the Auger features relative to the smoothly varying background. The spectrum allows surface contaminants to be identified. The shape of the Auger features is determined by the density of filled states.

identifying small quantities of surface impurities is evident; indeed, it is no exaggeration to state that the development of electron-excited Auger electron spectroscopy has revolutionized the field of surface science.

Initially, scientists were content to let Auger spectroscopy serve as a monitor of the cleanliness of experimental surfaces,[44] and it is this application that has had the greatest impact on surface science. Clearly, however, if Auger spectroscopy could be used to quantitatively measure elemental abundances in the surface region, rather than simply identify contaminants, it would open up whole new fields of inquiry. Indeed, references to "quantitative Auger analysis" are fairly common.[45] Unfortunately, they signify only that the authors do not comprehend the problem.

As a case in point, consider a recent paper dealing with the use of "elemental sensitivity factors".[46] The accuracy of this approach was tested on clean MgO samples which were believed to be homogeneous in composition out to the surface. Applying elemental sensitivity factors to the Auger spectrum, it was verified that the composition was made up of roughly equal numbers of magnesium and oxygen atoms which demonstrates only that, if the composition of a sample is completely homogeneous, Auger electron spectroscopy can be used to perform a <u>bulk</u> analysis.

The difficulty is that unless the structure of the surface is known, its composition is not defined. We cannot simply attach a percentage to each element as we do for homogeneous bulk samples, since the surface is necessarily inhomogeneous along its normal. Thus, except in cases where the surface structure is known, as in certain examples of monolayer formation, Auger spectroscopy can give us only a qualitative notion of elemental abundance. The same limitation applies to the other core level spectroscopies we have discussed.

Interest in the use of Auger electron spectroscopy to study surface elemental composition has to some extent led to its neglect for the study of the electronic structure in the surface region, although Lander called attention to this aspect in his 1953 paper.[42] The Auger spectrum is, of course, considerably more complex than either the core level appearance potential spectrum or the core level characteristic loss spectrum. The complexity results from the fact that an Auger line represents term differences between three levels. For example, a transition labelled KL_1L_2 refers to an initial vacancy in the K shell (1s) that undergoes a transition to holes in the L_1 (2s) and L_2 ($2p_{1/2}$) shells plus an Auger electron. For a heavy element, the L-shell Auger spectrum alone consists of hundreds of lines, the energies of which are not susceptible to precise first-principle calculations. It is generally possible to identify spectral groups, but frequently, individual features within those groups cannot be unambiguously labelled except for the light elements. Elemental identification is, therefore, based on matching spectra against "standard" plots taken from samples of known composition.

In those cases where the transitions giving rise to a particular feature can be identified, it should be possible to use the lineshape to reveal something about the electronic structure. For example, in cases where one or both of the final state holes lies in the valence band, as shown in Figure 10, it should be possible to interpret the Auger lineshape in terms of the occupied density of states.[47,48] This information is complementary to that obtained by the appearance potential and core level characteristic

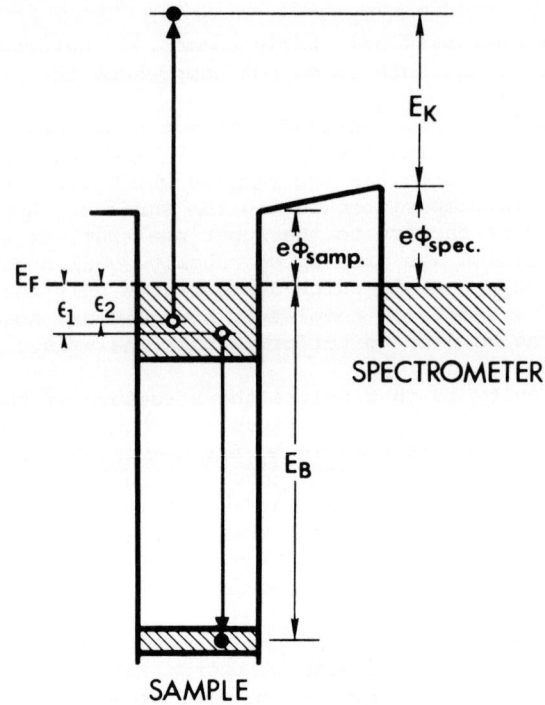

Fig. 10. Energy level diagram of the Auger electron emission experiment. The initial state is taken as including a vacancy in a core hole. The energy liberated when the core hole recombines with a valence electron may be transferred to another valence electron with an energy $E_B - \varepsilon_1 - \varepsilon_2$. The measured kinetic energy must be corrected for the work function of the spectrometer, $e\phi_{spec}$. The shape of the core-valence-valence Auger line is determined by the self-convolution of the valence band density of states broadened by the lifetime of the core hole. The maximum kinetic energy of an Auger electron corresponds to the case in which both final state holes are at the Fermi level ($\varepsilon_1, \varepsilon_2 = 0$).

loss experiments, both of which examine the unfilled states.

Problems

It is an inherent flaw of brief surveys that they provide little motivation for further research. The impression is left with the reader that the overall phenomena are well understood and only the details remain to be worked out. As the remaining thickets are hacked away, however, we often discover that they have been obscurring our view of whole jungles of ignorance.

Thus, it has been with the interaction of electrons with solid surfaces. Every attempt to understand one spectroscopy seems only to spawn other equally poorly understood spectroscopies.

Much of the difficulty results from the very inelastic damping processes that enable us to restrict our view to the surface region in the first place. We have, for example, stressed the relationship between the lineshapes observed in the various surface spectroscopies and the chemical environment of the atoms. For each electron which conveys information about a particular excitation, however, there are perhaps 10^3 or 10^4 electrons which experience other inelastic scattering events before escaping from the sample. Thus, a single sharp spectral line becomes a complete secondary emission spectrum resembling the plot shown in Figure 5. The secondary emission spectrum is a sort of response function describing what Robinson[49] called electron straggling. It represents the response of the crystal to a mono-energetic beam of electrons. To properly interpret the shape of a particular spectral feature in terms of the electronic states of the surface region requires that we understand this response, but at this point we do not have a reliable theory of even its gross features.

In the core level spectroscopies, we must also be wary of problems resulting from the screening response of the conduction electrons to the core hole that is suddenly created or annihilated.[50] For surface spectroscopies, the predicted consequences of this dynamic screening include singular behavior near thresholds,[51,52] strong plasmon satellites,[53,54] skewed lineshapes,[55] and a screening energy correction to the core electron binding energy.[56] Until the problem of electron straggling is better understood, however, it cannot be asserted unequivocally that these effects are actually observed.[57]

ACKNOWLEDGEMENTS

This work was supported by the Office of Naval Research under grant N00014-75-C-0292, and the National Science Foundation under grant DMR 75-01096. The author is also grateful to the Center of Materials Research for its assistance.

REFERENCES

1. R.L. Park, Physics Today, **28**, 52 (April 1975).
2. G.E. Laramore and W.J. Camp, Phys. Rev. B **9**, 3270 (1974).
3. K. Codling, Rep. Prog. Phys., **36**, 541 (1973).
4. R.L. Park and M. den Boer, CRC Critical Reviews of Sol. State Phys. (to be published).
5. J.E. Henderson and R.E. Badgley, Phys. Re., **38**, 590 (1931).

6. R.W. Strayer, W. Mackie and L.W. Swanson, Surface Science, 34, 225 (1973).
7. R.E. Thomas, J. Appl. Phys., 41, 5330 (1970).
8. H.E. Farnsworth, Phys. Rev., 25, 41 (1925).
9. J.J. Quinn, Phys. Rev., 126, 1453 (1962).
10. D. Penn, J. Vacuum Sci. and Technol., 13, 221 (1976).
11. C.J. Powell, Surface Science, 44, 29 (1974).
12. K.D. Sevier, *Low Energy Electron Spectrometry* (Interscience, New York, 1972), Chapt. 8.
13. A.J. Braundmeier, Jr., M.W. Williams, E.T. Arakawa, and R.H. Ritchie, Phys. Rev. B 5, 2754 (1972).
14. C.V. von Koch, Phys. Rev. Letters, 25, 792 (1970).
15. E.A. Stern and R.A. Ferrell, Phys. Rev., 120, 130 (1960).
16. H.E. Farnsworth, Phys. Rev., 31, 405 (1928).
17. O.W. Richardson, Proc. Roy. Soc. (London), A119, 531 (1928).
18. R.L. Gerlach, Surface Science, 28, 648 (1971).
19. J.E. Houston and R.L. Park, Phys. Rev. B 5, 3808 (1972).
20. R.L. Park and J.E. Houston, J. Vac. Sci. Technol. 11, 1 (1974).
21. See for example: E.J. McGuire, Phys. Rev. A 3, 1801 (1971).
22. O.W. Richardson and C.B. Bazzoni, Philos. Mag., 42, 1015 (1921).
23. R.L. Park, Surface Science, 48, 80 (1975).
24. A.M. Bradshaw, Surface and Defect Properties of Solids, 3, 153 (1973).
25. S. Anderson, H. Hammarquist and C. Nyberg, Rev. Sci. Instrum. 45, 877 (1974).
26. S. Andersson and C. Nyberg, Surface Science, 52, 489 (1975).
27. J. Kanski and P.O. Nilsson, Physica Scripta, 12, 103 (1975).
28. O. Hachenberg and W. Brauer in *Electronics and Electron Physics*, Vol. XI, edited by L. Marton (Academic Press 1959).
29. J.E. Houston and R.L. Park, Bull. Am. Phys. Soc., 14, 769 (1970).
30. F.M. Propst and T.C. Piper, J. Vac. Sci. Technol., 4, 53 (1967).
31. H. Ibach, J. Vac. Sci. Technol., 9, 713 (1972).
32. L.H. Little, *Infrared Spectra of Adsorbed Species*, (Academic Press, New York, 1966).
33. S. Ohtani, K. Terada and Y. Murata, Phys. Rev. Lett., 32, 415 (1974).
34. R.L. Gerlach, J.E. Houston, and R.L. Park, Appl. Phys. Lett., 16, 179 (1970).
35. R.L. Park and J.E. Houston, J. Vac. Sci. Technol., 10, 176 (1973).
36. R.L. Park, J.E. Houston and G.E. Laramore, Japan J. Appl. Phys., Suppl. 2, Part 2, 757 (1974).
37. P.A. Wolff, Phys. Rev., 95, 56 (1954).
38. A.J. Dekker in *Solid State Physics*, ed. by F. Seitz and D. Turnbull (Academic Press, New York, 1958).
39. C.B. Duke, Adv. Chem. Phys., 27, 1 (1974).
40. V.E. Henrich, Phys. Rev. B 7, 3512 (1973).
41. L.J. Haworth, Phys. Rev., 48, 88 (1935).
42. J.J. Lander, Phys. Rev., 91, 1382 (1953).

43. L.A. Harris, J. Appl. Phys., 39, 1419 (1968).
44. N.J. Taylor, J. Vac. Sci. Technol., 6, 241 (1969).
45. C.C. Chang in *Characterization of Solid Surfaces*, P.F. Kane and G.B. Larrabee, eds., (Plenum, New York, 1974).
46. P.W. Palmberg, J. Vac. Sci. Technol., 13, 214 (1976).
47. J.E. Houston, J. Vac. Sci. Technol., 12, 255 (1975).
48. J.W. Gadzuk, Phys. Rev. B 9, 1978 (1974).
49. H. Robinson, Proc. Roy. Soc. (London), A104, 455 (1923).
50. G.D. Mohan, Phys. Rev., 163, 612 (1967).
51. M. Natta and P. Joyes, J. Phys. Chem. Solids, 31, 447 (1970).
52. G.E. Laramore, Phys. Rev. Lett., 27, 1050 (1971).
53. D.C. Langreth, Phys. Rev. Lett., 26, 1229 (1971).
54. G.E. Laramore, Solid State Commun., 10, 85 (1972).
55. S. Doniach and M. Sunjić, J. Phys. C 3, 285 (1970).
56. G.E. Laramore and W.J. Camp, Phys. Rev. B 9, 3270 (1974).
57. R.L. Park, J.E. Houston and G.E. Laramore, Japan J. Appl. Phys., Suppl. 2, Part 2, 757 (1974).

Surface Characterization by Electron Microprobe

Ian M. Stewart

Walter C. McCrone Associates, Inc.
Chicago, Illinois   60616

The electron microprobe analyzer is an analytical tool utilizing the principles of x-ray fluorescence induced by electron bombardment. As the bombarding electrons have at most a penetration depth of about a micrometer, the method is particularly applicable to the analysis of surfaces. As well as determining what the bulk composition of a surface may be, it is possible to focus the electron beam to a small probe which allows an identification of inhomogeneities on the surface such as may result from the presence of contaminant particles or debris. Not only may thin films be analyzed chemically but in many instances their thickness may also be determined by monitoring either the intensity of their fluorescent signal or the attenuation of the fluorescent signal originating from their substrates. Identification of polymer films on surfaces, however, is severely limited and is generally dependent on the presence of characteristic ions either in the organic or in a filler used with the polymer. Some examples of the application of the electron microprobe to surface characterization problems will be described.

The development of the electron microprobe analyzer[1,2,3,4] over the last two decades has resulted in a very sophisticated tool for the determination of the chemical elements present in or near the surface of bulk material with a spatial resolution of the order of 1 micrometer. There is no other instrument that can give such information on small microvolumes combined with such accuracy.

To understand both the potential of the instrument and its limitations, it is necessary to consider the underlying principles of the operation of the instrument. When a high energy electron beam is stopped by a target material several important reactions take place (Figure 1). One of the most important of these reactions is the ejection of x-rays whose wavelengths or energies are characteristic of the chemical elements of which the target material is composed. Thus, if it were possible to measure either the wavelength or the energy of these x-rays, it would be possible to determine what chemical species are present in the target. Methods do exist to determine both the wavelength and the energy of the x-rays and hence, it is possible to use such determinations as a basis for chemical analysis. If it were further possible to concentrate the electron beam into a small spot, it would be possible to obtain this analysis from a very small area on the sample. Again, such a possibility is within the range of existing technology. Electron beams may be focused electrostatically or electromagnetically, and generally the latter is chosen because of its greater flexibility, to produce a very finely focused probe at the specimen surface. It was Castaing[5] who pioneered this approach in 1951 and thus opened up the whole realm of electron microprobe analysis. Duncumb and Melford[6] added to the technique the capability to sweep the electron beam over the sample surface and at the same time monitor the x-ray beams being generated. They thus put the scanning into scanning electron microprobes.

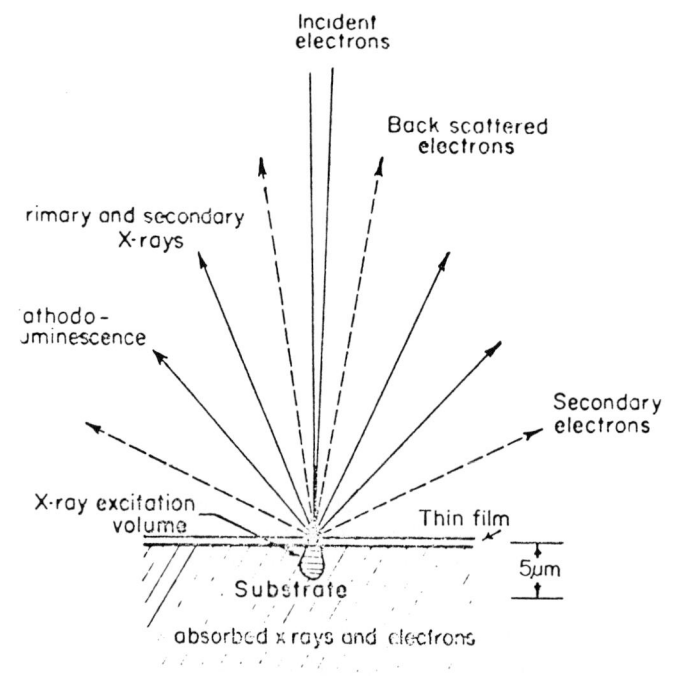

Fig. 1. Electron beam - sample interactions for a thin film.

A typical schematic layout of an instrument embodying these principles is shown in Figure 2. Electrons generated from a hot tungsten filament by thermionic emission are accelerated down the electron optical column at voltages generally in the range of 20-40 kV and are focused by a condenser lens and objective lens. Scanning coils are placed in or near the objective lens to enable the beam to be deflected on the specimen surface. Analysis of the x-ray energies generated is accomplished by a solid-state detector consisting of a lithium silicon diode whose output is related to the energy of the impinging x-rays. Analysis of x-ray wavelength is performed in a crystal spectrometer in which the x-rays are allowed to strike a crystal whose interatomic spacings are known, hence, by measuring the angle of the diffracted beams and using Bragg's law, $n\lambda = 2d \sin \theta$, where n is the order of the wavelength, d is the spacing of the crystal and θ is the angle of reflection; the wavelength λ may be calculated. Generally such spectrometers are semifocusing or fully focusing and are usually linear, the advantage of a linear spectrometer being that a constant take-off angle may be obtained at the specimen surface. The physical appearance of an electron microprobe analyzer is shown in Figure 3.

There are several areas in which one may consider the electron microprobe analyzer has made a major impact in the understanding of materials. The facility to provide area maps of elemental distribution has been employed in studying the reactions occuring at bonded junctions, particularly in semiconductor applications, and in understanding the structure and behavior of complex coatings such as are produced by flame spraying. In almost all the instances described, however, the coating layers have been of the order of several micrometers thick. What can be done in the electron microprobe for thin coatings?

The success of the study of thin coatings in the electron microprobe might almost be proportional to the mean atomic number of the coating. With heavy metal layers such as gold, platinum, etc., one can determine not only that the metal is present on the surface at thickness as low as 10-20 nm but one can also, after calibration with suitable standards, determine the thickness of the coating with some accuracy. To achieve this, the accelerating voltage of the electron microprobe is gradually decreased while monitoring the signals both of the coating and of the substrate. In this way, an intensity profile vs voltage is obtained which can be correlated with thickness.

In many instances, it is not always known whether an element detected is present as a surface coating or whether it is uniformly dispersed in small amounts in the subsurface layers. Here, quite a simple approach can be applied. The sample surface is simply lightly scratched and an x-ray scanning photograph prepared of the scratched area. This light scratching will disrupt

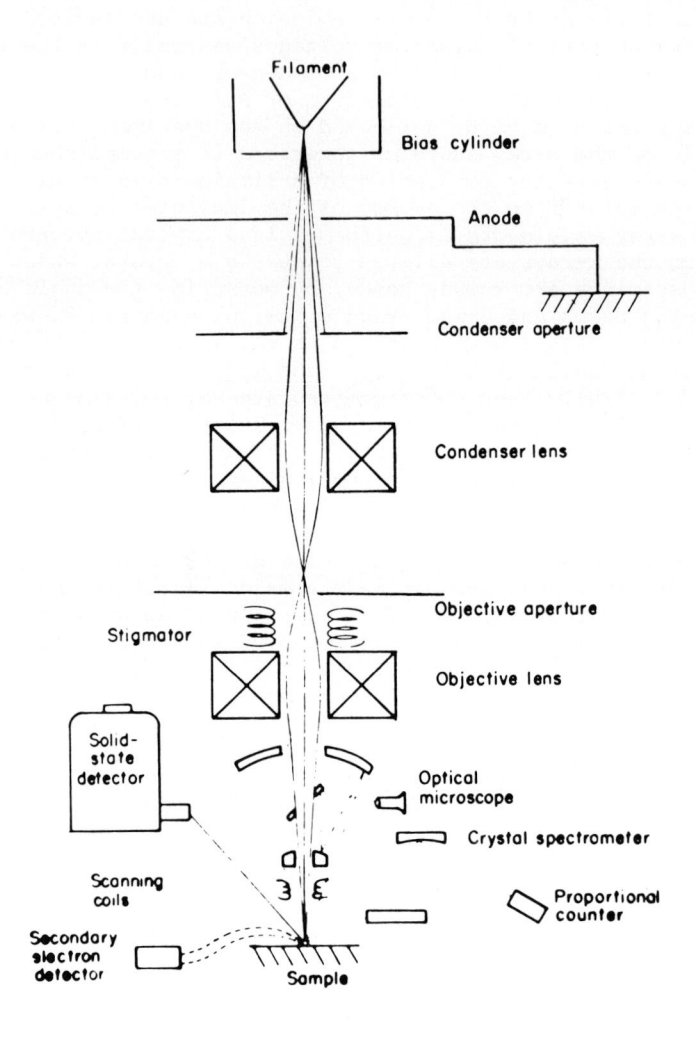

Fig. 2. Schematic of an electron microprobe analyzer.

any surface film allowing the base material to show through, Figure 4.

Fig. 3. A.R.E. Electron microprobe analyzer.

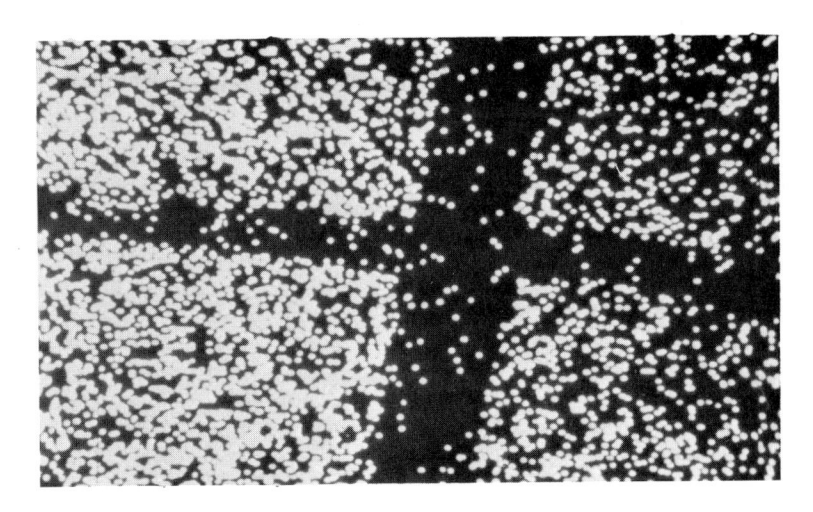

Fig. 4. Al signal from thin film of Al with two near orthogonal scratches through to substrate.

The application of the electron microprobe analyzer to organic coatings is unfortunately somewhat less successful. Although the microprobe can certainly detect the presence of carbon and oxygen, hydrogen, having no outer electron shell, can produce no x-rays while the x-ray signal from nitrogen is a particularly difficult one to detect. Chlorine, sulfur and fluorine, present in many hydrocarbons or polymers may be readily detected and thus may assist in making a somewhat speculative identification of the particular plastic coating used. If, however, some information is available on what might be expected and good standards are available of the suspect materials then the identification may be made more certain. The presence of filler materials in a polymer coating may also aid the positive identification of a coating.

In summary, then, the electron microprobe analyzer is an extremely powerful tool in surface analysis yielding information on the surface distribution of elements with a spatial resolution unequaled by any other analytical technique. The method does, however, yield information to some appreciable depth in the sample surface and is, therefore, not applicable to mono-molecular layers. In conjunction with other surface techniques which may indicate the presence of thin films on the surface, however, the electron microprobe can frequently indicate whether or not this material is uniformly distributed. Estimation can be made of the thicknesses of many inorganic films. Organic films and layers, however, are less amenable to study by this technique.

REFERENCES

1. W.C. McCrone and J.G. Delly, Ed., *The Particle Atlas*, Ed. II, Vol. II, pp. 155-168, Ann Arbor Science Publishers, Inc., 1973.
2. A.J. Tousimis and L. Marton, Ed., "Electron Probe Microanalysis" in *Advances in Electronics and Electron Physics*, Supplement 6, Academic Press, New York, 1969.
3. J.I. Goldstein and H. Yakowitz, Ed., *Practical Scanning Electron Microscopy*, chapter on Electron and Ion Microprobe Analysis, Plenum Press, New York, 1975.
4. G.A. Hutchins, "Electron Probe Microanalysis" in *Characterization of Solid Surfaces*, P.F. Kane and G.B. Larrabee, Ed., Plenum Press, New York, 1974.
5. R. Castaing, Thesis, University of Paris, Paris, France, 1951; Publ. ONERA, No. 55.
6. P. Duncumb and D.A. Melford, Proc. First Nat. Conf. on Electron Microprobe Analysis, 1966.

Auger Electron Spectroscopy of Solid Surfaces

J.T. Grant

Universal Energy Systems, Inc.
3195 Plainfield Road
Dayton, Ohio   45432

Auger electron spectroscopy (AES) is a powerful method for studying the composition of the few outermost atomic layers of a solid. Besides being capable of identifying all elements other than H and He, information about chemical environment can often be obtained using AES. The variation in composition with distance from the surface can be determined by depth profiling, and the variation across a surface can be obtained by electron beam scanning. Progress in quantification of AES is reviewed and some recent advances in AES discussed. Several applications of AES to studying solid surfaces are given.

I. INTRODUCTION

The technique of Auger electron spectroscopy (AES) had its experimental beginnings in 1923 when Pierre Auger was studying the photoelectric effect in gases using a cloud chamber.[1] He wanted to "visualize the whole story of an atomic photoexcitation": (i) the production of a photoelectron, (ii) the subsequent emission of a radiation quantum, and (iii) the absorption of this quantum with the production of yet another photoelectron, all in the same cloud chamber.[2] However, after the initial production of a photoelectron he observed that additional electrons emerged from the same point as the photoelectron, and he thought they might be due to the re-absorption of the radiation quantum by the same atom that produced it. However, by the time he published his thesis in 1926, Auger[3] had realized that these extra electrons were not due to re-absorption but were due to a non-radiating transition, which had been predicted by Rosseland.[4] These electrons emitted by non-radiative transitions have energies characteristic of the emitting atom and are now referred to as Auger electrons.

With solids, "humps" in the curves of secondary electrons that were independent of primary electron beam energy were reported as early as 1935[5], but it was not until 1953[6] that certain peaks in secondary electron distributions were identified as due to Auger transitions. It was also pointed out that most elements could produce characteristic Auger electrons in the low electron energy range where their absorption coefficients are very high (a few atomic layers) and that the excitation of Auger peaks would be an interesting technique for determining the composition of solid surfaces.[6] Although the technique was used in some surface studies,[7] the difficulty in detecting the Auger peaks due to their superposition on a large varying background of secondary electrons detracted from its usefulness.

It was not until 1967 when Harris showed that the detectability of Auger peaks could be enhanced enormously by electronically differentiating the electron energy distribution, that the technique became viable for surface analysis.[8] Within a few years, Auger electron spectroscopy was in use in many laboratories, particularly after Weber and Peria[9] showed that these derivatives could also be obtained using a three-grid low energy electron diffraction (LEED) system, typical of the type already in widespread use for LEED studies. The technique was further advanced in 1968 with the addition of a high current grazing incidence electron gun and the addition of a fourth grid to the standard three-grid LEED system.[10] The measurement of Auger spectra was made much easier with the introduction of the cylindrical mirror analyzer (CMA) in 1969.[11] The improved signal-to-noise ratio of the CMA enhanced the sensitivity of the technique and made it possible to display Auger spectra covering a range of up to 1000 eV on an oscilloscope, thereby allowing rapid acquisition of data. For many elements, the detectability limit is of the order of 0.001 monolayers.

Besides identifying the elements present at solid surfaces, AES can sometimes be used to obtain information about the chemical environment and bonding of surface species.[12] Coupling AES measurements with simultaneous sputtering to remove surface layers provides the composition as a function of depth from the original surface (called a depth profile).[13] The lateral distributions of elements at the surface can also be measured,[14] and when combined with sputtering provides a three-dimensional measurement of the elemental distribution in near surface regions of solids.

In this paper, Auger studies of solid surfaces will be briefly reviewed. Where appropriate, the reader is referred to other articles for further information.

II. THE AUGER PROCESS

When an atom is ionized in an inner shell it may relax to a less energetic state by one of two processes: the emission of a characteristic X-ray or the emission of an Auger electron. The Auger process may be thought of as the filling of the initial inner shell vacancy by an electron from an outer shell, with the energy produced being transferred to another electron (in one of the outer shells) allowing it to escape the atom.[15] If the atom is initially singly ionized in the W level, it will be doubly ionized following Auger emission and if these two ionized levels are denoted by X and Y, the Auger electron emitted is referred to as a WXY Auger electron (e.g. KL_1L_3, $L_3M_3M_5$, $M_5N_3N_5$, etc. Auger electrons). When the initial vacancy occurs in one of the inner subshells, there is often sufficient energy available so that one of the two final vacancies occurs in an outer subshell of the initially ionized shell, e.g., an $L_1L_3M_3$ transition, and such transitions are then referred to as Coster-Kronig transitions.

With solids, if none of the electrons involved in the Auger process are valence electrons similar labelling of Auger transitions is used. When a valence electron is involved the letter V is often used to indicate this, e.g. $M_{4,5}N_1V$, L_3VV, etc., Auger transitions. More complicated Auger transitions may also occur, e.g., due to initial multiple ionization,[16] auto-ionization,[17] or cross transitions.[18]

The energy $E_Z(WXY)$ of a WXY Auger electron emitted from an element of atomic number Z in a solid can be estimated from the one electron binding energies $E_Z(W)$ by:[19]

$$E_Z(WXY) = E_Z(W) - E_Z(X) - E_{Z+1}(Y) \quad (1)$$

This equation has met with some success but is of an empirical nature. A more precise calculation of the Auger electron can be made from the following equation:[20,21]

$$E_Z(WXY;S) = E_Z(W) - E_Z(X) - E_Z(Y) - \xi(XY;S) + R(XY) \quad (2)$$

where S denotes the final state with holes in the X and Y levels, $\xi(XY;S)$ is the interaction energy between the X and Y holes in the final state S, and $R(XY)$ is the total relaxation energy. Such theoretical calculations agree quite well with measurements on transition metals and their salts.[21]

Hydrogen and helium cannot be detected by Auger electron spectroscopy as they do not have sufficient electrons for the process. However, following suitable inner shell ionization all other elements produce Auger electrons in the energy range 20-2000 eV--see for example, the Auger charts in reference 15(a) or reference 22. For such low energies, the Auger yield is near unity[23,24] and this means that light elements can be detected as easily as heavy elements using Auger spectroscopy. This contrasts with X-ray fluorescence analysis where the fluorescent yield is very small for light elements.

Because the energies of Auger electrons emitted from atoms in solids are characteristic of the emitting atom, measurement of Auger electron energies can be used to identify the atoms present. In order for the Auger electrons to be detected, they must be able to escape from the solid without losing energy by scattering. As the escape depth of electrons in the energy range 20-2000 eV is $\sim$1 nm,[25] only Auger electrons produced within the few outermost atomic layers of a solid will contribute to the Auger signal--hence the surface sensitivity of Auger electron spectroscopy.

III. INSTRUMENTATION

Several commercial Auger electron spectroscopy systems are available.[26] Because of the surface sensitivity of AES, measurements should be made in an ultra-high vacuum system ($\sim 10^{-7}$ Pa and below) to prevent contamination of the specimens while being studied. Auger electrons are usually excited with an electron beam (typically 3 to 10 keV at 1 to 30 μA) because such beams are easily obtained, focussed, and produce a relatively large number of Auger electrons.[27] X-ray[28,29], proton[30] and in some cases, heavy ion[31,32] excitation can be used but in most cases the Auger signals produced are much weaker. A number of energy analyzers can be used but the most common type presently in use is the cylindrical mirror analyzer.[33] The commercial systems include devices for sample manipulation and sample treatment, e.g. heating, cooling, fracture, gas adsorption or inert gas sputtering. Using electron excitation, Auger analysis can be made in a number of modes, e.g., point analysis (with spatial resolution <1μm), line scan analysis and two-dimensional mapping.[34] When coupled with inert gas sputtering to continuously remove surface layers, changes in composition with depth can also be measured.[13] For many specimens, some sample treatment is necessary before analysis can be made due to environmental contamination. Analysis is rapid and several specimens can be examined in one day. Many specimens are usually mounted on a carrousel and suitable vacuum pumpdown is obtainable overnight.

IV. ASPECTS OF AUGER SPECTROSCOPY

Auger Spectra

Auger spectra are usually measured as the first derivative of the electron energy distribution, $dN(E)/dE$ or $\mathcal{N}_m^{(1)}(E)$. Such spectra are normally obtained by modulating the energy analyzer with a small sine wave and using phase sensitive detection techniques.[8,9,35] An example of such a spectrum is shown in Fig. 1, being that from the surface of a heat treated Al alloy. No other treatment to the specimen was made before Auger analysis. The main Auger transitions observed are due to C, O and Mg, and are identified in the Figure. For such derivative spectra, the energies of the Auger transitions are conventionally measured as the energies at which the peak minima are detected.[36] Note the lack of any strong Al Auger transition (which would occur near 1400 eV) from the spectrum shown in Fig. 1. The bulk concentration of Mg in the alloy is 2.5% but it can be seen that after heat treatment Mg, O and C completely cover the surface of

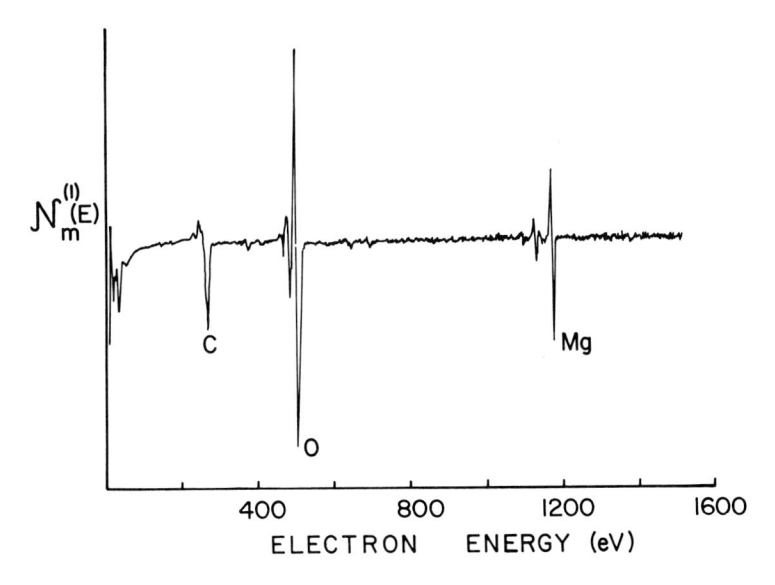

Fig. 1. Auger spectrum (in derivative form) of the surface of a heat-treated 7075 Al alloy (bulk composition includes 2.5% Mg). Note the lack of any significant Al Auger peak near 1400 eV. This spectrum was taken using a 5 keV electron beam for excitation, with a sinusoidal modulation of 4 eV peak-to-peak, and a 100 ms time constant.

the Al alloy. This illustrates the surface sensitivity of the technique. The Auger transitions of C and O are KVV transitions, while those of Mg are KLL as no bonding electrons are involved in these Mg transitions. Specifically, the largest Mg Auger feature is due to $KL_{2,3}L_{2,3}$ transitions while the next largest feature (on its low energy side) is due to $KL_1L_{2,3}$ transitions.

Depth Profiling

A depth profile of this specimen is shown in Figure 2 which shows the variation in the peak-to-peak heights of C, O, Mg and Al Auger transitions from a particular point (∼10μm diameter) on the specimen as a function of sputtering time. Such peak-to-peak height measurements are commonly used to determine changes in the concentration of elements with depth, but are not reliable if Auger peak shape changes occur, e.g., due to changes in bonding (see section on quantitative analysis). Such depth profiles can be obtained automatically using commercially available multiplexing equipment. For accurate sputtering rate determination, suitable standards are needed as the sputtering rate depends on element, ion,

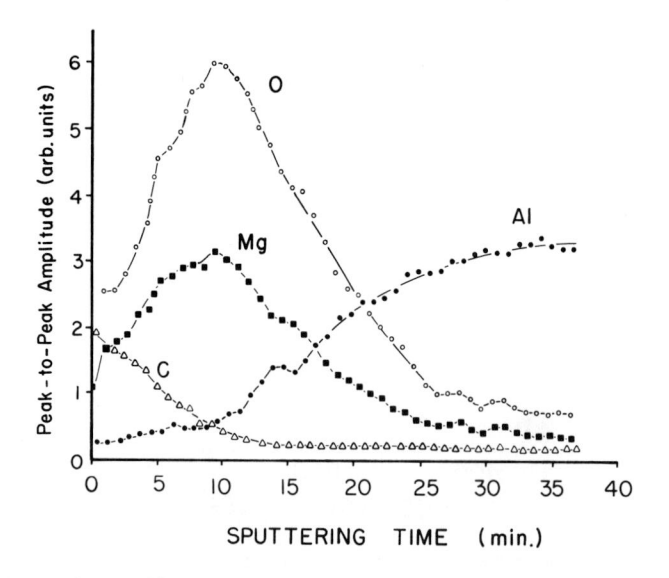

Fig. 2. Depth profile of heat treated 7075 Al alloy, showing the variation in C, O, Mg($KL_{2,3}L_{2,3}$) and Al($KL_{2,3}L_{2,3}$) Auger peak-to-peak signal strengths as a function of sputtering time. Experimental conditions are the same as for Figure 1. Ar was used for sputtering.

ion energy[37] and matrix.[38] In this case, the sputtering rate was 4 nm/min so it can be seen that the Al alloy was covered with about 75 nm of carbon and magnesium oxide. Carbon was present mainly in the outermost layers, the carbon Auger signal decreasing monotonically with time. After sputtering away 150 nm the Auger spectrum shown in Figure 3 was obtained, the main Auger features being due to Al. Argon was used for sputtering and note that some argon was embedded in the specimen during profiling. Typical depth resolution is ∼5% of the amount removed.

Scanning Auger Microscopy

The variation in concentration of particular elements across the surface of a specimen can be individually determined by scanning the electron beam across the specimen and simultaneously monitoring the appropriate Auger signal strength. This technique was discussed during this Symposium.[34]

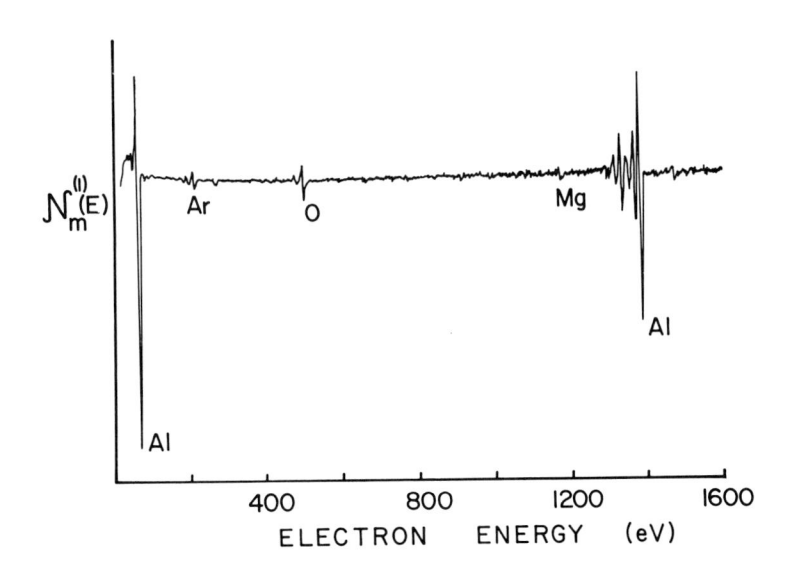

Fig. 3. Auger spectrum of heat-treated 7075 Al alloy after the removal of 150 nm by sputtering. The Auger spectrum consists mainly of Al Auger transitions. Note that some argon was embedded in the alloy during sputtering. Experimental conditions are the same as for Figure 1.

Reference Spectra

A handbook of reference Auger spectra in first derivative form for most of the pure elements is available[22] and is often used to identify peaks in Auger spectra, thereby allowing determination of the elements present at a particular surface. The handbook also contains relative sensitivity factors for the different elements and allows semiquantitative estimates of surface concentrations to be made from peak-to-peak height measurements. Care must be exercised when using such a handbook for quantitative analysis to ensure that the experimental conditions used to obtain Auger spectra are similar to those used in obtaining the reference spectra, as relative Auger signal strengths often depend on the modulating voltage used[39] (for phase sensitive detection), the incident electron beam energy[15,33], surface roughness[40], geometrical effects, etc. As electron binding energies and relaxation energies change with chemical bonding, shifts in Auger energies for a particular element can occur for changes in surface bonding and sometimes need to be taken into account in identifying Auger transitions.[12] Similarly, with insulating specimens, any sample charging will cause a corresponding change in measured Auger kinetic energies. Further, when valence electrons are involved in the Auger process, or when electron scattering is affected by a change in specimen matrix, marked changes in Auger electron energy distributions can take place, often making identification of elements present somewhat more difficult.[12] Of course, such changes in Auger spectra are useful in obtaining information about changes in surface bonding or electron scattering. When such changes in Auger electron energy distributions occur, accurate quantitative measurements cannot be made directly from the raw derivative data unless suitable standard spectra are available for the various Auger line shapes. These problems in quantitative analysis due to changes in modulating voltage and changes in Auger line shape can be largely overcome by signal processing or by using special modulation waveforms and are discussed later.

Chemical Effects

As mentioned above, important information about surface bonding can be obtained from Auger spectra when valence electrons are involved in the Auger process.[12] Interpretation is simpler when only one valence electron is involved in the Auger process,[41] as when two valence electrons participate convolution products of density of states functions are involved.[42] Good agreement between Auger data and band structure was found.[41,42] However, some LVV Auger spectra (of solids) show distinct characteristics of free-atom spectra and do not reflect the band structure, possibly because of electron localization due to increased screening.[43] However, the line shapes of Auger spectra are often useful as a method for fingerprinting the chemical form of surface atoms, e.g., in the case of carbon, to distinguish between graphite-like overlayers or surface carbides on Mo[44] and Si[45], or the different bonding mechanisms of CO to clean metal surfaces.[46] As an example, the Auger line shapes of carbon and oxygen following the exposure of clean Nb, Co and Pt surfaces to CO are shown in Figure 4. Chemical effects are sometimes useful in distinguishing metal oxides from pure metals, e.g., using Mg, Al or Si LVV spectra,[47] or in studying electron energy loss mechanisms.[48]

When only core level electrons are involved in the Auger process, changes in bonding may be reflected as a shift in the Auger energy[49] and such measurements may also be useful as an aid to compound identification when large energy shifts ($\sim$ few eV) occur.[50,51]

Unwanted Effects

As with any technique, there are a number of problems that might occur during analysis, e.g., electron beam induced decomposition of the surface,[15] electron beam desorption or adsorption,[15,52] localized heating, surface charging (with insulating surface layers), variation in sputtering yield with matrix[38] or ion mobilization in insulating films[53] (in depth profiling), ion excited Auger transitions during depth profiling,[54] surface roughness effects,[40] and problems due to Auger peak overlap.[55] Electron beam effects can usually be reduced by changing the beam energy, defocussing or rastering, and further details about these problems can be obtained from the references cited.

V. QUANTITATIVE ANALYSIS

From a theoretical standpoint, the current produced by a particular Auger transition of an element present in a homogeneous solid surface using electron excitation depends on: (a) the ionization cross-section of the inner shell, which depends on the primary electron beam energy; (b) the probability for relaxation by Auger emission; (c) the primary beam current; (d) the angle of incidence of the primary beam; (e) surface roughness; (f) electron backscattering; (g) the escape depth of the Auger electrons; (h) the angle of emission; and (i) the concentration of the particular element. For a non-uniform lateral distribution of an element, the Auger current will, of course, depend on the region probed by the electron beam, and for a non-uniform depth distribution the current produced will also depend on the attenuation of the primary electron beam in the solid.[56] With crystalline surfaces, the Auger current produced has a structured angular dependence which is related to the crystal orientation (and azimuth).[57]

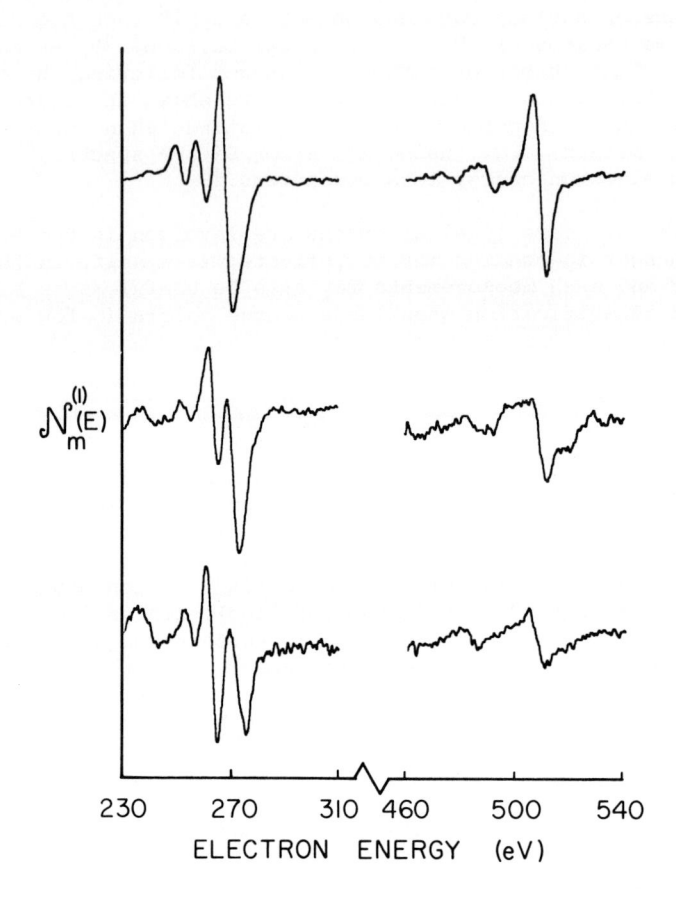

Fig. 4. Auger spectra of C (left hand side) and O (right hand side) following exposure of CO to clean surfaces of Nb (top), Co (middle) and Pt (bottom). Note the diferences in Auger line shapes between C spectra, and between O spectra. Such spectra provide information about the bonding of CO to metals. These spectra were obtained using a 0.3µA, 1.5 keV electron beam for excitation, the sinusoidal modulation was 5 eV peak-to-peak, the time constant was 100 ms, and the averages of 8 scans are shown. The C Auger spectrum for CO on Pt had to be obtained using spectrum subtraction techniques.[46]

In measuring the Auger current, other factors have to be taken into account, e.g., (a) the type of energy analyzer used; (b) the solid angle detected by the analyzer; (c) the transmission and resolution of the analyzer and (d) the orientation of the analyzer relative to the solid surface.

With solids, an Auger current cannot be measured directly because it is quite small relative to the background current of secondary and backscattered electrons.[8] Several methods have been used to quantify measured Auger signals, which can then be related to surface concentrations by theoretical models or by comparison with standard Auger spectra from clean metal surfaces, e.g., (i) by fitting and subtracting backgrounds from measured spectra and then appropriately integrating the result, (ii) by using peak-to-peak height measurements in first derivative spectra, $\mathcal{N}_m^{(1)}(E)$; (iii) by using dynamic background subtraction (DBS), where the background is removed by taking derivative spectra and then integrating them, or (iv) by using tailored modulation techniques (TMT) to obtain Auger current values directly from the phase sensitive detector.

Curve Fitting

Two similar approaches have been used. In one method[58], the electron energy distribution $\mathcal{N}_m(E)$ is measured directly, and then the background is fit by a suitable polynomial and digitally subtracted from the data. The resultant Auger peaks are then integrated and compared with integrals of known electron currents in order to obtain absolute values for the measured Auger currents.

In the other method[59], adsorbates were studied and backgrounds measured before adsorption were subtracted from first derivative spectra, $\mathcal{N}_m^{(1)}(E)$ {and in some cases, $\mathcal{N}_m(E)$}. The areas under the Auger peaks were then determined by graphical integration, and corrections were applied if large modulation voltages were used to obtain the original spectra.

Both these methods have met with success and the reader is referred to the work cited for further details.

Peak-to-Peak Heights

The use of peak-to-peak height measurements from first derivative Auger spectra is the most common method presently in use for quantitative analysis. This method can be used in either of three ways: (i) by comparing an Auger signal with that from a pure elemental standard, (ii) by comparing the signals with that from a pure silver specimen and then using known sensitivity factors relative to silver or (iii) when a pure silver standard is not used, by using relative sensitivity factors and appropriate summation over each element present at the surface.[22] Silver was chosen as a standard to measure the relative sensitivity factors for the pure elements, but any other element could be used so long as the sensitivities of the other elements can be conveniently related to it.

This method has met with a lot of success particularly in studying metals and their alloys.[60] When changes in Auger line shape occur, the accuracy of this method decreases, the decrease depending on the degree of line shape change. In studying metals and their oxides, for example, such inaccuracies can easily be a factor of two or more.[17,61] Such errors would be reduced if appropriate standard Auger spectra (having the same line shape) were available but this is not always possible, particularly if mixtures of metals and their oxides are present in the surface. Auger line shape changes due to using relatively large modulation voltages can be accurately allowed for, if the dependence of peak-to-peak height on modulation has been measured for the relevant Auger transitions.[39]

Dynamic Background Subtraction (DBS)

DBS involves multiple differentiation followed by multiple integration of some experimental variable.[62] It is applied to Auger spectroscopy to remove the large background signals and it is generally found that double differentiation of electron current distributions followed by double integration is sufficient to remove these backgrounds.[61,63] Differentiation can be conveniently carried out by using phase sensitive detection techniques[8,9] while the integration can be done digitally[39,61,63] or using analog methods.[64,65] An example of analog integration is shown in Figure 5 where the three derivative Auger features shown in part (a) are individually integrated twice, with the outputs shown in part (b). The flat region on the high energy side of the peak is set at zero and integration is then carried out by sweeping the analyzer pass energy through the peak of interest, the output at any particular pass energy being the double integral down to that particular energy. Of course, the electron energy distribution can also be obtained from such derivative spectra by single integration as can be seen in Figure 6, where single integration over a wide energy range is illustrated. Note that the largest features in the derivative Auger spectra do not necessarily correspond to the biggest Auger currents.

Note that tails are present on the low energy side of Auger peaks following the first integration. This means that double integrals do not converge as the energy range of integration increases,[64] necessitating that double integrals be measured down to specified energies[39,61] or through specified energy ranges.[64]

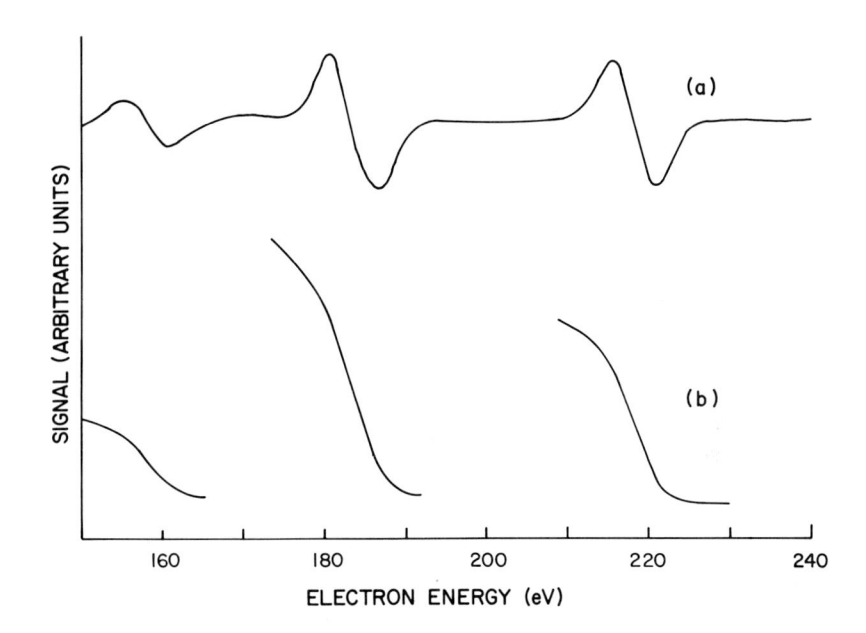

Fig. 5. (a) Part of the Auger spectrum of Mo, shown in first derivative form; (b) double integrals of the individual Mo Auger features obtained using two analog integrators operated in series. The integrals shown in part (b) are inverted for clarity.

This non-convergence of the double integral has not caused any great problems in using DBS for quantitative Auger analysis, good results having been obtained when Auger line shape changes are present, e.g., (i) in studying metals and their oxides, where particular Auger transitions are monitored,[39,61] and (ii) in determining the relative carbon to oxygen atomic ratio following the exposure of clean Mo(110)[64] and Ni(110)[66] to CO, where different Auger transitions are monitored.

DBS has also proven useful in quantitative Auger analysis as exact corrections for the modulation voltage can be made, no matter how large the voltage is or how distorted the Auger peaks might be.[39,65,67,68] DBS can also automatically correct data for small misalignments in sample positioning when a retarding potential energy analyzer (e.g. a retarding grid system) is used.[68]

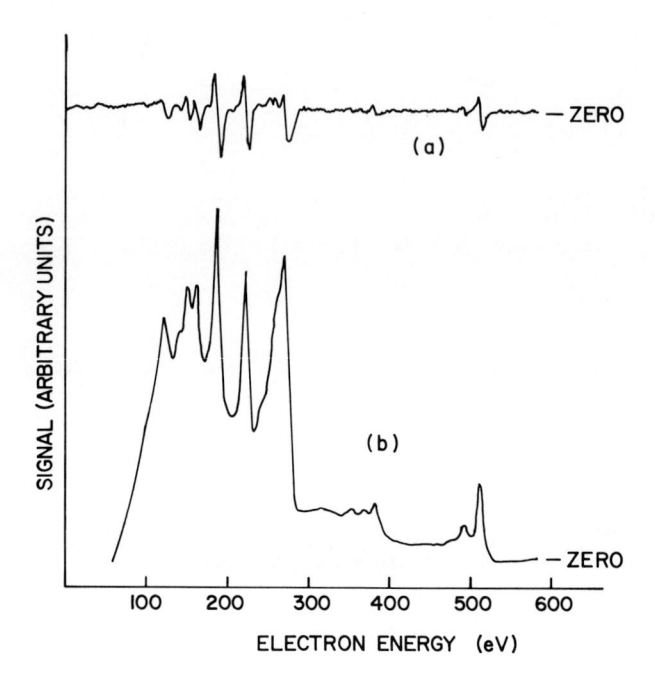

Fig. 6. (a) An Auger spectrum from a contaminated Mo surface, shown in first derivative form; (b) single integral of this spectrum showing the Auger electron energy distribution over a wide energy range. This integral was obtained using analog integration.

Tailored Modulation Techniques (TMT)

In TMT the modulation waveform is tailored to the instrument response function and the degree of background subtraction required, allowing the areas under Auger peaks to be measured directly at the output of the phase sensitive detector.[69] A direct comparison of results obtained using DBS and TMT to measure Auger signal strengths shows nearly perfect proportionality between the results.[69]

The application of TMT to scanning Auger microscopy and depth profiling to overcome problems due to line shape changes or energy shifts in Auger spectra has been successful, and shows much promise for the future.[70]

VI. APPLICATIONS

Hundreds of articles have been written on Auger electron spectroscopy and it would be impossible to refer to all its applications.[71] Some applications are listed below and references are provided for further reading:

(a) detection and identification of surface contaminants, particularly in studying methods for preparing clean surfaces, e.g. for use in LEED studies;[72,73]

(b) studies of chemical reactions at surfaces, e.g., oxidation of surfaces,[17] corrosion and weathering,[74] gas adsorption[17,64] and catalysis.[75,76] Such studies are often coupled with other techniques such as LEED, ultra-violet photoelectron spectroscopy, X-ray photoelectron spectroscopy, appearance potential spectroscopy, ion scattering spectroscopy, secondary ion mass spectroscopy, etc.;

(c) segregation at surfaces[77] or grain boundaries[78] (e.g., during heat treatment), or ion migration in insulators;[53]

(d) temper embrittlement of steels;[74,79,80]

(e) friction and wear studies;[81]

(f) adhesion and bonding studies, e.g., the adhesion of paint to surfaces,[74] the brazing of alloys,[74] the transfer characteristics of polymers in sliding contact with metals,[82] and in studying biomaterials;[83]

(g) studies of coatings on metals[84] and glass;[85]

(h) semiconductor technology, e.g., in identifying contaminants introduced during processing, rinsing, etc., that lead to device failure;[86] in studying contact failure mechanisms;[87] and in studying thin film devices (by depth profiling).[88]

VII. OTHER RECENT ADVANCES

There are at least two areas where recent advances have been made and deserve mention: (i) the production of Auger spectra (in the electron energy distribution mode) using TMT and (ii) the use of spectrum subtraction techniques (SST) to reduce problems due to Auger peak overlap.

Auger Spectra Using TMT

There are two ways of using TMT to obtain Auger electron energy distributions, $\mathcal{N}_m(E)$: (a) the use of relatively simple waveforms to obtain such spectra over limited energy ranges, the range depending on the amplitude of the waveform,[89] and (b) the use of more complicated special waveforms to obtain such spectra over the entire energy range (although with slope artifacts above the Auger peaks), coupled with modest real time spectrum deconvolution.[90]

<u>Simple Waveforms</u>. If a CMA is used to obtain spectra, square wave modulation will produce a first derivative spectrum $\mathcal{N}_m^{(1)}(E)$ for small modulation amplitudes and an $\mathcal{N}_m(E)$ spectrum for large modulation amplitudes.[89] This is illustrated in Figure 7 for a Co surface contaminated with C and O. Figure 7(a) shows a first derivative Auger spectrum (inverted relative to conventional display) while parts (b), (c) and (d) show $\mathcal{N}_m(E)$ type spectra. $\mathcal{N}_m(E)$ type spectra are obtained over energy ranges equal to the amplitude of the square wave and the Auger spectra move up in energy by half the amplitude--compare the results shown in Figure 7(b), (c) and (d) with the derivative spectrum shown in Figure 7(a). It can be seen from Figure 7(d) that with a modulation amplitude of 200 eV almost all the Co LMM Auger features are shown as $\mathcal{N}_m(E)$. Note that problems due to peak interference can occur. Some background is present in the $\mathcal{N}_m(E)$ spectra shown but it could be further reduced by using a different waveform.[69]

<u>Special Waveforms</u>. Special waveforms can be used to obtain $\mathcal{N}_m(E)$ type spectra over the entire energy range but problems due to slope artifacts arise, e.g. in determining areas under Auger peaks.[90] However, it is claimed that with such waveforms the maximum possible sensitivity and resolution are simultaneously available, peak broadening due to potential modulation having been completely eliminated. Problems due to Auger energy shifts with large square wave modulations are also eliminated.

Some real-time spectrum deconvolution can also be achieved by using special waveforms and the reader is referred to reference 90 for more details.

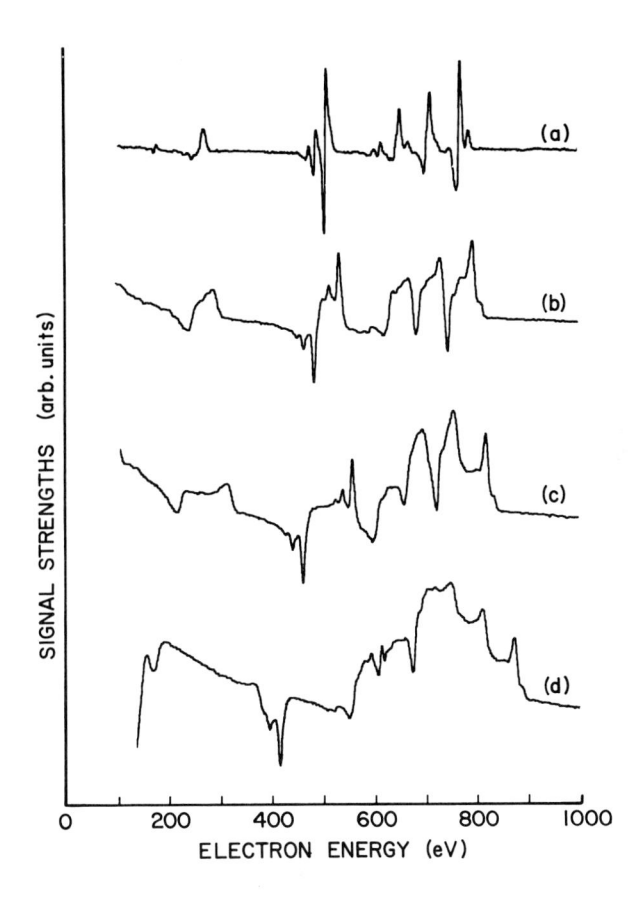

Fig. 7. Auger spectra from a contaminated Co surface using square wave modulation of amplitudes (a) 5 eV, (b) 50 eV, (c) 100 eV and (d) 200 eV. Spectra were obtained using a 5μA, 5 keV electron beam for excitation, with a time constant of 40 ms.

Spectrum Subtraction Techniques (SST)

Sometimes in Auger spectroscopy problems arise due to the overlap of Auger peaks from different elements. This is particularly troublesome if an element has only one strong Auger transition, e.g., S, and it falls on top of another transition, e.g., Mo $M_{4,5}N_{2,3}N_{2,3}$, making the detection of low concentrations of S directly from the data virtually impossible.[77] As Mo has other nearby Auger transitions, the Mo Auger spectrum can be removed

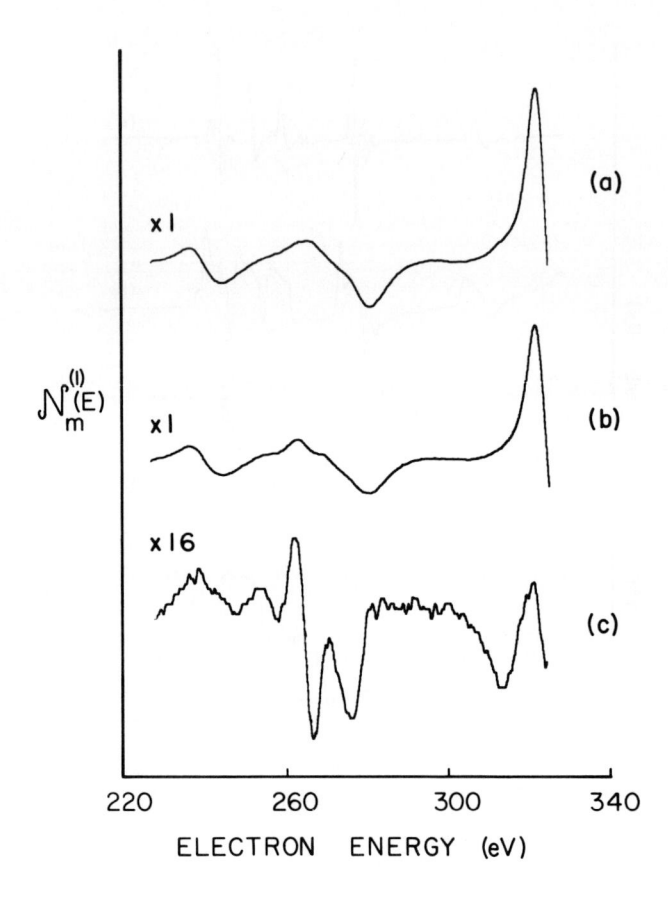

Fig. 8. Illustration of the application of SST to retrieve the carbon Auger line shape for CO adsorbed on clean Pd; (a) part of the Auger spectrum from clean Pd; (b) that obtained following exposure to CO at room temperature, and (c) the resulting spectrum following the application of SST,[55] shown at a magnification of 16X relative to part (b). The spectra were obtained using a 0.3μA, 1.5 keV electron beam for excitation, the sinusoidal modulation was 5 eV peak-to-peak, the time constant was 100 ms and the averages of 8 scans are shown.

by applying SST, thereby enhancing the detectability of S on Mo surfaces.[55]

The technique can also be used to obtain the Auger line shape of a particular element when overlap problems occur.[55] An excellent example of this is illustrated in Figure 8 where severe overlap problems occur between C and some of the minor Pd Auger transitions. Part of the Auger spectrum from clean Pd is shown in Figure 8(a), and the corresponding part following CO adsorption is shown in Figure 8(b). A change in peak shape due to C is detected around 270 eV but it is impossible to directly deduce the C Auger line shape. However, after applying SST to minimize the size of the Pd feature near 320 eV, the C Auger line shape emerges and is shown in Figure 8(c). It can be seen that the C spectrum is essentially a doublet with two minor features on the low energy side. This C Auger line shape from CO on Pd is essentially identical with that observed from CO on Ni[66] (where no peak overlap problems exist) implying that C has a similar chemical environment when CO is adsorbed on Pd and Ni.

VIII. ACKNOWLEDGEMENTS

Thanks are extended to M.P. Hooker and R.G. Wolfe for their assistance in taking some of the data, to J.R. Miller for technical assistance and to R.W. Springer for discussions regarding square wave modulation.

This work was sponsored by the Air Force Materials Laboratory, Air Force Systems Command, United States Air Force, Contract F33615-74-C-4017.

REFERENCES

1. P. Auger, Compt. Rend. (Paris), 177, 169 (1923).
2. P. Auger, Surface Sci., 48, 1 (1975).
3. P. Auger, Ann. Phys. (Paris), 6, 183 (1926).
4. S. Rosseland, Z. Physik, 14, 173 (1923).
5. L.J. Haworth, Phys. Rev., 48, 88 (1935).
6. J.J. Lander, Phys. Rev., 91, 1382 (1953).
7. See for example, G.A. Harrower, Phys. Rev., 102, 340 (1956); L.N. Tharp and E.J. Scheibner, J. Appl. Phys., 38, 3320 (1967).
8. See L.A. Harris, J. Appl. Phys., 39, 1419 (1968) and L.A. Harris, J. Vac. Sci. Technol., 11, 23 (1974).
9. R.E. Weber and W.T. Peria, J. Appl. Phys., 38, 4355 (1967).
10. P.W. Palmberg, Appl. Phys. Letters, 13, 183 (1968).
11. P.W. Palmberg, G.K. Bohn and J.C. Tracy, Appl. Phys. Letters, 15, 254 (1969).

12. T.W. Haas, J.T. Grant and G.J. Dooley, J. Appl. Phys., **43**, 1853 (1972).
13. P.W. Palmberg, J. Vac. Sci. Technol. **9**, 160 (1972).
14. N.C. MacDonald and J.R. Waldrop, Appl. Phys. Letters, **19**, 315 (1971).
15. See for example (a) N.J. Taylor in *Techniques of Metal Research*, ed. by R.F. Bunshah (Interscience, New York), 1972, Vol. VII, Part 1, Chapter 2; or (b) T.E. Gallon and J.A.D. Matthew, Review of Physics in Technology, **3**, 31 (1972).
16. J.T. Grant and T.W. Haas, Surface Sci., **23**, 347 (1970).
17. M.P. Hooker, J.T. Grant and T.W. Haas, J. Vac. Sci. Technol., **13**, 296 (1976).
18. T.E. Gallon and J.A.D. Matthew, Phys. Status Solidi, **41**, 343 (1970).
19. T.W. Haas, J.T. Grant and G.J. Dooley, Phys. Rev. B, **1**, 1449 (1970).
20. S.P. Kowalczyk, R.A. Pollack, F.R. McFeely, L. Ley and D.A. Shirley, Phys. Rev. B **8**, 2387 (1973).
21. S.P. Kowalczyk, L. Ley, F.R. McFeely, R.A. Pollack and D.A. Shirley, Phys. Rev., B **9**, 381 (1974).
22. P.W. Palmberg, G.E. Riach, R.E. Weber and N.C. MacDonald, *Handbook of Auger Electron Spectroscopy*, Physical Electronics Industries, Inc., Edina, 1972.
23. E.J. McGuire, J. Physique **32**, C4 (1971).
24. K. Siegbahn et. al., *ESCA - Atomic, Molecular and Solid State Structure Studied by Means of Electron Spectroscopy*, North-Holland, Amsterdam, 1967.
25. See, for example: C.J. Powell, Surface Sci., **44**, 29 (1974).
26. See the list in: C.A. Evans, Anal. Chem., **47**, 855A (1975).
27. For details regarding electron excitation in AES see reference 15.
28. L.I. Yin, E. Yellin and I. Adler, J. Appl. Phys. **42**, 3595 (1971).
29. T.E. Gallon and J.A.D. Matthew, J. Phys. D: Appl. Phys., **5**, L69 (1972).
30. R.G. Musket and W. Bauer, Appl. Phys. Letters, **20**, 455 (1972).
31. J.F. Hennequin and P. Viaris de Lesegno, Surface Sci., **42**, 50 (1974).
32. J.T. Grant, M.P. Hooker, R.W. Springer and T.W. Haas, J. Vac. Sci. Technol., **12**, 481 (1975).
33. See for example: C.C. Chang in *Characterization of Solid Surfaces*, ed. by P.F. Kane and G.B. Larrabee (Plenum Press, New York), 1974, Chapter 20.
34. See N.C. MacDonald, A paper presented to this Symposium.
35. N.J. Taylor, Rev. Sci. Instrum., **40**, 792 (1969).
36. H.E. Bishop and J.C. Rivière, Surface Sci., **17**, 462 (1969).
37. N. Laegreid and G.K. Wehner, J. Appl. Phys. **32**, 365 (1961).
38. M.L. Tarng and G.K. Wehner, J. Appl. Phys., **43**, 2268 (1972); J.W. Coburn and E. Kay, Crit. Rev. Solid State Sci., **4**, 561 (1974).

39. J.T. Grant, M.P. Hooker, and T.W. Haas, J. Colloid Interface Sci., in press.
40. P.H. Holloway, J. Electron Spectrosc., 7, 215 (1975).
41. D.R. Arnott and D. Haneman, Surface Sci., 45, 128 (1974).
42. G.F. Amelio, Surface Sci., 22, 301 (1970).
43. L. Yin, I. Adler, T. Tsang, M.H. Chen and B. Craseman, Phys. Lett., 46A, 113 (1973).
44. T.W. Haas and J.T. Grant, Appl. Phys. Letters, 16, 172 (1970); J.T. Grant and T.W. Haas, Surface Sci., 24, 332 (1971).
45. J.T. Grant and T.W. Haas, Phys. Lett., 33A, 386 (1970).
46. J.T. Grant and M.P. Hooker, J. Electron Spectroscopy, in press.
47. See for example, D.T. Quinto and W.D. Robertson, Surface Sci., 27, 645 (1971).
48. See for example, P.W. Palmberg in *Electron Spectroscopy*, ed. by D.A. Shirley (North-Holland, Amsterdam, 1972), p. 835; M.P. Seah, Surface Sci., 40, 595 (1973).
49. T.W. Haas and J.T. Grant, Phys. Lett., 30A, 272 (1969).
50. J.T. Grant and T.W. Haas, Surface Sci., 26, 669 (1971).
51. F.J. Szalkowski and G.A. Somorjai, J. Chem. Phys., 56, 6097 (1972).
52. S.V. Pepper, Rev. Sci. Instrum., 44, 826 (1973).
53. R.A. Kushner, D.V. McCaughan, V.T. Murphy and J.A. Heilig, Phys. Rev. B 10, 2632 (1974).
54. M.P. Hooker and J.T. Grant, Surface Sci., 51, 328 (1975).
55. J.T. Grant, M.P. Hooker, and T.W. Haas, Surface Sci., 51, 318 (1975).
56. J.J. Vrakking and F. Meyer, Surface Sci., 35, 34 (1973).
57. L. McDonald, D.P. Woodruff, and B.W. Holland, Surface Sci., 51, 249 (1975).
58. P. Staib and J. Kirschner, Appl. Phys., 3, 421 (1974).
59. F. Meyer and J.J. Vrakking, Surface Sci., 33, 271 (1972).
60. L.E. Davis and A. Joshi in *Surface Analysis Techniques for Metallurgical Applications*, ASTM STP 596 (American Society for Testing and Materials, Philadelphia), in press.
61. J.T. Grant, T.W. Haas, and J.E. Houston, J. Vac. Sci. Technol., 11, 227 (1974).
62. J.E. Houston, Rev. Sci. Instrum., 45, 897 (1974).
63. J.T. Grant, T.W. Haas, and J.E. Houston, Japan J. Appl. Phys., Suppl., 2, Pt. 2, 811 (1974).
64. J.T. Grant, M.P. Hooker and T.W. Haas, Surface sci., 46, 672 (1974).
65. L.C. Isett and J.M. Blakely, Rev. Sci. Instrum., 45, 1382 (1974).
66. M.P. Hooker and J.T. Grant, Surface Sci., in press.
67. J.T. Grant, T.W. Haas and J.E. Houston, Surface Sci., 42, 1 (1974).
68. J.T. Grant and T.W. Haas, Surface Sci., 44, 617 (1974).
69. R.W. Springer, D.J. Pocker, and T.W. Haas, Appl. Phys. Letters, 27, 368 (1975).

70. R.W. Springer, T.W. Haas and J.T. Grant, to be presented at the 12th Annual Symposium of the New Mexico Chapter of the American Vacuum Society, Albuquerque, March 1976; manuscipts in preparation.
71. See for example the bibliographies: T.W. Haas, G.J. Dooley, J.T. Grant, A.G. Jackson and M.P. Hooker, Progress in Surface Science, 1, 155 (1971); D.T. Hawkins, Bell Laboratories Bibliography No. 303 (1975).
72. A.M. Horgan and I. Dalins, J. Vac. Sci. Technol., 10, 523 (1973).
73. J.C. Tracy and P.W. Palmberg, J. Chem. Phys., 51, 4852 (1969).
74. N.A. Gjostein and N.G. Chavka, Journal of Testing and Evaluation, 1, 183 (1973).
75. Y. Takasu and H. Shimizu, J. Catalysis, 29, 479 (1973).
76. M.M. Bhasin, J. Catalysis, 34, 356 (1974).
77. T.W. Haas, J.T. Grant and G.J. Dooley, J. Vac. Sci. Technol., 7, 43 (1970).
78. M.P. Seah and E.D. Hondros, Proc. R. Soc. Lond. A, 335, 191 (1973).
79. H.L. Marcus and P.W. Palmberg, Trans. TMS-AIME, 245, 1664 (1969).
80. A. Joshi, Scripta Metallurgica, 9, 251 (1975).
81. D.H. Buckley, Japan. J. Appl. Phys. Suppl., 2, Pt. 1, 297 (1974).
82. S.V. Pepper, J. Appl. Phys., 45, 2947 (1974).
83. C.G. Pantano, D.B. Dove and G.Y. Onoda, Appl. Phys. Letters, 26, 601 (1975).
84. R.J. Sunderland, Japan. J. Appl. Phys. Suppl., 2, Pt. 1, 347 (1974).
85. G.Y. Onoda, D.B. Dove, and C.G. Pantano, in *Materials Science Research*, Plenum Press, New York, Vol. 7, 1974, p. 39.
86. M.G. Yang, K.M. Koliwad, J. Electrochem. Soc., 122, 675 (1975).
87. C.A. Haque, IEEE Trans., Vol. PHP-9, 58 (1973).
88. J.M. Morabito, Thin Solid Films, 19, 21 (1973).
89. R.W. Springer, personal communication; manuscript in preparation.
90. D.J. Pocker, R.W. Springer, F.E. Ruttenberg and T.W. Haas, J. Vac. Sci. Technol., 13, in press.

A Study of the Passive Film Using Auger Electron Spectroscopy

C.E. Locke, J.H. Peavey[‡], O. Rincon[1], M. Afzal[2]

School of Chemical Engineering
University of Oklahoma
Norman, Oklahoma 73069

The passive film formed on Nickel-200 and Inconel X-750 has been studied with Auger Electron Spectroscopy. The film was formed in H_2SO_4 containing concentration levels of NaCl ranging from 0→5.8% by weight at various electrode potentials within the passive region of the polarization curve. The elemental composition of the film changed with the electrode potential and chloride ion content. The film composition as a function of thickness was studied by sputtering away the film using 500 eV argon ions. The film on Nickel-200 is an oxide containing some sulfur. The film on Inconel X-750 is an oxide of nickel and chromium and is chromium enriched over the composition of the bulk alloy. The film on Inconel-750 is thicker than on Nickel-200. The film contains chloride at a concentration somewhat independent of solution concentration for Nickel 200, but somewhat dependent on solution concentration for Inconel X-750. Auger Electron Spectroscopy is a new and useful tool in the continuing effort to elucidate the fundamental nature of the passive film.

INTRODUCTION

Passive metals are widely used commercially but the phenomena of passivity has not been completely understood.[1,2] In fact, there is some disagreement even in defining passivity. The

---

[1]University of Zulia
Maracaibo, Venezuela

[2]Electrocast Steel Foundry
Cicero, Illinois 60650

[‡]Motorola Semiconductor
Products Division
Phoenix, Arizona 85008

definition considered valid for this study states that passivity is the loss of chemical reactivity experienced by some metals in special environments. These metals and alloys, when passive, behave like the noble metals. This noble-like behavior is attributed to the presence of a thin film which is formed on the surface of the metal. There continues to be disagreement concerning the mechanism of the formation of this film. Two theories have been advanced and are vigorously defended by the proponents. Uhlig[1] states that the film is formed by a chemisorbed layer of oxygen which displaces the water molecules from the metal surface. The oxygen layer may subsequently react with the base metal to form an oxide reaction product.

Bockris and Reddy[3] however argue that the film is an oxide reaction product. The dissolution of the metal occurs first, then a salt or hydroxide with limited solubility is formed. This precursor subsequently undergoes some sort of change that allows it to be an electronic conductor and it is then the passive film. This solid state theory has, in the works of Bockris and Reddy, "...seemed to have gained the day".

The mechanism theory arguments have been in vogue for over 25 years and have not been completely resolved. The work presented here does not, as yet, support either theory. It is hoped, however, that it will help in the understanding of the nature of the passive film which is formed, regardless of the mechanism.

The passive film on nickel has been studied by a number of investigators and film formation mechanisms have been debated using data obtained from nickel. Bockris, et al[4] used ellipsometry to study the film formed on nickel in 0.5M K_2SO_4 + 0.1N H_2SO_4. They found a porous "pre-passive" film up to 60 Å thick was formed in the active region which converted to a passivating film at the passivation potential. This passivating film had high electronic conductivity. Hoar[5] thinks their interpretation is in error. He states the oxide film is formed under a porous nickel sulfate layer. Sato and Okamoto[6] by using potential time decay experiments with anodic polarization, postulated the film to be Ni_3O_4 which was formed by successive reactions of nickel with hydroxyl ions. Kunze and Schwabe[7] state that a porous nickel oxide was formed in the active region and passivation was caused by oxygen chemisorption. Tokuda and Ives[8] found that the film thickness on nickel was 1 or 2 monolayers of NiO when the Ni was placed directly at a passive potential rather than polarizing relatively slowly from active to passive. They believe that chemisorption of oxygen must be the initial step in passivation.

The passive film on Inconel X-750 has not been studied but a few investigators have examined other alloys such as 300 series stainless steels. Okamoto[9] has studied the composition and structure of the passive film on an 18-8 stainless steel using several analytical methods. He varied the potential of the electrode and observed film characteristics under these known electrochemical conditions. He found that the composition and thickness of the film was potential dependent. It also contained solution anions at specific potentials. Chromium was enriched in the passive film with nickel and iron being depleted at potentials below 0.40V.

Lumsden and Staehle[10] have used Auger Electron Spectroscopy (AES) to evaluate the passive film on 316 SS. They studied the composition as a function of thickness by sequentially removing the film by argon ion bombardment followed by AES. They found that the film composition was different than the bulk metal, being enriched chromium with iron and nickel being depleted.

Chloride ions have a detrimental effect on the passive film on nickel and Inconel alloys. Pascoe,[11] and Trout and Daniels[12] have shown that chloride ions increase the passive currents and decrease the passive potential region on Inconel X-750. Piron, et al.[13] studied nickel-200 and Inconel-600 in acid solutions containing chloride ions. They found that the results with these alloys were very similar to those obtained with Inconel X-750 discussed above. Nickel was more sensitive than the Inconel alloys to the presence of the chlorides. Various theories of the role of chloride ions in the disruption of the passive film[8,14] have been proposed. This is an area that is also not completely understood.

Auger Electron Spectroscopy is a tool newly applied to the study of the passive film. The Auger effect is a secondary electron phenomenon resulting from a radiationless transition between energy levels. The atom is first excited by bombardment with a 2-5 KeV electron beam which ionizes one of the inner electron levels. An outer electron subsequently decays into the vacancy. If this is an Auger transition, energy is conserved by the emission of a secondary electron; otherwise, an x-ray is emitted. These two events are mutually exclusive - the total probability of their occurrence is unity. The Auger electrons are unique since the transitions occur between the quantized electronic energy levels. The elements from which they originated can be identified by analyzing the electron energies.

This technique has not been widely used for the study of passivity. Revie et al.[15] studied the passive film formed on iron with a unique electrochemical cell coupled to the AES apparatus. Lumsden and Staehle et al.[10,16,17] have also used AES to examine passive films on various alloys. Their work involves removing the sample from the electrochemical cell to transfer it to the AES apparatus. This undoubtedly changes the film from the state in the solution.

The work presented in this paper also involves transferring the samples from the electrochemical cell to the AES vacuum system. It is difficult to justify that the film studied is identical to the film in place when the metal is in the corrosive environment. However, the transfer procedure should be a reasonable technique for comparing properties of the films formed under varying electrochemical conditions. This was the primary thrust of the study initially, but as will be shown, the data gives some insight into the nature of the passive film. The complete elucidation of the passive film will probably be a result of using several experimental approaches. AES is one such approach.

EXPERIMENTAL

Commercially available nickel-200 and Inconel X-750 manufactured by International Nickel, Huntington Alloys Product Division were used. The chemical compositions of these alloys are given in Table 1.

Table 1: Chemical Composition in Weight Percent

	Nickel-200	Inconel X-750
Ni	99.5	74.02
Cr		14.82
Fe	0.15	6.33
Ti		2.50
Mn	0.25	0.12
S	0.005	0.007
Cu	0.05	0.03
C	0.06	0.03
Al		0.88
Cb		0.87
Si	0.05	0.27

The Inconel X-750 was solution heat treated for one hour at 1204°C followed by water quenching. The surfaces of both metals were polished with 180 to 600 grit silicon wet grinding paper followed by sequential washes in trichloroethylene, acetone and distilled water prior to the electrochemical treatments.

The polarization studies and electrochemical treatments were conducted potentiostatically. The polarization curves were obtained by the manual step procedure whereby the potential was shifted in 50 mv steps after 3 minutes at each potential. All electrochemical experiments were conducted at 25°C in 1 normal sulfuric acid. Chloride ions were added as sodium chloride in concentrations ranging from 0.05 - 5.8 weight percent NaCl. All electrochemical experiments were conducted in hydrogen purged solution. The hydrogen deaeration was started 30 minutes prior to the experiment and was continued throughout the polarization experiments. The metal samples were polarized to the preset potentials within the passive region for 15 minutes in preparation for the AES experiments. The samples were removed from the electrochemical cell, rinsed with distilled water and stored in a chamber at approximately 10^{-2} Pa for transfer to the Auger system. Nitrogen was used to bring the chamber to atmospheric pressure for sample transfer. The samples were transferred from the vacuum chamber to the AES spectrometer which was in a nitrogen atmosphere to prevent contamination of the sample with air.

The experimental system used in this study was a sputter ion pumped stainless steel chamber, OFHC copper gasket sealed, bakeable to 250°C, and capable of achieving ultra high vacuum. Figure 1 is a schematic diagram of the system used. Clyndrical Auger electron optics which contained a coaxial electron gun and a thirteen-stage electron multiplier were used for the Auger electron analysis. The Auger spectrum was recorded as $dN(e)/dE$ on an x-y recorder. The spectra were obtained in the energy range of 0 to 1000 ev. An electron gun focusing a 40μa beam of 3 KeV electrons on an area of less than 1 mm^2 was used to irradiate the specimen at an angle of incidence of 90°. The residual gas pressure was approximately 1×10^{-7} Pa. The AES spectrum was taken of the passive film and then the specimen was rotated to face the sputtering ion gun. The argon ion current density was 3×10^{-5} μamp/cm^2 at 500 ev. The argon pressure was 4×10^{-4} Pa. After sputtering, the specimen was rotated back to the electron analyzer for an Auger spectrum. This procedure was repeated until the maximum sputtering of approximately 70×10^{-4} coul/cm^2 was reached.

RESULTS

Anodic Polarization. Figures 2 and 3 contain the anodic polarization curves for nickel-200 and Inconel X-750 respectively

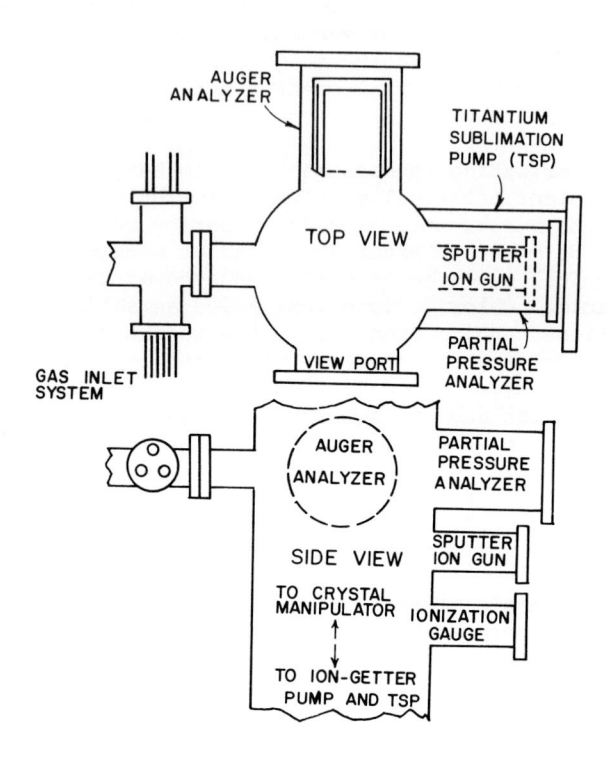

Fig. 1. Arrangement of apparatus in vacuum system.

in hydrogen purged 1N H_2SO_4 with various levels of sodium chloride. Both metals have the polarization characteristic of active-passive metals. These data also indicate that chloride ions increase the currents in the passive region and the currents necessary to achieve passivity. The potential range of passivity is decreased by chloride ions. Nickel is more sensitive to chloride ions than Inconel X-750 since only 0.5% NaCl was sufficient to completely eliminate the passive region on nickel. A level of 5.8% NaCl was required for the Inconel X-750. These data were in agreement with previously published results for these alloys.[12,13]

Auger Electron Spectroscopy. Typical spectra for nickel-200 and Inconel X-750 are given in Figures 4 and 5. The nickel spectrum has peaks characteristic of nickel, oxygen, carbon, sulfur, nitrogen, and argon. The Inconel X-750 spectrum has peaks characteristic of nickel, iron, chromium, oxygen, carbon, sulfur, nitrogen, argon, and titanium. These data were used to calculate relative amounts of the elements observed by comparing the response

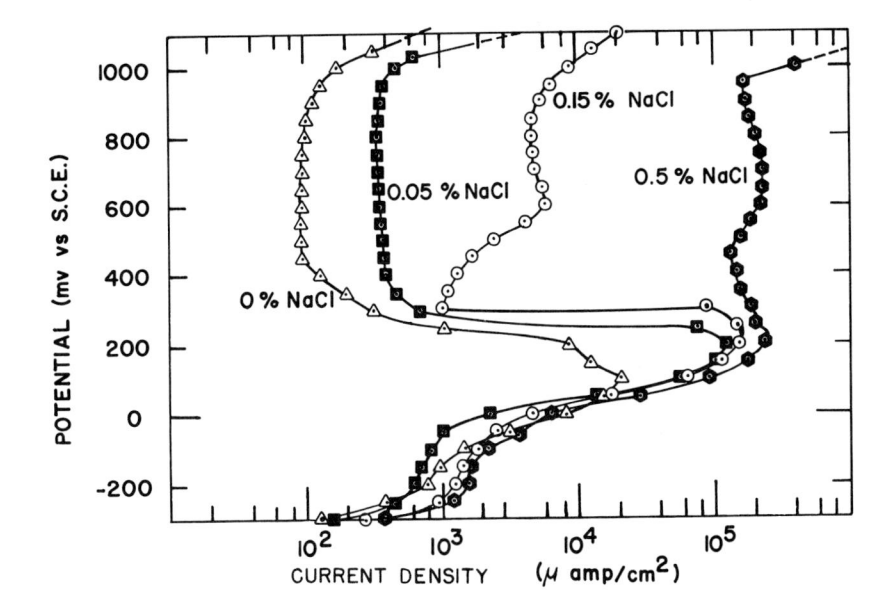

Fig. 2. Potentiostatic anodic polarization curves of nickel-200 in 1N H_2SO_4 containing 0, 0.5, 0.15, and 0.5% NaCl (wt).

of the element in the alloy to that of the pure element. Complete details of the calculations are described by Rincon.[18] All calculated data are reported on a carbon free basis since the carbon source was undoubtedly in the vacuum system. Revie, et al.[15] reported similar carbon contamination in their AES experiments. They subsequently purposely exposed their iron sample to CO and found there was no change in the AES spectra.

<u>Chloride Free Solutions</u>. Table 2 summarizes the data for nickel-200 polarized at +300, 500, and 900 mv (SCE). One of the numbers for each element is the relative amount at the surface of the passive film, and the second number is the relative amount after approximately 70×10^{-4} $coul/cm^2$ sputtering was done. Notice that potassium and chloride were found even though none was added to the solution. Possibly, these species could have diffused through the Luggen capillary probe.

Figures 6-8 contain the relative amounts of nickel, sulfur, and oxygen as a function of sputtering for nickel-200 polarized at +300, +500 and +900 mv (SCE). From these figures, it is apparent the relative amount of the base metal is reached after

Table 2. The relative amounts of elements in atomic percentages for nickel-200. On the left, values of AES at passive film. On the right, values taken at last sputtering depth. (Approximately 70 × 10⁻⁴ coul/cm²)

Potential mv (SCE)	S		Ni		O		N		Cl		Ar		K	
+300	20.1	3.0	43.4	85.7	19.2	3.0	13.6	4.8	3.7	0.0	0.0	2.4	0.0	1.1
+500	15.8	2.8	46.8	82.9	14.0	2.4	16.9	5.6	6.6	3.0	0.0	2.3	0.0	1.0
+900	16.7	1.3	53.7	89.4	13.8	1.7	12.4	4.8	3.4	0.0	0.0	1.9	0.0	0.9

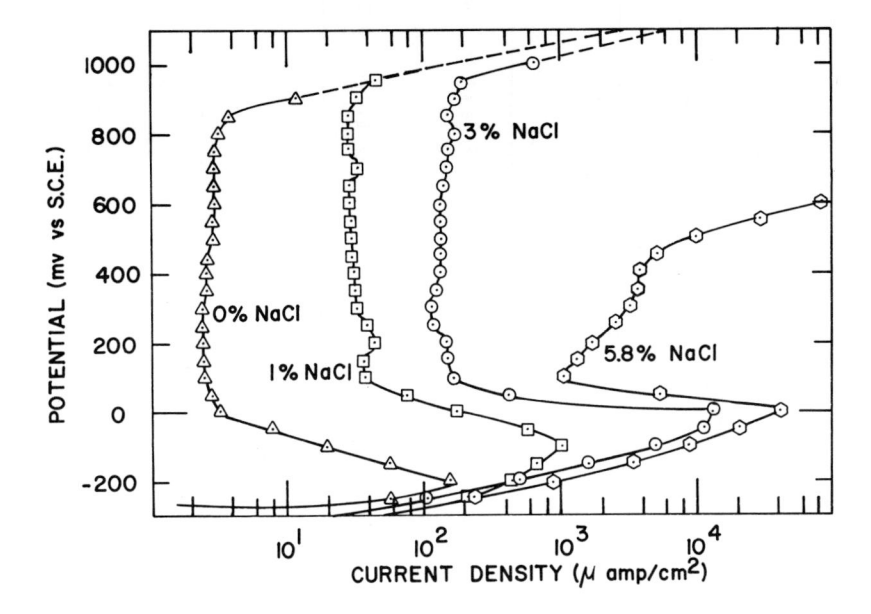

Fig. 3. Potentiostatic Anodic Polarization Curves of Inconel X-750 in 1N H_2SO_4 containing 0, 1, 3, and 5.8% NaCl (wt).

sputtering 30 to 40 x 10^{-4} coul/cm^2. At the passive surface, nickel, sulfur, and oxygen are the important constituents of the film. The relative amounts of these three elements at the surface of the passive film as a function of potential are shown in Figure 9. The nickel content increases with potential and the oxygen and sulfur decrease.

Table 3 contains the relative amounts of the elements (on a carbon free basis) found on the passive surface of Inconel X-750 and after sputtering approximately 70 x 10^{-4} coul/cm^2 for materials polarized at +200, +500, and +800 mv (SCE). The concentrations of sulfur, nickel, oxygen, and chromium changed from the passive surface to that exposed by sputtering. Iron and titanium remained essentially constant. Nitrogen and argon are not significant since they came from the gaseous environment as contaminants in the sample handling and sputtering operations. The potassium and chlorides are surprising contaminants, probably coming inadvertently from the electrochemical cell. Figures 10-12 are plots of the relative amounts of nickel, chromium, oxygen, and sulfur as a function of sputtering amount for samples polarized at 200, 500, and 800 mv (SCE). The chlorine amount is included in

Table 3. The relative amounts of elements in atomic percentages for Inconel X-750. On the left, values recorded from AES at the passive film. On the right, values taken at last sputtering depth. (Approximately 70 x 10^{-4} coul/cm^2).

Potential mv (SCE)	S	Ni	O	Cr	Fe	Ti	N	Cl	Ar	K
+200	4.1 1.7	35.3 59.1	25.6 4.7	13.3 11.4	5.0 5.6	2.2 2.4	11.4 10.9	2.7 1.3	0.0 2.4	0.0 0.6
+500	2.5 1.1	25.9 56.4	29.4 7.8	16.0 12.7	6.1 5.8	2.7 2.3	10.7 11.7	6.7 1.6	0.0 0.7	0.0 0.0
+800	3.7 1.2	23.7 55.6	33.1 7.2	18.2 12.9	4.9 5.3	2.7 2.7	11.2 11.0	2.5 1.6	0.0 1.7	0.0 0.8

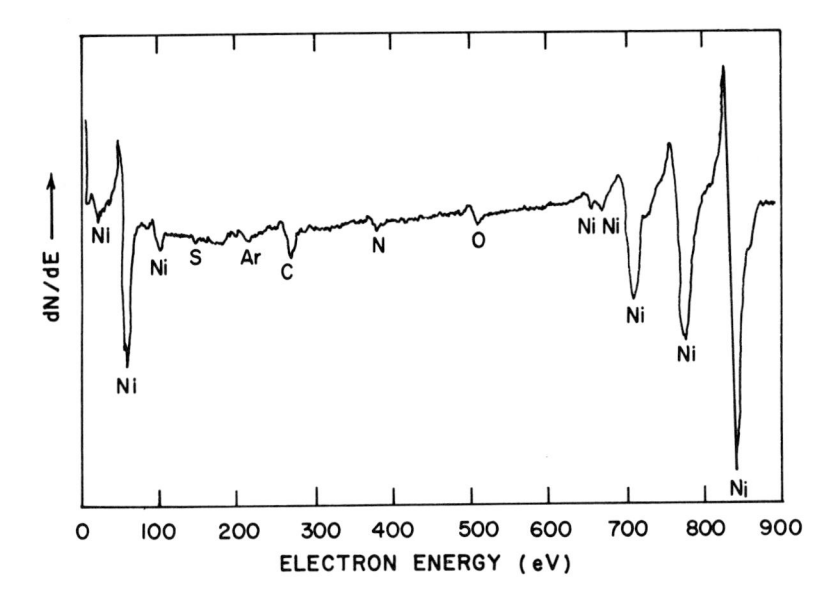

Fig. 4. Auger electron spectroscopic spectra for nickel-200.

Figure 12 for the sample polarized at +800 mw. The nickel content at each of these potentials is low at the surface then increases with an increase in amount of sputtering. The oxygen content decreases with sputtering while chromium and sulfur are approximately constant as the film is sputtered away. The passive film composition is depleted in nickel and therefore is enriched in chromium. This is best seen from the data presented in Figure 13 where the nickel to chromium relative amount ratio is plotted against the sputtering amount for samples polarized at the three electrode potentials. This ratio is dependent on the sputtering amount and electrode potential. The Ni/Cr ratio for the bulk alloy is about 5. Therefore, the bulk alloy was reached for the sample polarized at 200 mv when sputtering of 70×10^{-4} coul/cm^2 was accomplished. The bulk alloy evidently had not been reached for the samples polarized at 500 and 800 mv. Therefore, the film thickness must vary with electrode potential.

Chloride Containing Solutions. The chloride content of the 1N H$_2$SO$_4$ solution was varied for samples of nickel-200 and Inconel X-750 each polarized at +500 mv (S.C.E.). Nickel was found to pit severely in solutions containing 0.5% NaCl which was not surprising from the anodic polarization results for metal in this

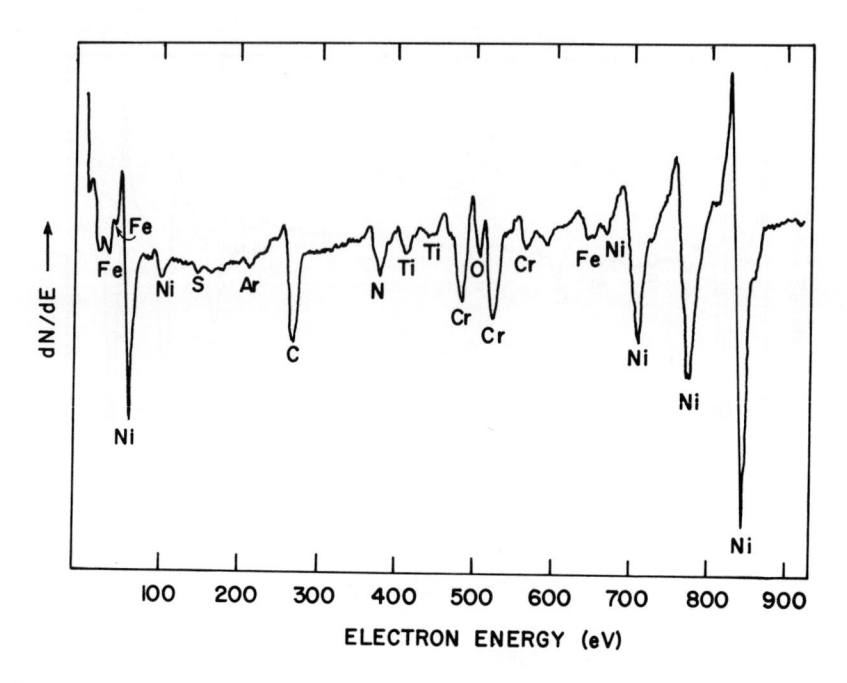

Fig. 5. Auger electron spectroscopic spectra for Inconel X-750.

solution. Data is therefore presented for nickel-200 in solutions containing 0.05 and 0.15% NaCl. Figures 14 and 15 are plots of the relative amounts of nickel, oxygen, sulfur and chlorine as a function of sputtering amount for nickel-200 polarized at 500 mv in 1N H_2SO_4 containing 0.05 and 0.15% NaCl. Oxygen, sulfur and chlorine decrease with sputtering. Nickel increases with sputtering, reaching a maximum at about 60 to 70 x 10^{-4} coul/cm^2 for both NaCl concentrations. The relative amounts of sulfur, nickel, oxygen, chlorine, nitrogen, argon and potassium for nickel-200 polarized at +500 mv(SCE) in 1N H_2SO_4 solutions with NaCl contents of 0 to 0.15% are shown in Table 4. The chlorine contents are independent of NaCl solution content, oxygen and sulfur contents are changed in the solutions containing chloride ions from the chloride free solutions. The sources of nitrogen, argon and potassium are as discussed above for nickel.

There was some difficulty in reproducing the AES data for nickel samples polarized in the chloride containing solutions. These samples were pitted somewhat and the reproducibility difficulties might be due to the presence or absence of pits in the area subjected to AES analysis.

Table 4. The relative amounts of elements in atomic percentages for nickel-200. On the left, values recorded from AES at the passive film. On the right, values taken at last sputtering depth. (Approximately 70 x 10^{-4} coul/cm^2 at 500 mv (SCE))

NaCl wt.%	S		Ni		O		Cl		N		Ar		K	
0	15.8	2.8	46.8	82.9	14.0	2.4	6.6	3.0	16.9	5.6	0.0	2.3	0.0	1.0
0.05	7.8	2.3	48.2	88.2	32.5	2.6	6.4	1.3	5.2	2.8	0.0	2.7	0.0	0.0
0.15	9.3	3.1	45.6	81.2	34.6	7.3	6.5	2.3	4.0	4.0	0.0	2.1	0.0	0.0

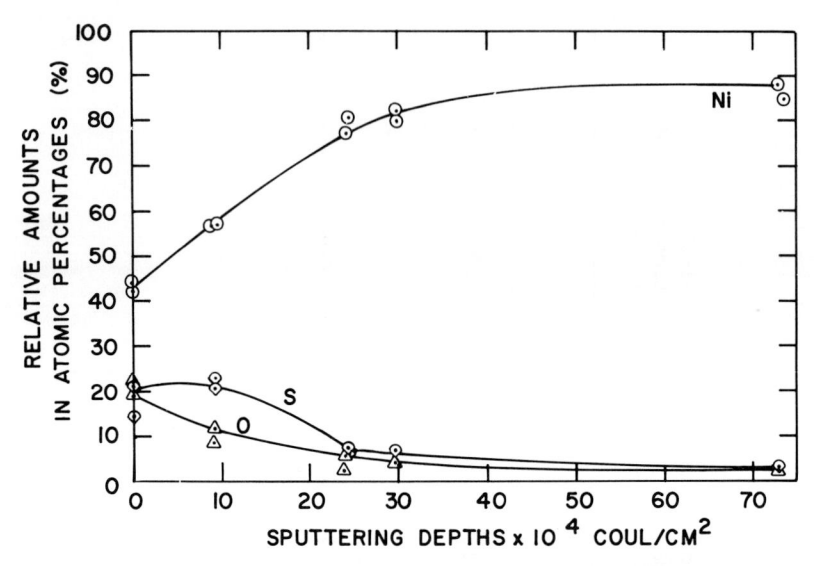

Fig. 6. Relative amounts of elements versus sputtering depth for nickel-200 in 1N H_2SO_4 at 300 mv (S.C.E.).

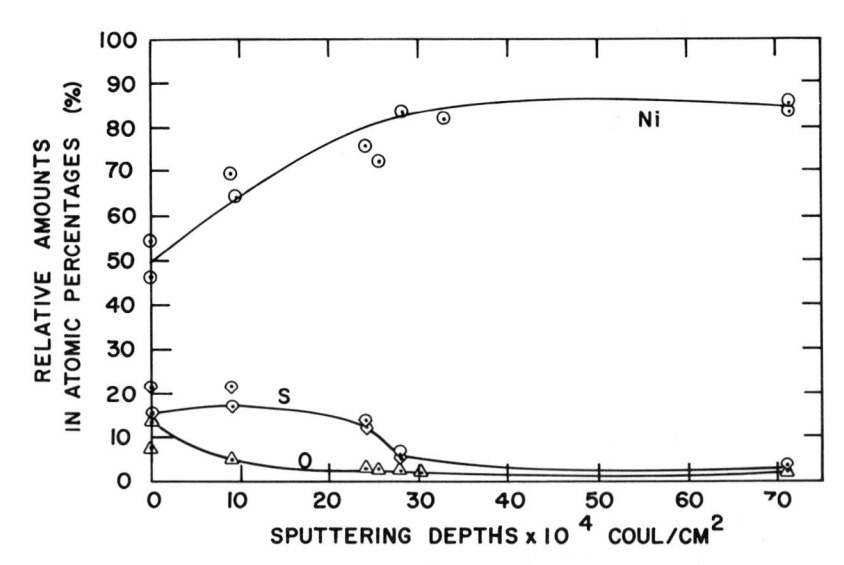

Fig. 7. Relative amounts of elements versus sputtering depth for nickel-200 in 1N H_2SO_4 at 500 mv (S.C.E.).

Passive Film   169

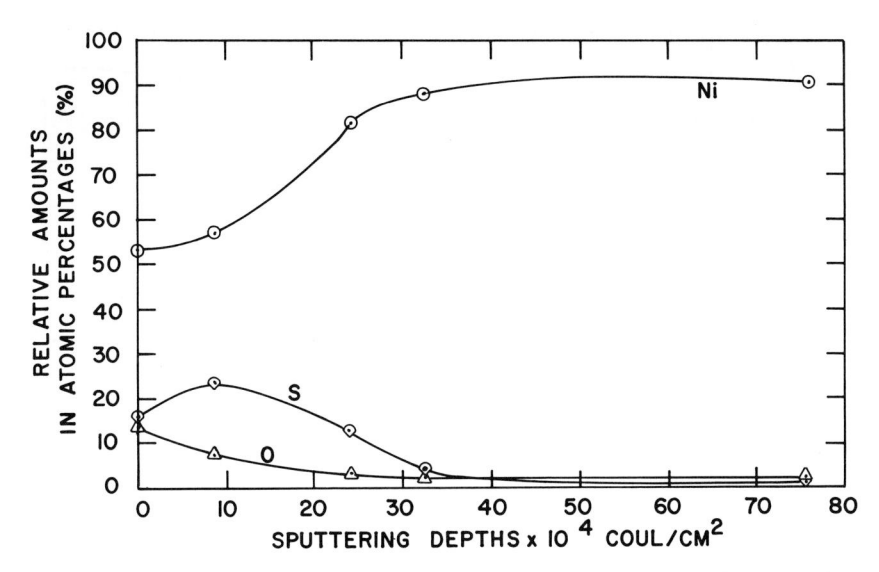

Fig. 8. Relative amounts of elements versus sputtering depth for nickel-200 in 1N H_2SO_4 at 900 mv (S.C.E.).

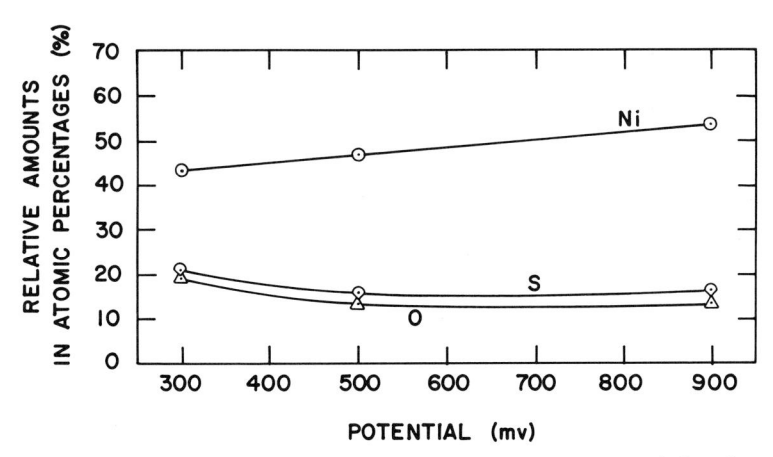

Fig. 9. Relative amounts of elements versus potentials for nickel-200 in 1N H_2SO_4 at passive film (not sputtered).

170   C. E. Locke et al.

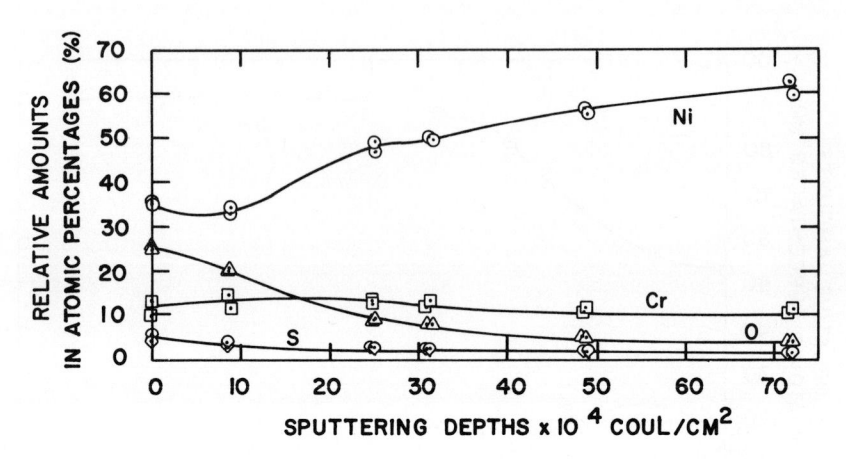

Fig. 10. Relative amounts of elements versus sputtering depth for Inconel X-750 in 1N H_2SO_4 at 200 mv (S.C.E.).

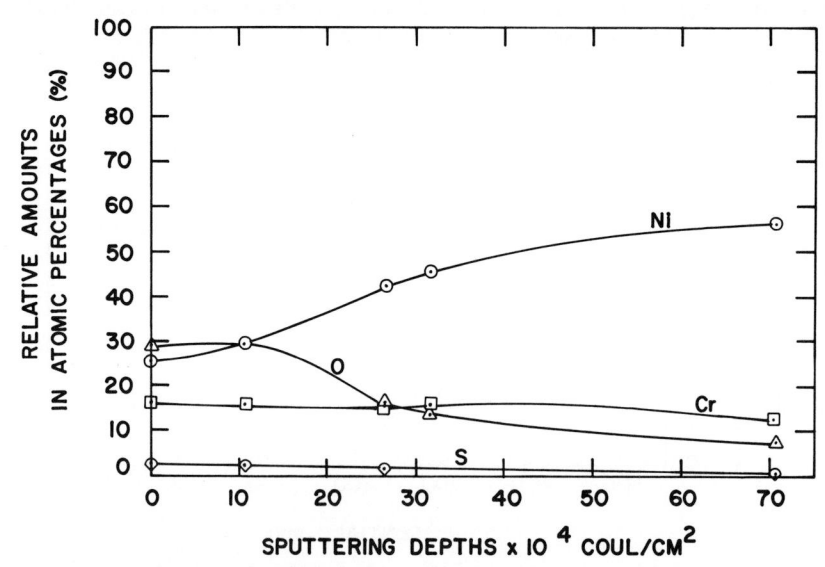

Fig. 11. Relative amounts of elements versus sputtering depth for Inconel X-750 in 1N H_2SO_4 at 500 mv (S.C.E.)

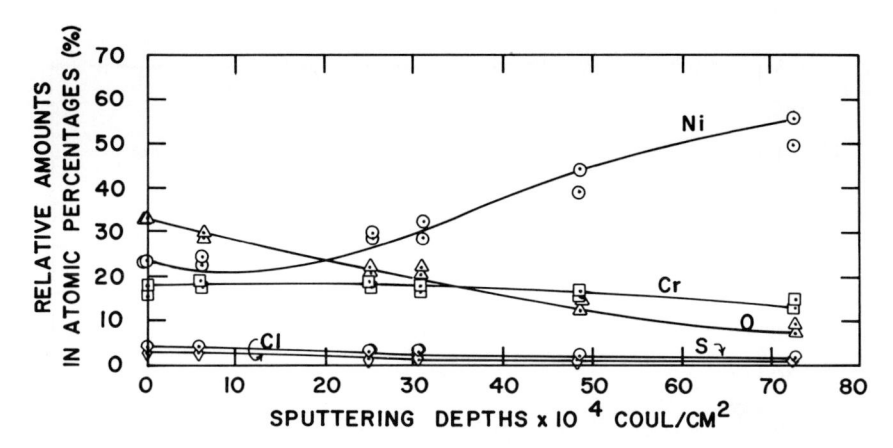

Fig. 12. Relative amounts of elements versus sputtering depth for Inconel X-750 in 1N H_2SO_4 at 800 mv (S.C.E.).

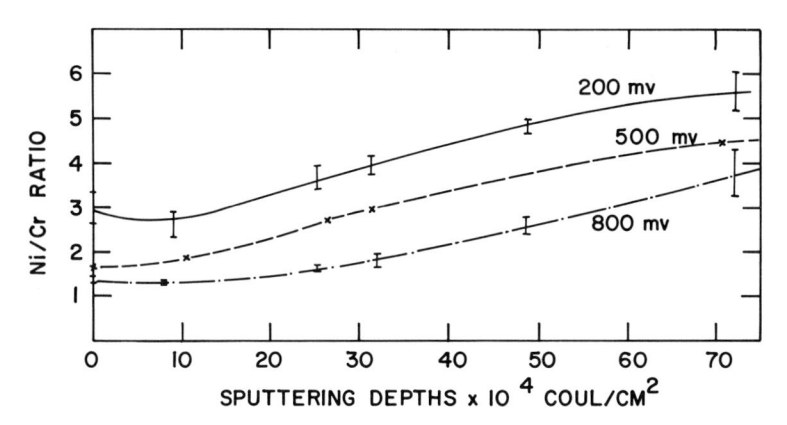

Fig. 13. Variation of Ni/Cr ratio as function of sputtering depth at different electrode potentials for Inconel X-750 in 1N H_2SO_4.

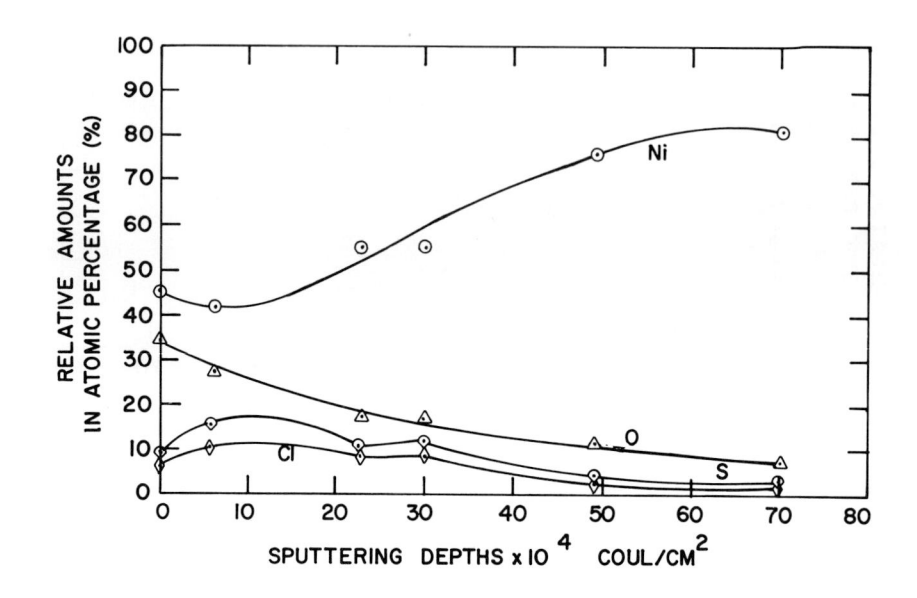

Fig. 14. Relative amounts of elements versus sputtering depth for nickel-200 in 1N H_2SO_4 containing 0.15% NaCl (wt) at 500 mv (S.C.E.).

Figures 16 and 17 contain the relative amount data for nickel, chromium, oxygen and chlorine in Inconel X-750 polarized to 500 mv in 1N H_2SO_4 containing 1% and 3% NaCl. The nickel content increases, oxygen and chlorine decrease while chromium is essentially constant with increasing amounts of sputtering. Table 5 summarizes the data for relative amounts of sulfur, nickel, oxygen, chlorine, iron, titanium, nitrogen, argon and potassium for Inconel X-750 in 1N H_2SO_4 with 0, 1 and 3% NaCl. Data at the surface of the film and after sputtering of 70 x 10^{-4} coul/cm^2 are shown. The chlorine content of the film increased when chloride ions were added to the solutions. However, there is no substantial difference in chlorine content in the samples polarized in 1 and 3% sodium chloride solutions. Nickel, chromium and titanium contents at the passive surface were unaffected by chloride ion in the solution. The iron content was lowered when the chloride ion content increased. Figure 18 illustrates that the Ni/Cr ratio is affected very little by the presence of chloride ions in the sulfuric acid. The variations seen in this figure are probably mostly experimental error and not substantial differences.

Table 5. Relative amounts of elements in atomic percentages for Inconel X-750. On the left, values recorded from AES at the passive film. On the right, values taken at last sputtering depth. (Approximately 70 x 10^{-4} coul/cm^2 at 500 mv (SCE))

NaCl wt.%	S		Ni		O		Cr		Cl		Fe		Ti		N		Ar		K	
0	2.5	1.1	25.9	56.4	29.4	7.8	16.0	12.7	6.6	1.6	6.1	5.8	2.7	2.3	10.7	11.7	0.0	0.7	0.0	0.6
1	3.7	1.2	24.2	47.7	30.3	9.6	15.1	14.6	9.8	1.5	4.5	4.6	2.8	3.4	9.6	15.3	0.0	1.6	0.0	0.4
3	4.0	0.9	24.2	53.1	28.9	7.7	15.8	13.2	9.6	1.0	3.8	5.1	2.2	3.1	11.6	13.6	0.0	1.9	0.0	0.3

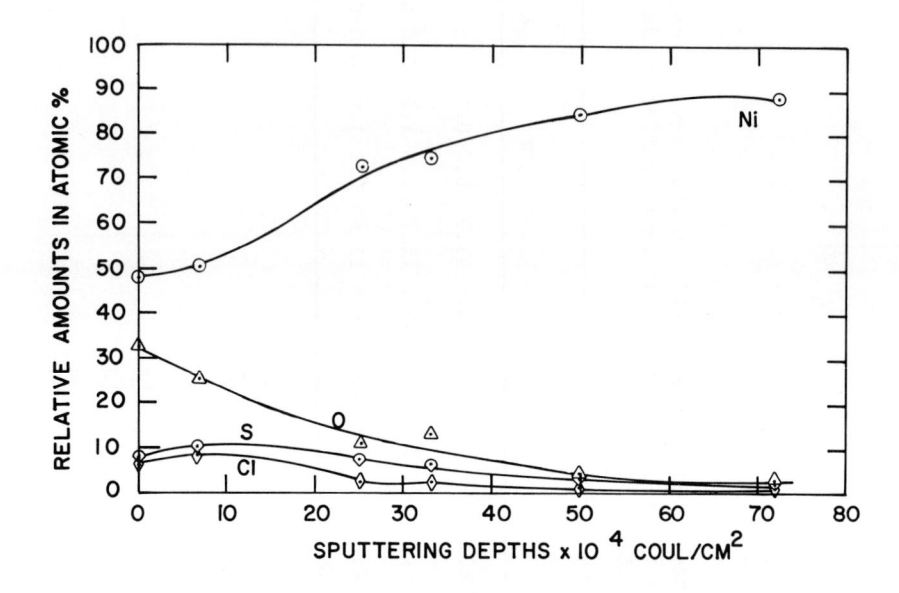

Fig. 15. Relative amounts of elements versus sputtering depth for nickel-200 in 1N H_2SO_4 containing 0.05% NaCl (wt) at 500 mv (S.C.E.).

DISCUSSION OF RESULTS

The results of the Auger Electron Spectroscopy of the passive films on nickel-200 and Inconel X-750 will be discussed in this section. The results obtained with chloride-free solutions will be discussed first, followed by the results from the chloride containing solutions.

Chloride Free Solutions

<u>Nickel-200</u>. The passive film on nickel-200 formed in 1N H_2SO_4 contains nickel, oxygen and sulfur as the primary constituents. There does not seem to be any stoichiometric ratio of these elements. Bockris et al.[4] postulated that the film was made up of a non-stoichiometric nickel oxide based on calculations of polarizability of these compounds compared to the ellipsometric data. The composition of the film varied with potential which was also observed for Inconel X-750 as will be discussed below.

Passive Film   175

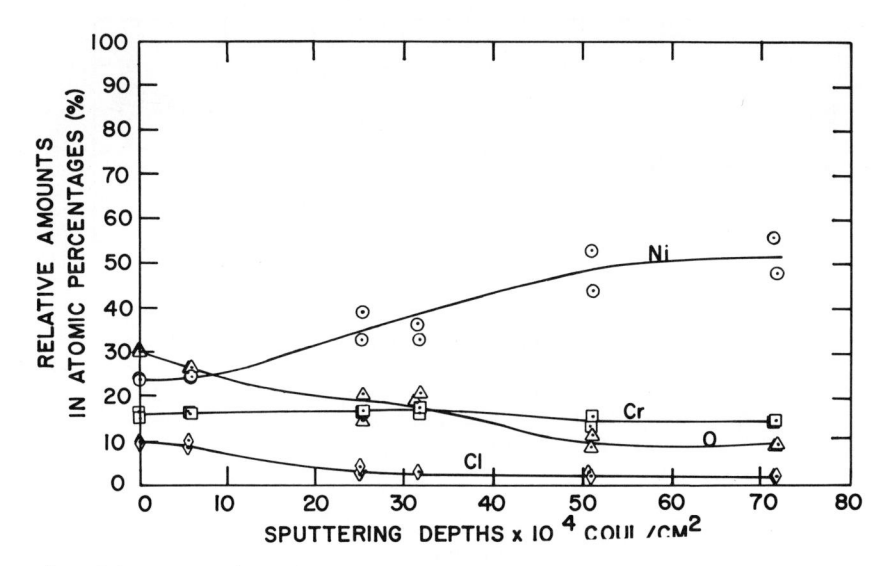

Fig. 16. Relative amounts of elements versus sputtering depth for Inconel X-750 in 1N H_2SO_4 containing 1% NaCl (wt) at 500 mw (S.C.E.).

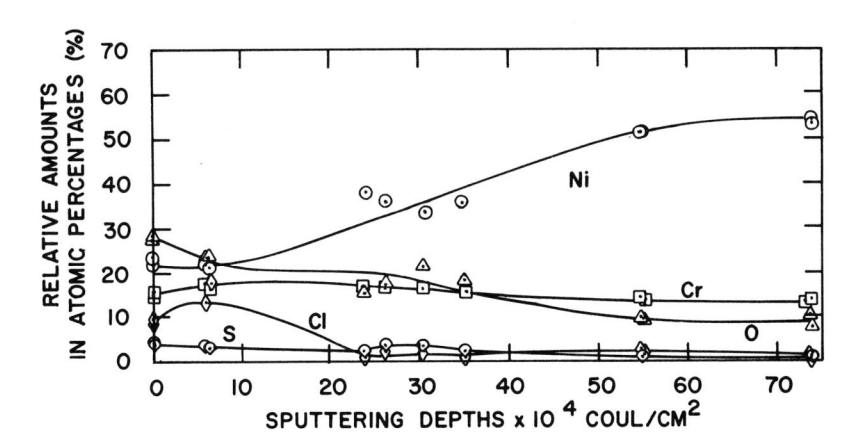

Fig. 17. Relative amounts of elements versus sputtering depth for Inconel X-750 in H_2SO_4 containing 3% NaCl (wt) at 500 mw (S.C.E.).

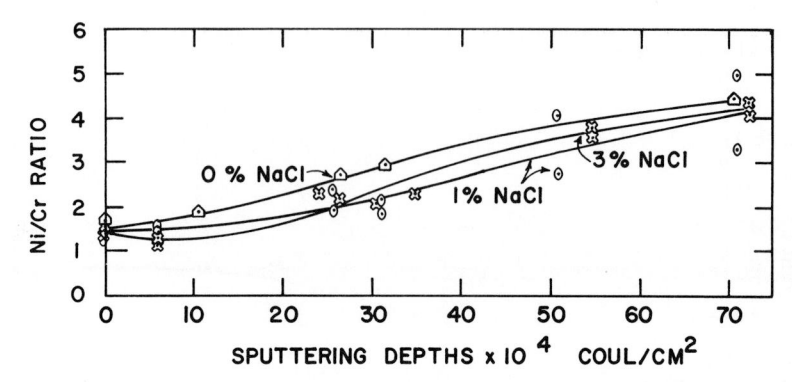

Fig. 18. Variation of Ni/Cr ratio at function of sputtering depth at different NaCl concentration for Inconel X-750 in 1N H_2SO_4 at 500 mw (S.C.E.).

The thickness of the film was estimated from the data obtained from the sputtering. The base metal was reached when the nickel content did not vary with further sputtering. The thickness of the film was calculated assuming the rate of removal of the film to be the same as nickel. As discussed by Afzal[19], 1×10^{-4} coul/cm^2 sputtering will remove 0.5271 monolayer of nickel. It was calculated then that 1×10^{-4} coul/cm^2 removes 1.312 Å of nickel. This is an estimation since the film removal rate may be different from that of the pure nickel. However, using these estimations with the data obtained in this study, the passive film on nickel-200 was 40 to 60 Å. This is in agreement with the ellipsometric data found by Bockris et al.[4] Tokuda and Ives[8] postulate the film formed on nickel when it is placed directly at the passive potential, as was done in this study, was relatively thin. They state that it is 1 to 2 monolayers of NiO. This does not agree with the data in this study since 1 to 2 monolayers would have a thickness of 3 to 6 Å.

Chlorine and potassium appeared in the "chloride-free" results. Possibly, these could have been introduced into the cell by the potential measuring circuit. A saturated calamel electrode was used to measure the electrode potential. It was connected to the electrochemical cell by a Luggen probe filled with 1N H_2SO_4 which provide a long diffusion path between the electrode and cell. Even with this path, it is possible that the KCl bridge solution might have diffused into the cell and was incorporated into the film.

Inconel X-750. The composition and thickness of the passive film are affected by the electrode potential for Inconel X-750, polarized in 1N H_2SO_4. The nickel content decreases with potential while the chromium and oxygen contents increase. Sulfur, iron and titanium do not change greatly with potential. Okamoto[9] found that the composition of the passive film on 18-8 SS also varied with potential which he explained by superimposing anodic polarization curves for the pure elements in the alloy with the polarization curve for the alloy. This procedure can help predict that the film should be enriched in chromium at potentials lower than +400 mv (SCE). Data from this study does not completely agree with this approach. However, at all potentials the passive film Ni/Cr ratio is lower than the bulk alloy ratio which may indicate chromium is enriched. The composition may be a mixture of chromium and nickel oxides, the stoichiometry of which may be potential dependent.

The apparent film thickness is potential dependent. If the same assumptions used for nickel-200 are used for Inconel X-750, the film is estimated to be about 90 Å when formed at +200 mv and thicker than 90 Å at the higher potentials. These data indicate the film to be much thicker than found by Lumsden and Staehle[10] and Okamoto[9] for stainless steel. Titanium and iron contents vary little with electrode potential and sputtering. Titanium may strengthen or stabilize the passive film. Comparison of the polarization results of Inconel X-750 and Inconel-600 indicates that X-750 can tolerate higher levels of chloride ion than 600. Titanium is the only essential difference between these two alloys.

The sulfur content was low and essentially constant with sputtering. However, oxygen content decreased with sputtering. This indicates that the passive film on Inconel X-750 may be an oxide rather than a sulfate. It is possible that the film is a chromium oxide mixed with a nickel oxide. It is not possible to assign any specific stoichiometric oxides as the main constituents.

Chloride Containing Solutions

Nickel-200. The surprising result of the data found with nickel in chloride containing solutions is that the chloride is not present to any great extent in the film. Chlorine is found in the film but notice that the amount does not vary with solution composition. The samples polarized in the chloride containing solutions were pitted. The pitting theory presented by Vijh[14] states that the chloride ions would not necessarily be included in the film but would form a metal-chloro complex. This complex would decompose in solution forming the metal ion and releasing the chloride ion for further metallic attack. It may be possible that that mechanism can explain these data. Also, the film is apparently thicker when formed in chloride containing solutions than when formed in "chloride free" solutions. The oxygen content is also higher. These data were difficult to reproduce whereas in all other metal-solution combinations, reproducibility was no problem. Possibly, the presence of pits made it difficult to examine the surface as discussed in the results section.

<u>Inconel X-750</u>. The composition of the film formed on Inconel X-750 was not changed substantially by chloride ions in the 1N H_2SO_4. The chloride content of the film increased when chloride ion was added but did not vary with solution chloride ion content. The concentration of the other elements in the film did not vary with the chloride content in the solution. This is apparent from Figure 18 where it is seen that the Ni/Cr ratio did not change as the chloride ion content varied from 0 to 3%.

The ionic conductivity of the film must increase by increased chloride ion content as evidenced by the polarization data. It is difficult, however, to assign this phenomenon to penetration of the film by the chloride ion as predicted as Hoar.[5] There was no pitting at these concentrations so therefore the film may be stable and resistant to these levels of chloride ions.

CONCLUSIONS

This paper can be considered to be a progress report of a study of the passive film. The data do indicate interesting facts concerning the film on nickel and Inconel X-750. Additional work must be done to fully elucidate the nature of the film. Several conclusions can be made from this study.

1. The passive film on nickel-200 formed in 1N H_2SO_4 was 40 to to 60 Å as determined by AES. When chloride ion was added to the 1N H_2SO_4, the film was apparently thicker.

2. The film formed on Inconel X-750 when in 1N H_2SO_4 was somewhat greater than 90 Å and was dependent on potential. It was thicker as the potential was made more noble. There was little difference in film thickness due to chloride ion in the sulfuric acid.

3. The composition of the film on nickel-200 was a nonstoichiometric compound of nickel, oxygen and sulfur. It is difficult to call this a sulfate; chlorides were not included in the film as a function of solution composition.

4. The composition of the film on Inconel X-750 was dependent on potential. The nickel/chromium ratio varied with potential, and was lower than the ratio in the bulk alloy. The film may therefore

consist of a mixture of nickel and chromium oxides. Chloride ion seems to be included in the film as a function of solution composition. The thickness and composition of the film were not changed by the chlorides when at concentrations that did not cause pitting of the metal.

REFERENCES

1. H.H. Uhlig, *Corrosion and Corrosion Control*, 2nd Ed, John Wiley, New York (1971).
2. L.L. Shreir, *Corrosion*, V.1, G. Newnes, London, 1962.
3. J.O'M. Bockris and A.K.N. Reddy, *Modern Electrochemistry*, V.2, Plenum Press, New York, 1970.
4. J.O'M. Bockris, A.K.N. Reddy and B.Rao, J. Elec. Soc., $\underline{113}$, 1133 (1966).
5. T.P. Hoar, Corr. Sci., $\underline{7}$, 341 (1967).
6. N.Sato and G. Okamoto, J. Elec. Soc., $\underline{110}$, 605 (1963).
7. E. Kunze and K. Schwabe, Corr. Sci., $\underline{4}$, 109 (1964).
8. T. Tokuda and M.B. Ives, Corr. Sci., $\underline{11}$, 297 (1971).
9. G.Okamoto, Corr. Sci., $\underline{13}$, 471 (1973).
10. J.B. Lumsden and R.W. Staehle, Scripta Metalurgia, $\underline{6}$, 1205 (1972).
11. R.F. Pascoe, *Induction of Localized Corrosion on a Nickel Alloy*, Ph.D. Dissertation, The University of Oklahoma (1973).
12. B.L. Trout and R.D. Daniels, Corrosion, $\underline{28}$, 9 (1972).
13. D.L. Piron, E.P. Koutsoukos and K. Nobe, Corrosion, $\underline{25}$, 156 (1969).
14. A. Vijh, Corr. Sci., $\underline{11}$, 161 (1971).
15. R.W. Revie, B.G. Baker and J.O'M. Bockris, J. Elec. Soc., $\underline{122}$, 1460 (1975).
16. J.B. Lumsden and R.W. Staehle, "Application of AES to Commercial Iron Base Alloys," Private Communication.
17. M. Seo, J.B. Lumsden and R.W. Staehle, Surface Sci., $\underline{42}$, 337 (1974).
18. O. Rincon, *Effect of Chloride Ion on the Passive Film on Nickel-200 and Inconel X-750*, Masters Thesis, University of Oklahoma, (1975).
19. M. Afzal, *The Study of the Passive Film on Nickel-200 and Inconel X-750*, M.S. Thesis, The University of Oklahoma, (1975).

Discussion

On the Paper by R.L. Park,
M. den Boer and Y. Fukuda

J.D. Andrade (*University of Utah*): Please comment on the use of low energy electrons and photons to probe the density of states or molecular orbitals at a surface.

R.L. Park (*University of Maryland*): Low energy electrons and photons probe essentially the same states. The principal advantage of photons as a probe is in the study of fragile surfaces where electron beams may be too destructive.

D.E. Williams (*Dow Corning Corp.*): You have commented that peak intensities are affected both by the surface elemental composition and by the depth distribution of those elements and that hence one should not expect ESCA or Auger to give meaningful surface compositions from empirical "sensitivity factors". Is it your view that peak intensities are of no value to obtain some insight on surface composition unless the depth distribution is known?

R.L. Park: We all use peak intensities to give us some notion of elemental abundance. If we are really interested in chemical analysis, however, we should be trying to understand the secondary features of the spectrum. Line shapes and chemical shifts provide information on chemical bonding, which may be more informative than peak intensities. The straggling tail on the low energy side of the peak may provide information on depth distributions.

D.M. Hercules (*University of Georgia*): It seems to me that your remarks about the non-quantitative character of surface techniques need some qualification. You are correct that absolute analyses using ESCA and AES are difficult but not so for relative measurements. Few spectroscopic technique yield quantitative analyses without prior calibration. Perhaps, the problems here revolves around the meaning of the word "quantitative". Is a measurement of ± 100% (relative) quantitative if it cannot be performed at all by other techniques? If so, then I submit these techniques are quantitative.

R.L. Park: I do not believe the meaning of "quantitative" is what is at issue. Rather, it is the definition of "surface composition". To describe surface composition in terms of elemental percentages does not make sense unless we can agree on what region we are talking about. If a substrate of element B is covered by a monolayer of element A, is the surface 100% element A? That is not what my instruments will tell me. The problem is that the surface is necessarily inhomogeneous along its normal and its composition cannot be given independent of its structure.

On the Paper by I.M. Stewart

Participant: Can Auger analysis be performed in the electron microprobe as well as X-Ray analysis?

I.M. Stewart (*W.C. McCrone Associates*): Auger analysis is not a practical possibility in a conventional electron microprobe as the vacuum system of the latter is not normally good enough to prevent contamination forming on the surface of the sample. Auger analysis has been added to some SEMs with high vacuum systems with some measure of success from areas down to about 1μm.
Comment from the floor (as I recall it). Scanning Auger has been demonstrated with resolution of better than 1000Å

On the Paper by J.T. Grant

L.H. Lee (*Xerox Corp.*): I would like to thank Dr. Grant for this thorough review of the subject matter. There have been many review papers on Auger spectroscopy, but Dr. Grant has tried especially to bring the survey up to date in a readily readable manner.

PART III

Low-Energy Electron Diffraction

Introductory Remarks:

David L. Allara

Bell Telephone Laboratories
Murray Hill, New Jersey   07974

The premier research tool for characterization of the geometrical arrangement of atoms on a solid surface is Low Energy Electron Diffraction (LEED). In principle, a LEED experiment can provide quantitative structural information. However, in practice only very simple surface structures can be analyzed and the interpretation of spectra, particularly intensity patterns, usually involves exceedingly difficult model calculations. Because of the recent explosion of interest in surface structure and chemisorption, it is clear that the development of techniques for interpreting LEED data is of great importance. The papers presented in this session address this problem.

Of particular significance, these papers demonstrate current approaches to dealing with intensity patterns and include both direct interpretation and model calculations. These recent developments leave one hopeful that eventually LEED spectroscopy will be useful in mapping out surface structure in more complex systems such as catalysts, but at present it is clear that a great deal of effort is needed to make LEED a more routine tool for quantitative analysis.

Plenary Lecture: LEED Studies of Surface Layers*

Peder J. Estrup

*Department of Chemistry and Department of Physics
Brown University
Providence, R. I. 02912*

At the present time low energy electron diffraction, LEED, is probably the most important experimental technique for the investigation of surface structures. LEED is primarily a research tool in fundamental studies since it requires ultra-high vacuum conditions and single-crystal samples. However, if these requirements are met, LEED has the potential of determining the positions of all the atoms in the topmost layers of the crystal. A complete structure analysis has so far been carried out for relatively few surfaces, but a determination of just the two-dimensional periodicity provides very useful information about phenomena such as surface reconstruction, surface phase transitions, and surface adsorption and reactions.

INTRODUCTION

During the last two decades, a series of experimental techniques have been developed for the study of solid surface phenomena on the atomic and molecular level. Many of them are the subject of reviews or surveys in this volume. In favorable circumstances, the application of these techniques permits the characterization of a surface to an extent comparable to that attainable for bulk materials, i.e., they give a detailed description of the chemical composition, the crystallographic structure, the electronic and thermal properties, etc. Genuine progress in the fundamental understanding of surface phenomena has resulted from the experimental study of surfaces which have been defined in this manner.

---

*Supported by the Materials Research Program at Brown University, funded through the National Science Foundation.

It is well known that the physical properties of a surface depend strongly on its <u>structure</u>. Surface planes of different crystallographic orientation usually show differences in, for example, work function, vibrational amplitude of the surface atoms, and electron density of states. It is not surprising, therefore, that in studies of surface chemical bonding and surface reactions each crystal plane must be considered a distinct chemical reagent. Since defects and impurities also may have a profound effect on the behavior, a complete elucidation of a process occurring on an ordinary surface represents a formidable problem. Of necessity most of the current research therefore involves surfaces of single crystals, prepared to have a particular orientation and to contain a minimum of imperfections. The macroscopic structure, i.e., the surface morphology, is thus considered to be eliminated as a variable, and the structural investigations can be directed towards the determination of atomistic models of the surface geometry.

The most widely used method for surface studies of this type is low energy electron diffraction, LEED. In many respects, this method is comparable to x-ray diffraction of a bulk material; in a crystal the regular arrangement of atoms represents a grating to the electrons and diffracted beams appear at angles which depend on the interatomic spacings. However, unlike x-rays, electrons with a wavelength $\lambda \sim 1$ Å can penetrate only a few atomic layers without undergoing inelastic collisions, and LEED is therefore specific to the surface of a solid. The observed diffraction pattern is thus determined by the two-dimensional (2D) periodicity of the surface plane. A number of theoretical and experimental difficulties have delayed the development of LEED to its full potential but the method, nevertheless, is the most important source of data on the structure of surface layers. A bibliography of LEED studies up to 1970 lists more than 700 papers[1], and this number has at least doubled since then. The field has been reviewed frequently in recent years[2].

Early LEED investigations[2h,2i] were done without the benefit of the complementary methods which later became available, and the interpretation of the results were in many cases rather speculative. In current work, it is considered essential to apply several measuring techniques simultaneously to a given surface; for example, Auger Electron Spectroscopy (AES)[3] is an indispensable adjunct to LEED to monitor the surface composition. Photoemission, in particular by means of ultra-violet radiation (UPS)[4], work function measurements, inelastic electron spectroscopy and flash desorption are among the other methods which are readily combined with LEED. They provide information about the electronic structure and the bonding of the surface atoms.

The discussion in the following sections centers on LEED but it should be noted that a number of other experimental probes are available for surface structural studies, e.g., high energy electrons, ions and neutrons. Generally speaking, they do not match the versatility, flexibility and relatively low cost of LEED, at the present time, but they offer a very valuable alternative in special cases.

The technique of RHEED (reflection high energy electron diffraction)[5] employs electrons in the 50 keV range at glancing incidence (a few degrees) to the surface. Since the penetration into the solid depends on the electron momentum perpendicular to the surface, the surface sensitivity can be made comparable to that of electrons in a LEED beam which typically have an energy of only about 100 eV but which are normal to the surface. The 2D lattice parameters can be determined quite precisely from a RHEED pattern[5], and the theoretical treatment of the diffracted intensity[6] is likely to be less complicated than in the case of LEED. However, these advantages tend to be offset by stringent requirements regarding surface perfection; when the primary electron beam is incident at glancing angle the effects of surface steps and facets become very important. It has been suggested[7] that such difficulties can be avoided by use of electrons with intermediate energy and larger angle of incidence (MEED).

Measurements of the backscattering of mono-energetic ions from a solid can, in principle, give information about both the chemical identity and the geometrical arrangement of the outermost surface atoms. Noble-gas ions of an initial energy of $\sim$ 1 keV are suitable for this purpose. The energy spectrum of the ions which have undergone binary scattering provides a mass-analysis of the target atoms[8], and by observation of the angular dependence of the scattering it is possible to investigate the relative position of different surface species on single-crystal samples[9]. Ions in the MeV range have also been used with some success in surface studies[10]; the scattering can be treated as a "channeling" process, as is done in bulk studies.

Very recently it has been demonstrated that neutron diffraction can be used to study the structure of surfaces[11]. Neutrons are strongly penetrating particles and will not ordinarily be sensitive to surface properties. However, diffraction from the surface becomes detectable when an overlayer of strongly scattering atoms is deposited on a relatively transparent adsorbent. Among the advantages of a neutron beam as a surface probe is the absence of any significant attenuation in the gas phase even at quite high pressures and the potential for studies of the dynamic properties of the surface layer by inelastic scattering. So far the method has been applied only to physisorbed layers on graphite (Grafoil)[11].

Field ion microscopy[12] and other experimental techniques are discussed elsewhere in this volume.

ELEMENTS OF LEED

The elements of a LEED experiment are illustrated in Figure 1. The primary (incident) electrons are given a kinetic energy E eV, numerically equal to the acceleration voltage V. The corresponding electron wavelength is obtained from the deBroglie relationship and can be calculated by

$$\lambda \simeq \sqrt{\frac{150}{E}} \text{ Å}$$

Thus, acceleration by a potential of 150 volts leads to a wavelength of 1 Å.

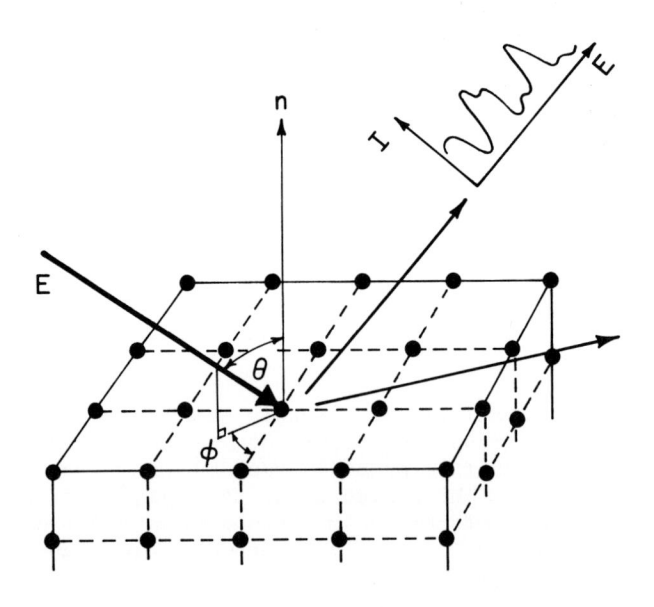

Fig. 1. Schematic diagram of the LEED geometry. The primary electron beam has energy E and its direction is defined by the angles θ_i and ϕ. Two diffracted beams are shown and the I-E curve for one of them is indicated.

The direction of the incident beam is defined by the polar angle θ_i relative to the surface normal and by the azimuthal angle ϕ relative to a major crystallographic axis in the surface plane. The incident electrons are scattered by the surface atoms, and the scattered waves interfere constructively along certain directions to produce diffracted beams. The resulting diffraction pattern is determined by the two translation vectors which characterize the 2D surface net[2f], in the same way that an x-ray diffraction pattern is determined by the geometry of the bulk lattice via Bragg's law. For example, if the primary beam is at normal incidence ($\theta_i = 0$), the diffracted beams appearing at $\phi' = 0$ satisfy the condition

$$a \sin \theta_r' = n\lambda$$

where a is the appropriate 2D lattice parameter, θ_r' is the angle of "reflection" and n is an integer. (In the general case, the diffraction conditions are found most conveniently by an Ewald sphere construction[2f].) A change in the periodicity of the clean surface (the "substrate") will obviously cause a change in the diffraction pattern. Adsorption of foreign atoms ("adatoms") frequently has this effect; if, for example, an adatom is placed on every other substrate atom, the magnitude of the translation vector is doubled and the new diffraction condition is

$$2a \sin \theta_r' = n\lambda$$

This condition is satisfied by twice as many values of $\sin \theta_r'$, and the adsorption thus produces a diffraction pattern with twice as many beams. A more detailed discussion of such changes is given below, but it is important to realize that in general a diffraction pattern can give only the dimensions of the surface unit cell and not the position of the atoms inside the cell. Thus, if all the adatoms were shifted by the same distance relative to the substrate, the diffraction pattern would not change. The *intensity* of the diffracted beams would change, however, and by means of an intensity analysis it should therefore be possible to determine the surface structure completely.

For a given structure, the intensity I of a diffracted beam varies with the electron energy E, the angles θ and ϕ, and the temperature. Experimental intensity data may be recorded in different ways[2d], but they are usually presented in the form of I-E curves (see Figure 1) which give I as a function of E at fixed angles of incidence. Whereas a diffraction pattern ordinarily can be interpreted without difficulty, the I-E curves (or "intensity profiles") are much harder to analyze. If weakly interacting radiation - such as x-rays - were used, the shape of the I-E curves could be predicted from kinematical theory, i.e., the intensity could be calculated by summing the single-scattering amplitudes over the atoms in the crystal. An I-E curve would then consist of sharp spikes, each spike being produced when waves from different layers were in phase, and thus the interlayer spacing could be found from the location of the spikes. In actual fact, the LEED electrons are turned around in collisions with atom cores in the first few layers near the surface, and since the cross-sections for elastic and inelastic scattering both are large, the intensity is greatly affected by multiple scattering and attenuation of the electrons[2d]. A substantial effort has been given to the development of an adequate theoretical treatment of these processes so that the structural information can be extracted from LEED intensity data. Considerable success has been achieved recently[2a], and a number of surface structures have now been worked out in detail. However, the procedures require sophisticated computer calculations and cannot yet be done routinely. The problems involved are discussed further in other articles in this volume[13].

A schematic diagram of a LEED apparatus is shown in Figure 2. The ultra-high vacuum (UHV) chamber can be maintained at a base pressure of about 10^{-10} torr so that rapid adsorption of impurities by the crystal surface can be prevented. The instrumentation includes a sputtering gun for cleaning by ion bombardment, and either the LEED gun or a separate, glancing incidence electron gun may be used to excite Auger electrons from the sample for analysis of the surface composition.

Bombardment of the crystal by mono-energetic electrons causes the emission of "secondary" electrons with an energy spectrum which may be obtained by measuring the transmitted current as a function of the retarding voltage applied to one of the grids. The spectrum typically has a smooth background with peaks corresponding to the elastic electrons, to electrons that suffered characteristic losses, and - if the primary energy is sufficiently large - to Auger electrons. In LEED experiments, only the elastic component is transmitted, and the electrons are post-accelerated for display on the fluorescent screen. Figure 3 shows a photograph of such a display system. The diffraction pattern can be observed directly, and the intensity of an individual diffracted beam is obtained by measuring the spot brightness with a spot-photometer placed outside the chamber.

Another way of detecting the electrons is to use a Faraday cup which can be moved to any position in front of the screen. If this device is available, the angular distribution of the inelastic electrons and of the Auger electrons may also be measured[14].

LEED systems of the type shown in Figure 3 have been available commercially for about ten years and have been used in the majority of LEED studies carried out to date. It seems likely that future work will require some modifications in the equipment in at least two respects. First, an absolute determination of a 2D lattice parameter depends on values of the diffraction angles, and in the present systems these are difficult to measure with an accuracy better than ∿ 5%. It is usually assumed that in the plane of the

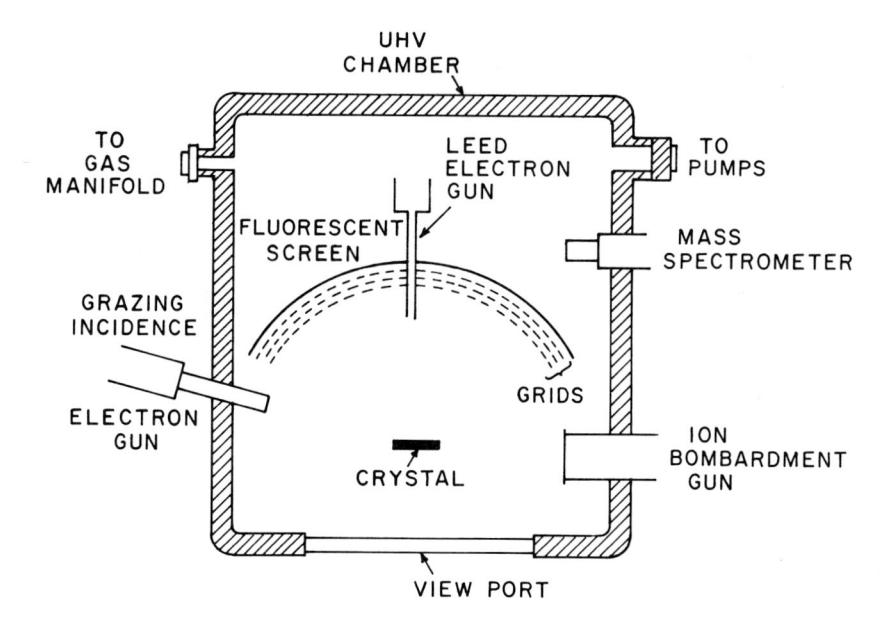

Fig. 2. Schematic drawing of the LEED apparatus. Other components, such as a UV lamp for photoemission studies, may be added.

surface the lattice parameters are identical to those of a plane in the bulk crystal (or simple multiples hereof) and that x-ray values therefore can be used. However, some careful experiments[15] for surfaces with an adsorbate indicate that this assumption is not always justified. Second, the standard procedure for acquiring intensity data is difficult and time consuming; it may take weeks to get the I-V curves needed for the analysis of just a single surface structure. New methods using image intensifiers, vidicon tubes, channel plates and photographic techniques are therefore under investigation[16].

INTERPRETATION OF LEED PATTERNS

As an example of the interpretation of a LEED pattern a discussion will be given of one of the patterns produced when oxygen adsorbs on a molybdenum (100) surface. It is the spot pattern

Fig. 3. Photograph of the LEED apparatus. The sample is mounted on a vertical crystal holder and can be rotated from outside the UHV chamber. The sample may be cooled by liquid nitrogen or heated resistively. A diffraction pattern is displayed on the phosphor screen behind the crystal. The attachment on the upper left is the grazing incidence gun for AES. On the upper right is a quadrupole mass spectrometer and below it is the sputtering gun for sample cleaning.

appearing on the screen in Figure 3 and also shown in the diagram of Figure 4b.

Diffraction from a clean Mo(100) surface gives a pattern which contains only the "normal" spots, i.e., the pattern is that expected from a surface with the same periodicity as a parallel layer in the bulk crystal. Some of the beams are indicated in Figure 4a, namely the specular or 00 beam surrounded by the first order beams with indices as shown. This figure also represents a part of the

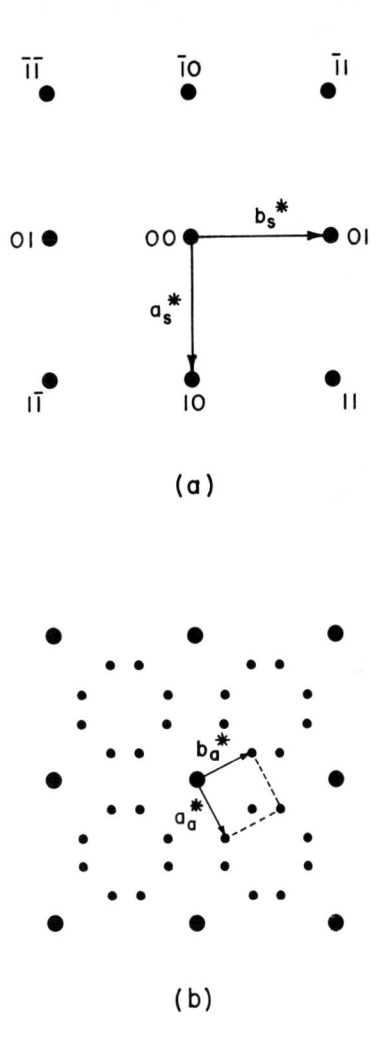

Fig. 4. (a) Diagram of a spot pattern. The filled circles represent the diffraction beams from a (100) surface of a cubic crystal. a_s^* and b_s^* are the reciprocal lattice vectors for the clean substrate. (b) Diagram of the spot pattern appearing on the screen in Figure 3. The extra spots of 1/5 order are due to an overlayer of adsorbed oxygen. Half of these spots can be generated by translation of the vectors a_a^* and b_a^*. The remaining spots may be generated by reflection through a vertical line.

reciprocal net for the substrate under consideration, a net which is generated by the reciprocal translation vectors a_s^* and b_s^*. The corresponding lattice in real space is a 2D square net defined by the vectors a_s and b_s which obey the relations[17]

$$a_s \cdot a_s^* = b_s \cdot b_s^* = 1$$

$$a_s \cdot b_s^* = b_s \cdot a_s^* = 0$$

The magnitude of the vectors is given by

$$a_s^* = \frac{b_s}{A} \quad \text{and} \quad b_s^* = \frac{a_s}{A}$$

where A is the area of the unit cell, i.e.,

$$A = |a_s \times b_s|$$

For molybdenum (100), the table value of the lattice parameter (measured by x-ray diffraction) is $a_s = b_s = 3.15$ Å, and hence $a_s^* = b_s^* = 0.32$ Å^{-1}.

The adsorption of oxygen leads to a new surface periodicity as evidenced by the appearance of "extra" spots in the LEED pattern, Figures 3 and 4b. The extra spots can be indexed n/5 m/5, where m and n are integers, and clearly the new reciprocal net is related to the original in a relatively simple manner. The presence of fifth order spots might suggest a new unit cell with lattice vectors $a_a = 5a_s$ and $b_a = 5b_s$, i.e., a (5x5) overlayer structure[2f]. If the new unit cell were primitive, this structure would contain one adsorbed species for every twenty-five substrate atoms and the coverage would therefore be $\theta = 0.04$ monolayers. AES data indicate that θ is much larger. The model can also be rejected on the basis of the LEED pattern alone; many of the expected fifth order spots are absent in the pattern for all values of E (for example $0\,\frac{1}{5}$, $\frac{1}{5}\,\frac{1}{5}$ and $\frac{1}{5}\,0$ are missing), and the unit cell would therefore have to be non-primitive and have the atoms inside the unit arranged in a manner to cause the observed extinctions[2a,f]. Although the possibility of finding a unit cell with the required geometrical structure factor cannot be ruled out immediately, there is a much simpler way to account for the pattern.

Quite often more than one structure is present on a surface, and the LEED screen will then show the superposition of the corresponding patterns. In the present case, the assumption of a twinned structure appears to be consistent with the data. As illustrated in Figure 4b, half of the spots in the observed pattern correspond to a reciprocal net generated by the vectors a_a^* and b_a^*. a_a^*

connects the origin 00 with the point $\frac{2}{5}\frac{1}{5}$, while $\underline{b_a}^*$ terminates at $\frac{1}{5}\frac{2}{5}$. Thus, the magnitude of the reciprocal translation vectors for the adsorbate layer is given by

$$a_a^* = b_a^* = \frac{\sqrt{5}}{5} a_s^*$$

where, as before, the subscript s refers to the substrate net.

The area of the reciprocal unit cell is

$$A_a^* = |a_a^* \times b_a^*| = \frac{1}{5}(a_s^*)^2 = \frac{1}{5} A_s^*$$

In real space, the overlayer translation vectors have magnitude

$$a_a = b_a = \frac{b_a^*}{A_a^*} = \sqrt{5}\, a_s$$

Since a_a is perpendicular to b_a^* and b_a is perpendicular to a_a^*, the overlayer net in real space can be drawn as shown on the left in Figure 5. The structure is denoted $(\sqrt{5} \times \sqrt{5})$ R (26°.6) since $\frac{a_a}{a_s} = \frac{b_s}{b_a} = \sqrt{5}$, and since the overlayer net is rotated by 26°.6 relative to the substrate net. It is clear that if the adsorbed oxygen has a tendency to produce this geometry, the twin structure $(\sqrt{5} \times \sqrt{5})$ R (-26°.6), shown on the right in Figure 5, is equally likely to be produced. The incident LEED beam has a cross-section of the order of 1 mm^2 and if within this area there exists many nucleation centers during the formation of the overlayer, the two structures will be produced in equal abundance.

It may be noted that the translation vectors are related by

$$\begin{bmatrix} \underline{a_a} \\ \underline{b_a} \end{bmatrix} = \begin{bmatrix} 2 & 1 \\ -1 & 2 \end{bmatrix} \begin{bmatrix} \underline{a_s} \\ \underline{b_s} \end{bmatrix}$$

and that the overlayer structure equally well can be specified by means of the matrix $\begin{bmatrix} 2 & 1 \\ -1 & 2 \end{bmatrix}$. The twin can similarly be specified by the matrix $\begin{bmatrix} 2 & -1 \\ 1 & 2 \end{bmatrix}$.

The proposed model (Figure 5) may be tested and refined if additional data can be obtained. Oxygen is believed to dissociate

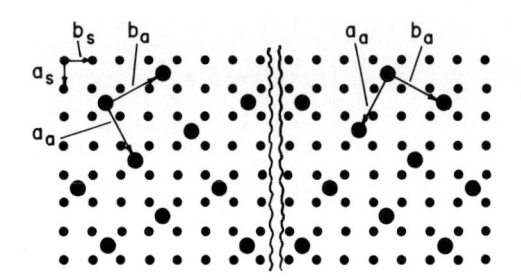

Fig. 5. Model of a surface structure giving the pattern shown in Figure 4b (and in Figure 3). The small filled circles represent substrate atoms (Mo); the large open circles are the adatoms (oxygen). $\underline{a}_s$ and $\underline{b}_s$ are the translation vectors of the substrate lattice; $\underline{a}_a$ and $\underline{b}_a$ are the translation vectors of the overlayer. The two domains which are shown are "twins" and are related by a reflection. A super-position of the diffraction patterns from these two structures produces all the observed spots.

on most metals, and if the unit cell is correctly identified the coverage should be 0.2 monolayer O atoms. (In the case of molecular adsorption, the apparent value of θ would be twice as large.) The consistency of the model can therefore be checked by AES measurements of the coverage[18]. As mentioned earlier, with rare exceptions[21], the location of an individual adatom relative to the substrate atoms cannot be deduced from the LEED pattern. It is reasonable to suppose that an O atom will occupy a site of high symmetry[19], e.g. co-ordinated to four surface Mo atoms as shown in Figure 5, directly on top of a Mo atom, or in a bridged position between two Mo atoms. For oxygen adsorption, it might be possible to distinguish between these possibilities by means of electron loss spectroscopy[20], by electron impact desorption[21], and by photoemission[22], but a complete structure determination, including the Mo-O bond lengths, would require a LEED intensity analysis. Such an analysis should also allow for a possible rearrangement of the surface Mo atoms, and many different structural models would therefore have to be examined.

Detailed computations of this type have so far been carried out only for structures characterized by a small unit cell[2a,13]. The corresponding LEED patterns are simple, and the surface periodicity is obtained almost by inspection. Figure 6 gives some examples, all involving overlayers on the (100) surface of a cubic crystal. The arrangement in Figure 6a may be described by the matrix $\begin{bmatrix} 2 & 0 \\ 0 & 2 \end{bmatrix}$ but is usually called a p(2x2) structure, p being included to emphasize that the unit cell is primitive. The LEED pattern from this surface, Figure 6b, has extra, half order spots. Figure 6e shows a frequently occuring overlayer geometry which almost always is referred to as c(2x2) structure; the letter c means that the unit cell has an additional scatterer in the center. It would be more correct, however, to identify the structure by ($\sqrt{2}$ x $\sqrt{2}$) R(45°) or by $\begin{bmatrix} 1 & 1 \\ -1 & 1 \end{bmatrix}$. It gives rise to ½ ½ spots in the diffraction pattern, Figure 6f. Similarly, the $\begin{bmatrix} 2 & 1 \\ 0 & 2 \end{bmatrix}$ arrangement in Figure 6c is described as a c(4x2) structure in the literature. In this case, the overlayer has a lower symmetry than the substrate, but nevertheless, the LEED pattern is expected to exhibit fourfold symmetry since most probably the c(2x4) structure will be formed along with the c(4x2).

It is characteristic of all the structures discussed above that the overlayer is in _registry_ with the substrate net. Even when the site remains undetermined, it can be concluded from the LEED pattern that the bonding at the surface creates a distinct geometrical relationship of the adatom position to the position of the substrate atoms[19]. This is not always the case, however. Surface layers may be observed for which the registry is weak ("coincidence lattices") or even absent ("incommensurate" or "incoherent" structures). In the latter case, the extra spots due to the overlayer cannot be indexed by rational fractions[2f].

A LEED pattern is relatively insensitive to structural imperfections[2f,23]. The resolution of the LEED instrument sets a lower limit for the angular width of a diffracted beam, i.e. on the size of a spot. To cause an additional broadening of the beam profile, there must be many defects in an area having dimensions comparable to the _coherence zone_[24]. This is the region over which the primary electron wave is effectively coherent, and for low energy electrons it has a radius of 10^2-10^3 Å. Certain types of disorder are readily identified, however. For example, thermal vibration of the surface atoms[24] makes the diffracted beams weaker (Debye-Waller effect), increases the background intensity, and changes the peaks in the I-E curves. In the case of overlayer structures, a sharp LEED pattern can be expected only within a certain range of coverage

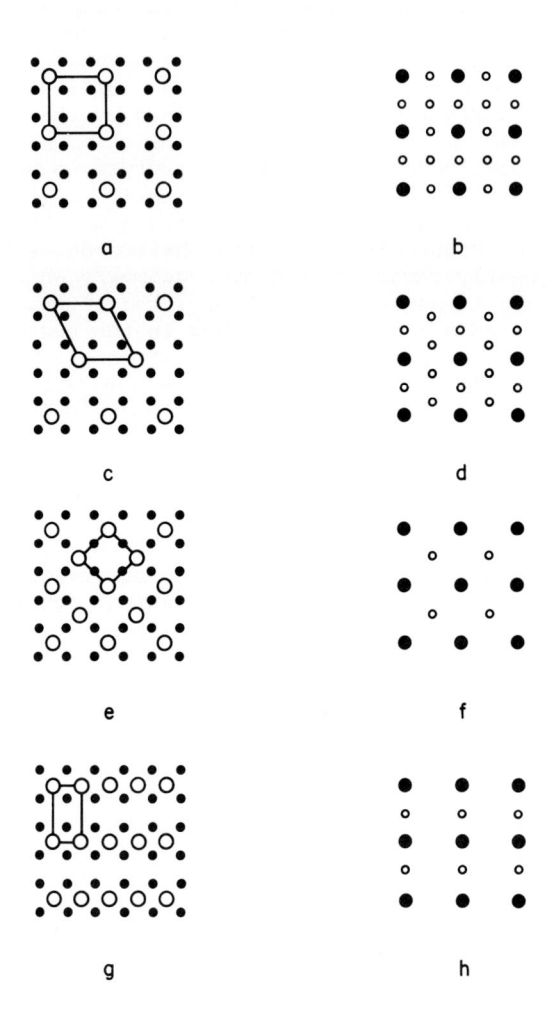

Fig. 6. Examples of simple overlayers on a square net (i.e. on a (100) surface) and the corresponding spot patterns.
- (a) p(2x2) structure. The dots represent the substrate atoms, the open circles are the adatoms.
- (b) Spot pattern from a p(2x2) structure. The filled and empty circles represent normal and extra spots respectively.
- (c) c(4x2) structure.
- (d) c(4x2) pattern.
- (e) c(2x2) structure.
- (f) c(2x2) pattern.
- (g) (2x1) structure
- (h) (2x1) pattern.

and temperature. Outside this range, the pattern may show features such as diffuse (or split) spots or streaks. The patterns may be interpreted by the methods used in the study of disordered solids by x-ray diffraction[2,25,26], and quantitative measurements of the beam profile (as well as the intensity) can therefore be useful in characterizing the nature of the interactions[27].

APPLICATIONS OF LEED

LEED should not be considered as a technique for routine surface analysis but rather as a tool applicable to a variety of investigations of a fundamental nature[2]. Among the problems studied by LEED is surface reconstruction and surface atom vibrations for a clean substrate, and the long-range ordering of the surface of an alloy or a ferromagnetic material. The most important application, however, is to the development of models for adsorption and surface reactions. Only a brief discussion of some of these problems will be given here.

Clean Surfaces

Because of the inherent asymmetry of the surface, the atoms in the topmost layers (the "selvedge"[28]) cannot be expected to occupy exactly the same positions they would have in an infinite crystal. LEED studies have shown that changes both in the 2D (lateral) periodicity and in the interlayer distance (perpendicular to the surface) may occur. The formation of a new 2D periodicity is called <u>reconstruction</u> in the LEED literature, and it is readily observed in the corresponding spot pattern. The phenomenon appears to be common among covalent materials, and it is of particular interest in studies of semiconductor surfaces. In the case of Si(111), for example, it has been found that, depending on the thermal treatment of the sample, the surface may have (2 x 1), (7 x 7) or (1 x 1) periodicity, and that there is a strong correlation between the surface structure and the electronic properties[29]. Definitive models of the atomic arrangement are not yet available, and the nature of the structural transformations is still under investigation[30]. Reconstruction has also been observed for at least two metals, namely Au and Pt; a (5 x 20) structure is observed on the (100) face of both of these substrates after cleaning[31]. It has been speculated that small amounts of impurities are needed to stabilize the superstructure, but AES and other techniques have failed to detect any foreign species.

Data concerning the interlayer distance have accumulated more slowly since, as mentioned earlier, the determination of this parameter requires a LEED intensity analysis. Nickel is one of the substrates for which detailed studies have been made[2a-d]. The comparison of model calculations with experimental I-E curves indicates[32] that at the (100) surface the interlayer spacing is close to the bulk value but that a contraction takes place at both the (110) and the (111) surface, amounting to 5% and 2.5%, respectively, of the bulk spacing. These values may be typical for face-centered cubic metals[2a]. On the other hand, at the (100) surface of molybdenum, which is body-centered cubic, a contraction of 11-12% has been reported[33]. Considerably more work is needed, however, before any trends can be firmly established.

By studying the effect of temperature on the diffracted intensity, information may be obtained regarding the vibrational amplitude of the surface atoms[2b,2d,24]. The temperature dependence enters via the Debye-Waller factor, and kinematic (single-scattering) diffraction theory[26] provides an explicit expression for I(T) which shows that

$$\ln I \propto \left[\frac{T}{\theta_D^2}\right] \propto \langle u^2 \rangle$$

where θ_D is the Debye temperature. This parameter can, in turn, be related to $\langle u^2 \rangle$, the mean square vibrational amplitude of the scatterer in a direction determined by the diffraction geometry. In the case of LEED, complications arise because of the multiple scattering of the electrons and because of the increase of the penetration depth with electron energy. The qualitative results are clear, however; θ_D(surface) - determined by LEED - is significantly smaller than θ_D(bulk) - determined for example by x-ray diffraction. Nickel is again a substrate for which detailed studies have been performed[24,32], and the analysis of LEED data show that θ_D(surface) $\sim$ 300°K as compared to θ_D(bulk) $\simeq$ 420°K, which means that the mean square vibrational amplitude is approximately twice as large at the surface.

Few investigations of these types have been carried out with substrates other than elemental crystals. The properties of compounds are of no less interest, but the difficulty of obtaining a well-defined surface is considerably greater for these systems since the surface composition must be controlled as well. Studies of alloys[34] have indicated some of the problems involved, even when techniques as powerful as AES are available.

Although LEED is most effective in studies of low index crystallographic planes, it is noteworthy that the method has been used successfully in the characterization of surfaces containing a regular array of steps[35]. This is of considerable importance in studies of chemisorption and catalysis, since it appears that the sites at the edges of a step may endow the surface with special chemical properties[36]. In general, work on stepped surfaces should help to show if the "idealized" surfaces, used in LEED, are representative of "real" surfaces.

Adsorbed Layers

Adsorption phenomena are of fundamental importance in the physics and chemistry of surfaces, and LEED is playing a major role in the development of a microscopic description of adsorbed layers.

Some of the characteristic features of adsorption may be discussed with the aid of Figure 7 which shows the potential energy of an atom as it approaches the (clean) solid. The atom becomes trapped in a potential well of depth E_d; since this process is spontaneous and lowers the entropy, it must be exothermic. The potential energy variation may of course be considerably more complicated than indicated here[37]; the adsorption may have an activation energy and a precursor state may be formed initially. Additional changes must be made in the curve if the adsorption involves a molecule which dissociates. However, for the present purpose the important feature is that the adsorbate is confined to a layer parallel to the surface by barriers against both desorption and sorption (i.e. penetration into the solid).

The adsorbed state will be stable at low temperature. If the energy barriers are as shown in Figure 7a, sorption will be important at intermediate temperatures. An example is the interaction of carbon with a Ni(111) surface for which $E_d \sim 7.5$ eV and $E_s \sim 7$ eV[38]. Below $\sim 1200°K$, carbon segregates at the surface, but above this temperature the adsorbed state is depleted and a dilute solid solution is formed. However, most adsorption studies have been done on systems which are better described by Figure 7b. To a good approximation sorption can be ignored, and the surface coverage θ is determined only by the rates of adsorption and desorption. The adsorption rate is given by

$$r_a = \left[\frac{sp}{\sqrt{2\pi mkT}}\right]$$

where p is the gas phase pressure; T the temperature; s the sticking probability, and k Boltzmann's constant. For a gas like nitrogen at 10^{-6} torr and $300°K$ and with $s \sim 1$, the rate is of the order of 1 monolayer per second (1 monolayer $\simeq 10^{15}$ atoms/cm^2). The desorption rate may be written as

$$r_d = \nu \theta^\alpha \exp\left[-\frac{E_d}{kT}\right]$$

where ν is a frequency factor and α is the order of the kinetics. It is seen that if $E_d \gg kT$ the adsorption will appear to be irreversible; the process will continue until the surface is saturated with adsorbed species. On the other hand, if $E_d \ll kT$, no adsorbed layer will build up. Using typical values of the parameters ($\nu = 10^{13}$ sec^{-1}, $\theta = 0.5$, $\alpha = 1$, $s = 0.5$, $p = 10^{-8}$ torr), the

(a)

(b)

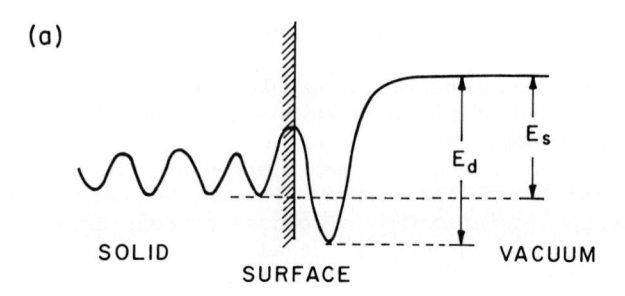

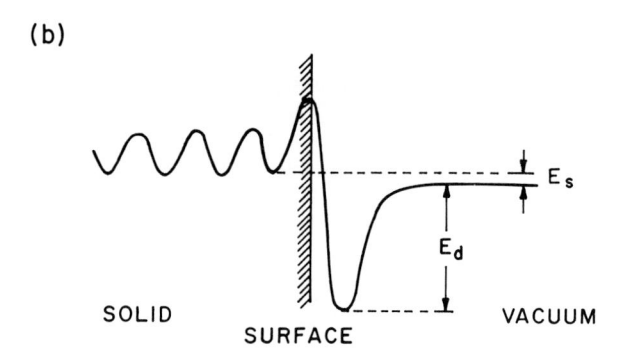

Fig. 7. Simplified diagrams illustrating the potential energy of an atom at the solid-gas interface. E_d is the barrier against desorption. In (a) the formation of a solid solution is expected at intermediate temperatures. In (b) sorption by the bulk can be ignored.

temperature at which the equilibrium $r_a = r_d$ obtains is estimated to be $T_e \sim 700°K$, assuming E_d to be 1 eV. An adsorbate for which E_d is of this magnitude is sometimes called "weakly chemisorbed", and in this case the equilibrium region is readily explored experimentally. This makes it possible to determine the adsorption isostere (i.e. the set of (p,T) values which gives the same coverage θ) and hence to find the (isosteric) heat of adsorption, q

$$\left[\frac{d\ln p}{dT}\right]_\theta = -\frac{g}{RT^2} \simeq \frac{E_d}{RT^2}$$

where θ is monitored by AES, work function measurements and LEED. The procedure is exemplified by results for CO on Ni(100) for which E_d has been measured over the whole range of θ[39].

Similar experiments are difficult to do if E_d is much smaller or much larger. For a physisorbed layer, e.g. a noble gas adsorbed on graphite, $E_d \lesssim 0.2$ eV so that temperatures below 100°K are required. At the other extreme, strongly chemisorbed species, such as oxygen on molybdenum, may have $E_d > 5$ eV and a T_e of several thousand degrees can be expected. This is usually too high for equilibrium measurements to be made and E_d must instead be estimated from thermal desorption spectra, i.e. from a trace of the partial pressure of the desorbed species versus crystal temperature[37,40].

The details of the potential energy surface will of course depend on the identity of the substrate and the adatoms, but for a given system one typically finds a range of values for E_d, which suggests that the adsorbed layer is not homogeneous even when the substrate is a perfect single-crystal face. One reason may be an "intrinsic" heterogeneity, due to the occupation of sites with different coordination. Another effect, which may be termed "induced" heterogeneity, arises because the bonding of an adatom will be influenced by the presence of another adatom nearby, i.e. because of lateral adatom-adatom interactions. If this interaction is repulsive for example, E_d will decrease with coverage.

The overlayer geometry depends on both the adatom-substrate and the adatom-adatom forces[19]. For the reasons discussed earlier, a complete structure analysis is available for only a few overlayers[2a,13] but LEED observations of the 2D periodicity alone may provide both qualitative and quantitative information about the surface interactions. A few examples will illustrate the point.

The adsorption of hydrogen on tungsten(100) is one of the most widely studied adsorption systems. The LEED pattern which appears initially is that of a c(2x2) structure (see Figure 6e and 6f), and evidently the H atoms avoid occupation of adjacent sites. There must be an effective repulsion between the adatoms, but this repulsion does not reach beyond the nearest neighbor because the c(2x2) pattern is present even when θ is 0.25 or less[41], i.e. the overlayer tends to form <u>islands</u> with a c(2x2) structure. If the lateral interaction were a monotonically decreasing repulsion, a more stable structure at $\theta = 0.25$ would be the c(4x2) structure shown in Figure 6c; if the repulsion acted on nearest and next nearest neighbors only, the structure to be expected might be the p(2x2) (see Figure 6a). Thus, the LEED data show that the lateral interaction oscillates, being repulsive at the nearest neighbor distance and attractive at the next nearest neighbor distance[42]. It appears that the strength of the repulsion also can be obtained from LEED measurements. An increase in the temperature induces an order-disorder transition of the c(2x2) structure, and the intensity of the (½ ½) spot decreases as the overlayer is randomized[27]. Statistical mechanics gives the connection between the observed critical temperature and the adatom pair interaction which turns out to be ∿ 0.1 eV. This value agrees with the repulsion energy estimated from an analysis of the thermal desorption spectrum[43]. Unfortunately, many systems do not show structural changes suitable for this type of analysis. For example, a thorough study of Mo(100) + H has been carried out[44], and the phase diagram[19] shows numerous transitions but no order-disorder transformations.

An interesting example of the interplay between adsorbate-adsorbate and adsorbate-substrate interactions is provided by the Ni(100) + CO system[39]. At a coverage of half a monolayer, a c(2x2) structure is formed (Figure 6e), and there are reasons to believe that each CO molecule occupies a site of 4-fold coordination. However, as the adsorption progresses and the coverage increases, the overlayer is gradually "compressed" towards an approximately hexagonal structure. There is no longer a unique coordination of the adsorbed molecules relative to the substrate; the overlayer structure is incommensurate and is governed by the CO density. Thus, the arrangement is _not_ determined by some preferred bond geometry. A dominance of the surface structure by adatom-adatom interactions is known to occur in systems as diverse as Ni(100) + K[45] and Pd(100) + Xe[46]. However, the first system involves ionic bonding and the second van der Waals bonding, and strongly directional forces are therefore not expected. It is surprising that the same is true for Ni + CO, since the bond appears to be covalent and according to UPS data[47] is similar to that in the nickel carbonyl molecule.

Strongly chemisorbed adsorbates such as oxygen, nitrogen and carbon monoxide on tungsten and molybdenum produce surface structures in registry with the substrate, as was illustrated in Figures 3 and 5. Apparently the adatom-substrate interactions dictate the use of specific sites in these cases, and a convincing demonstration of this has been given for nitrogen on tungsten[48]. Studies of the LEED patterns and of the saturation coverage on the (100), (310), (210) and (110) faces indicate that in order for chemisorption to occur adjoining sites must be available, each consisting of four W atoms in a square. These are the sites found on the W(100) face which readily adsorbs nitrogen. The (310) and (210) faces contain terraces with this configuration, and some adsorption takes place. Sites of this type are absent on the (110) face, however, and little or no chemisorption of nitrogen is observed.

That the adatom-adatom interactions remain important also for these systems is obvious from the tendency of the adsorbates to form overlayers with long-range order at coverages less than a monolayer. (Since strongly chemisorbed species usually have relatively low surface mobility, the layer may not order at room temperature, however.) The lateral interactions modify other surface properties as well. For example, the so-called β-state of CO on Mo(100) is believed to be dissociated and to consist of carbon and oxygen atoms which occupy equivalent sites[49]. At half the saturation coverage, the adsorbate tends to form a c(2x2) structure, and at this stage the adatom-adatom distance is thus $\sqrt{2}\ a_s$ = 4.5 Å. As the adsorption proceeds, the adatoms (C or O) are forced to occupy neighboring sites, a distance of a_s = 3.15 Å apart. This has the drastic effect of reversing the surface dipole and of lowering the binding energy by ∿ 1 eV. It is interesting that the distance between the adatoms is much larger than a chemical bond and that the lateral interaction therefore must be indirect, involving the metal substrate.

CONCLUDING REMARKS

No attempt has been made here to give a thorough review of the many interesting and important LEED experiments which have been carried out in recent years. The discussion has been mainly confined to very simple surface layers. These give the clearest illustration of the basic principles of LEED, and they represent the structures for which the technique so far has been most successful. Furthermore, they serve as model systems in the development and testing of theoretical approaches to surface phenomena. It may be noted, however, that LEED to an increasing degree is being applied to more complex problems, such as adsorption of large molecules[50], catalytic oxidation[51], the poisoning of catalysts[52], the formation of 3D oxides[15,53], the study of surface reactions under conditions other than UHV[54], and even to solid-liquid interface phenomena[55]. This trend is likely to continue in the future.

REFERENCES

1. T. W. Haas, G. J. Dooley, A. G. Jackson and M. P. Hooker, Progress in Surface Science 1, 155 (1971).
2. a) J. A. Strozier, D. W. Jepsen and F. Jona, Chapter I of *Surface Physics of Materials*, J. M. Blakely, ed., Academic Press (1975).
   b) J. B. Pendry, *Low Energy Electron Diffraction*, Academic Press (1974).
   c) C. B. Duke, Adv. Chem. Phys., 27, 1 (1974).
   d) M. B. Webb and M. G. Lagally, Solid State Physics 28, 301 (1973).
   e) G. A. Somorjai and H. H. Farrell, Adv. Chem. Phys. 20, 215 (1971).
   f) P. J. Estrup and E. G. McRae, Surface Sci. 25, 1 (1971).

g) E. N. Sickafus and H. P. Bonzel, Progress in Surface and Membrane Science 4, 115 (1971).
h) J. W. May, Adv. Catalysis 21, 151 (1970).
i) J. J. Lander, Progress in Solid State Chemistry 2, 26 (1965).
3. See, for example, the articles by J. T. Grant and by R. L. Park in this volume.
4. See, for example, the article by T. N. Rhodin in this volume.
5. P. E. Hojlund Nielsen, Surface Sci. 35, 194 (1973).
6. N. Masud and J. B. Pendry, J. Phys. C (in press).
7. A. R. Moon and J. M. Cowley, J. Vac. Sci. Technol. 9, 649 (1972).
8. H. H. Brongersma and T. M. Buck, Surface Sci. 53, 649 (1975).
9. W. Heiland, F. Iberl, E. Taglauer and D. Menzel, Surface Sci. 53, 383 (1975). H. H. Brongersma, J. Vac. Sci. Technol. 11, 231 (1974).
10. D. M. Zehner, B. R. Appleton, T. S. Noggle, J. W. Miller, J. H. Barrett, L. H. Jenkins and O. E. Schow III, J. Vac. Sci. Technol. 12, 454 (1975).
11. J. K. Kjems, L. Passell, H. Taub, J. G. Dash and A. D. Novaco, Phys. Rev. B13, 1446 (1976).
12. See the article by E. W. Müller in this volume.
13. See the articles by D. Adams, by S. Y. Tong, and by P. M. Marcus in this volume.
14. A recent example is given in J. R. Noonan, D. M. Zehner, and L. H. Jenkins, J. Vac. Sci. Technol. 13, 183 (1976).
15. C. Leygraf and S. Ekelund, J. Vac. Sci. Technol. 11, 189 (1974); Surface Sci. 40, 609 (1973).
16. M. G. Lagally, J. C. Buchholz and G. C. Wang, J. Vac. Sci. Technol. 12, 213 (1975); P. C. Stair, T. J. Kaminska and G. A. Somorjai, Phys. Rev. B11, 623 (1975); T. E. Felter and P. J. Estrup, Rev. Sci. Instr. 47, 158 (1976).
17. A discussion of reciprocal lattice vectors is given in most texts on crystallography. See, for example, H. D. Megaw, *Crystal Structures: A Working Approach,* W. B. Saunders, Philadelphia (1973).
18. Figure 3 is taken from unpublished work by T. E. Felter, C. H. Huang and P. J. Estrup. At the time of this writing only preliminary AES data are available.
19. P. J. Estrup, Physics Today 28, No. 4, 33 (1975).
20. H. Froitzheim, H. Ibach and S. Lehwald, Phys. Rev. Lett. 36, 1549 (1976); Phys. Rev. B (July 15, 1975) (in press).
21. T. E. Madey, J. J. Czyzewski and J. T. Yates, Surface Sci. 49, 465 (1975); J. I. Gersten, R. Janow and N. Tzoar, Phys. Rev. Lett. 36, 610 (1976); T. E. Felter and P. J. Estrup, to be published.
22. J. W. Gadzuk, Phys. Rev. B10, 5030 (1974); A. Liebsch, Phys. Rev. Lett. 32, 1203 (1974); E. W. Plummer, private communication.
23. R. L. Park, J. E. Houston and D. G. Schreiner, Rev. Sci. Instrum. 42, 60 (1971). J. E. Houston and R. L. Park, Surface Sci. 21, 209 (1970), 18, 213 (1969).

24. M. G. Lagally, in *Surface Physics of Materials*, Vol. 2, J. M. Blakely, ed., Academic Press (1975).
25. P. J. Estrup and J. Anderson, Surface Sci. 8, 101 (1967).
26. A. Guinier, *X-Ray Diffraction*, Freeman, San Francisco (1963), B. E. Warren, *X-Ray Diffraction*, Addison-Wesley (1969).
27. P. J. Estrup, in *The Structure and Chemistry of Solid Surfaces*, 19-1, Wiley (1969); J. C. Buchholz and M. G. Lagally, Phys. Rev. Lett. 35, 442 (1975); G. Doyen, G. Ertl and M. Plancher, J. Chem. Phys. 62, 2957 (1975).
28. E. A. Wood, J. Appl. Phys. 35, 1306 (1974).
29. W. Mönch, in Festkörperprobleme 13, 241 (1973).
30. E. Tosatti and P. W. Anderson, Japan. J. Appl. Phys. Suppl. 2, Pt. 2, 381 (1974). J. A. Appelbaum and D. R. Hamann, Rev. Mod. Phys. 48, 479 (1976); J. C. Philips, Surface Sci. 53, 474 (1975).
31. D. G. Fedak and N. A. Gjostein, Surface Sci. 8, 77 (1967); A. E. Morgan and G. A. Somorjai, Surface Sci. 12, 405 (1968); F. Grönlund and P. E. Höjlund Nielsen, J. Appl. Phys. 43, 3919 (1972).
32. J. E. Demuth, P. M. Marcus and D. W. Jepsen, Phys. Rev. B11, 1460 (1975).
33. A. Ignatiev, F. Jona, H. D. Shih, D. W. Jepsen and P. M. Marcus, Phys. Rev. B11, 4787 (1975).
34. See, for example, J. M. McDavid and S. C. Fain, Surface Sci. 52, 161 (1975); C. R. Helms, J. Catalysis 36, 114 (1975).
35. W. P. Ellis and R. L. Schwoebel, Surface Sci. 11, 82 (1968); M. Henzler and J. Clabes, Japan. J. Appl. Phys. Suppl. 2, Pt. 2, 389 (1974).
36. G. A. Somorjai, Catal. Rev. 7, 87 (1972).
37. See, for example, D. O. Hayward, in *Chemisorption and Reactions on Metallic Films*, J. R. Anderson, Ed., Academic Press, New York (1971).
38. J. C. Shelton, H. R. Patil and J. M. Blakely, Surface Sci. 43, 493 (1974).
39. J. C. Tracy, J. Chem. Phys. 56, 2736 (1972).
40. L. D. Schmidt, Catal. Rev. 9, 115 (1974).
41. P. J. Estrup and J. Anderson, J. Chem. Phys. 45, 2254 (1966).
42. The nature of this interaction has been considered theoretically by T. L. Einstein and J. R. Schrieffer, Phys. Rev. B7, 3629 (1973); J. R. Schrieffer and P. Soven, Physics Today 28, No. 4, 24 (April 1975).
43. D. L. Adams, Surface Sci. 42, 12 (1974).
44. C. H. Huang and P. J. Estrup, to be published.
45. S. Andersson and V. Jostell, Solid State Commun. 13, 829 (1973).
46. P. W. Palmberg, Surface Sci. 25, 598 (1971).
47. G. E. Becker and H. D. Hagstrum, J. Vac. Sci. Technol. 10, 31 (1973).
48. D. L. Adams and L. H. Germer, Surface Sci. 27, 21 (1971); S. P. Singh-Bopari, M. Bowker and D. A. King, Surface Sci. 53, 55 (1975).
49. T. E. Felter and P. J. Estrup, Surface Sci. 54, 179 (1976).

50. J. L. Gland and G. A. Somorjai, Surface Sci. 41, 387 (1974); J. L. Gland, K. Baron and G. A. Somorjai, J. Catalysis 36, 305 (1975).
51. G. L. Ertl and J. Koch, in *Proceedings of Vth Intern. Congress Catalysis*, J. W. Hightower, ed., North-Holland, p. 969 (1973).
52. H. P. Bonzel and R. Ku, J. Chem. Phys. 58, 4617 (1973); 59 1641 (1973), Y. Berthier, M. Perdereau and J. Oudar, Surface Sci. 44, 281 (1974).
53. P. H. Holloway and J. B. Hudson, Surface Sci. 43, 141 (1974).
54. G. Rovida, F. Pratesi, M. Maglietta and E. Ferroni, Surface Sci. 43, 230 (1974).
55. A. T. Hubbard, private communication.

The Use of Direct Methods in the Analysis of LEED

David L. Adams and Uzi Landman*

Institute of Physics
University of Aarhus
DK 8000 Aarhus C, Denmark

*Institute for Fundamental Studies
Department of Physics and Astronomy
University of Rochester
Rochester, N. Y. 14627

The conventional approach to surface structure determination by analysis of LEED consists of comparison between experimental intensities and intensities calculated on the basis of a model of the diffraction process and a trial model of the surface structure. Although this model calculation approach has been used with some success in the case of simple structures, it is unlikely that it can be easily extended to more complex systems unless a reasonable approximation to the structure is known or can be determined by another method. In this article, therefore, we consider the possible adaption and application to LEED of the methods of x-ray crystallography in which direct use is made of the experimental data without prior assumption of a model structure. Since these methods are based upon single-scattering theory, difficulties can be anticipated to arise from the known importance of multiple-scattering processes in LEED. In practice, however, more severe difficulties result from the paucity of data available from a LEED as opposed to an x-ray diffraction experiment, and from other characteristics of the scattering of low-energy electrons in solids. A method for solution of some of these problems is described and applications to experimental data analysis are presented. The results suggest that direct methods of analysis of LEED may indeed provide a means for obtaining a good approximation to surface structures.

CONTENTS

I. INTRODUCTION

II. SOME PROCEDURES AND PROBLEMS IN THE INTERPRETATION OF DIFFRACTION MEASUREMENTS

III. PATTERSON FUNCTIONS OF LEED INTENSITIES

1. Data Truncation

2. Energy Dependence of the Atomic Scattering Factor and Debye-Waller Effect

3. Differences in Atomic Scattering Factors

4. Inner Potential

5. Multiple Scattering

IV. TRANSFORM-DECONVOLUTION METHOD

V. APPLICATION OF THE TRANSFORM-DECONVOLUTION METHOD TO OVERLAYER SYSTEMS

VI. SUMMARY AND CONCLUSIONS

ACKNOWLEDGEMENTS

REFERENCES

I. INTRODUCTION

The importance of a knowledge of the atomic arrangement at the surface of a solid for the understanding of many physical and chemical properties of surfaces is very evident. That there exist methods capable of quantitative determination of surface structure in the general case of an arbitrary structure is less obvious. In recent years, sensitivity to surface structure has been demonstrated or suggested for a number of techniques. Examples include ion scattering[1] and atom scattering[2] methods, and angular-resolved photoemission spectroscopy[3]. At the present time, however, low-energy electron diffraction (LEED) is the only method to have attained a degree of sophistication in terms of structural interpretations. The technique is limited, of course, to single-crystal substrates.

Following intensive efforts over the last decade, theoretical models of LEED and, particularly, their numerical implementation on high-speed computers, have reached the stage where tolerable agreement between calculated and experimental LEED intensities has been achieved for some simple structures. This is despite rather considerable uncertainties which remain in the theory, and difficulties in obtaining precise experimental measurements.

As is well known, a source of major difficulty in describing the scattering of low energy electrons is the inadequacy of the first Born approximation. An accurate treatment of both atomic and interatomic multiple-scattering processes is necessary. The most accurate numerical treatments of this problem, by Marcus, Jepsen and their co-workers[4], in which multiple scattering processes are accounted for through the use of a multiple-beam representation of the layer KKR method, involve computational requirements beyond the reach of many laboratories. Accordingly, a major preoccupation in the field has been the investigation of schemes for approximate description of multiple-scattering. Perturbation expansion schemes in which interatomic multiple-scattering processes are summed to <u>finite</u> order have been developed, for example by Pendry[5], and refined by Tong and Van Hove[6]. The physical basis for such schemes is the low contribution of higher order scattering processes due to the high probability for inelastic scattering.

An alternative approach to LEED analysis based on single-scattering theory in conjunction with angular-averaged experimental data has been investigated following the work of Ngoc, Lagally, and Webb[7]. The physical basis for this method is the proposition that multiple-scattering contributions to the diffracted intensities should, in general, be aperiodic with normal momentum transfer. Angular averages made at constant normal momentum transfer should then be dominated by the single-scattering contribution to the intensities. The difficulty with this approach is that residual effects of multiple-scattering due to incomplete averaging may be of the same order as effects associated with the structure of an adsorbed layer[8]. To date the method has not been successfully used for cases other than clean surfaces.

In conclusion, at the present time it appears that of the model calculation approaches, the perturbation method, and in particular the algorithm of Tong and Van Hove, constitutes the most practical method for LEED analysis. The fact that tolerable agreement has been achieved, for some simple overlayer systems, both with the experimental data and with the complete calculations of Marcus and Jepsen[4] suggests that the main ingredients of a theoretical treatment can now be handled in a fairly accurate and practical way. It seems appropriate, therefore, to turn now to consideration of other fundamental questions which bear upon the general applicability of analysis of LEED for surface structure determination, and which have been largely held in abeyance to this time.

The main intention in this paper is to review the few attempts which have been made to develop <u>direct</u> methods for analysis of LEED, but we consider that this is most usefully viewed from the perspective of a more general discussion of the applicability of the technique, which raises issues of concern to both model calculation and direct approaches. In attempting to provide such a perspective in Section II, we draw upon the experience gained in the development of x-ray crystallography, from which we conclude that development of direct methods for analysis of LEED is probably essential if the technique is to develop into a crystallographic tool. In Section III, we consider the problems in adaptation of the direct methods of x-ray crystallography to the analysis of LEED and review the few previous attempts by other workers in this direction. In Section IV we turn to a recapitulation of the transform-deconvolution method, which we have described elsewhere[9,10] and discuss its application to the determination of the interlayer spacings in the surface of Al(100), Al(111), Ni(100), and Cu(100). Extension of the transform-deconvolution method to the case of overlayer systems is discussed in Section V. Conclusions of this work and possible future directions are presented in Section VI.

II. SOME PROCEDURES AND PROBLEMS IN THE INTERPRETATION OF DIFFRACTION MEASUREMENTS

In this section we are concerned with the determination of the structure of the unit cell from analysis of measurements of diffracted intensities. It is assumed that the lattice periodicity is known from analysis of the angular distribution of diffracted beams (diffraction pattern).

In x-ray crystallography, the magnitude of the structure factor $|F_{\vec{H}}|$ is obtained from the measured intensity after appropriate corrections for absorption, extinction, and various experimental factors. In the absence of dynamic effects and anomalous dispersion, the structure factor, $F_{\vec{H}}$ is related to the electron density, $\rho(\vec{r})$ by Fourier transformation:

$$\rho(\vec{r}) = \int |F_{\vec{H}}| e^{i\Phi} e^{-2\pi i \vec{H} \cdot \vec{r}} d\vec{H}, \qquad (2.1)$$

where $\vec{H}$ is a general reciprocal lattice vector and $\vec{r}$ the vector from some arbitrary origin in the crystal. The "phase problem" of diffraction analysis which prevents straightforward use of Equation 2.1 in structure determination is that only the magnitude $|F_{\vec{H}}|$ is available from the measurements; the associated phase, Φ, is not measured. Thus in the early days of x-ray crystallography, structure determination was entirely a trial and error procedure, based on the use of calculated phases in Equation 2.1, or based on comparison of calculated and experimental structure factor amplitudes. In either case, a model of the structure had to be assumed in order to begin the analysis.

The successful use of x-ray crystallography in the determination of increasingly complex structures has resulted from the introduction since the 1940's of "direct" methods of analysis. Partial solutions to the phase problem were achieved by Patterson[11] and by Harker and Kasper[12].

Patterson showed that the Fourier transformation of the diffracted <u>intensities</u> yields the self-convolution of the electron density:

$$\begin{aligned} P(\vec{r}) &= \int |F_{\vec{H}}|^2 e^{-2\pi i \vec{H} \cdot \vec{r}} d\vec{H} \\ &= \int \rho(\vec{r}) \rho(\vec{r} + \vec{r}') d\vec{r}'. \end{aligned} \qquad (2.2)$$

In this case, structure determination devolves to finding the convolution square-root of the Patterson function $P(r)$. Image-seeking procedures for performing this operation have been proposed and discussed in detail by Buerger[13]. It should be noted that the occurrence of overlapping peaks in the Patterson function, resulting from the self-convolution of the electron density, is a source of difficulty, the severity of which increases with increasing numbers of atoms in the unit cell.

Harker and Kasper[12] and later Hauptman and Karle[14] demonstrated the existence of certain statistical relationships between phases and intensities, formulated as a set of inequalities. These relationships allow the calculation of phases for use in equation 2.1 to be constrained by knowledge of the relative intensities of hkl reflections.

The development of the direct methods served to illuminate the fundamental problem of diffraction analysis, which is that the experimental diffracted intensities are not uniquely related to a real-space structure[15]. Determination of the correct physical solution requires that the structure determination procedure be suitably constrained by incorporation of a priori known physical and chemical information.

Incorporation of constraints occurs at the lowest level in the model calculation approach by construction of physically sensible structures as models for calculation of diffracted intensities, taking into account, for example, known dimensions of the atomic constituents and known or calculated scattering factors. The problem with this approach, however, is the need to range sufficiently widely over the geometric parameters of the model to ensure that best possible agreement is obtained with the experimental data. In practice, the degree of confidence to be placed in a model solution depends upon the range of experimental data over which comparison is made, in addition of course, to the accuracy of both data and calculations. Inaccuracies in the parameters of the physical model of the diffraction process or deficiencies in the physical model itself are potential sources of uncertainty. The amount of labor required in parameter variation to achieve best fit between experiment and theory can be considerably reduced if a good approximation to the structure is known beforehand. Thus the main use of model calculations in x-ray crystallography is in structure refinement after the structure has been determined by one of the direct methods.

The real-space, Patterson function method and the reciprocal-space, phase-determining procedures introduce a higher level of constraint into the process of structure determination, with the additional advantage that the results of the analysis, being derived from the experimental data, are by construction consistent with the data. However, in the direct methods, the diffraction process is not represented to the same accuracy as in a model calculation, particularly for multi-component systems, insofar as scattering factors and temperature corrections are concerned. It should be emphasized that in x-ray crystallography the direct methods are complementary to the model calculation approach, being used to provide a good first approximation to the structure for subsequent refinement.

It is instructive to compare the current situation in the analysis of LEED with that in x-ray crystallography. With the exception of our own work, and a few earlier investigations described in the next section, current analysis of LEED intensities is carried out by the trial and error, model calculation approach. That this approach is subject to the uncertainties referred to above is amply demonstrated in the conflicting interpretations of LEED data published in recent years[16]. Andersson and Pendry[17], for example, have noted the occurrence of "multiple coincidence" solutions; that is, equally acceptable fits between experimental and calculated intensities based on <u>different</u> structural models. While some particular issues of conflict appear now to be resolved[16], and while it should be noted that some of the conflicting interpretations of recent years can most probably be attributed to different standards of accuracy in the calculations of different authors, the possibility of ambiguous interpretation clearly exists, even for the relatively simple structures under current investigation.

The problem of uniqueness involved in variation of a number of model parameters is much more severe in the case of LEED than in x-ray crystallography. Calculations which properly describe the dynamics of the interaction are considerably more demanding than the single-scattering, kinematic theory applicable in the x-ray case. Furthermore, it is not certain at the present time that the scattering of electrons from individual atoms, the effect of thermal vibrations, and the effects of both elastic and inelastic electron-electron interactions are adequately treated. These difficulties are well known[18] and it suffices here to note that the demanding nature of the computations has precluded a proper variation over the model parameters. Comparison between experimental and calculated intensities is carried out in a rudimentary fashion, and is made uncertain by the lack of generally accepted criteria for assessing the extent of agreement.

Apart from the difficulties associated with the extra complexity of the diffraction process, an equally important difference between x-ray crystallography and LEED is the relatively restricted region of reciprocal space accessible to investigation in the latter case, and hence the relative paucity of data against which comparison with calculations can be made. Whereas in a typical x-ray structure determination a few thousand reflections may be measured, in LEED a number of factors combine to limit the available data to the equivalent of several tens of reflections.

The severity of this truncation problem in the case of LEED can be illustrated by a specific example. In the case of the specular diffracted beam, for which the most extensive measurements are possible, comparison between experimental and calculated intensity spectra is rarely made over a range of incident electron energy greater than 250 eV. (Measurement over several times this range is usually possible but with attendant decrease of surface sensitivity due to increasing electron mean free paths.) This energy range corresponds to a diffraction vector of about 2.5 Å^{-1} which, in turn, corresponds to a sampling of the real-space structure with a resolution of about 0.4 Å in the surface normal direction. At first glance it might appear that this poor resolution could be improved by including in the comparison between theory and experiment, intensity spectra for the <u>same</u> diffracted beam taken at different angles of incidence and azimuth, since multiple-scattering processes will lead to differences between such spectra. It is quite evident, however, that whilst such extended comparison is desirable, it can only lead to improved statistics of comparison. The region of reciprocal space explored is not extended, and hence the resolution is not changed.

Thus, ignoring errors in the experimental data, comparison between experimental and calculated intensities over typical energy ranges corresponds to comparison between the correct structure and the model structure on a grid of about 0.4 Å. Clearly this procedure can only lead to unambiguous and accurate structure determination, say of the order of ± 0.01 Å in interatomic distances if, inter alia, the spatial variation of the atomic scattering potential used in the calculations is very accurately known.

In summary, the intrinsic problems associated with the model calculation approach to LEED, and the crucial role of the direct methods in the development of x-ray crystallography lead us to conclude that progress in LEED beyond the analysis of the most simple surface structures will be very difficult, if possible at all, unless analogous direct methods can be devised to obtain a good approximation to the structure. In the remainder of this paper, we shall consider the feasibility of applying such methods to LEED.

III. PATTERSON FUNCTIONS OF LEED INTENSITIES

In comparison with the large number of studies aimed at the development of an adequate theory of LEED for use in model calculations, there have been rather few attempts to examine the feasibility of direct methods of analysis. There is no doubt that the importance of multiple-scattering processes in LEED is one of the obstacles to be surmounted in application of the direct methods. In x-ray crystallography, the direct methods depend upon the Fourier-pair relationship between the diffracted amplitudes and the real-space scattering potential, Eq. (2.1), which is valid only in the first Born approximation. As we have discussed elsewhere[9,10], however, the first problem to be solved in application of the direct methods to LEED is that associated with the severe truncation of the data and effects due to the energy dependence of the scattering potential and the Debye-Waller factors. In this section, therefore, we focus mainly on those aspects of the problem which would exist even in the absence of multiple scattering. Since the possible applications of phase-determining methods to LEED does not appear to have been discussed, in this article we shall consider only the use of the Patterson function.

Before discussing the significance of the Patterson function in the case of LEED, it is useful to outline briefly the meaning of the function in x-ray crystallography[13]. In the latter case, the Patterson function, Equation (2.2), is written as a summation over the Miller indices hkℓ:

$$P(xyz) = \sum_h \sum_k \sum_\ell I_{hk\ell} e^{-2\pi i(hx+ky+\ell z)}, \qquad (3.1)$$

where $I_{hk\ell}$ is the intensity of the hkℓ reflection. In the absence of anomalous dispersion, Friedel's law, $I_{hk\ell} = I_{\bar{h}\bar{k}\bar{\ell}}$, applies and so Equation (3.1) can be written as a real, cosine transform. For convenience, we retain the exponential form in the following.

As mentioned earlier, the Patterson function represents the self-convolution of the scattering potential of the structure. Thus the transformation given by Equation (3.1) defines a 3-D vector space. Peaks in the Patterson function define the positions of interatomic vectors in the real-space structure after translation to a common origin. It is convenient to illustrate this relationship using reduced forms of the full 3-D function. These reduced forms will later be shown to be relevant in the analysis of LEED data.

By substituting a constant value p for the variable z in Equation 3.1, we obtain a 2-D section through the 3-D function:

$$P(xyp) = \sum_h \sum_k \sum_\ell I_{hk\ell} e^{-2\pi i(hx+ky+\ell p)}. \qquad (3.2)$$

The 1-D line projection, P(z) is given by:

$$P(z) = \sum_\ell I_{oo\ell} e^{-2\pi i \ell z}. \qquad (3.3)$$

In Figure 1, the significance of the 2-D section and the line projection is schematically illustrated with respect to the bcc unit cell shown in Figure 1a. 2-D sections corresponding to values of p = 0 and p = d are shown in Figure 1b. The general section, P(xyp), contains peaks defining interatomic vectors between atoms lying in planes separated by p in the z direction, with the vectors shifted to a common origin and projected onto the xy plane. In the special case of the P(xyO) section, peaks correspond to vectors between atoms in the same xy plane.

In the P(xyO) section shown in Figure 1b, the peak positions are completely defined by the structure of a single xy plane, but we emphasize that in general, the P(xyO) section will contain vector sets from each non-equivalent plane. Similarly, the P(xyd) section shown in Figure 1b contains a single peak corresponding to

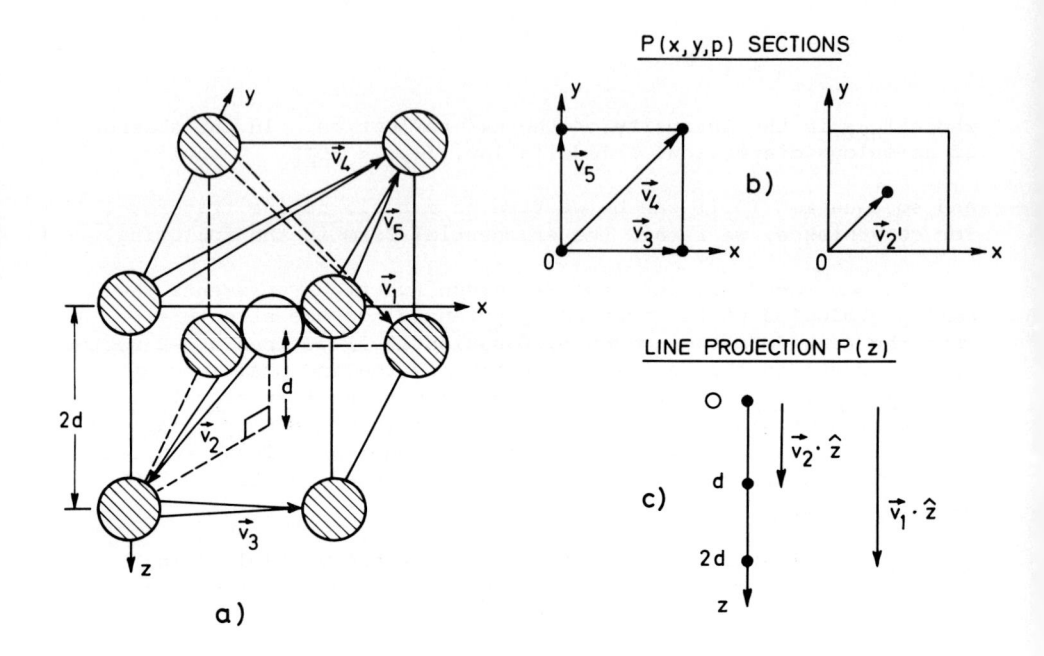

Fig. 1. Schematic illustration of the physical significance of reduced forms of the x-ray Patterson function, based on the bcc unit cell. a) bcc unit cell with body-centered atom indicated by open circle. Various interatomic vectors are indicated by arrows. b) P(xyp) sections for p = 0 (left-hand) and p = d (right-hand). c) P(z) line projection, showing peaks at multiples of layer spacing d.

the degenerate set of interatomic vectors between successive planes in the bcc example considered.

The P(z) line projection contains peaks corresponding to <u>all</u> the interatomic vectors shifted to a common origin and projected onto the z axis. Thus, in the example shown in Figure 1c, the P(z) function contains a single vector set of peaks at multiples of the layer spacing in the z direction. In general, P(z) contains a vector set associated with each different interlayer spacing.

Although the reduced forms of the Patterson function described above have obvious advantages compared to the 3-D function in terms of both computation and interpretation, they have the disadvantage of poorer resolution. In the case of a more complex unit cell than the simple example shown in the figure, it can easily be imagined that overlap of peaks associated with non-equivalent vector sets might prevent unambiguous location of peak positions. This particular problem, however, can be expected to be less troublesome in LEED than in x-ray crystallography, since the very limited penetration of low-energy electrons into solids should result in significant contributions to the Patterson function from only the first few atomic planes.

Accordingly, if it could be assumed that the Patterson functions of LEED intensities were subject to the straightforward interpretation possible in x-ray crystallography, then construction of the reduced forms described above would be sufficient to determine the surface structure. The interlayer spacings would be determined from a P(z) function, and use of these spacings in P(xyp) sections would enable the interlayer registries to be found. The structures of the layers would be determined from a P(xy0) section.

Before the significance of the Patterson function can be evaluated for the case of LEED, certain modifications to the form of the function must be made. The large scattering cross sections for low-energy electrons lead to strong attenuation of the electron beam as it penetrates the solid. Consequently, the third Von Laue condition is not properly established and the intensity of an hk diffracted <u>beam</u> varies continuously along the ℓ direction of reciprocal space, corresponding to the surface normal direction. Accordingly, Equations (3.1) to (3.3) must be rewritten for the case of LEED as:

$$P(xyz) = \sum_h \sum_k \int I_{hk}(s) e^{-2\pi i(hx + ky)} ds, \qquad (3.4)$$

$$P(xyp) = \sum_h \sum_k \int I_{hk}(s) e^{-2\pi i(hx + ky)} ds \qquad (3.5)$$

$$P(z) = \int I_{oo}(s) e^{-2\pi isz} ds, \qquad (3.6)$$

where the third Miller index ℓ has been replaced by the continuous variable s, with 2πs equal to the normal component of the diffracted wave-vector. In the case of LEED, Friedel's law does not apply.*

---

*From symmetry considerations[10] the following relation holds.

$I_{hk}(s) = I_{hk}(-s)$.

The problems which prevent straightforward interpretation of the Patterson function of LEED intensities were first identified by Landman and Adams[9a], and by Woodruff, Mitchell, and McDonnell[19]. In the case of the former authors[9a,b], solutions to these problems were also proposed, based on an explicit analysis of the physical significance of the Patterson function described later in this section.

It is convenient to illustrate some of the difficulties in interpretation, by constructing the Patterson functions of diffracted intensities calculated using single-scattering theory for assumed model structures. Figures 2a and 2b contain intensity spectra calculated for the specular diffracted beam from a clean surface with interlayer spacing d_s, and from the same surface containing half a monolayer of an adsorbate with spacing $d_s/2$ between the adsorbed layer and the substrate. The corresponding P(z) line projections are also shown in the figure. The real part, $Re[P(z)]$ is shown in Figures 2c and 2d, and the modulus, $|P(z)|$ is shown in 2e and 2f. Considering first $Re[P(z)]$, it can be seen that the functions do contain peaks at about the positions expected from the interlayer spacings of the model structures, but there are also additional peaks of no structural significance. Quite clearly a straightforward assignment of the peak positions to interatomic vectors would lead to an incorrect structure. In fact, even if a distinction between significant and spurious peaks could somehow be made, structural assignment would still be hazardous. Firstly, the structure-related peaks are in general shifted to higher z. Secondly, the peaks decay in amplitude with increasing z, making difficult the identification of higher-order components of the vector sets. Finally, it should be noted that in the particular example shown here, the ratio of one-half between the adsorbate layer spacing and the substrate layer spacing corresponds to the most favorable case as far as resolving the different structure-related peaks in Figure 2d is concerned. In general, resolution of the different vector sets might be difficult, particularly in the case of experimental data analysis, when some degradation of the P(z) function due to the effects of multiple-scattering and experimental errors could be expected.

It can be seen from Figures 2e and 2f that $|P(z)|$ also contains peaks at roughly the positions expected from the structural model, and that the additional spurious peaks are less pronounced than in the real part of P(z). However, the peaks in $|P(z)|$ are very broad and the shifts in peak positions quite large. In general, therefore, straightforward interpretation based on $|P(z)|$ is at best inaccurate, and in most cases the poor resolution prevents unambiguous determination of the vector sets.

In Figure 3 are shown $|P(xyp)|$ sections based on single-scattering intensities calculated for two different structural arrangements of the adsorbed layer. In each case, the layer had

Analysis of LEED 223

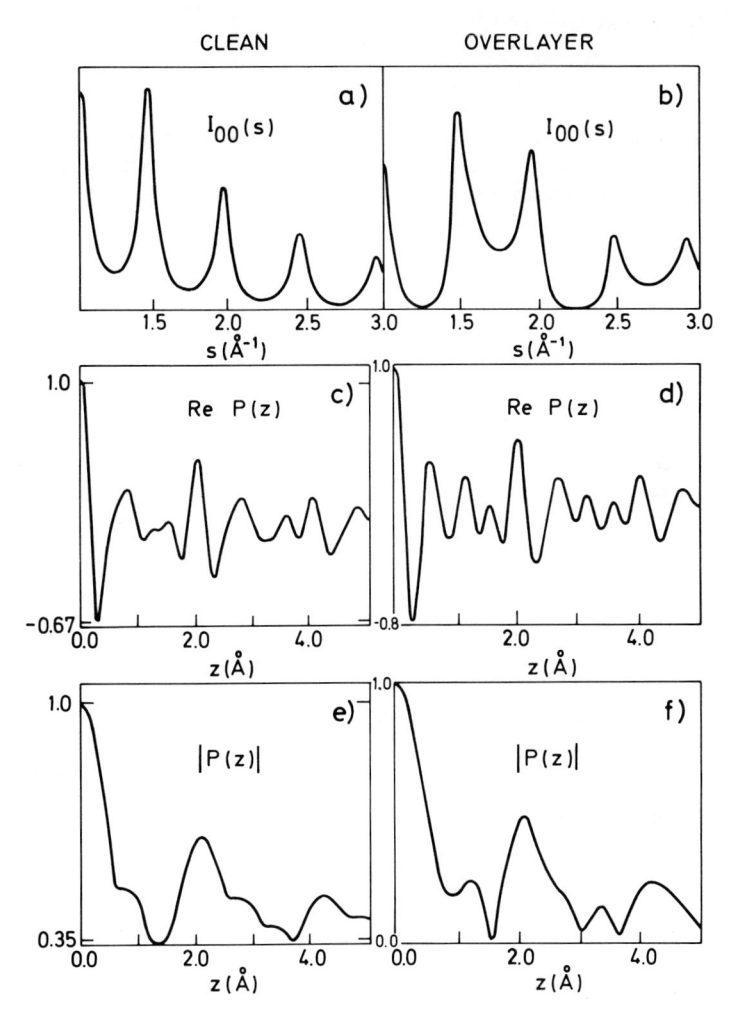

Fig. 2. Calculated intensity spectra for the specular diffracted beam and corresponding Fourier transforms, P(z).
a) Intensity spectrum from clean surface with uniform layer spacing d_S = 2.02 Å, calculated using s-wave scattering factor with scattering phase shift $\delta_S = \pi/2$
b) Intensity spectrum for same surface containing half a monolayer of adsorbed atoms in a layer $d_S/2$ above the substrate, calculated using a scattering phase shift $\delta_S = \pi/8$ for the adsorbed atoms.
c) and d) Real part of the complex Fourier transforms P(z) of the intensity spectra shown in a) and b), respectively. Note that in c), peaks are found at approximately m d_S, where m is an integer, and that in d), peaks occur at approximately m d_S and ($d_S/2$+m d_S). Note also, however,

Fig. 2. (cont'd)

the occurrence of extra, non-structure-related peaks and negative excursions.
e) and f) Modulus of the Fourier transforms of the intensity spectra in a) and b), respectively. Note the considerable broadening of the structure-related peaks as compared to c) and d).

a P(2 x 2) planar periodicity at a spacing of $d_s/2$ above the substrate, but the arrangements differed in the location of the adsorbed atoms. In the first arrangement, shown in Figure 3a, the adsorbed atoms occupied the sites of four-fold symmetry on the square substrate lattice. In the second arrangement, shown in Figure 3b, the adsorbed atoms occupied the two-fold symmetric sites. As can be seen from Figures 3a and b, the P(xyp) sections constructed for $p = d_s/2$ contain dominant peaks at the positions expected from the two different interlayer registries of the two model structures and distinction between the different atomic arrangements can clearly be made. As can be seen from the figures, however, the structure-related peaks are quite broad, as was the case for $|P(z)|$, and the $|P(xyp)|$ sections also contain spurious, subsidiary maxima. Nevertheless, it would seem at first sight that structural assignments could be made with rather less ambiguity than was the case for the P(z) line projections. Our experience with the analysis of P(xyp) sections suggests that this conclusion is often valid in the case of simple structures where peak overlap is not a significant problem. Unfortunately, one crucial proviso has to be made, which is that the interlayer spacing, $p = d_s/2$ in the example, must first be accurately known. P(xyp) sections constructed for arbitrary values of p are, in general, dominated by spurious peaks.

From the above discussion, it is evident that the problem of spurious peaks in the Fourier transforms of LEED intensities cannot be side-stepped by reliance on P(xyp) sections. The problem has to be tackled first for the case of the P(z) line projections.

In the following, we first itemize what we consider to be the main sources of difficulty in the appliation of Patterson functions in the analysis of LEED, then present an explicit analysis[9] of the significance of the P(z) function.

1. Data Truncation

In practice, the volume of reciprocal space which is accessible to investigation by LEED is contained within limiting hemispheres[20] of radii defined by the inner potential of the solid and by the highest incident electron energy used. The consequent truncation of the sums over h and k, and the finite limits of the integrals

TRANSFORM SECTIONS FOR (2×2) OVERLAYER ON (100) PLANE

$|P(x,y,p)|$ 4 fold SITE

$|P(x,y,p)|$ 2 fold SITE

$|P(x,y,0)|$ FOR EITHER 4 fold OR 2 fold SITE

a)

b)

c)

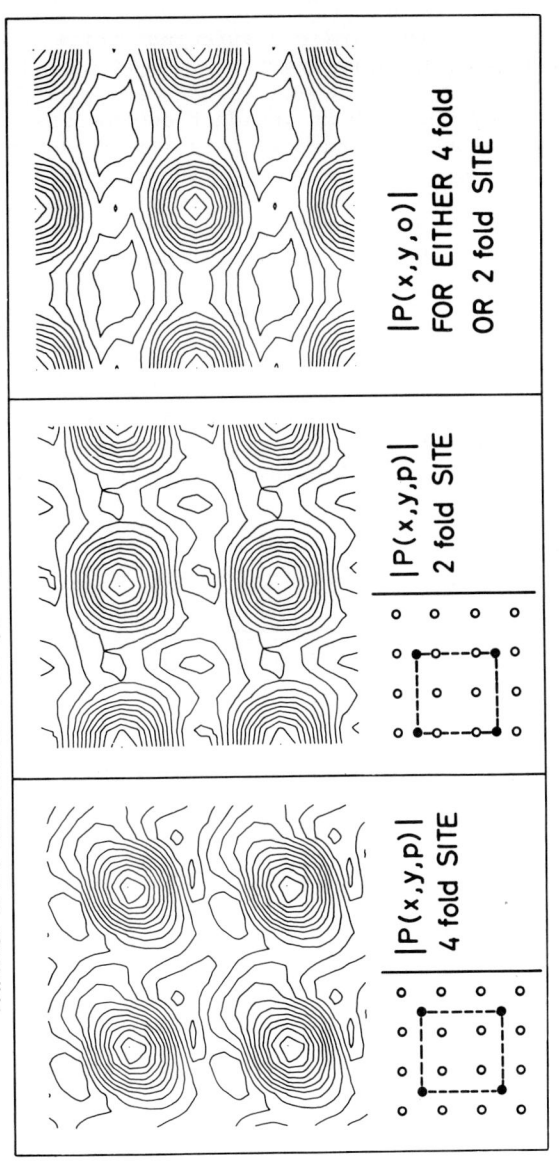

Fig. 3. P(xyp) sections based on intensities calculated for half a monolayer of adsorbed atoms located $d_s/2$ above the (100) plane of a uniform fcc substrate with layer spacing d_s, but for two different layer registries between the substrate and overlayer. Intensity spectra were calculated for nine diffracted beams in each case (h,k = 0,1,2 indexed according to the substrate unit mesh), with scattering factors as in the caption to Figure 2.
a) $|P(xyp)|$ section for $p = d_s/2$, for the case of adsorbed atoms lying in the 4-fold site, as shown in the underlying sketch. The section contains major peaks at 1/4,1/4; 1/4,3/4; 3/4,1/4; 3/4,3/4 with respect to the unit mesh, corresponding to all possible interlayer vectors for $p = d_s/2$.
b) $|P(xyp)|$ section for $p = d_s/2$, for the case of adsorbed atoms lying in the 2-fold site. Major peaks at 0,1/4; 0,3/4; 1/2,1/4; 1/2,3/4 correspond to all possible interlayer vectors.
c) $|P(xyp)|$ section for $p = 0$, obtained for both adsorlate locations. Major peaks at 0,0 (adsorbate) and 0,0; 0,1/2; 1/2,0; 1/2,1/2 (substrate) corresponding to the superposition of intralayer vectors from both substrate and overlayer.

over s in Equations 3.4 to 3.6 lead to the spurious peaks (Gibbs oscillations) in the Fourier transforms of LEED intensities.

2. Energy Dependence of the Atomic Scattering Factor and Debye-Waller Effect

For practical purposes, the highest incident energy used is limited by the decrease in ion-core scattering factor and by the Debye-Waller effect, which combine to attenuate the diffracted intensities at higher electron energies. The form of the energy dependence of the scattering factor and the Debye-Waller factor have a pronounced effect upon the line-shape of the Gibbs oscillations.

3. Differences in Atomic Scattering Factors

The difference in phase between electron waves scattered from different atoms in the surface depends not only upon the interatomic vectors, but also upon the scattering factors of the different atoms.

4. Inner Potential

In order to perform the Fourier transforms given by Equations 3.4 to 3.6, experimental measurements of diffracted beam intensities, I_{hk}, as a function of incident electron energy, E must be converted to tables of I_{hk} versus s, where $2\pi s$ is the normal component, $\Delta \vec{k}_\perp$ of the diffracted wave-vector. This is carried out using the conditions for conservation of energy and parallel momentum:

$$\frac{h^2 \vec{k}^2}{2m} = \frac{h^2 \vec{k}^{12}}{2m} = E + V_o , \qquad (3.7)$$

$$\vec{k}_{//}^1 = \vec{k}_{//} + \vec{g}_{//} . \qquad (3.8)$$

Thus

$$2\pi s = \Delta \vec{k}_\perp = |\vec{k}_\perp| + |\vec{k}_\perp^1| = |\vec{k}_\perp| + \sqrt{|\vec{k}|^2 - |\vec{k}_{//} + \vec{g}_{//}|^2} . \qquad (3.9)$$

In these equations, primed quantities refer to the outgoing diffracted beam and unprimed quantities to the incident beam. $\vec{k}$, $\vec{k}_{//}$ and $\vec{k}_\perp$ are the total, parallel, and perpendicular components, respectively of the electron wave-vector. E is the incident electron energy and V_o the inner potential of the crystal. $\vec{g}_{//}$ is the reciprocal lattice vector in the place parallel to the surface. Thus conversion from the experimental scale of E to a scale of s involves known quantities, except for the inner potential V_o. Approximate values for V_o can be obtained by summation of the work function and Fermi energy, but in general the value of V_o and, particularly, its energy dependence is not accurately known. Clearly inaccuracies in the scale of s resulting from inaccuracies in V_o will cause systematic errors in the Fourier transform.

5. Multiple Scattering

The consequences of multiple-scattering processes for the Fourier transform of LEED intensities are not fully delineated at this time. Heuristic arguments have been made[9,21] that multiple-scattering features in intensity measurements will be smeared out in the Fourier transform since they are expected to occur aperiodically with s, whereas single-scattering features (Bragg peaks) will add in phase to produce peaks in the transform, as we demonstrate below. These arguments receive general support from the results of Landman[22] of analytical Fourier transformation of an expression for dynamic scattering. It was found[22] that the consequences of multiple scattering was to introduce a slowly-varying modulation of the single-scattering features of the transform. The magnitude of this effect has not yet been evaluated for specific cases.

To illustrate the physical significance of the Fourier transform of LEED intensities, we derive below the analytical transform[9] $P(z)$ of an expression for the single-scattered intensity in the specular diffracted beam. We take as a model structure the case of an adsorbed layer with scattering factor $f_o(s)$, taken to be renormalized with respect to adsorbate coverage, at a distance d_o above the outermost plane of the substrate. The substrate scattering factor is $f_s(s)$ and the uniform interlayer spacing in the substrate is d_s. The atomic scattering factors are taken to be renormalized with respect to thermal vibrations by multiplication of the rigid lattice scattering factors by a Debye-Waller factor. Attenuation of the electron beam inside the solid by inelastic processes and by destructive interference with elastically scattered electrons is simulated by assuming an energy-independent mean free path, λ. Defining attenuation factors $\alpha_o = e^{-\mu d/\cos\theta}$ and $\alpha_s = e^{-\mu d_s/\cos\theta}$, where θ is the angle of incidence, $\mu = 2/\lambda$, and defining phase angles $\beta_o = 2\pi s\, d_o$ and $\beta_s = 2\pi s\, d_s$, then the diffracted intensity in the specular beam can be written in terms of a sum over the amplitudes scattered from each layer:

$$I_{oo}(s) = \left| f_o(s) + f_s(s)\alpha_o \sum_{\nu=0}^{\infty} \alpha_s^{\nu} e^{i(\beta_o + \nu\beta_s')} \right|^2$$

$$= f_{oo}(s) + f_{ss}(s)\alpha_o^2(1-\alpha_s^2)^{-1} \sum_{\nu=0}^{\infty} \alpha_s^{\nu}(e^{i\nu\beta_s} + e^{-i\nu\beta_s} - 1)$$

$$+ \alpha_o \sum_{\nu=0}^{\infty} \alpha_s^\nu [f_{os}(s) e^{-i(\beta_o+\nu\beta_s)} + f_{so}(s) e^{i(\beta_o+\nu\beta_s)}] \quad (3.10)$$

where $f_{oo}(s) = f_o(s)f_o^\dagger(s)$, $f_{ss}(s) = f_s(s)f_s^\dagger(s)$, $f_{os}(s) = f_o(s)f_s^\dagger(s)$, $f_{so}(s) = f_s(s)f_o^\dagger(s)$, and $\dagger$ indicates complex conjugate. The summations in Equation 3.10 can be easily carried out to give a closed expression for the diffracted intensity, but the Fourier transformation is more transparent, using the form of the equation given.

In carrying out the Fourier transformation of Equation 3.10, we use the Convolution Theorem of Fourier analysis, which may be written as:

$$\int_{-\infty}^{+\infty} f(s)g(s)e^{-2\pi i s z} ds = F(z)*G(z), \quad (3.11)$$

where $F(z)*G(z) = \int_{-\infty}^{+\infty} F(z)G(z-z')dz'$ and $F(z) = \int_{-\infty}^{+\infty} f(s)e^{-2\pi i s z} ds$ and

$$G(z) = \int_{-\infty}^{+\infty} g(s)e^{-2\pi i s z} ds.$$

Ignoring for the moment the fact that experimental intensities are available only for a truncated range of s, Fourier transformation of Equation (3.10) yields:

$$P(z) = F_{oo}*\delta(z) + F_{ss}(z) * \alpha_o^2(1-\alpha_s^2)^{-1} \sum_{\nu=0}^{\infty} \alpha_s^\nu [\delta(z-\nu d) + \delta(z+\nu d) - \delta(z)]$$

$$+ \alpha_o \sum_{\nu=0}^{\infty} \alpha_s^\nu [F_{os}(z) * \delta(z-d_o-\nu d_s) + F_{so}(z) * \delta(z+d_o+\nu d_s)], \quad (3.12)$$

where F_{oo}, F_{ss}, F_{os}, and F_{so} are respectively the Fourier transforms of f_{oo}, f_{ss}, f_{os}, f_{so}, * denotes the convolution operation, and $\delta(\chi)$ is the Dirac delta function.

Two particular cases may be obtained directly from Equation (3.12). In the case of a clean surface with first-interlayer spacing d_o different from subsequent layer spacings d_s, then by substitution in Equation (3.12) for

$$F_{oo}(z) = F_{os}(z) = F_{so}(z) = F_{ss}(z), \text{ we obtain:}$$

$$P(z) = F_{ss}(z) * \{\delta(z) + \alpha_o^2(1-\alpha_s^2)^{-1} \sum_{\nu=0}^{\infty} \alpha_s^{\nu}[\delta(z-\nu d_s)+\delta(z+\nu d_s)-\delta(z)]$$

$$+ \alpha_o \sum_{\nu=0}^{\infty} \alpha_s^{\nu}[\delta(z-d_o-\nu d_s) + \delta(z+d_o+\nu d_s)]\}. \qquad (3.13)$$

In the case of a clean surface with uniform layer spacing d_s, substitution of $\alpha_o = \alpha_s$, and $d_o = d_s$ into Equation (3.13) yields:

$$P(z) = F_{ss}(z) * (1-\alpha_s^2)^{-1} \sum_{\nu=0}^{\infty} \alpha_s^{\nu}[\delta(z-\nu d_s) + \delta(z+\nu d_s)-\delta(z)] \qquad (3.14)$$

The significance of the P(z) functions in Equations (3.12) to (3.14) is quite clear. For example, from Equation (3.14) it can be seen that in the case of a clean, uniform substrate, the P(z) function consists of a single set of delta functions positioned at multiples of the interlayer spacing d_s, convoluted with the Fourier transform F_{ss} of the atomic scattering factor. The delta-function amplitudes decay exponentially with increasing $|z|$ due to the factor α_s^{ν} associated with the attenuation of the incident and diffracted beams in the crystal. Evidently, the broadening of the structure-related peaks and the occurrence of extra, spurious peaks in the P(z) functions shown in Figure 2 are related to the form of F_{ss}.

It is now appropriate to consider the truncation of the Fourier integral in the case of experimental data analysis. As noted earlier, the minimum value of s is defined by the inner potential of the solid, to be: $s_1 = \frac{1}{\pi}(\frac{2m}{\hbar})^{1/2} V_o^{1/2}$, since electrons incident upon the solid at zero energy increase in energy by V_o upon entering the solid. In practice, s_1 is usually larger than the value given above because of the difficulty in working with electron beams of energy less than about 10 eV. The maximum value of s, s_2 is usually chosen arbitrarily by the experimentalist, but in general, the choice is very much influenced by the strong attenuation of diffracted intensity with increasing s, due to the decrease in scattering factor and to the Debye-Waller effect.

Thus, the Fourier integral of Equation (3.6) must be written as:

$$P^1(z) = \int_{s_1}^{s_2} I_{oo}(s) e^{-2\pi i s z} ds .$$

If we define a box-car window ω_B according to:

$\omega_B = 1; \quad s_1 \leq s \leq s_2$
$ = 0; \quad \text{elsewhere,}$

then the above integral can be written as:

$$P^1(z) = \int_{-\infty}^{+\infty} \omega_B(s) I_{oo}(s) e^{-2\pi i s z} ds ,$$

which, by application of the Convolution Theorem, leads to:

$$P^1(z) = \int_{S_1}^{S_2} e^{-2\pi i s z} ds * \int_{-\infty}^{+\infty} I_{oo}(s) e^{-2\pi i s z} ds, \text{ that is}$$

$$P^1(z) = W_B(z) * P(z), \qquad (3.15)$$

where

$$W_B = \int_{S_1}^{S_2} e^{-2\pi i s z} ds = \frac{(e^{-2\pi i s_1 z} - e^{-2\pi i s_2 z})}{2\pi i z} \qquad (3.16)$$

Thus the expressions for P(z) given in Equations (3.14) to (3.16) must be convoluted with $W_B(z)$ in the case of a truncated range of s. Equations (3.14) to (3.16) can be applied directly if the Fourier transforms $F_{ss}(z)$, etc., are understood to be truncated, that is:

$$F_{ss}(z) = \int_{S_1}^{S_2} f_s(s) f_s^\dagger(s) e^{-2\pi i s z} ds . \qquad (3.17)$$

The consequences of truncation of the Fourier integral, as derived above, are most simply illustrated for the case of a clean, uniform substrate. First of all, we further simplify by considering the case of a rigid lattice of unit, point scatterers, that is, with scattering factor independent of s. In the absence of truncation, the limits of integration of Equation (3.17) are infinite and $F_{ss}(z)$ is a delta function at the origin $\delta(z)$. P(z) in Equation (3.14) reduces simply to the set of delta functions at multiples of the interlayer spacing d_s. This P(z) function and the corresponding infinite intensity spectrum are shown respectively in Figures 4b and a.

In the case of truncation, still considering unit scatterers, $F_{ss}(z)$ is simply the Fourier transform $W_B(z)$, shown in Figure 4d, of the box-car window shown in Figure 4c. For this case, convolution of $W_B(z)$ with the set of delta functions, according to Equation 3.14, gives the P(z) function shown in Figure 4f. The equivalent procedure in reciprocal space is multiplication of the infinite spectrum (Figure 4a) with the box-car window (Figure 4c) to give the truncated intensity spectrum shown in Figure 4e. Thus the broadening of the delta functions and the extra non-structural

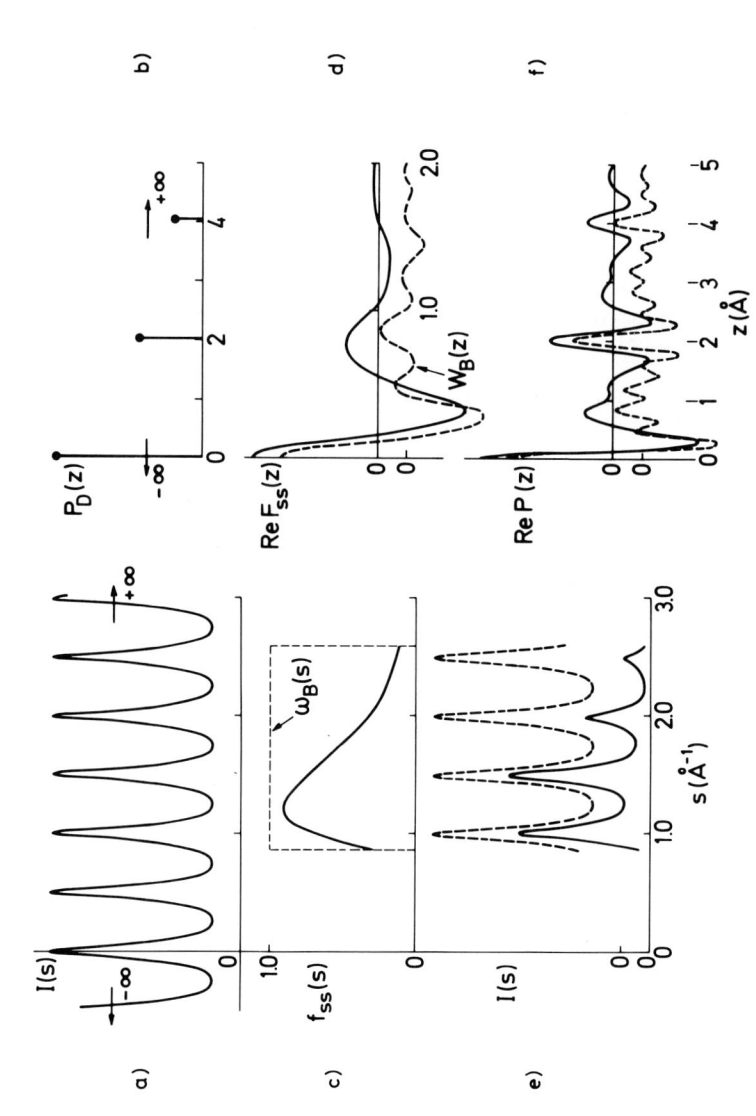

Fig. 4. Illustration of the consequences of data truncation and the energy dependence of the atomic scattering factor upon the Fourier transform P(z) of diffracted intensities for the specular beam.
a) Infinite intensity spectrum calculated for clean substrate with uniform interlayer spacing $d_s = 2$Å, and for unit, point scatterers.
b) Delta function set $P_D(z)$ obtained by Fourier transformation of the intensity spectrum of a).
c) Realistic, truncated atomic scattering factor $f_{ss}(s)$, calculated using 14 scattering phase shifts appropriate for Al and renormalized for thermal vibrations (see Section 4). Box-car window $\omega_B(s)$ shown as dashed line.
d) Fourier transforms of the functions shown in c). Note the pronounced effect of the energy dependence of $f_{ss}(s)$ upon the line-shape of Re[$F_{ss}(z)$].
e) Truncated intensity spectra obtained by multiplication of the infinite spectrum a) by the truncated atomic scattering factor and by the box-car window (dashed line).

231

Fig. 4. (cont'd)

>f) Real part of P(z) functions formed by convolution of $P_D(z)$ with $F_{SS}(z)$ and $W_B(z)$ (dashed line), or equivalently by Fourier transformation of truncated intensity spectra shown in e).

peaks in the Fourier transform (Figure 4f) of the truncated intensity spectrum are seen to be related to the line-shape of $W_B(z)$.

Finally, the P(z) function of Equation 3.14 in the case of a realistic scattering factor is also shown in Figure 4f. As can be seen, the energy dependence of the scattering factor (Figure 4c) leads to pronounced changes of the line-shape of its Fourier transform (Figure 4d) as compared to the box-car case, leading to corresponding changes in P(z).

In the case of a clean substrate, the P(z) functions given by Equations (3.13) and (3.14) contain the Fourier transform $F_{SS}(z)$ of the substrate scattering factor alone. Since $f_s(s)f_s^\dagger(s)$ is real, it follows from Equation (3.17) that $\text{Re}[F_{SS}(z)]$ is even and thus symmetric about the origin. Consequently, peak shifts in $\text{Re}[P(z)]$ from the delta-function positions result solely from the overlap of structure-related peaks with the oscillatory side-lobes of adjacent peaks, and will only be severe if the delta functions are closely spaced. $\text{Im}[F_{SS}(z)]$ is odd and thus antisymmetric about the origin. Consequently all peaks in $\text{Im.}[P(z)]$ are shifted to higher $|z|$. The shift in peak positions between $\text{Re}[P(z)]$ and $\text{Im.}[P(z)]$ is the cause of the very broad peaks in $|P(z)|$ shown in Figure 2.

In the case of an adsorbed layer, the P(z) function given by Equation (3.12) contains the Fourier transforms $F_{OS}(z)$ and $F_{SO}(z)$ of products of substrate and overlayer scattering factors. These products are complex so that $\text{Re}[F_{OS}(z)]$ and $\text{Re}[F_{SO}(z)]$ are not symmetric about the origin. Thus, it follows from Equation (3.12) that peaks in $\text{Re}[P(z)]$ which derive from the vector sets $\delta[z^\pm(d_0+\nu d_s)]$ will be subject to larger shifts than peaks which derive from the sets $\delta[z^\pm\nu d_s]$.

The derivations and discussion above provide a complete explanation for the occurrence of spurious peaks in the Fourier transforms of calculated, single-scattering intensities for the specular diffracted beam (e.g., Figure 2) and should provide an approximate basis for the analysis of experimental data since the effects of multiple scattering do not lead to pronounced new features in the transforms (see also Section IV). Although a detailed analysis will not be given here for the case of the P(xyp) sections (Figure 3), it is evident that truncation of the sums over h and k leads to the broadening of the structure-related peaks, and that the occurrence of spurious peaks in sections taken at arbitrary values of p is due to the truncation of the integral over s.

In the remainder of this section, we discuss previous uses of the Patterson functions of LEED intensities in the light of the above discussion. The first, brief, application to LEED appears to be that of Tucker[23], later followed by Tucker and Duke[24], but a detailed discussion of this work is made difficult by the omission of any reference to the particular form of the Patterson function which was used. From our reading of the latter paper[24], it appears that a P(xyO) section was constructed, for diffracted intensities measured for oxygen adsorbed on Rh(100). No consideration was given in this work to the consequences of data truncation, scattering potential, and vibronic effects, and it is not clear that the authors understood the physical significance of a P(xyO) section since the contribution of the planar structure of the substrate to the section was not considered.

The first detailed studies of the use of Patterson functions in LEED were made by Clarke, Mason, and Tescari[21]. These authors presented analyses of experimental LEED data from Si(100)[21a], carbon monoxide adsorbed on Pt(100)[21b], and clean Pt(100)[21c]. In the latter two papers, P(z) and P(xyp) functions were constructed using Equations (3.4) to (3.6). It was assumed that the detailed structural interpretation could be made in the same manner as in x-ray crystallography by considering only the real part of these complex functions. The potential problem of multiple scattering was raised, and the suggestion made that the effects of multiple scattering might be averaged out upon Fourier transformation.

The work of Clarke, Mason, and Tescari[21] created a considerable interest in this new approach to the analysis of LEED and stimulated a number of other studies, including our own[9,10,22]. It should be pointed out, however, that although Clarke et al. recognized that data truncation might be a problem in their analysis, their treatment of this problem was inadequate, and we believe that their conclusions regarding surface structures should be viewed with considerable skepticism. Since a detailed criticism of their work has been given by Woodruff, Mitchell, and McDonnell[19], we restrict the present discussion to consideration of two procedures proposed by Clarke et al.[21b,c] for identifying spurious peaks in Patterson functions of LEED intensities.

In their second paper[21b], Clarke et al. suggested that distinction between structure-related and spurious peaks could be made by requiring that the former constitute a self-consistent vector set. Whilst this, in principle, is a valid requirement, applying in the hypothetical case of point-scatterers and no data truncation, in practice the structure-related peaks are shifted in position, as discussed earlier in this section. Accordingly, the structure-related peaks can only form a vector set if some uncertainty in the position of each component of the set is allowed. This relaxation of the requirement for a vector set is, apart from leading to further inaccuracy in structural assignment, in our experience always sufficient to allow identification of further, spurious sets which include truncation peaks as members. Structural interpretation is therefore ambiguous.

In their latest paper[21c], Clarke et al. proposed an additional procedure for identification of spurious peaks based on comparison of Fourier transforms of the experimental data with transforms of single-scattering intensities calculated assuming a model of the surface structure, the surface structure itself being chosen by assuming major peaks in the experimental transform to have structural significance. Although this procedure obviously cannot take into account the effects of multiple scattering in the experimental transforms, nevertheless, in principle we consider it to be a useful concept, related in some aspects to the transform-deconvolution method[9] which we first described shortly afterwards. There are, however, several major problems which were not addressed by Clarke et al.[21c]. Firstly, as discussed earlier in this section, any procedure based on straightforward interpretation of the Patterson function, is necessarily inaccurate because of shifts in the positions of structure-related peaks. The ambiguity in identification of structure-related peaks means that alternative model structures must be considered, and proper criteria for assessing their validity in terms of comparison of the Fourier transforms of experimental and calculated intensities should be established. Clearly, the inaccuracy of the model structures leads to inaccuracy in the calculated intensities which would be a cause of uncertainty in any comparison process. Secondly, the calculation of single-scattering intensities must take into account the energy dependencies of the atomic-scattering factors and the Debye-Waller factor since these factors have a pronounced effect on the content of the Fourier transform of the intensities. In addition, the calculations must accurately account for inelastic scattering of electrons in the crystal since, as we have shown above, this causes a strong attenuation of peak amplitudes in the transform with increasing z, where z is the surface-normal direction. Finally, the inner potential of the crystal must be known for the calculation of both the intensities and the Fourier transforms.

The failure of Clarke et al.[21c] to provide a satisfactory treatment of the problems described above casts considerable doubt on the significance of their conclusions based on comparison of Fourier transforms of calculated and experimental intensities. The transform-deconvolution method[9] described in the next section, although addressing these problems from a rather different perspective, constitutes an attempt to provide a systematic procedure for their solution.

Finally, in this section we consider a different approach to the question of data truncation, proposed by Buchholtz and Lagally[25]. These authors have used an optical-transform method in the analysis of both individual and angle-averaged specular-beam-intensity spectra from Ni(100). The optical transform is a modified form of Equation (3.6) in which $|P(z)|$ is formed after pre-multiplication of the intensity spectra by a smoothing function K(s) ("Kaiser window"[26]):

$$|P(z)|_{opt} = \left| \int_{s_1}^{s_2} I_{oo}(s) K(s) e^{-2\pi i s z} ds \right| . \qquad (3.18)$$

The purpose of this procedure is to eliminate the Gibbs oscillations which result from truncation of the Fourier integral. This is achieved because the Fourier transform of the Kaiser window consists essentially of a single peak at the origin, unlike the Fourier transform of the box-car window (Figure 4e) which contains oscillatory side-lobes.

The crucial disadvantage of this procedure, which in our judgment makes it of little use in structure determination from LEED intensities, is that suppression of the Gibbs oscillations is achieved only at the expense of considerable broadening of the main peak. Thus the peak widths in an optical transform are even wider than in $|P(z)|$ (e.g., Figures 2e and f), and so peak resolution is very poor. In addition, the optical-transform procedure allows only for the fact of data truncation. The important consequences of the energy dependence of the scattering factor and the Debye-Waller factor, and of the angular dependence of the scattering factor are not taken into account.

IV. Transform-Deconvolution Method[9,10]

As described in the previous section, in the single-scattering approximation the Fourier transform P(z) of specular beam intensity spectra contains convolution products of functions of the required structural parameters with functions of the atomic scattering factors. The transform-deconvolution method consists of a procedure for deconvolution of the structural and non-structural content of the Fourier transform. In this section, we discuss the principles of the method and describe its application to experimental data analysis in the case of one-component systems. The case of overlayer systems is discussed in the next section.

For a clean substrate, allowing for the possibility of expansion or contraction of the interlayer spacings at the surface, Equation 3.13 can be compactly written as:

$$P(z) = P_D(z) * F_{ss}(z) , \qquad (4.1)$$

where $P_D(z)$ contains sets of delta functions associated with each different interlayer spacing. Given a knowledge of the inner potential, $V_O(s)$, $P(z)$ is obtained by Fourier transformation of an experimental intensity spectrum:

$$P(z) = \int_{S_1}^{S_2} I_{OO}(s) e^{-2\pi i s z} ds \:. \qquad (4.2)$$

$F_{ss}(z)$ is the truncated Fourier transform of the atomic scattering factor, $f_{ss}(s)$:

$$F_{ss}(z) = \int_{S_1}^{S_2} f_{ss}(s) e^{-2\pi i s z} ds \qquad (4.3)$$

where $f_{ss}(s)$ is calculated as described below. Thus ignoring for the moment any consequences of multiple-scattering and any uncertainties in $V_O(s)$ and $f_{ss}(s)$, determination of interlayer spacings requires solution of Equation 4.1 for $P_D(z)$ given $P(z)$ and $F_{ss}(z)$.

The atomic scattering factor in the absence of thermal vibrations is obtained from the partial wave expansion:

$$f_{ss}^1(s) = \left| \left(\frac{-i\cos\theta}{2\pi s}\right) \sum_{\ell=0}^{\infty} (2\ell+1)[e^{2i\delta_\ell}-1]P_\ell(\cos\Phi) \right|^2, \qquad (4.4)$$

where δ_ℓ is the ℓth partial wave phase shift obtained by integration of the radial Schrödinger equation, typically using an APW band-structure potential. In the equation, θ is the angle of incidence, Φ is the diffracted angle, and $P_\ell(\cos\Phi)$ is the ℓth Legendre polynomial.

It should be noted that in the context of a given model potential, the scattering factor $f_{ss}^1(s)$ can be obtained to arbitrary accuracy in our calculations by inclusion of a sufficient number of phase shifts. Calculation of the scattering factor is a relatively minor part of the analysis so that inclusion of a large number of phase shifts does not significantly increase computational time or storage, unlike the case of dynamic model calculations of diffracted intensities.

The effect of thermal vibrations is included approximately[27] by multiplying the rigid-lattice scattering factor $f_{ss}^1(s)$ by a Debye-Waller factor[20]:

$$f_{ss}(s) = f_{ss}^1(s) e^{-\gamma(s)s^2}, \qquad (4.5)$$

where

$$\gamma(s) = \frac{3h^2}{2Mk_B\theta_D}\left[\frac{1}{4} + \left(\frac{T}{\theta_D}\right)^2 \int_0^{\theta_D/T} \frac{x}{e^x-1}\,dx\right] \quad (4.6)$$

and θ_D is an effective Debye temperature.

Evaluation of Equations 4.2 and 4.3 must be carried out by numerical integration. This is done using the trapezoidal rule which is known to be quite accurate for integrals of this kind. In addition, the relationship of the trapezoidal rule to the discrete Fourier transform allows the use of the known convolution properties of such transforms[28].

Application of the trapezoidal rule to Equations 4.2 and 4.3 yields aliased[28] versions of $P(z)$ and $F_{ss}(z)$, which are periodic with period $T = 1/\Delta s$:

$$P(z) \simeq P''(z) = \frac{1}{T} e^{-2\pi i s_1 z} \sum_{j=0}^{N-1} I_{oo}(j\Delta s) e^{-2\pi i j z/T} \quad (4.7)$$

and

$$F_{ss}(z) \simeq F_{ss}''(z) = \frac{1}{T} e^{-2\pi i s_1 z} \sum_{j=0}^{N-1} f_{ss}(j\Delta s) e^{-2\pi i j z/T}, \quad (4.8)$$

where

$$N = (s_2 - s_1)/\Delta s \quad \text{and} \quad P''(z) = \sum_{n=-\infty}^{+\infty} P(z+mT).$$

Thus by choosing a small enough value for Δs, T can be made sufficiently large that the errors in approximating $P(z)$ by $P''(z)$ and $F_{ss}(z)$ by $F_{ss}''(z)$ are negligible in the range $|z| = 0$ to $T/2$, because of the rapid attenuation of $P(z)$ and $F_{ss}(z)$ with increasing $|z|$.

From Equations 4.7 and 4.8, $P(z)$ and $F_{ss}(z)$ can be obtained on a discrete grid of z at intervals of Δz over the range $|z| = 0$ to $T/2$, as sequences of $2M+1$ values where $M = T/2\Delta z$. The convolution integral of Equation 4.1 is written now in discrete form as:

$$P(k\Delta z) = \Delta z \sum_{j=-M}^{M} P_D(j\Delta z) F_{ss}[(k-j)\Delta z] \quad (4.9)$$

with the indices interpreted modulo $2M$. Defining $P_k = P(k\Delta z)$, $W_k = \Delta z F_{ss}(k\Delta z)$, $q_k = 2P_D(k\Delta z)$ for $|k| > 0$, and $q_k = P_D(k\Delta z)$ for $k = 0$, and taking advantage of the even nature of $P_D(z)$, Equation 4.9 can be written as:

$$P_k = \sum_{j=0}^{M} q_j (W_{k-j} + W_{k+j}). \quad (4.10)$$

Since $P(-z) = P^\dagger(z)$, we need consider only positive values of k, so in discrete form the convolution equation can be expressed as a set of simultaneous equations:

$$\sum_{k=0}^{M} P_k = \sum_{k=0}^{M} \sum_{j=0}^{M} q_j A_{k,j}, \quad (4.11a)$$

or as the vector-matrix product:

$$\vec{p} = \vec{q}(A), \quad (4.11b)$$

where $A_{k,j} = W_{k+1} + W_{k-j}$ and indices are interpreted modulo 2M.

Having written the convolution equation in a form suitable for numerical solution, there are many well-known procedures which could be used. Most directly, $\vec{q}$ could be obtained by matrix inversion:

$$\vec{p}(A)^{-1} = \vec{q}.$$

Unfortunately, however, solution of Equation 4.1 or 4.11 to yield a unique mathematical solution is not possible, even in principle, in the present case. This can be readily appreciated by considering the application of a standard procedure[28] for solution of convolution equations. Defining the normal and inverse Fourier transforms by the operators F and F^{-1}, inverse Fourier transformation of Equation 4.1 gives:

$$F^{-1}[P(z)] = F^{-1}[P_D(z) * F_{ss}(z)] \quad (4.12a)$$

or

$$I_{OO}(s) = f_{ss}(s) F^{-1}[P_D(z)](s) \omega_B(s), \quad (4.12b)$$

where ω_B is the box-car window defined in the previous section. In principle, $P_D(z)$ could be obtained from:

$$P_D(z) = F\left[\frac{I_{OO}(s)}{f_{ss}(s)\omega_B(s)}\right] \quad (4.13)$$

but in the present case this equation is undefined for values of s outside the range of the box-car window, s_1 to s_2. If the range of s is restricted to s_1 to s_2, then the solution obtained is not $P_D(z)$ but $P_D(z) * W_B(z)$, where $W_B(z)$ is the Fourier transform of ω_B defined by Equation 3.16. Thus, both $P_D(z)$ and $P_D(z) * W_B$ are possible solutions of Equation 4.12. In general, there are an infinite number of solutions which are identical to $F^{-1}[P_D(z)](s)$

in the range of s between s_1 and s_2, but which are arbitrary different outside that range. An alternative statement of the problem is that a unique determination of $P_D(z)$ requires a knowledge of $I_{OO}(s)$ over the complete range of s, and can only be achieved if $I_{OO}(s)$ can be uniquely constructed for all s from its known values in the range s_1 to s_2.

The problem of nonuniqueness in the interpretation of limited experimental data is not uncommon. As noted previously, it may be regarded as perhaps the fundamental problem of diffraction analysis. In the field of surface physics, another well-known example is the process of unfolding ion-neutralization spectra[29] to obtain information regarding densities of states. In general, the approach to solving problems of this kind must be to incorporate in the solution algorithm as much information, known a priori, concerning the nature of the solution as is possible. In the following we examine ways in which physical constraints can be built into a solution procedure for the problem at hand.

The main feature of the correct physical solution, $P_D(z)$ is that it contains vector sets of delta functions, as derived in Section 3. Thus, in our first attempt[9a] to implement the transform-deconvolution approach for experimental data analysis, for clean Ni(100), the procedure adopted was to convolute trial sets of delta functions with $F_{ss}(z)$ according to Equations 4.1 and 4.11, and compare the results with the experimental transform P(z). Reasonable correspondence was achieved between experimental and calculated P(z) functions for a single set of delta functions at positions close to the bulk interlayer spacing in Ni(100), indicating that the effects of multiple-scattering in the experimental intensities did not interfere too severely with the analysis. It was evident, however, that manual variation of the parameters characterizing the sets of delta functions, and visual comparison of the experimental and calculated P(z) functions would be an inadequate procedure in the case of more complicated systems. In addition, in this first work the atomic scattering factor was crudely described using s-wave phase shifts only.

In our efforts to refine this preliminary deconvolution scheme, we carried out computer experiments based on the Fourier transforms of calculated, single-scattering intensities. Various modifications to conventional algorithms for solution of systems of linear equations such as Equation 4.11 were investigated. Somewhat surprisingly we found that a number of such algorithms could be used to obtain solutions to Equation 4.1, in which the structural parameters of the deconvolution coincided with those used in the original calculation of the single-scattering intensities. However, with the exception of one method described below, the algorithms were numerically unstable, reflecting the ill-posed nature of the problem. The addition of extremely small amounts of random noise to the calculated intensities led to mathematical solutions of the convolution equation which bore no relation to the proper physical solution.

After a frustrating period during which the computer resolutely refused to solve our problem, we discovered a relaxation procedure, attributed to Southwell[30,31], which quite fortuitously turned out to be considerably biased toward producing solutions of the required form.

The method begins in the general manner of relaxation schemes with definition of a <u>residual</u> vector:

$$\vec{r} = \vec{p} - \vec{q}(A) \qquad (4.14)$$

where $\vec{p}$, $\vec{q}$, and (A) are defined in Equation 4.11. Clearly, if a correct solution for $\vec{q}$ is used in Equation 4.14, $\vec{r} = 0$. Thus an iterative procedure is used to reduce $\vec{r}$:

0th iteration: $\vec{r}^{(0)} = \vec{p}; \quad \vec{q}^{(0)} = 0.$

1st iteration: $\vec{r}^{(1)} = \vec{r}^{(0)} - \Delta q_k^{(0)} \vec{a}_k^{(0)}; \quad q_k^{(1)} = q_k^{(0)} + \Delta q_k^{(0)} \cdot \qquad (4.15)$

νth iteration: $\vec{r}^{(\nu)} = \vec{r}^{(\nu-1)} - \Delta q_k^{(\nu-1)} \vec{a}_k^{(\nu-1)}; \quad q_k^{(\nu)} = q_k^{(\nu-1)} + \Delta q_k^{(\nu-1)}.$

The main feature of the Southwell method is that only one component of the solution vector $\vec{q}$, say q_k, is adjusted in each iteration by Δq_k. In the νth iteration the component of $\vec{q}^{(\nu-1)}$ to be adjusted is determined by the position of the maximum component of $\vec{r}^{(\nu-1)}$, that is $r_k^{(\nu-1)}$, at $z = k\Delta z$. The value of $\Delta q_k^{(\nu-1)}$ is then determined from the approximation:

$$r_k^{(\nu-1)} \cong q_k^{(\nu-1)} A_{kk}, \qquad (4.16)$$

which amounts to taking the dominant term only of the expansion given in Equation 4.11. This choice of $\Delta q_k^{(\nu-1)}$ leads to reduction to zero of $r_k^{(\nu)}$ by subtraction of the convolution product $\Delta q_k^{(\nu-1)} \vec{a}_k^{(\nu-1)}$ from $\vec{r}^{(\nu-1)}$ according to Equation 4.15, where $\vec{a}_k^{(\nu-1)}$ is the kth column of the matrix (A), that is, $\vec{a}_k = \sum_{j=0}^{N} A_{jk}$.

The progress of the Southwell scheme is illustrated by the example shown in Figure 5, based on the Fourier transform of calculated, single-scattering intensities for the specular diffracted beam from a clean substrate with uniform interlayer spacing. In the present application, the most important characteristic of the method is that it necessarily starts correctly by placing a delta function at the origin. From Equations 4.2 and 4.15 it can be seen that the maximum value of $\vec{r}(0)$ must occur for z = 0, since $\vec{r}(0)$=P(z) and the maximum value of P(z) is P(0). Thus in the first iteration, with an accuracy depending upon the approximation made in calculating the delta function amplitude from Equation 4.16, the truncation oscillations associated with the origin peak of $P_D(z)$ are removed from the residual. Accordingly, the possibility of correctly locating the next delta function and obtaining a good approximation to its amplitude from the maximum component of $\vec{r}^{(1)}$ in the second iteration is much increased.

In the example shown in Figure 5, the positions of the delta functions are exactly consistent with the values of the interlayer spacing used in calculating the diffracted intensities. In addition, the relative amplitudes decay exponentially with an exponent close to the value of the layer attenuation exponent used in the intensity calculation. In general, however, for less ideal cases the approximations necessary in choosing the delta function positions and amplitudes can lead to errors and to the occurrence of additional noise peaks in the deconvolution. The Southwell procedure is self-correcting, in principle, in the limit of a large number of iterations, but in practice it is expedient to terminate the iteration sequence when the residual has been reduced to a predetermined level characterized by the value of an error indicator R, given by:

$$R^{(\nu)} = \sum_{k=0}^{M} |r_k|^{(\nu)} \Big/ \sum_{k=0}^{M} |P_k| . \qquad (4.17)$$

The terminal value of R was typically taken to be 0.05.

In order to remove the errors discussed above, two modifications to the Southwell scheme were made. Briefly, the first consists of choosing the component of $\vec{q}$ to be adjusted, $q_k^{(\nu-1)}$, according to the criterion that $R^{(\nu-1)} - R^{(\nu)}$ should be a maximum. The second modification is a scheme in which the iteration sequence is oscillated back and forth, so that the value of $q_k^{(\nu)}$ can be used to correct the values $q_\ell^{(\nu-1)}, q_m^{(\nu-2)}$, etc. obtained in previous iterations.

With these modifications, the Southwell method was found to give rapid convergence to accurate deconvolutions in all cases involving P(z) functions formed from calculated, single-scattering intensities. The method has some limitations, however, in application to experimental data analysis, stemming essentially from the fact that the only physical constraint built into the procedure is that $P_D(z)$ must contain a delta function at the origin. Nevertheless, the method was used quite successfully in the analysis of experimental data from Al(100)[9b-d], as described below, and its discussion here at some length has been warranted by the fact that

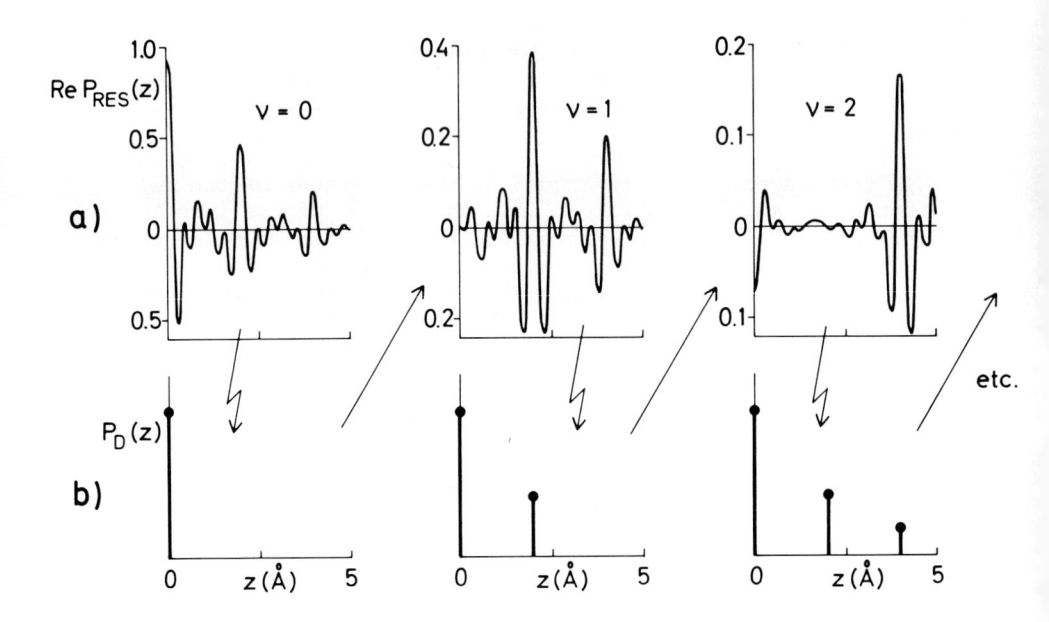

Fig. 5. Illustration of Southwell deconvolution scheme, using P(z) function of calculated intensities from a uniform, point lattice with layer spacing 2Å. In the zeroth iteration the residual function $P_{RES}(z)$ is set equal to P(z). The maximum value of $P_{RES}^{(0)}(z)$ occurs at z = 0, so a delta function is placed at the origin in the solution $P_D^{(1)}(z)$, with amplitude $P_{RES}^{(0)}(z)/F_{ss}(0)$. Convolution of $P_D^{(1)}(z)$ with $F_{ss}(z)$ and subtraction from $P_{RES}^{(0)}(a)$ yields $P_{RES}^{(1)}(z)$. The maximum value of $P_{RES}^{(1)}(z)$ occurs at z = 2Å so a delta function of amplitude $P_{RES}^{(0)}(2.0)/F_{ss}(0)$ is placed at this position in $P_D^{(2)}(z)$, and so on.

it forms an essential first step in a more refined procedure to be described later in this section.

In Figure 6 are shown experimental intensity spectra[9] for the specular diffracted beam for Al(100) at six different angles of incidence. The corresponding P(z) Fourier transforms are shown in

Analysis of LEED 243

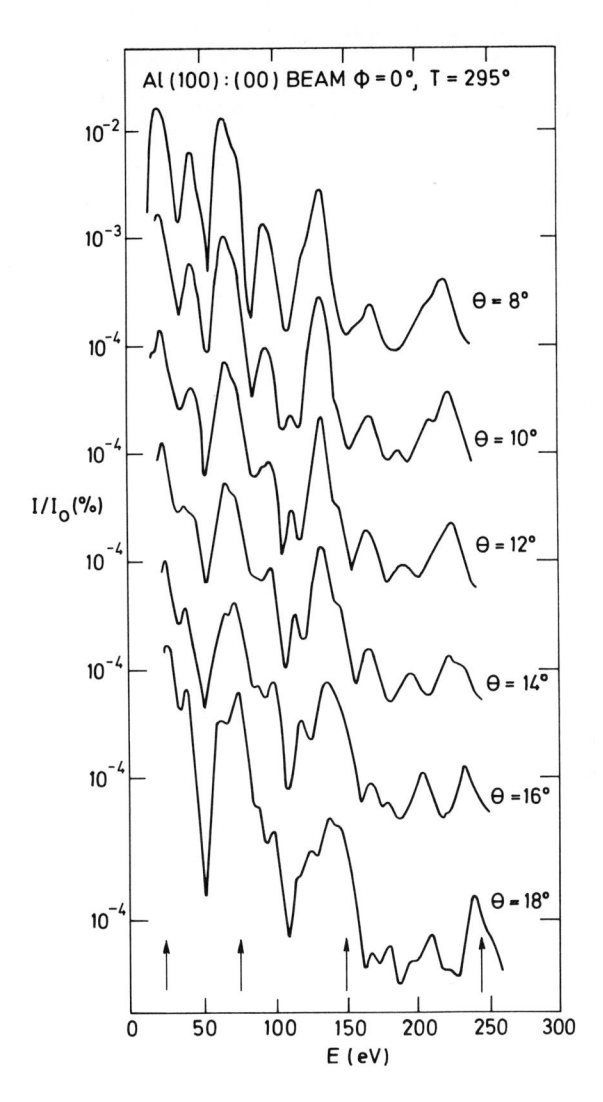

Fig. 6. Normalized intensity spectra for the specular diffracted beam from clean Al(100) at six angles of incidence, $\theta = 8°$ to $18°$ in $2°$ increments, and azimuthal angle $\Phi = 45°$. The intensities are plotted on a logarithmic scale. Single-scattering, Bragg peak positions are indicated by arrows on the $\theta = 18°$ spectrum. Note the strong effects of multiple-scattering in producing extra peaks and fine structure in the peak line-shapes. Note also the strong attenuation

Fig. 6. (cont'd)

of the spectra with increasing electron energy, due to the decrease in atomic scattering factor and Debye-Waller effect.

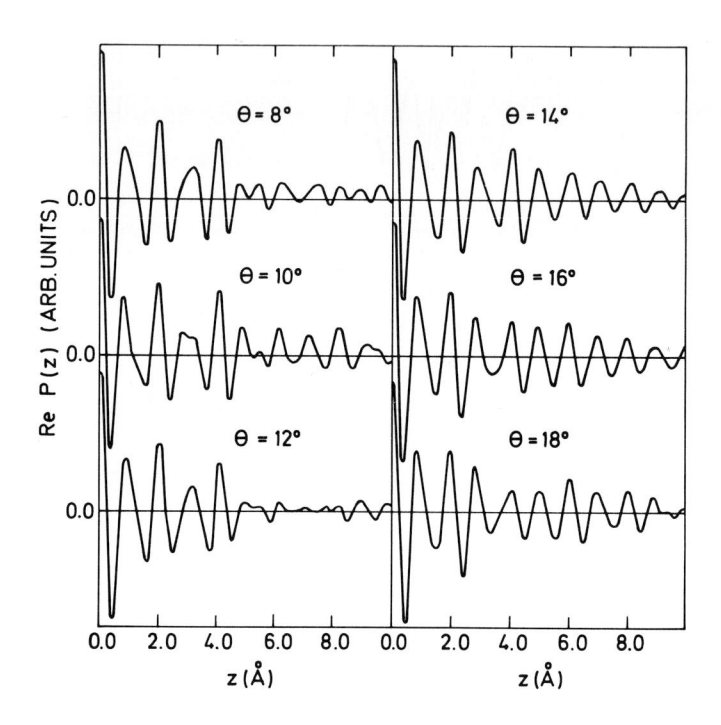

Fig. 7. Real parts of the P(z) functions of the intensity spectra of Figure 6.

Figure 7. In Figure 8 are shown the deconvolutions obtained using the modified Southwell method of the P(z) functions of Figure 7. In performing this analysis, the inner potential was taken to be 14 eV. The scattering factor $f_{ss}(s)$, and hence its Fourier transform $F_{ss}(z)$, was obatined using 14 phase shifts in the partial wave expansion, Equation 4.4. The phase shifts were derived from Snow's potential[31] for Al. The scattering factor was renormalized for thermal vibrations according to Equations 4.5 and 4.6 using a Debye temperature of 340°K.

As shown in Figure 8, the deconvolutions all contain dominant peaks corresponding to a consistent vector set with spacing 2.05 Å.

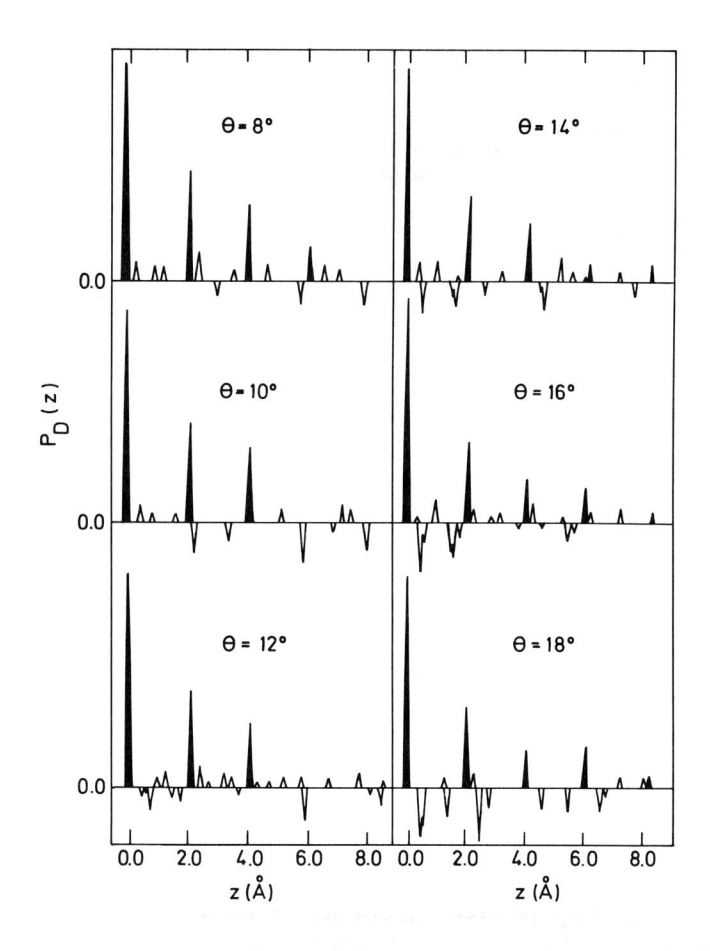

Fig. 8. Southwell deconvolutions of the P(z) functions of Figure 7. $F_{ss}(z)$ used in the deconvolution was derived from an atomic scattering factor for Al calculated using 14 scattering phase shifts and a Debye temperature of 340°K. The deconvolutions were carried out on a grid of $\Delta z = 0.05$ Å. Peaks forming a consistent vector set at $z = \nu 2.05$ Å, where ν is an integer have been filled. Note the occurrence of additional random noise peaks, and the failure to observe all the components of the vector sets.

This value may be compared with the bulk interlayer spacing along the [001] direction of 2.025 Å. However, as can be seen from the figure, the deconvolutions also contain random noise peaks. An important question is the relative contribution of various factors to this noise component, apart from experimental errors in the intensities.

Approximations made in the input to the analysis include the use of a spherical model potential in deriving the scattering phase shifts, the assumption of an energy-independent inner potential, and the use of a Debye-Waller correction for thermal vibrations including the use of a single Debye temperature. These assumptions are also usually made in dynamic model calculations and the lack of quantitative agreement between such calculations and experimental data does not inspire great confidence in their accuracy. The remaining source of uncertainty in the present case is the extent to which multiple-scattering processes contribute to the noise in the deconvolution, and affect the parameters characterizing the set of delta functions.

The relative importance of the approximations discussed above is difficult to judge. Some useful insight was obtained, however, by studying the effects of deliberately introduced errors in the scattering factor used in the deconvolution. In an attempt to isolate possible effects of multiple-scattering, comparison was made of the deconvolution of $P(z)$ functions obtained from both the experimental intensities and intensities calculated using single-scattering theory. In the intensity calculation the complete scattering factor was used, but in the deconvolutions of $P(z)$ functions from both experimental and calculated intensities the complete scattering factor was replaced by its factors $f_{ss}^1(s)$ and $\gamma(s)$, which are shown together with $f_{ss}(s)$ in Figure 9.

Thus in the deconvolutions shown in Figures 10b-d, $f_{ss}(s)$ was approximated by respectively, the box-car window ω_B, the rigid-lattice scattering factor $f_{ss}^1(s)$, and the Debye-Waller factor $\gamma(s)$. In Figure 10e the complete scattering factor $f_{ss}(s)$ was used. As is evident in Figures 10b-d, the signal-to-noise level is about the same from both experimental and calculated intensities. Since the calculated intensities did not take into account multiple-scattering, and were based on the assumptions of an energy-independent inner potential and a single Debye-Waller correction, the noise level in the corresponding deconvolutions results solely from the inaccurate representation of the scattering factor. The rather similar character of the noise in the experimental deconvolutions suggests that this results largely from the same cause.

In the case shown in Figure 10e, where the complete scattering factor was used, a noise-free deconvolution is obtained in the case of the calculated intensities, as is expected since the values of V_0, θ_D, and $f_{ss}(s)$ used in the deconvolution were those used in calculating the intensities.

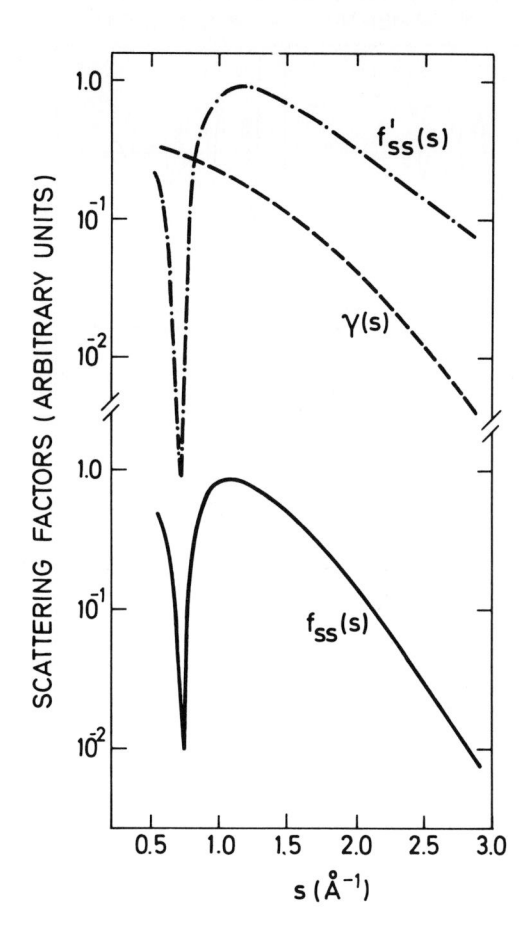

Fig. 9. Components of the atomic scattering factor for Al used in obtaining the deconvolutions shown in Figure 10. $f'_{ss}(s)$ is the rigid-lattice scattering factor calculated using 14 phase shifts. $\gamma(s)$ is the Debye-Waller factor for $\theta_D = 340°K$. $f_{ss}(s) = \gamma(s)f'_{ss}(s)$ is the complete scattering factor. Calculations are for $\theta = 8°$.

In the light of the comparisons shown in Figures 10b-d, it seems reasonable to infer that the residual noise in the experimental deconvolution shown in Figure 10e is at least in part attributable to remaining inaccuracies in the scattering factor $f_{ss}(s)$. For this particular system, the obvious importance of the Debye-Waller correction, evident upon comparison of Figures 10c and d, suggests

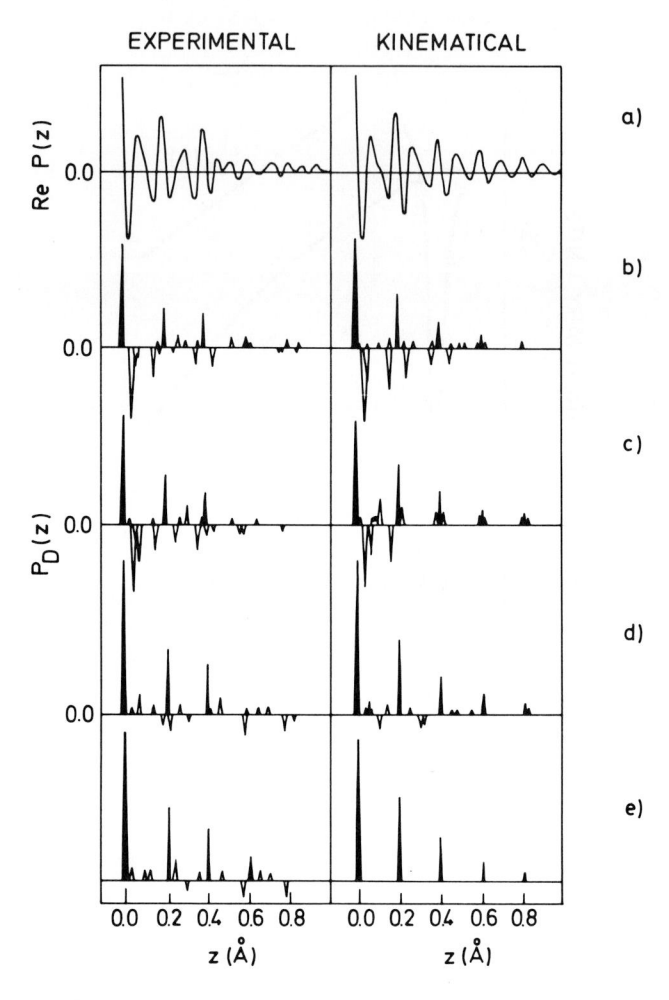

Fig. 10. Comparison of Southwell deconvolutions of P(z) functions of experimental and calculated intensity spectra for Al(100). θ = 8°.
a) P(z) functions. Note the similarity of the functions, which occurs despite the strong multiple-scattering features in the experimental intensities (Figure 6).
b) Deconvolutions in which $f_{ss}(z)$ was approximated by the box-car window only.
c) Deconvolutions in which $f_{ss}(s)$ was approximated by the truncated rigid-lattice scattering factor $f'_{ss}(s)$.
d) Deconvolutions in which $f_{ss}(s)$ was approximated by the truncated Debye-Waller factor $\gamma(s)$.
e) Deconvolutions using the full scattering factor $f_{ss}(s)$.

that inaccuracy in the description of thermal vibrations might be the dominant cause of noise in the deconvolution. Certainly, it seems reasonable to conclude that multiple-scattering is probably not the major cause of noise.

In carrying out the analysis of experimental data, as described above, a variation was performed over the values of V_O and θ_D in order to obtain a maximum signal-to-noise in the deconvolutions. A major deficiency of the Southwell method in this respect is the fact that the "signal", which we define as a consistent vector set of delta functions, is not unequivocally distinguished from the noise, since nothing in the procedure guarantees that the solution will contain a consistent vector set. In general, errors in the analysis can lead to displacements of delta functions from their correct positions in a vector set or, in more extreme cases, to failure to find the complete set. A related, practical difficulty with the method is the need to carry out the analysis on a finite grid of z. In the results described above, an interval, $\Delta z = 0.05 \text{\AA}$ was used in order to keep the number of operations necessary in performing the convolution $P_D(z)*F_{SS}(z)$ to a reasonably low value, when $P_D(z)$ and $F_{SS}(z)$ were described for the range 0 - 10 Å. Whilst a finer grid of z could be used without requiring excessive computation, in general some error is introduced into the analysis unless the actual value of the interlayer spacing falls fortuitously on the grid of z which is used.

In the light of our experience in applying the modified Southwell deconvolution scheme to experimental data analysis, we have evolved a considerably refined procedure, as described below. The main features are as follows:

A. The convolution Equation 4.1 is modified to explicitly account for the occurrence of errors in the input to the analysis:

$$P(z) = [P_D(z) + P_N(z)]*F_{SS}(z), \qquad (4.18)$$

where $P_N(z)$ is the noise component of the solution.

B. The solution algorithm is constrained to produce vector sets of delta functions, $P_D(z)$ and random noise, $P_N(z)$, which are unambiguously distinguished.

C. The interlayer spacings are obtained on a continuous grid of z.

D. A variation over V_O and θ_D is carried out to maximize signal-to-noise in the deconvolution.

The main points of an algorithm incorporating these features are listed below:

1. $P_{ex}(z)$ is constructed from the experimental intensities using a first guess for V_o. $F_{ss}(z)$ is constructed using a calculated scattering factor, including first guess for θ_D.
2. The Southwell deconvolution scheme is used, but is terminated after the first 2-4 iterations, yielding first approximations to the interlayer spacings, $d_i^{(0)}$, and giving a first approximation to the layer attenuation exponent $\mu^{(0)}$, calculated using the relative amplitudes of the delta functions, inserted in Equation (3.13).
3. Based on $d_i^{(0)}$, $\mu^{(0)}$, and a scaling constant $c^{(0)}$, a complete $P_D^{(0)}(z)$ function is constructed using Equation (3.13). $P_D^{(0)}(z)$ is convoluted with $F_{ss}(z)$ to give:

$$P_{calc}^{(0)}(z) = [P_D^{(0)}(z) + P_N^{(0)}(z)] * F_{ss}(z) \qquad (4.19)$$

4. Defining an error indicator:

$$R^{(\nu)} = \frac{\sum |P_{ex}(z) - P_{calc}^{(\nu)}(z)|}{\sum |P_{ex}(z)|} \qquad (4.20)$$

d_i, μ, c, and $P_N(z)$ are varied until $R \leq 0.1\%$.

5. Steps 1 to 4 are iterated with an outer variation over V_o and θ_D until a maximum of signal-to-noise is obtained in the deconvolution.

In interpreting the results of step 2, we have typically restricted consideration of the number of different interlayer spacings, d_i, to $i \leq 3$, which is justified in general by the limited penetration of low-energy electrons into solids. In practice, in a particular analysis of a given experimental intensity spectrum, it is convenient, although not essential, to definitely constrain the value of i. Thus the algorithm is constrained to produce a solution with a single interlayer spacing, or with first layer spacing different from subsequent uniform layer spacings, or so on.

It is emphasized that the treatment of errors in the analysis as resulting in random noise in the deconvolution is empirical. As mentioned earlier, there are grounds[22] for expecting multiple-scattering processes to contribute a slowly-varying modulation of $P_D(z)$, rather than a random noise component. In addition, the assumption has been made that the best approximation to the correct structural parameters is obtained from a deconvolution corresponding to a local maximum of signal-to-noise.

Evaluation of the procedure must rest upon the results of experimental data analysis. For this reason, we have concentrated to date upon the analysis of diffracted intensities from clean metal surfaces whose first interlayer spacings are believed, from the results of dynamic model calculations, to lie within ± 10% of the bulk values. Apart from requiring sensible output values of the structural parameters, however, additional consistency checks can be applied. In particular, reasonable output values of the non-structural parameters, V_o, θ_D, and μ must be obtained. This criterion is, in fact, used to discriminate between deconvolutions corresponding to different local maxima of signal-to-noise. In addition, analysis of intensity spectra taken for different angles of incidence and azimuth must produce internally self-consistent results.

Since the procedure involves a parameter variation to maximize signal-to-noise, S/N, an obvious practical requirement is that plots of both the structural and non-structural parameters versus S/N should exhibit well-defined maxima. To demonstrate that this is in fact the case, plots of S/N versus d, V_o and θ_D are shown in Figures 11 to 13 respectively, from an analysis[10] of an experimental intensity spectrum for Al(100). In these plots and in the analyses described later in this section, S/N was determined according to:

$$S/N = P(0)/\Sigma_N |P_N(z)| . \qquad (4.21)$$

In Figure 11 is shown a plot of S/N versus d, for fixed V_o and θ_D, where d is the uniform interlayer spacing in the Al(100) surface (see below). The figure contains the results of a large number of deconvolutions, for the case of which d was held fixed at a particular value. As can be seen from the figure, the result is not quite unambiguous since a small peak is found at 4.02 Å, but it is obvious that a well-designed variation procedure will converge to the correct result, since the first approximation to d obtained from the Southwell procedure was 2.05 Å in this case.

In Figure 12 is shown a plot of S/N versus V_o for fixed θ_D. The interlayer spacing d was allowed to vary in obtaining the deconvolutions used to construct this figure, and the corresponding plot of d versus V_o is shown as the dashed line. As can be seen, the S/N exhibits a well-defined maximum as a function of V_o, although the dependence is weaker than upon d, shown in Figure 11. Nevertheless, the dependence of d upon V_o shown in Figure 12 indicates the need for an accurate choice of V_o to obtain an accurate value of d.

A plot of S/N versus θ_D for fixed V_o is shown in Figure 13. Again a well-defined maximum is obtained. The value of d was allowed to vary in the deconvolutions, but the plot of d versus θ_D, also shown in Figure 31, indicates that d depends less critically upon θ_D than upon V_o, as would be expected.

Having established that the signal-to-noise depends critically upon the value of the structural parameter and strongly upon the values of the non-structural parameters, the deconvolution procedure

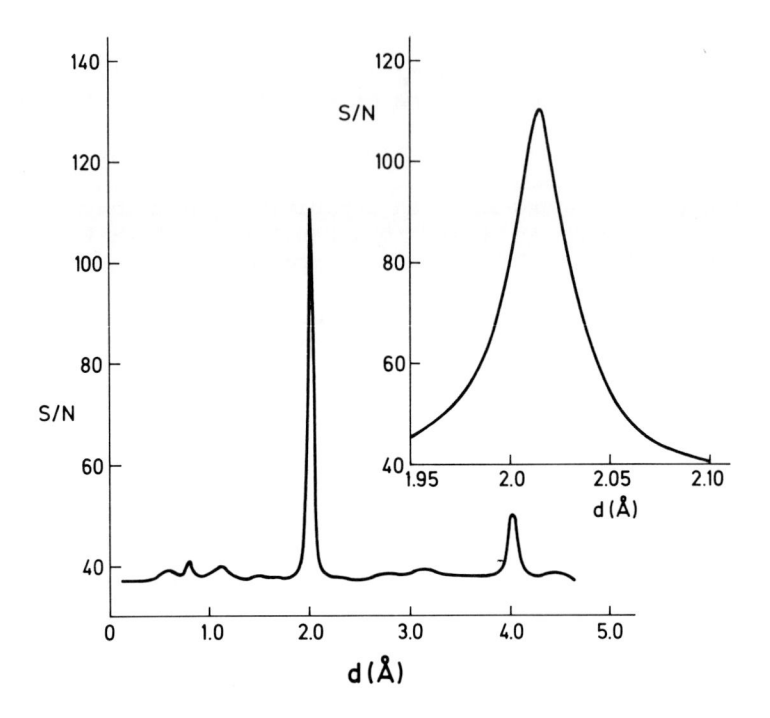

Fig. 11. Signal-to-noise versus d, the uniform interlayer spacing in the Al(100) surface. Plot constructed from many deconvolutions of the P(z) function of the intensity spectrum for $\theta = 10°$, in which d was held fixed. All deconvolutions were carried out with fixed V_o = 16.6 eV and θ_D = 369°K. Inset shows the main peak at 2.01 Å on an expanded scale of z.

was applied[10] to the analysis of experimental specular beam intensity spectra from Cu(100)[33], Ni(100)[34], Al(111)[35], and Al(100)[9]. In comparing the quality of the results obtained for these systems, it should be noted that the accuracy of the input data varied widely. For Al(100) and Cu(100), an accurate representation of the original experimental data was available, in the former case from our own work[9], and in the case of Cu(100) through courtesy of Jim Burkstrand[33] of General Motors. The data for Ni(100) and Al(111), however, were obtained from photographic enlargements of figures published in the literature, and hence were a less faithful representation of the original spectra.

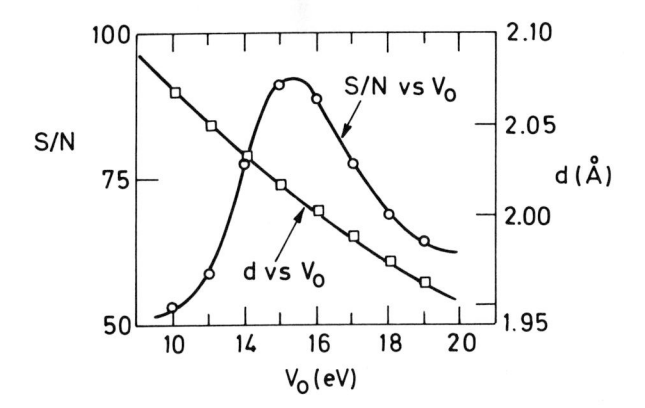

Fig. 12. Signal-to-noise versus V_o for fixed $\theta_D = 380°K$. Dashed line shows variation of d with V_o. Again based on data for Al(100), $\theta = 10°$.

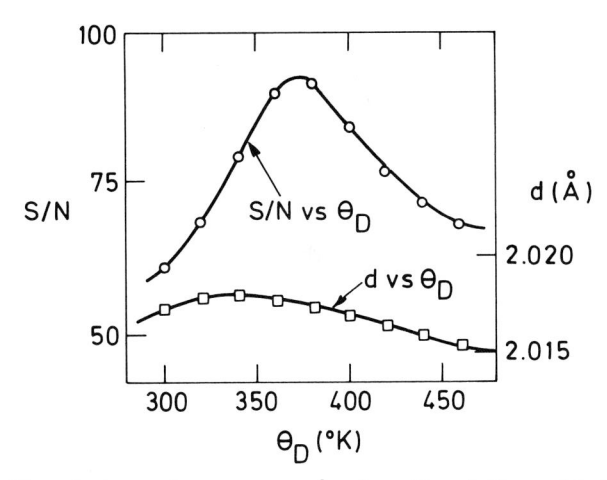

Fig. 13. Signal-to-noise versus θ_D for fixed $V_o = 16.0$ eV. Dashed line shows corresponding variation of d with θ_D. Again obtained from deconvolutions of P(z) function of Al(100) $\theta = 10°$ intensity spectrum.

In the first analyses of the above systems, the deconvolution procedure was constrained to produce a <u>single</u> interlayer spacing. The propriety of this constraint is discussed later.

The results of the deconvolutions are summarized in Tables 1 to 4, which contain the output values for d, V_o, θ_D, and μ and S/N for a number of different intensity spectra from each surface. Computer deconvolutions corresponding to the best and worst cases in terms of S/N for each system are shown in Figures 14-17.

To a large extent the content of Tables 1-4 and Figures 14-17 is self-explanatory and we forego a detailed discussion, which can be found in our original article[10]. Several points deserve emphasis, however. As can be seen from Figures 14-17, even in the cases of worst signal-to-noise, the delta function signal is well distinguished from the noise. From Tables 1-4, it can be seen that analyses of different intensity spectra from the same surface gave very consistent values for the interlayer spacing, d, and for the inner potential, V_o. The spread in output values of θ_D and μ is generally higher, and systematic trends in the values of these parameters with angle of incidence occur in some cases. A completely consistent explanation for these trends cannot be found, but in the cases of Ni(100) and Al(100), for example, it is notable that θ_D decreases uniformly with increasing angle of incidence. This behavior might be attributable to the increasing sensitivity to the outermost layers of the surface with their larger thermal vibrations.

The mean values of the output parameters for the four systems, weighted according to the signal-to-noise, are given in Table 5, which also contain values of the bulk interlayer spacings, and the values of the volume-averaged inner potentials determined in the band-structure calculations of the crystal potentials[32,36] which we used to derive the scattering phase shifts. It can be seen from the table that for the (100) planes, the determined values of the surface interlayer spacing are within ± 0.02 Å of the corresponding bulk spacings. The output values of the inner potentials are within ± 1 eV of the calculated, volume-averaged values in the cases of Al(100), Al(111), and Cu(100). The output value for Ni(100) is 2 eV higher than the calculated value. Finally, the output values for θ_D are generally similar to values used in model calculations for those systems[4], and the values of the mean-free path λ are consistent with experimental measurements[37].

As mentioned previously, in the analyses described above, the deconvolution procedure was constrained to produce a single interlayer spacing. In the case of the (100) planes, the results summarized in Table 5 indicate that this constraint was appropriate. In the case of Al(111), however, the mean value of d corresponds to a 3% contraction of the surface layer spacing relative to the bulk value. It is of interest, therefore, to determine if this contraction is genuine, or if it reflects systematic errors in the

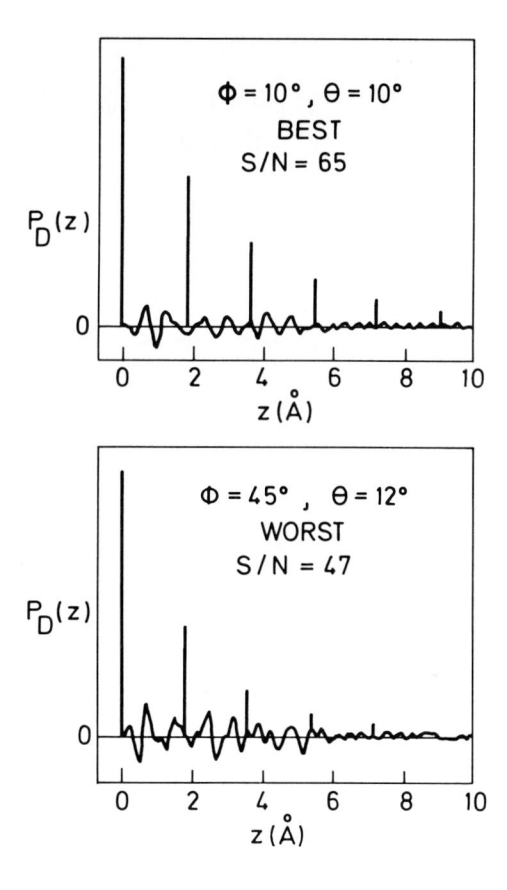

Fig. 14. Deconvolutions of P(z) functions of intensity spectra for Cu(100), corresponding to the best and worst cases in terms of signal-to-noise of the seven intensity spectra which were analyzed (see Table 1). The deconvolutions contain delta functions at multiples of the uniform interlayer spacing in the Cu(100) surface, together with a random noise component. The exponential decay of the delta function series is due to the attenuation of the electron flux in the crystal, as characterized by the exponent μ.

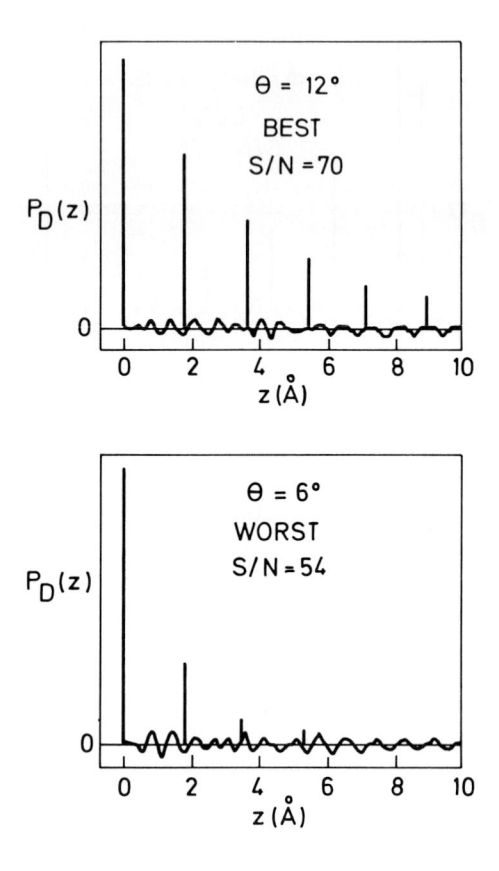

Fig. 15. Deconvolutions of P(z) functions of intensity spectra for Ni(100), corresponding to the best and worst cases in terms of signal-to-noise of the five intensity spectra analyzed (see Table 2).

experimental data or the analysis. Accordingly the data for Al(111) and also, for the purpose of a consistency check, the data for the (100) planes were re-analyzed with the deconvolution procedure constrained to produce a first interlayer spacing possibly <u>different</u> from subsequent spacings which were fixed at the bulk value. For this case the deconvolution contains two sets of delta functions, at $z = d_o + \nu d_s$ and at $z = \nu d_s$, where d_o is the first layer spacing.

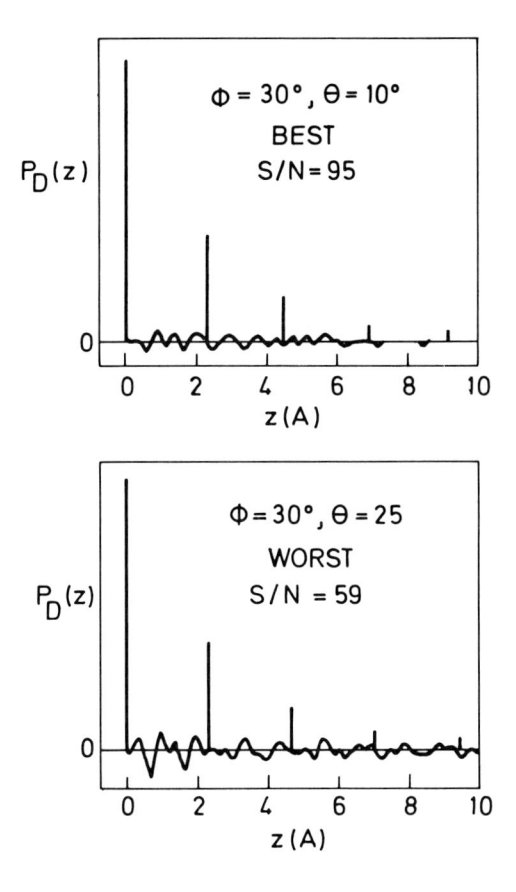

Fig. 16. Deconvolutions of P(z) functions of intensity spectra for Al(111), corresponding to the best and worst cases in terms of signal-to-noise of the eight intensity spectra analyzed. For this surface, the constraint of a uniform interlayer spacing is probably not justified since the derived value of 2.27 Å is ∼3% less than the bulk value of 2.34 Å (see Table 3 and following text).

A typical result for Al(111) is shown in Figure 18. The value of the first layer spacing in this example is 2.24 Å compared to the value of 2.25 Å obtained previously (Table 3) on the assumption of a uniform layer spacing. A complete analysis of the Al(111) intensity spectra confirmed this result; the mean value of the first

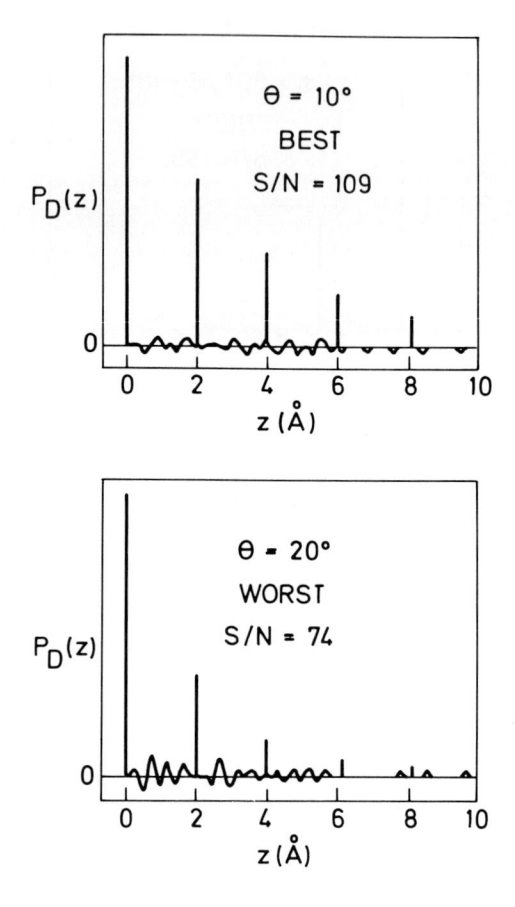

Fig. 17. Deconvolutions of P(z) functions of intensity spectra for Al(100), corresponding to the best and worst cases in terms of signal-to-noise of the seven intensity spectra analyzed (see Table 4).

layer spacing being 2.26 Å. However, the S/N values obtained in the second analysis were not significantly better than in the first. Also, the strong attenuation of the delta function amplitudes with increasing z led to less than unambiguous distinction between the νd_s set and the noise (see Figure 18). Thus, while the second analysis of the Al(111) data does tend to confirm the occurrence of a contraction of the first layer spacing of 3%, it also indicates that resolution of differences in structural parameters of less

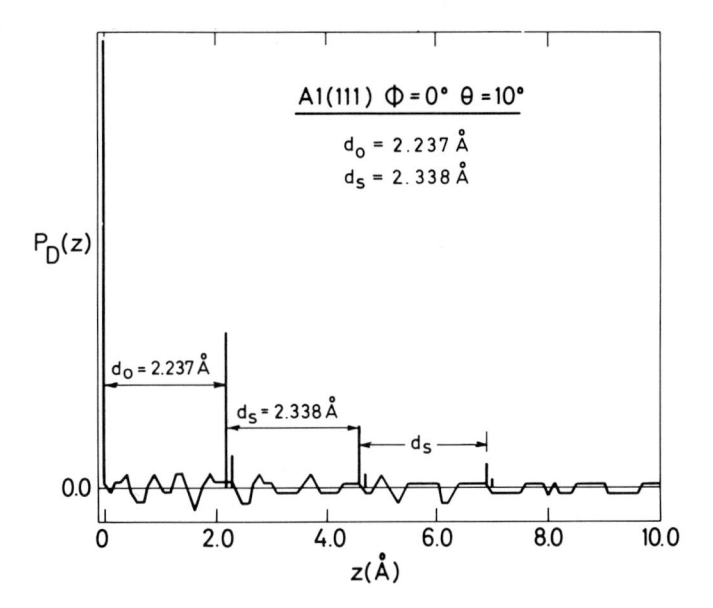

Fig. 18. Typical deconvolution of P(z) function of an intensity spectrum for Al(111), in which the deconvolution was constrained to produce sets of delta functions at $z = \nu d_s$ and $z = d_o + \nu d_s$. d_s was fixed at the bulk value of 2.338 Å, and the first layer spacing d_o was allowed to vary freely.

than about 0.1 Å may be difficult to achieve without some ambiguity.

Finally, the re-analysis of the data for the (100) planes confirmed the result that the first layer spacing is within ± 0.02 Å of the bulk value. A typical deconvolution is shown in Figure 19 for Al(100). As shown in the figure, the two sets of delta functions are virtually coincident, although the value of the first layer spacing was allowed to vary freely. The values of S/N were essentially unchanged, as were the output values of V_o, θ_D, and μ.

In our judgment, the internal consistency of the results described above, and the reasonable correspondence with the results of other studies where comparison is possible, indicate that the transform-deconvolution method can be used to provide an accurate and rapid structure determination in the case of clean surfaces. In the following section, we consider the extension of the analysis

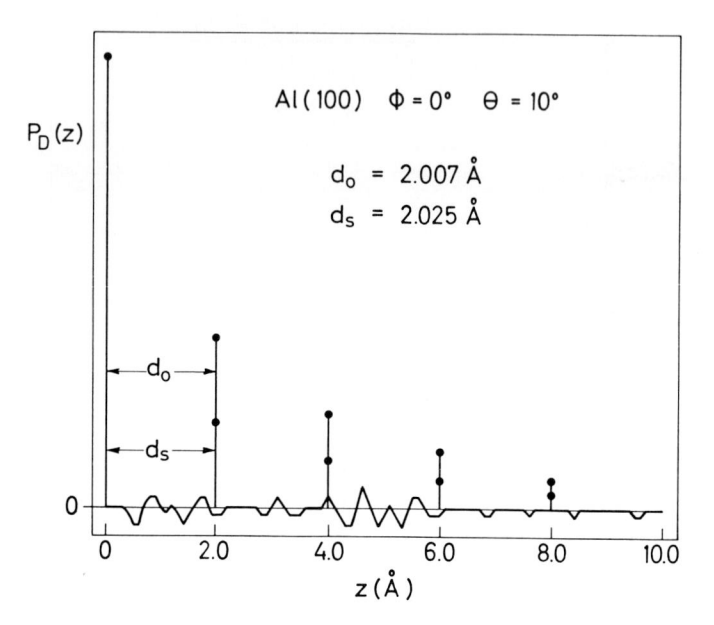

Fig. 19. Typical deconvolution of P(z) function of an intensity spectrum for Al(100), in which the deconvolution was constrained to produce two sets of delta functions, as in the caption to Figure 18, with d_s fixed at the bulk value of 2.025 Å.

to the case of surfaces with adsorbed overlayers.

V. APPLICATION OF THE TRANSFORM-DECONVOLUTION METHOD TO OVERLAYER SYSTEMS

In this section we consider the application of the transform-deconvolution method to be the more difficult problem of systems containing more than one kind of surface atom. Since the analysis has been to date tested only with calculated, single-scattering intensities, we restrict ourselves here to a brief discussion of the main problems.

For convenience, we consider the case of an adsorbed layer at a distance d_O from the first layer of the substrate which has a uniform layer spacing d_S. This model contains the basic elements of a multi-component system and can be discussed without loss of generality. The single-scattering intensity in the specular diffracted beam is given by Equation 3.10 and the corresponding Fourier transform P(z) is given by Equation 3.12. The latter equation may be written compactly as:

$$P(z) = [P_O(0) * F_{oo}(z)] + [P_D(z) * F_{ss}(z)] + \\ + [P_o(+z) * F_{os}(z)] + [P_o(-z) * F_{os}^\dagger(z)] , \qquad (5.1)$$

where $P_O(z)$ contains a set of delta functions at $z = \pm(d_O + \nu d_S)$, and $P_D(z)$, as in Equation 4.1, contains a set at $z = \pm \nu d_S$. $F_{oo}(z)$ and $F_{ss}(z)$ are respectively the Fourier transforms from $s = s_1$ to s_2 of the overlayer and substrate scattering factors, and $F_{os}(z)$ is similarly the Fourier transform of the mixed overlayer-substrate scattering factor $f_o f_s^\dagger$. Equation 5.1 can be written in vector-matrix form as:

$$\vec{p} = \vec{q}_o(A_o) + \vec{q}_s(A_s) , \qquad (5.2)$$

where the symbols have the same connotation as in Equation 4.11, and the subscripts distinguish overlayer and substrate contributions to $\vec{p}$.

The basic problem in analysis of multi-component systems is immediately apparent upon comparison of Equations 4.11 and 5.2. Whereas in the case of a one-component system, the structural and non-structural variables $\vec{q}$ and (A) are separable, such a separation cannot, in general, be made in the multi-component case. Equation 5.2 could not be uniquely inverted to obtain $\vec{q}_o$ and $\vec{q}_s$ given $\vec{p}$, (A_o) and (A_s) even in the absence of the additional truncation problem.

In a previous article[9b], we suggested a "substrate-subtraction" procedure for solution of Equation 5.2 in the case where intensity spectra for both the clean substrate and substrate plus overlayer are available. On the assumption that the adsorption process does not lead to a change in the substrate structure, then $\vec{q}_s(A_s)$ in Equation 5.2 is known to within a scaling factor g which can be found by a variational procedure[9b]. Thus, defining the Fourier transform of the intensities from the clean substrate as $\vec{P}_s$, a residual function $\vec{p}_r$ can be determined as:

$$\vec{p}_r = \vec{p} - g\vec{P}_s = \vec{q}_o(A_o) . \qquad (5.3)$$

Equation 5.3 can be solved for $\vec{q}_o$ using the methods of the previous section.

It must be emphasized, however, that the substrate-subtraction procedure is compromised by the assumption that the substrate structure is unchanged upon adsorption. Whilst this assumption may be justified in many cases, its validity cannot be independently assessed. Clearly, a more general approach to the analysis of multi-component systems is needed.

Recently we have found[10] that Equation 5.2 can be solved in essentially the same manner as described in the previous section for one-component systems, by a variation of the parameters of trial values of $\vec{q}_O$ and $\vec{q}_S$ (trial sets of delta functions), but a modification to the Southwell procedure must be made in obtaining the first approximation to the solution. The modification consists of making an approximate separation of the structural and non-structural variables by replacing the individual atomic scattering factors by their average value. This approximation is similar to that used in obtaining "sharpened" Patterson functions of x-ray intensities[13]. Having obtained a first approximation to the structural parameters of $\vec{q}_O$ and $\vec{q}_S$ in this way, the individual atomic scattering factors are used in the subsequent refinement procedure.

VI. SUMMARY AND CONCLUSIONS

In this paper, we have described the problems involved in adaptation of the direct, Patterson function method of x-ray crystallography to the analysis of LEED. Studies of the Fourier transforms of both calculated and experimental LEED intensities have led us to conclude that the main difficulty in application of the method results from the severely truncated region of reciprocal space which is accessible to LEED. The extra complications of the scattering of low-energy electrons from solids, in particular the occurrence of multiple scattering, give rise to additional but lesser difficulties.

We have shown that the Fourier transforms of LEED intensities contain convolution products of structural and non-structural parameters, from which the structural parameters can be recovered by a deconvolution procedure which requires a priori knowledge of the atomic scattering factors, and good first approximations to parameters characterizing the inner potential of the solid and the effects of thermal vibrations. The deconvolution procedure is constrained to select the correct physical solution from an infinite number of possible mathematical solutions, with constraints based on the results of analytical Fourier transformation of expressions for diffracted intensities.

The transform-deconvolution method has been applied to experimental LEED intensities from clean Al(100), Al(111), Ni(100), and Cu(100). The results for a number of different intensity spectra in each case show a high degree of internal consistency, which argues against the introduction of systematic errors due to the approximate treatment of multiple-scattering. We believe that the successful analysis of these simple surfaces augers well for the planned future application to reconstructed clean surfaces and to overlayer systems.

At its present stage of development, the transform-deconvolution method applies only to the determination of interlayer spacings via analysis of Fourier transforms of intensity spectra for the specular diffracted beam. We believe, however, that its extension to the case of 2-D sections for the purpose of determination of layer structure and registry should be quite straightforward. The additional problem of truncation over h and k in this latter case is completely defined by the extent of the experimental data, and can be solved without the requirement of ancillary physical information concerning the system under study.

Finally, we should like to comment that the study of direct methods of analysis of LEED is a relatively new and largely unpopulated field. In this paper we have argued that the development of rapid and economical direct methods, to provide at least a first approximation to surface structure determination for subsequent refinement via model calculations, is vital if the full potential of LEED is to be realized. In discussing the transform-deconvolution method in some detail we have hoped to show that the use of direct methods is feasible and thereby to encourage their further study and development.

ACKNOWLEDGEMENTS

The authors are grateful to John F. Hamilton for many useful discussions. One of us (DLA) would like to thank the American Chemical Society of providing support for travel to the Centennial Meeting of the ACS in New York, at which a brief account of this work was presented. The assistance of Alice Grandjean and Inge Schmidt in preparing the manuscript and of Svend Olesen, V. Blak Nielsen and Tove Asmussen in constructing figures has been greatly appreciated.

Table 1

Deconvolution Output Parameters for Cu(100)

θ	Φ	d (Å)	V_o (eV)	θ (°K)	μ (Å^{-1})	S/N (Arb)
10	0	1.81	12.0	181	0.24	59
10	5	1.81	11.8	180	0.25	62
10	10	1.80	13.0	198	0.32	65
10	45	1.78	10.0	190	0.57	50
12	0	1.81	12.0	170	0.32	57
12	10	1.81	12.0	176	0.40	54
12	45	1.78	12.1	195	0.49	47
*		1.80	11.9	184	0.36	
		±0.01	±0.01	±0.8	±0.11	

*Weighted mean values

Table 2

Deconvolution Output Parameters for Ni(100)

θ	Φ	$\overset{\circ}{d}$ (Å)	V_o (eV)	θ (°K)	μ (Å^{-1})	S/N (Arb)
6	0	1.77	16.0	328	0.69	54
8	0	1.79	14.1	284	0.43	56
10	0	1.78	16.0	320	0.32	61
12	0	1.79	15.9	280	0.24	70
14	0	1.78	15.8	246	0.34	56
*		1.78	15.6	291	0.39	
		±0.01	±0.7	±29	±0.15	

*Weighted mean values

Table 3

Deconvolution Output Parameters for Al(111)

θ	Φ	d_0 (Å)	V_o (eV)	θ (°K)	μ (Å^{-1})	S/N (Arb)
10	0	2.25	18.0	322	0.43	86
15	0	2.27	14.8	360	0.42	72
20	0	2.25	17.0	310	0.40	73
25	0	2.25	20.8	338	0.51	85
10	30	2.27	16.0	327	0.43	95
15	30	2.26	18.0	369	0.42	79
20	30	2.31	16.0	325	0.38	87
25	30	2.31	20.9	350	0.44	59
*		2.27	17.6	337	0.43	
		±0.02	±2.0	±19	±0.04	

*Weighted mean values

Analysis of LEED

Table 4

Deconvolution Output Parameters for Al(100)

θ	Φ	d (Å)	V_o (eV)	θ (°K)	μ (Å^{-1})	S/N (Arb)
8	45	2.01	16.0	373	0.28	105
10	45	2.01	16.6	369	0.27	109
12	45	2.02	16.0	350	0.32	88
14	45	2.02	16.0	337	0.39	92
16	45	2.01	17.1	320	0.44	90
18	45	2.01	16.1	327	0.43	75
20	45	2.02	16.0	320	0.47	74
*		2.01	16.3	345	0.36	
		±0.01	±0.4	±21	±0.08	

*Weighted mean values

Table 5

Mean Values of Output Parameters

Surface	d (Å)	V_o (eV)	θ_D (°K)	λ^* (Å)	d_{BULK} (Å)	V_{Inner} (eV)
Al(100)	2.01	16.3	345	5.6	2.025	16.7
Al(111)	2.27	17.6	337	4.7	2.34	16.7
Ni(100)	1.78	15.6	291	5.1	1.76	13.6
Cu(100)	1.80	11.9	184	5.6	1.81	12.4

*Mean free path $\lambda = 2/\mu$

REFERENCES

1. See for example: E. Bogh, Chapter XV, Application to Surface Studies, in: *Channeling,* V. Morgan, Ed., John Wiley, New York (1974); W. Heiland and E. Taglauer, J. Vac. Sci. Technol. 9, 620 (1972); H. H. Brongersma and J. B. Theeton, Surf. Sci. 54, 519 (1976); J. J. Davies, D. P. Jackson, J. B. Mitchell, P. R. Norton, and R. L. Tapping, Phys. Letters 54A, 239 (1975).
2. J. P. Toennies, App. Phys. 3, 91 (1974).
3. See for example: J. S. Gadzuk, Phys. Rev. B 10, 5030 (1974); A. Liebsch, Phys. Rev. Letters 32, 1203 (1974); J. Anderson and G. J. Lapeyre, Phys. Rev. Letters 36, 376 (1976).
4. D. W. Jepsen, P. M. Marcus, and F. Jona, Phys. Rev. B 5, 3933 (1972); P. M. Marcus, this volume.
5. J. Pendry, J. Phys. C 1, 2501, 2514, 3095 (1971).
6. S. Y. Tong, this volume, and references therein.
7. M. G. Lagally, T. C. Ngoc, and M. B. Webb, Phys. Rev. Letters 26, 1557 (1971); J. Vac. Sci. Technol. 9, 645 (1972); T. C. Ngoc, M. G. Lagally, and M. B. Webb, Surf. Sci. 35, 117 (1973).
8. See for example: C. B. Duke and D. L. Smith, Surf. Sci. 23, 411 (1970); 29 237 (1972); J. B. Pendry, J. Phys. C 5, 2567 (1972); D. T. Quinto and W. D. Robertson, Surf. Sci. 34, 501 (1973).
9. a) U. Landman and D. L. Adams, J. Vac. Sci. Technol. 11, 195, (1974).
   b) D. L. Adams and U. Landman, Phys. Rev. Letters 33, 585 (1974).
   c) D. L. Adams, U. Landman, and J. C. Hamilton, J. Vac. Sci. Technol. 12, 206 (1975).
   d) U. Landman and D. L. Adams, Surf. Sci. 51, 149 (1975).
10. D. L. Adams and U. Landman, to be published; J. Vac. Sci. Technol. 13, 363 (1976) (abstract only).
11. A. L. Patterson, Phys. Rev. 46, 372 (1934); Z. Krist. (A) 90, 517 (1935).
12. D. Harker and J. S. Kasper, Acta Cryst. 1, 70 (1948).
13. M. J. Buerger, *Vector Space*, John Wiley, New York (1959).
14. J. Karle and H. Hauptman, Acta Cryst. 3, 181 (1950).
15. W. L. Bragg, Chemistry 22, 26 (1948).
16. T. N. Rhodin and S. Y. Tong, Physics Today, October (1975), p. 23.
17. S. Andersson and J. B. Pendry, Solid State Comm. 16, 563 (1975).
18. See for example References 4 and 6 and: S. Y. Tong, J. B. Pendry, and L. L. Kesmodel, Surf. Sci. 54, 21 (1976).
19. D. P. Woodruff, K. A. R. Mitchell, and L. McDonnell, Surf. Sci. 42, 355 (1974).
20. A. Guinier, *X-Ray Diffraction*, Freeman, San Francisco (1963).
21. a) T. A. Clarke, R. Mason, and M. Tescari, Surf. Sci. 30, 553, (1972);
    b) Proc. Roy. Soc., London, A 331, 321 (1972);
    c) Surf. Sci. 40, 1 (1973).
22. U. Landman, Discuss. Faraday Soc. 60, 230 (1976).
23. C. W. Tucker, Surf. Sci. 2, 516 (1964).

24. C. W. Tucker and C. B. Duke, Surf. Sci. 29, 237 (1972).
25. J. C. Buchholz, M. G. Lagally, and M. B. Webb, Surf. Sci. 41, 248 (1974).
26. J. F. Kaiser, in: *System Analysis by Digital Computer*, F. F. Kuo and J. F. Kaiser, Eds., John Wiley, New York (1966).
27. C. B. Duke and G. E. Laramore, Phys. Rev. B 2, 4765 (1970); G. E. Laramore and C. B. Duke, Phys. Rev. B 2, 4782 (1970).
28. J. W. Cooley, P. A. W. Lewis, and P. D. Welch, IEEE Trans. on Audio and Electroacoustics, AU-15, 79 (1967).
29. H. D. Hagstrum, Phys. Rev. 150, 495 (1966).
30. M. S. Patterson, Proc. Phys. Soc. 63, 477 (1950); R. V. Southwell, *Relaxation Methods in Engineering Science*, Oxford University Press (1946).
31. We are indebted to J. C. Hamilton for constructing the first version of the computer program based on the Southwell relaxation method.
32. E. C. Snow, Phys. Rev. 158, 683 (1967).
33. J. M. Burkstrand, G. G. Kleiman, and F. J. Arlinghaus, Surf. Sci., 46, 43 (1974).
34. J. E. Demuth and T. N. Rhodin, Surf. Sci. 42, 261 (1974).
35. F. Jona, IBM J. Res. Develop. 14, 444 (1970).
36. S. Wakoh, J. Phys. Soc. (Japan) 20, 1894 (1965).
37. C. J. Powell, Surf. Sci. 44, 29 (1974).

Surface Structure by Analysis of 'LEED' Intensity Measurements

P. M. Marcus

*IBM Research Center
Yorktown Heights, N. Y. 10598*

A brief review of the physical basis of LEED is followed by a discussion of why the determination of the surface structure of crystals by analysis of LEED intensity spectra is possible. The procedure for fixing geometrical parameters of the surface layers is illustrated by selected examples, and comments are made on ways of simplifying the structure determination and on the possible complications that can occur. Tables of structures of clean-metal surfaces and of ordered overlayers on metal surfaces are given and commented on. Overlayer systems with unusual features are singled out. Some trends in bonding on a surface are noted and represented as effective radii of adsorbed atoms.

INTRODUCTION

In the last few years, a number of structures of ordered surfaces have been convincingly determined by analysis of the LEED intensity-energy profiles or spectra. By "surface structure" is meant the positions of all atoms which deviate from bulk positions, hence can include several layers of atoms, not just the first layer, and necessarily includes any overlayers or layers which mix foreign atoms and substrate atoms. LEED is, in fact, highly suited to such determination, since it is sensitive to just the first 4 or 5 layers of a crystal. The results give information of great chemical interest and provide a microscopic description of the ordered regions of surfaces.

This review first briefly describes the physical basis of LEED, then discusses the reasons why the structure determination has been possible, and the theoretical and experimental limitations on the procedures used in the determination.

A tabulation of successful clean-metal and overlayer-on-metal-substrate structures is given and discussed with attention to three types of results: 1) the effective radii of overlayer atoms, 2) the occurrence of overlayer atoms in unexpected positions, i,e., positions different from the positions that would be occupied by the next layer of substrate atoms, 3) strains in the first substrate layer.

The review concludes with remarks on some important directions for further development of LEED as a tool for surface structure.

PHYSICAL BASIS OF 'LEED'

When electrons of a given energy and direction are incident on a surface, a small fraction of them are scattered elastically and coherently. If the surface is ordered, i.e., has translational symmetry over some finite area, the scattered electrons group themselves into a finite set of diffracted beams concentrated into particular directions (see Fig. 1). One of these beams is always in the specular direction; the other beams are defined by the geometry of the surface, i.e., the unit cell of the surface net, analogous to diffraction of a wave by a two-dimensional grating which leads to higher order diffracted waves with wave numbers differing by reciprocal lattice vectors of the grating. When formulated mathematically as the scattering of an incident plane wave by a semi-infinite crystal, the basic description of the electron wave function is achieved by a super-position of a discrete set of outgoing (reflected) plane waves and a discrete set of outgoing (transmitted) Bloch waves (characteristic modes in the crystal) all of which have the same energy and (reduced) wavenumbers parallel to the surface - including waves which attenuate strongly going away from the surface. The total wave function and its normal derivative are then continuous across the surface. When an ordered overlayer is present, ingoing and outgoing Bloch waves characteristic of the layer are introduced to carry the wavefunction and its normal derivative continuously across the layer.

The intensity in each beam (the electron flux) relative to the incident beam intensity is then a function of energy giving rise to the LEED spectrum or intensity profile (Fig. 2) which shows a series of broad smooth peaks with characteristic shapes (the upper curves of each beam). These peaks are then very different from the sharp lines of x-ray diffraction, primarily due to the much stronger scattering of electrons by atoms, including the strong inelastic scattering. The strong elastic scattering gives rise to much multiple scattering and a complex spectrum which differs substantially from the single-scattering or Bragg spectrum that is adequate for x-ray diffraction. Such complex spectra are shown in the lower curves of Fig. 2 calculated with the imaginary part of the potential V_i equal to zero and may be associated with energy gaps in the band

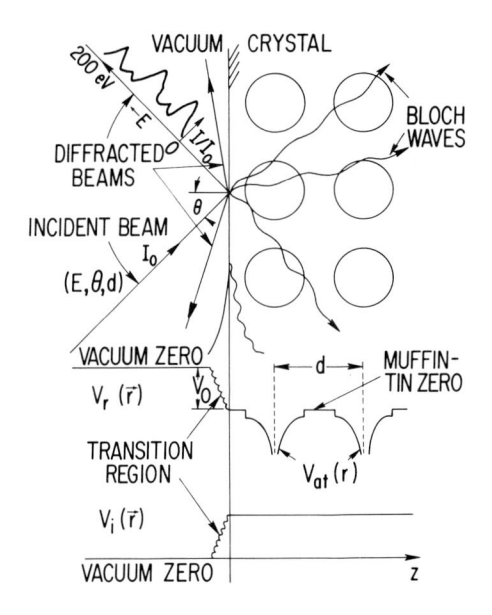

Fig. 1. Model of the semi-infinite crystal showing incident and diffracted electron beams and transmitted Bloch waves; model of the real part of the crystal scattering potential $V_r(\vec{r})$ in muffin-tin form showing spherical atomic potential $v_{at}^r(r)$, interlayer spacing d, and constant potential between atoms lying V_o below vacuum level (middle); model of imaginary part of the scattering potential $V_i(\vec{r})$ taken constant inside the crystal (bottom).

structure; sharp changes in intensity occur when the energy crosses a band edge, and a channel for carrying flux into the crystal is lost or gained (corresponding to increased and decreased reflection respectively). The model potential used for the elastic scattering calculation is shown in the center of Fig. 1, and consists of a regular array of spherical atomic potentials separated by a region of constant potential.

The model of the imaginary part of the potential, which approximately describes the inelastic scattering of the electrons, is shown at the bottom of Fig. 1 and is taken to be spatially constant. When a V_i of 0.3 Ry = 4.1 eV is included in the

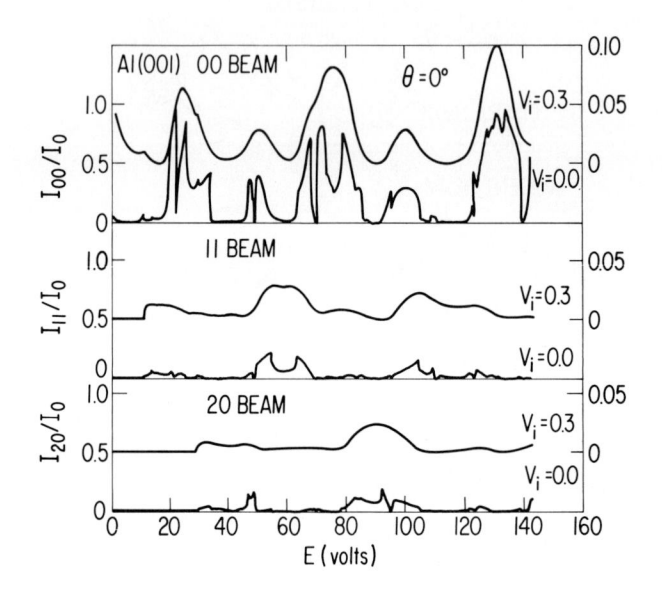

Fig. 2. Calculated LEED spectra for Al(001) at normal incidence for 00, 11 and 20 beams at typical absorption $V_i = 0.3$ Ry, and for zero absorption, $V_i = 0$.

calculation, the LEED spectra for Al(001) become the upper curves in Fig. 1, with an order of magnitude reduction in intensity (scale on the right), and a smoothing away of the sharp structure; however some structure still remains in the shapes of the peaks, which are sensitive to the geometry of the lattice and are used to determine the atomic geometry. Note that the widths of these broad smooth peaks correspond to the envelope of a succession of nearby sharp peaks for the elastic scattering calculation, merged and broadened by the inelastic scattering.

WHY THE STRUCTURE PROBLEM IS SOLVABLE

Features of the Analysis
Which Make Structure Determination Practicable

Determination of surface structures from LEED spectra, i.e., determination of the positions of atoms which are not in the lattice positions fixed by the bulk of the substrate, does not require a quantitative theory of the intensity spectra. However, the theory must be able to establish the positions of peaks and structural features well enough to discriminate against a random fit. Study of the correspondence of random spectra over the typical range of energy used in the LEED analysis, 20 to 200 eV, indicates that random spectra can show an average fit of $\pm$ 4 or 5 eV with fluctuations of $\pm$ 1 eV, e.g., comparison of two of the spectra in Fig. 4 for values of d_o separated by more than 0.2 Å.

The required discrimination against random fits can be achieved by focusing attention in the comparison of theory and experiment on particular features and allowing for uncertainties in the model of the scattering potential. Four aspects of the comparison are:

1) The comparison relies primarily on the relative positions and intensities of nearby peaks and gives up any attempt to compare absolute peak positions (relative to vacuum) and absolute intensities (diffracted beam intensities as fractions of incident beam intensity). These uncertainties reflect our lack of knowledge of the energy dependence of the real and imaginary parts of the potential, corresponding to the decrease in the magnitude of the correlation energy of a fast-moving electron, and to the change in the inelastic scattering with energy. Smoothly varying functions $V_o(E)$ and $V_i(E)$ are introduced and fitted to the measured peak positions and intensities; $V_{o2}(E)$ usually drops 4 or 5 eV over the energy range of interest (20 to 200 eV); the fitted behavior of V_i is shown for Ni in Fig. 3 along with values determined for Cu and Al by McRae by fitting optical data[3], and calculated values for the homogeneous electron gas (jellium) at the Al density. The curves of $V_i(E)$ are similar smooth curves with magnitudes of several eV. Choice of $V_i(E)$ permits adjusting the relative intensities in different energy ranges to fit experiment.

2) The comparison with experiment is made in an energy window whose lower bound is picked to permit use of a transferable scattering potential for each type of atom. Above this lower-bound energy, the scattering is largely ion-core scattering and is not sensitive to the environment of the atom. A conservative estimate of this energy is about 50 eV, and, in fact, potentials taken from muffin-tin band structure potentials appear to be satisfactory to substantially lower energies. However simple superposition potentials (obtained by superposing atomic charge distributions) and even free-ion potentials, are almost as good. The upper bound in energy is fixed by the complexity of the calculation, and is around 200 eV (more for simple structures, less for more complicated structures).

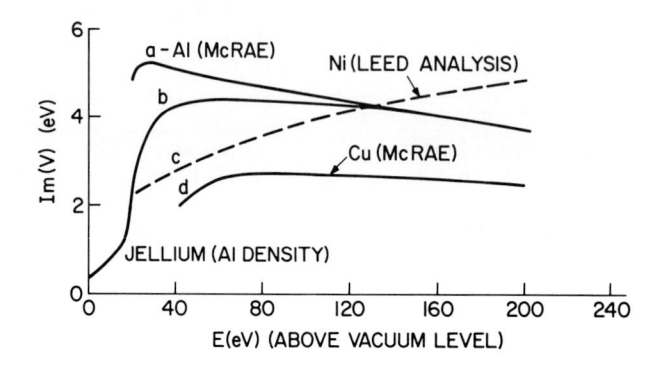

Fig. 3. Imaginary part of the scattering potential V_i as a function of energy: a) Al curve calculated by McRae from optical data, b) Al curve for imaginary part of self-energy of jellium at the Al valence electron density, c) $V_i(E)$ for Ni determined by fitting measured LEED spectra, d) $V_i(E)$ for Cu calculated by McRae.

3) The amount of useful data in the energy window is substantially expanded by using many beams and many angles of incidence, which provide independent data for the analysis. Expansion of the data in this way is necessary for many cases in which the fit to experiment of any one spectrum is not too close, hence discrimination among different structures requires expansion of the data base.

4) The LEED spectrum is not sensitive to disordered regions or to defects in the surface structure such as steps, kinks, vacancies or impurities, but only to sufficiently large patches of ordered surface layers. The presence of disorder will affect the intensity of the diffracted beam (and contribute to the diffuse background intensity) but will not significantly affect the shapes or positions of peaks, hence will not interfere with the surface structure determination. Lattice motion will have effects similar to disorder, and has a considerably smaller effect than the absorption described by V_i.

Procedure for Structure Determination

The features of the analysis of LEED spectra noted above are incorporated in the procedure used for structure determination, which will now be illustrated for the ordered overlayer structures in Figs. 4e and 4a. These figures show the positions of S atoms on Ni surfaces; in Fig. 4e S atoms occupy 1/4 of the positions in one set of three-fold hollows on the (111) surface of Ni, namely the set which has no atom below the hollow in the third layer (second Ni layer), just as in the actual FCC Ni structure; in Fig. 4a S is in 1/2 the four-fold hollows on the (001) surface of Ni, again as in the FCC lattice continuation. In these two cases, spectra are calculated for a specific model of the scattering potential, like that in Fig. 1, supplemented by a similar potential for the overlayer. The potentials in any layer are taken from band structure potentials in muffin-tin form, which essentially provide energy-dependent phase shifts that can be used to describe the scattering in a range of geometrical arrangements. The calculated spectra will be compared with measured spectra to fix the values of the geometrical parameters.

The functions $V_o(E)$ and $V_i(E)$ are also left to be fixed by fitting the data, i.e., they are respectively adjusted to fit slow trends in peak positions and intensities over the energy range of the data. The fitting is done first for the clean substrate LEED spectra, the $V_o(E)$, $V_i(E)$ are used again to analyze an overlayer on the same substrate but with separate values of V_o and V_i for the overlayer. Other, less important, parameters for substrate and overlayer structure determination are the effective Debye theta values θ_D for substrate and overlayer, which are used to describe the effects of temperature motion on the scattering; the values of θ_D are readily found by fitting the temperature dependence of measured LEED spectra and incorporated in the calculation.

The calculation of the theoretical spectra when the model of the scattering potential has been fixed - including appropriate ranges of geometrical parameters, and V_o, V_i θ_D - is similar to calculation of band structures. Both calculations are concerned with the energies and wave functions of multiple scattered electrons among ordered arrays of scattering centers. The LEED calculation is somewhat more elaborate, and differs in the following ways from the band calculation: 1) the multiple scattering is treated separately in each layer and then the interlayer scattering is treated, usually by a different method; this flexibility in treatment allows each ordered layer to be different and differently spaced from neighboring layers, and of course allows for the occurrence of a surface; 2) the LEED calculation goes to much higher energies than the usual band calculation, requiring many more phase shifts to describe the scattering at each atomic center: 3) at some stage it is necessary to convert spherical wave expansions into plane waves, since the outgoing beams in vacuum are plane waves;

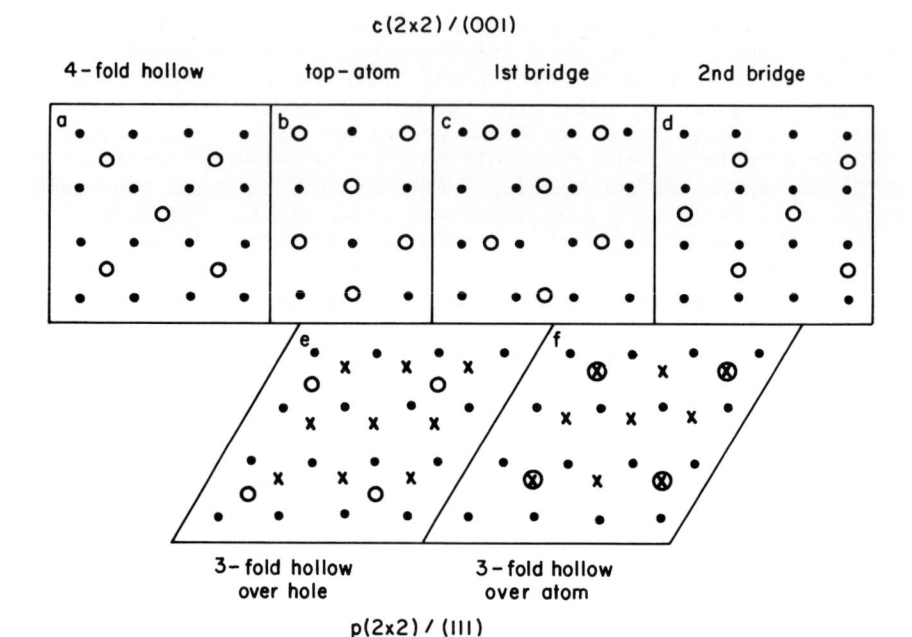

Fig. 4. Ordinary patterns for overlayers on (001) and (111) faces of an FCC crystal of overlayer (open circles) in the four-fold hollows of (001) surface of substrate (dots), b) c(2x2) overlayer in top-atom positions on (001) substrate surface, c) c(2x2) overlayer in horizontal bridge positions on (001) substrate, d) c(2x2) overlayer in vertical bridge position on (001) substrate, e) p(2x2) overlayer (circles) in three-fold hollows in the first substrate layer (dots) over holes in second substrate layer (crosses) for (111) substrate, f) p(2x2) overlayer in three-fold hollows over atoms in second substrate layer for (111) substrate.

4) strong attenuation of the electron in the crystal must be included usually in the form of an imaginary part to the scattering potential.

Tha main approaches to the calculation[4,5,6] now agree on the treatment of scattering within each layer, which is done accurately by summation of multiple scattering series to convergence. The approaches differ in the treatment of interlayer scattering, which can be done equally accurately by solving an eigenvector problem[4], or less accurately by treating a finite sequence of interlayer scatterings chosen to give adequate accuracy in the total scattering and relying on the attenuation to give convergence[5,6]. The former procedure can treat any amount of attenuation with full accuracy, the latter is more economical of computer time and space.

The variation of LEED spectra with variation of geometrical parameters is illustrated in Figs. 5 and 6 where the interlayer spacing between a layer of S atoms and the first layer of Ni atoms is systematically varied to determine the best fit to the experimental results below. In both cases, we see that the spectra are sensitive to values of d_o, and permit selection of a closest fit to ± 0.1 Å. In Fig. 5, the spectrum at $d_o=1.41$ Å or slightly smaller d_o appears to give closest overall agreement, although there are discrepancies in peak positions and intensities. The spectra with d_o deviating by ± 0.2 Å or more from 1.41 Å are clearly strongly discriminated against, as are spectra at other registrations of the S layer including the three-fold positions of Fig. 4f. In Fig. 6 is shown an expanded scale view of calculated and measured ½ ½ spectra for normal incidence on c(2x2)S/Ni(001) in the registration of Fig. 4a. This particular overlayer case provides the best agreement we have obtained between theory and experiment. Only the best fit value at $d_o=1.28$ and the adjacent d_o values differing by ± 0.105 Å are shown for comparison with the experimental spectrum on which ten features have been marked with vertical lines. The corresponding features are marked with vertical tics on the calculated spectra and it is striking that all the experimental features appear in the calculation. The best d_o appears to be fixed by these spectra to better than ± 0.1Å, perhaps to ± 0.05 Å, but it is difficult to give a closer estimate until the theory is further refined.

Limitations and Difficulties in Comparison of Theory and Experiment

The comparisons shown above give reasonable confidence that we are close to the right structure for those simple systems, but considerable refinement of the theory is needed before closer limits of accuracy can be fixed for structural quantities and before the theory can be extended to more complex systems. This section will note some of the refinements needed, and will also mention some of the complications encountered in the analysis, which might be expected to grow as more complex systems are studied.

Most structure determinations, like the ones above, fixed a single structural parameter by a best-fit criterion; in a few cases a best value of a second interlayer spacing has been found, as will be noted in the next section. However, in general, several

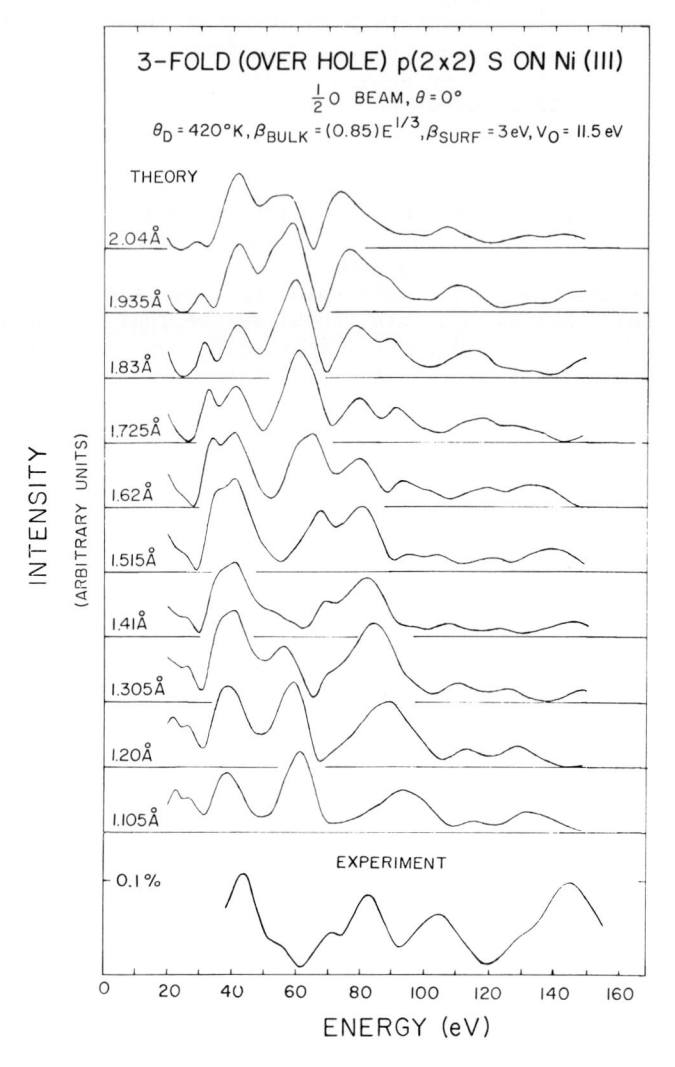

Fig. 5. Sequence of calculated LEED spectra for the ½ 0 beam, normal incidence, for p(2x2)S on Ni(111) in the three-fold hollows over holes in second substrate layer as the interlayer spacing d_o between S and the first layer of the Ni substrate is varied from 1.105 Å to 2.04 Å in steps of 0.2 atomic units (= 0.106 Å). The imaginary part of the potential in the crystal $V_i = \beta_{BULK} = 0.85\ E^{1/3}$ eV where E is the electron energy in eV with respect to vacuum level; in the overlayer $V_i = \beta_{SURF} = 3$ eV, $V_o = 11.5$ eV at all energies, effects of lattice vibration given by Debye parameter $\theta_D = 420°K$; experimental spectrum of Demuth and Rhodin, closest fit at $d_o = 1.41$ Å.

'LEED' Intensity Measurements 281

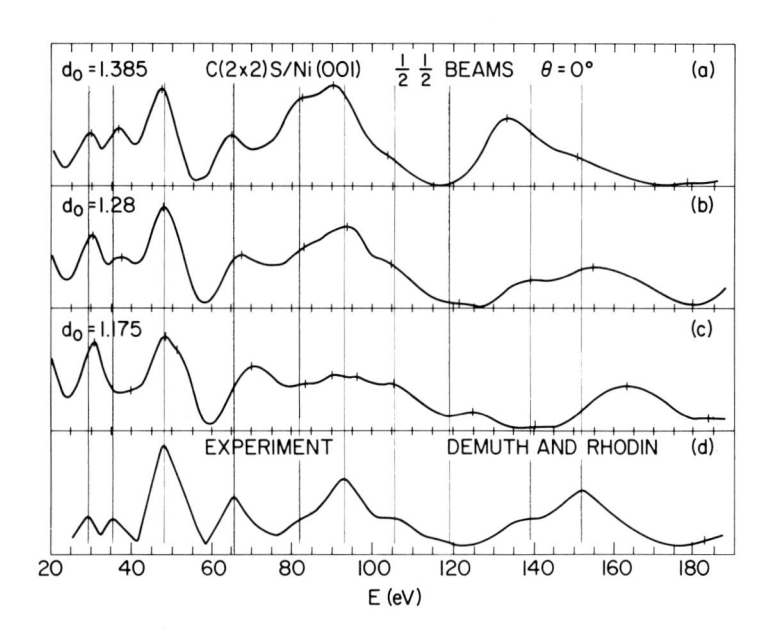

Fig. 6. Sequence of ½ ½ beam spectra at θ = 0° for c(2x2)S/Ni(001) for S in four-fold hollows of Ni substrate at 3 uniformly-spaced values of d_o calculated with V_o = 11.4 eV in substrate surface layer, $V_i(E) = 0.85E^{1/3}$ eV (E in eV), V_i (surface) = 3 eV, θ_D (bulk) = 420°K, θ_D (surface) = 335°K, T = 300°K, 8 phase shifts; best agreement with experiment of Demuth and Rhodin is at d_o = 1.28 Å. Vertical lines through experimental features to be compared with vertical tics at corresponding features of calculated spectra.

structural parameters need to be adjusted, and the single-parameter best fit will be modified. These modifications appear to be less than 0.1 Å for the cases given above, but have not been precisely fixed. For more complex structures, more parameters and greater modifications will be needed. Since it is very laborious to go beyond two parameters in the fitting process with present calculation procedures, residual uncertainties will remain until better fitting procedures are available.

The energy window within which the calculation is practicable has been mentioned above. Extension of this window would provide a greater data base for analysis. The upper limit is fixed by the increase in computational complexity as the number of phase shifts and the number of propagating waves in the crystal and vacuum (which fix the size of the representation) increase. There are no limitations in the formulation which prevent raising the upper limit, but a substantial increase in the power of the computational procedures is needed. A decrease in the lower limit, however, requires a conceptual improvement in the scattering potential, which will now be affected by the atomic environment. An improvement here would require detailed study of the construction of the scattering potential, and possibly adjustment of the potential for each configuration, i.e., abandonment of the convenient approximation of transferability.

The structures determined so far are largely monotomic overlayers on elemental substrates with small surface cells (up to p(2x2)). More complicated overlayer structures with large unit cells, unit cells containing several atoms, compound overlayers, compound substrates require more powerful procedures.

In the analysis, it is convenient to assume the overlayer atom had a position on the substrate with the same symmetry as the LEED pattern, thereby restricting the registration possibilities, e.g., with a LEED pattern of square symmetry, only the patterns in Figs. 4a and 4b are possible. However it is entirely possible for the overlayer to have less than the full LEED pattern symmetry and simulate a higher symmetry by the presence of several ordered domains, e.g., Figs. 4c and 4d show the two bridge position domains which, if evenly mixed, give the LEED pattern the appearance of square symmetry. Clearly four domains for a generally-positioned overlayer could also simulate square symmetry and greatly complicate the search for the correct structure.

Surfaces in general have steps one layer thick. Hence if the structures of successive layers are different, the LEED patterns are produced by an average over the different steps. This situation occurs with basal planes of close-packed hexagonal structures[8].

Surfaces may have patches of bare substrate or of overlayers with different structures. The LEED pattern may be produced by averaging over various patches, e.g., averaging with a bare patch of substrate appeared necessary[9] in reproducing the LEED pattern of c(2x2) Te/Ni(001).

REVIEW OF STRUCTURE DETERMINATIONS

Procedures like those illustrated above for ordered S layers on Ni surfaces have been applied to analysis of LEED spectra of some 25-30 overlayer-substrate systems, as well as to the clean surfaces of the corresponding substrates. Each of these analyses is an elaborate experimental and theoretical study. In particular, finding the structural, i.e., geometrical, parameters that best fit the measured spectra is a tedious process requiring many repetitions of the lengthy electron scattering calculation as the parameters are varied. The best fit is obtained within a rather narrow range of parameter values, outside of which the correspondence is random and inside of which the correspondence is frequently not very close. Most fits are less convincing than the fit shown in Fig. 6, and the case for a best fit is usually strengthened by using many beams and many angles of incidence. Although agreement in relative positions and amplitudes of nearby peaks that is better than random is achieved, discrepancies for some beams and some energy ranges frequently persist. Presumably, in these cases only part of the structure has been found, and the discrepancies would only be removed by adjustment of additional geometrical parameters or consideration of some of the other interfering effects mentioned above (steps, patches, domains, etc.).

The structures fixed by these analyses, partially or wholly, will not all be discussed in detail here, but the principal facts for surfaces of the clean-metal and overlayer-on-metal systems will be incorporated in tables, and some trends will be noted. Then the geometry of five overlayer systems which exhibit some interesting special features will be described and discussed in detail.

Clean Metal Surface Structures

Analysis of clean metal surface structures was carried out initially to test the theory, rather than to look for new structural information. The successful fit to experiment, with agreement of peak positions to better than ± 1 eV, showed that both the model and the method of calculation were adequate. However this success was not attained until scattering potentials with proper core contributions were used (such as self-consistent atomic calculations, or crystal potentials based on such atomic calculations), and until a sufficiently large number of phase shifts was used. The results (see Table 1) demonstrated that in general the surface layers were simply extensions of the bulk, with the same interlayer spacing and registration as the bulk layers, but some remarkable new structural information did appear. The analysis appeared to establish that FCC(110) faces and BCC(001) faces show about 5 to 10% compression of the first interlayer spacing, as noted in Table 1. All cases analyzed give this

Table 1

	Clean-Metal Surface Structures from LEED Spectra			
Metal	Lattice	Face	References	Special Features
Al	FCC, a=4.05*	001	12,13,14	
		111	12	
		110	12	5% contraction,** poorer fit***
Ni	FCC, a=3.52	001	2	
		111		
		110	2	5% contraction
Cu	FCC, a=3.61	001	4	
		111	15	
Ag	FCC, a=4.09	001	16	
		111	17	
		110	18	7% contraction, poorer fit
Pt	FCC, a=3.92	110	19	
Mo	BCC, a=3.15	001	20	11.5% contraction
Na	BCC, a=4.23	110	21	Crystal grown epitaxially on Ni(001)
W	BCC, a=3.16	110	22,23	Analyzed by constant momentum transfer averaging (Ref. 22)
Fe	BCC, a=2.87	001	24	1.4% contraction
Be	HCP, a=2.27, c=3.59†	0001	26	Analyzed by perturbation treatment of multiple scattering
Ti	HCP, a=2.95, c=4.68	0001	26	2% contraction
Zn	HCP, a=2.66, c=4.95	0001	27	Analyzed by constant momentum transfer averaging; 2% contraction
Cd	HCP, a=2.98, c=5.62	0001	28	Crystal grown epitaxially on Ti(0001); 4 layers or more = pure Cd.

* FCC and BCC a=side of cubic cell; all distances in Angstroms

** First interlayer spacing contracted from bulk value by 5%

*** Poorer fit theory to experiment compared to other surfaces

† HCP a,c = lengths of sides of unit cell (containing two atoms), in and out of basal plane

compression, and in some cases, e.g., Ni(110), the result is very clear. Less extensive results also suggest a smaller compression of the interlayer spacing of HCP(0001) faces. The other faces of FCC and BCC crystals either show no deviation from bulk spacing or a small expansion.

Later studies of clean metal surfaces were undertaken mainly to obtain appropriate parameters to use in the analysis of ordered overlayers on that metal as substrate, including $V_o(E)$ and $V_i(E)$. The level of agreement of peak positions in the best cases is of order ± 1 eV, but some cases, e.g., Al(110), Ag(110), have unexplained discrepancies.

Ordered Overlayer Structures

Extension of the analysis to systems with ordered overlayers on metal surfaces has succeeded with a moderate number of systems - Table 2 lists 29 - but in general is confined to monatomic overlayers on monatomic substrates, and cell sizes no larger than (2 x 2), i.e., four times the size of the substrate cell. Some systems are completely determined, others are only partially successful, but are included here because some new information is obtained, e.g., the interlayer spacing of the overlayer to the first substrate layer d_o is usually fixed reliably, but the registration is frequently less certain.

Enough information is now available on overlayer structures to note some significant trends. Column two of Table 2 shows that in the large majority of the systems analyzed, the overlayer atom sits in an "expected position", defined as a position that would be occupied by atoms of the next layer of substrate. Several exceptions have been found, however, which provide examples of novel bonding arrangements on surfaces. Some of these will be illustrated and discussed in more detail later. Column four gives the bond length to the nearest neighbor. A simple way to interpret the bond lengths is to introduce the touching radius of the substrate atoms, determined by bulk separations (the so-called metallic radius), and subtract this touching radius from the bond length to give the effective radius of the overlayer atom. Various overlayer effective radii are given in Table 3, where they may be compared with standard single-bond covalent radii. We note that most effective radii are close to the single-bond covalent radii, but can be significantly greater, as for O, suggesting a slight ionic character, whereas S appears to have an effective radius smaller than the single-bond covalent value. Finally, in column 6 of Table 2, a number of cases are noted in which strain in the substrate geometry has been detected by a refined analysis, i.e., O/Fe(001), C/Ni(001), N/Ti(0001), H/Ni(110). In the first three of these, the first substrate interlayer spacing is expanded; in the last case the substrate is compressed.

TABLE 2. Ordered Overlayer Structures from LEED Spectra

System[*]	Position[**]	Bond Length(Å)[***]	Ref.	Special Features
c(2x2)O/Ni(001)	E	1.98	32,1	
p(2x2)O/Ni(001)	E	1.98	33	
c(2x2)S/Ni(001)	E	2.18	32,1	
p(2x2)S/Ni(001)	E	2.19	33	
c(2x2)Se/Ni(001)	E	2.27	32	
p(2x2)Se/Ni(001)	E	2.34	33	
c(2x2)Te/Ni(001)	E	2.58	33,9	Overlayer spectrum averaged with patches of substrate spectrum improves fit
p(2x2)Te/Ni(001)	E	2.52	33	
(2x1)O/Ni(110)	U	1.91	1	O sits on top of short bridge
c(2x2)S/Ni(110)	E	2.17	34	S sits closer to 2nd layer of Ni.
(2x1)H/Ni(110)	-	-	35	H is not detected; Ni shows a double strain- [110] rows move together and into substrate
c(2x2)Na/Ni(001)	E	2.84	36	Fermi levels in Na and Ni matched
c(2x2)S, c(2x2)Na /Ni(001)	E	3.06 (Ni-Na)	37	Na moves out in presence of S to covalent bond distance
p(2x2)O/Ni(111)	-	1.88	1	Choice between 3-fold sites not fixed
p(2x2)S/Ni(111)	E	2.02	34	
p(2x2)C/Ni(001)	E	2.13	38	First Ni interlayer distance expanded 8.5%; structure not completely fixed
c(2x2)Se/Ag(001)	E	2.80	39	Fit not as good as on Ni
c(2x2)Cl/Ag(001)	E	2.67	40	Bond length varies from 2.6 to 2.7 Å depending on choice of Cl potential
($\sqrt{3}$x$\sqrt{3}$)30° I /Ag(111)	E	2.80	41	Atomic I potential is used,(better than ion potential).

TABLE 2. (Cont.)

System	Position	Bond Length (Å)	Ref.	Special Features
(1x1)O/Fe(001)	E	2.08	24	O sits closer to 2nd layer of Fe; first Fe interlayer distance expanded 7.5%
(1x1)S/Fe(001)	E	2.30	42	
(1x1)N/Ti(0001)	U	2.12	43	N sits in octahedral sites between first two layers of Ti
(1x1)Cd/Cd/Cd/Ti(0001)	U	3.08 (Ti-Cd)	28	1st Cd layer sits in FCC 3-fold sites; successive layers build an HCP Cd lattice
(2x1)O/W(110)	U	2.08	44	O sits in three-fold coordinated site and forms two domains
(1x1)Si/Mo(001)	E	2.51	45	
c(2x2)N/Mo(001)	E	2.45	46	Fit poor for some beams
c(2x2)Na/Al(001)	E	2.88 / 2.90	14 / 13	The bond length varies from 2.83 to 2.95 depending on the beam used for best fit
c(2x2)C$_2$H$_2$/Pt(111)	–	2.2-2.5 (C-Pt)	47	C$_2$ sits either centered and along main diagonal of unit cell (Fig. 11), or at 90° to that line and centered over 3-fold position.
c(2x2)N/Cu(001)	E	2.32	48	Analysis made by data averaging.

* The overlayer and its symmetry are given first, then the substrate.

** E = expected position (same as next layer of substrate)
U = unexpected position (different from next layer of substrate)
– = unknown or no expectation

*** internuclear distance overlayer atom to nearest substrate atom

TABLE 3. EFFECTIVE RADII r^A_{eff} (Å)

| Element | Single-Bond Covalent Radius* | \multicolumn{8}{c}{Substrate r^S_B (bulk radius)} |
		Ni 1.24	Fe 1.26	W 1.39	Pt 1.38	Ti 1.47	Ag 1.44	Al 1.43	Mo 1.36
O	0.66	0.74	0.83	0.72					
S	1.04	0.94	1.03						
C	0.77	0.89			0.99 (C$_2$)				
N	0.70					0.62†			
Se	1.13	1.03					1.36		
Te	1.39	1.34							
Cl	0.99						1.23		
I	1.33						1.36		
Si	1.17								1.15
Na	1.85**	1.60						1.43	

$r^A_{eff} \equiv b_{closest} - r^S_B$

$b_{closest} \equiv$ bond length to nearest substrate atom

* (Pauling 1960)
** Metallic radius
† underlayer

Five cases of unusual features or unexpected positions will now be illustrated and discussed in more detail. In Fig. 7 the (1 x 1)O/Fe(001) ordering pattern is above, then the top and side views are shown with touching radii drawn to scale. The small O atom is seen to sit so deeply in the four-fold hole (the expected position) that it rests on a second-layer substrate atom. The most noteworthy feature is the expansion of the first substrate interlayer distance by 8%, whereas the clean Fe surface is contracted 4%, approaching the configuration of flat planes of FeO.

In Fig. 8, the remarkable position of O in the structure (2 x 1)O/Ni(110) is shown, in which the O atom chooses to sit on

'LEED' Intensity Measurements 289

(a) (1x1)O/Fe(001)

(b) TOP VIEW

(c) SIDE VIEW

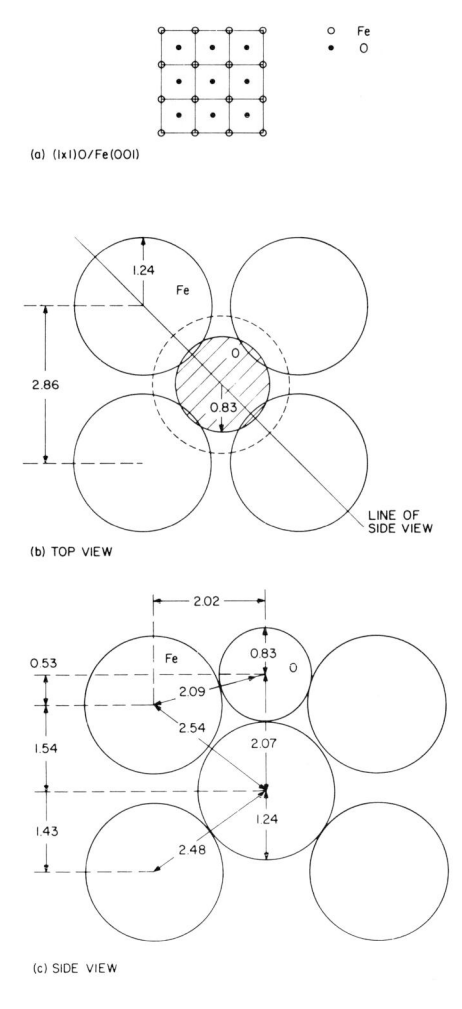

Fig. 7. Geometry of (1x1)O/Fe(001) a) ordering pattern with O in four-fold hollows of (001) surface of BCC Fe (expected position), b) top view showing three layers of atoms, atoms drawn to scale with touching radii (all distances in angstroms), O atom above (filled area), Fe atoms in first substrate layer (solid lines at bulk touching radius), Fe atom in second substrate layer (dashed line), intersection line of perpendicular plane in which side view is shown, c) side view showing O atom resting on second substrate layer Fe atom and 7.7% expansion of first interlayer spacing compared with bulk.

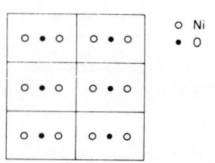

(a) (2x1)O/Ni(110)

(b) TOP VIEW

(c) SIDE VIEW

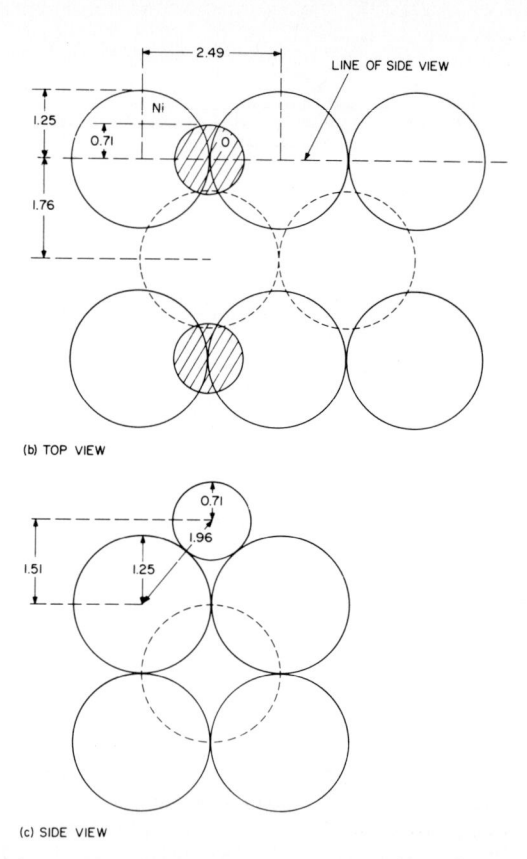

Fig. 8. Geometry of (2x1)O/Ni(110) a) ordering pattern with O on top of short bridge position between two touching Ni atoms on (110) face of FCC lattice, two Ni atoms to each O atom (in unexpected position), b) top view to scale with touching radii showing three layers of atoms and line of side view, c) side view showing two layers of atoms, O touching Ni atoms with bulk radius.

top of the short bridge of the rectangular unit cell of Ni(110) rather than down in the trough, in contrast to S which sits in the trough in the expected position on top of a second-layer Ni atom. The position of the O atom has been verified by ion-scattering experiments[29].

In Fig. 9, the unexpected position of O in the structure (1 x 2)O/W(110) is drawn with touching radii to show that the small O atom moves away from the central (expected) position above the unit cell (top view) to a three atom-coordinated position at one side of the unit cell. The analysis requires that two domains of ordered O atoms be averaged, the other domain corresponding to O atoms sitting in the other three-atom coordinated position in the unit cell, i.e., obtained by reflection in a vertical line in the plane of the figure.

In Fig. 10, the strange behavior of (1 x 1)N/Ti(0001) is shown in which the initial ordered structure formed on Ti(0001) by exposure to N is an underlayer of N, sitting in the octahedral holes of Ti HCP lattice. The first three layers essentially form a single layer of TiN with the same interlayer spacings as in bulk TiN, for which the spacing between Ti layers is 5% expanded from bulk Ti (interlayer spacing of close-packed planes 2.34 Å). The touching radius in the basal plane, from the lattice constant in the plane, should be 1.48 Å while the touching radius between planes, allowing for the 5% expansion, is 1.49 Å. (The 1.50 Å used in the figure corresponds to the touching radius of Ti in bulk TiN.) The analysis requires averaging over the two different surfaces exposed by the presence of monatomic steps.

In Fig. 11, the first example of a structure determination of an ordered molecular overlayer is shown for (2 x 2)C$_2$H$_2$/Pt(111). In this case, there are many more parameters to fix, since, in addition to the interlayer distance, fixing the registration requires fixing both translation and rotation of the molecule on the surface and also the C-C distance. Fortunately the H atoms have small enough scattering power to be ignored. In view of the difficulties, the results must be regarded as tentative. We note that in the publication the preferred interlayer distance is given as 1.95 Å (rather than the 1.8 Å in Fig. 11), the C-C distance is not fixed at all (and presumably could be anywhere between triple bond value 1.20 Å to single bond value of 1.54 Å), and another orientation is given equal preference with the one shown in Fig. 11, namely with the center of the molecule over the center of one of the equilateral triangles of Pt atoms, and the C-C bond parallel to a side of the trianble (so that the two C atoms have the same bonding environment). In each case, the analysis requires averaging over the three equivalent 120°-rotated domains. The shortest C-Pt distance for the former orientation is 2.47 Å, for the latter it is 2.25 Å, corresponding to an effective radius for C of 1.08 Å or 0.86 Å respectively; the size of this radius compared to even the single-bond C radius

292   P. M. Marcus

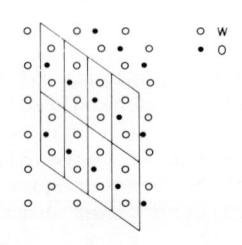

(a) (1x2) O/W (110)

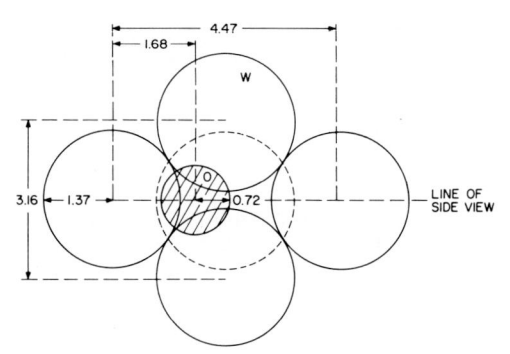

(b) TOP VIEW

(c) SIDE VIEW

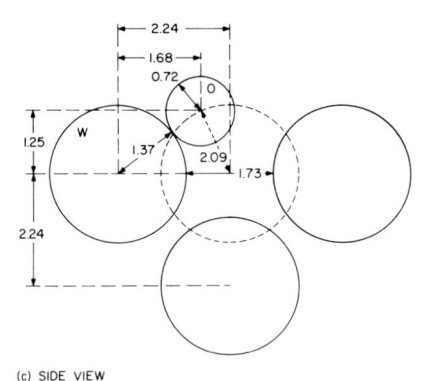

Fig. 9. Geometry of (1x2)O/W(110) a) ordinary pattern showing O in hollow touching 3 W atoms on (110) face of BCC lattice, one O atom to two W atoms (in unexpected position), b) top view showing three layers of atoms drawn with touching radii and line of side view, c) side view showing two layers of atoms.

'LEED' Intensity Measurements 293

(a) (1x1)N/Ti(0001)

(b) TOP VIEW

(c) SIDE VIEW

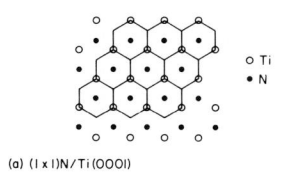

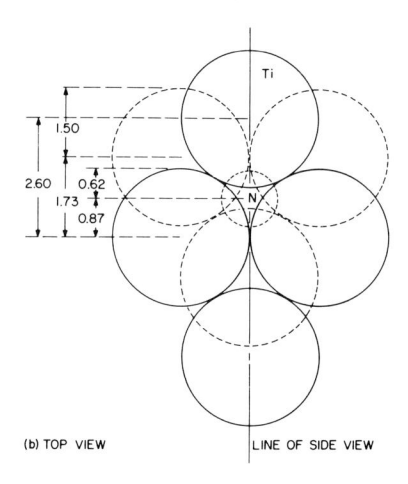

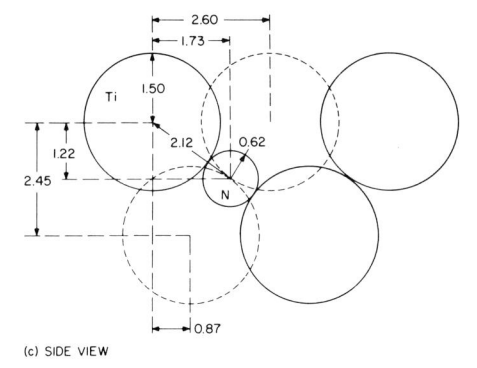

Fig. 10. Geometry of (1x1)N/Ti(0001) a) ordering pattern showing N in three-fold hollows of basal plane of HCP lattice of Ti, b) top view drawn to scale with touching radii showing three layers of atoms with N atom in second layer <u>beneath</u> first Ti layer (short dashes) and above second layer of Ti (long dashes) and line of side view, c) side view showing two layers of atoms with N in octahedral hollow of Ti atoms.

(a) (2x2)C₂H₂/Pt(111)

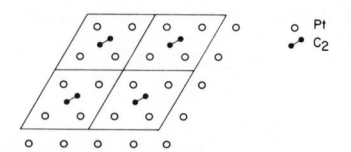

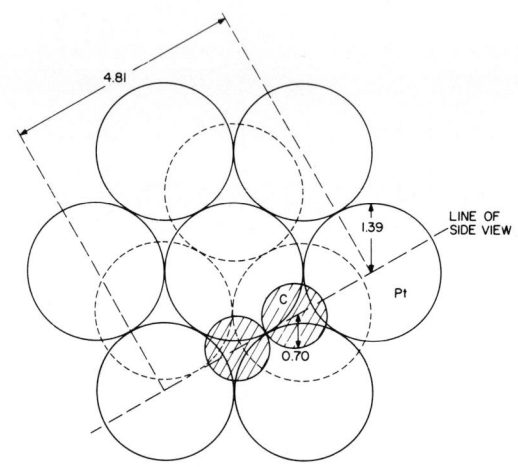

(b) TOP VIEW

(c) SIDE VIEW

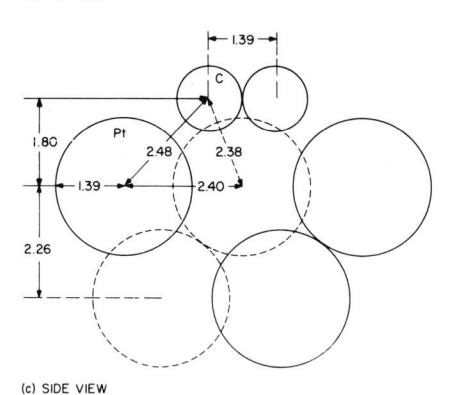

Fig. 11. Geometry of (2x2) C₂H₂/Pt (111) a) ordering pattern on (111) face of FCC Pt lattice, b) top view to scale with touching radii of three layers of atoms, carbon radii based on assumed C-C distance of 1.39 Å (not fixed by LEED analysis), H atoms not shown, c) side view showing two layers of atoms, interlayer spacing from C layer to first Pt layer of 1.8 Å fixed by LEED analysis, shortest C-Pt bond length of 2.38 Å, which goes out of plane, shown dashed

of 0.77 Å suggests a rather weak bond of C to Pt. An inherent difficulty in this work is the rapid deterioration of the ordered structure and the LEED pattern under the incident beam, which required that the entire spectrum be measured in two minutes.

CONCLUSIONS

Despite the exacting conditions required by both theory and experiment for systems for which the intensity analysis can be carried out, the success in fitting experiment achieved in the systems described above suggests that the field of surface crystallography by LEED intensity analysis is believably launched. In selected systems that meet the conditions for simplicity imposed by theory and ease of preparation by experiment, the positions of atoms in surface layers can be fixed to better than ± 0.1 Å, with a sensitivity that suggests an ultimate accuracy of the order of ± 0.01 Å when the theory is adequately refined. However, in many systems the analysis is only partially successful; in these systems the correspondence of theory to experiment is close enough to be substantially better than a random match, but the discrepancies point to complications that have not been found and require patient unraveling. Some of the many complications that can occur have been mentioned - aside from experimental difficulties, domains, steps, patches are possible - and more will undoubtedly appear as more complex systems are studied.

Some important developments of the theory will now be noted, developments which are needed to advance structure determination by LEED. Fitting the intensity spectra is now a tedious and chancy procedure, although when a good fit according to criteria described above is obtained, there is little doubt of its correctness. More rapid analysis and extension to more complex systems will require 1) more effective search procedures for correct geometrical parameters, 2) more powerful calculation procedures to find intensity spectra. These developments will now be discussed.

Before lengthy intensity calculations are carried out, the parameter ranges to be searched should be narrowed down by applying quicker and less costly procedures - chemical information on likely surface configurations of course, which will improve as knowledge of surface binding increases, and the results of other techniques, such as ion scattering[29]. But development of data handling techniques which automatically point to values of parameters promises to be very useful. The method of averaging LEED spectra over angle at constant scattering vector (defined as the difference between outgoing and incoming electron wave vectors) produces simpler, more kinematic spectra, i.e., spectra described well by single scattering, as in x-ray diffraction theory[30]. Such spectra can be fitted more quickly by the simpler kinematic diffraction formulas, and many structures can be quickly eliminated, although more of the burden of the structure determination is transferred to the experiment. Another data handling procedure is Fourier inversion of the LEED spectrum[31], leading to a distribution of atomic spacings (analogous to the Patterson function of x-ray diffraction theory). The chief difficulty in obtaining useful information by Fourier inversion arises from the short range of energy (or wavenumber) available in the measured spectrum, and the spurious structure produced by the truncation error associated with terminating the range for Fourier inversion. In fact, Fourier inversion can be carried out on the angle-averaged spectra, which reduces the amount of spurious structure. Systematic reliance on these data-handling methods is probably needed to analyze the more complicated and more interesting structures that lie ahead.

The current multiple scattering calculation procedures for LEED spectra are reaching a practical limit related to the number of beams (plane waves) used in the description of the wave function. Extension of the calculation to more complex structures or to higher energies requires a more powerful calculation procedure that avoids the beam description, e.g., both the wave function and scattering calculation could be described in spherical waves. It is clear that such a more powerful procedure would be very useful now in advancing surface structure determination by LEED, and essential in the near future.

ACKNOWLEDGEMENTS

The author is indebted to his colleagues, D. Jepsen and J. Demuth at IBM Research and F. Jona at the State University of New York at Stony Brook, for much discussion of LEED theory and its application to structure determination.

REFERENCES

1. P. M. Marcus, J. E. Demuth and D. W. Jepsen, Surface Sci., 53 501 (1975).
2. Results for Ni are given in J. E. Demuth, P. M. Marcus and D. W. Jepsen, Phys. Rev., B11, 1460 (1975).
3. E. G. McRae, private communication (to be published).
4. D. W. Jepsen, P. M. Marcus and F. Jona, Phys. Rev., B5, 3933 (1972).
5. J. B. Pendry, Low Energy Electron Diffraction, Academic Press, 1974, chapts. 4, 5.
6. S. Y. Tong, Progress in Surface Science, 7, 1 (1975).
7. J. E. Demuth and T. N. Rhodin, Surface Sci., 45, 249 (1974).
8. H. D. Shih, F. Jona, D. W. Jepsen and P. M. Marcus, J. Phys. C: Solid State Phys., 9, 1405 (1976).
9. J. E. Demuth, D. W. Jepsen and P. M. Marcus, J. Phys. C, 6, L307 (1973).

10. J. S. Strozier, Jr., D. W. Jepsen and F. Jona, Chapt. I in *Surface of Physics of Crystalline Materials*, J. M. Blakely, Ed., Academic Press, New York (1975).
11. S. Andersson, *Lectures at the NATO Advanced Study Institute on Electronic Structure and Reactivity of Metal Surfaces*, Namur, Belgium 1976, Ed. E. G. Derouane and A. A. Lucas, Plenum Press (1976).
12. D. W. Jepsen, P. M. Marcus and F. Jona, Phys. Rev., B6, 3684, (1972).
13. M. Van Hove, S. Y. Tong and N. Stoner, Surface Sci., 54, 259 (1976).
14. B. N. Hutchins, T. N. Rhodin and J. E. Demuth, Surface Sci., 54, 419 (1976).
15. G. E. Laramore, Phys. Rev., B9, 1204 (1974).
16. D. W. Jepsen, P. M. Marcus and F. Jona, Phys. Rev., 8, 5523 (1973).
17. D. W. Jepsen, P. M. Marcus and F. Jona, Surface Sci., 41, 223 (1974).
18. E. Zanazzi, F. Jona, D. W. Jepsen and P. M. Marcus (to be published).
19. L. L. Kesmodel and G. A. Somorjai, Phys. Rev., B11, 630 (1975).
20. A. Ignatiev, F. Jona, H. D. Shih, D. W. Jepsen and P. M. Marcus, Phys. Rev., B11, 4787 (1975).
21. S. Andersson, P. Echenique and J. B. Pendry (to be published) (1975).
22. M. G. Lagally, J. C. Buchholz and G. C. Wang, J. Vac. Sci. Technol., 12, 213 (1975).
23. M. A. Van Hove and S. Y. Tong (to be published).
24. K. O. Legg, F. Jona, D. W. Jepsen and P. M. Marcus, J. Phys. C, 8, L492 (1975) and to be published.
25. J. A. Strozier and R. O. Jones, Phys. Rev., B3, 3228 (1971).
26. H. D. Shih, F. Jona, D. W. Jepsen and P. M. Marcus, J. Phys. C, 9, 1405 (1976).
27. W. N. Unertl and H. V. Thapliyal, J. Vac. Sci. Technol., 12, 263 (1975).
28. H. D. Shih, F. Jona, D. W. Jepsen and P. Marcus, Communications on Physics, 1, 25 (1976).
29. W. Heiland and E. Taglauer, J. Vac. Sci. Tech., 9, 620 (1972), Surface Sci., 35, 381 (1973).
30. See, for example, the reivew by M. B. Webb and M. G. Lagally in Solid State Physics, 28, 301 (1973).
31. U. Landman and D. L. Adams, J. Vac. Sci. Technol., 11, 195 (1974).
32. J. E. Demuth, D. W. Jepsen and P. M. Marcus, Phys. Rev. Letters, 31, 540 (1973).
33. M. Van Hove and S. Y. Tong, J. Vac. Sci. Technol., 12, 230 (1975).
34. J. E. Demuth, D. W. Jepsen and P. M. Marcus, Phys. Rev. Letters, 32, 1182 (1974).
35. J. E. Demuth, D. W. Jepsen and P. M. Marcus, Bull. APS Ser. II, 20, 855 (1975).

36. J. E. Demuth, D. W. Jepsen and P. M. Marcus, J. Phys. C: Solid State Phys., 8, L25 (1975).
37. J. B. Pendry and S. Andersson, to be published.
38. M. A. Van Hove and S. Y. Tong, Surface Sci., 52, 673 (1975).
39. A. Ignatiev, F. Jona, D. W. Jepsen and P. M. Marcus, Surface Sci., 40, 439 (1973).
40. E. Zanazzi, F. Jona, D. W. Jepsen and P. M. Marcus, Bull. APS Ser. II, 21, 320 (1976); Phys. Rev., B14, 432 (1976).
41. F. Forstmann, W. Berndt, and P. Büttner, Phys. Rev. Letters, 30, 17 (1973); F. Forstmann, Proc. 2nd Intern'l. Conf. on Solid Surfaces, J. Appl. Phys. Suppl., 2, 657 (1974).
42. K. O. Legg, F. P. Jona, D. W. Jepsen and P. M. Marcus, Bull. APS Ser. II, 21, 319 (1976), to be published.
43. H. D. Shih, F. Jona, D. W. Jepsen and P. M. Marcus, Phys. Rev. Letters, 36 798 (1976).
44. M. A. Van Hove and S. Y. Tong, Phys. Rev. Letters, 35, 1092 (1975).
45. A. Ignatiev, F. Jona, D. W. Jepsen and P. M. Marcus, Phys. Rev., B11, 4780 (1975).
46. A. Ignatiev, F. Jona, D. W. Jepsen and P. M. Marcus, Surface Sci., 49, 189 (1975).
47. L. L. Kesmodel, P. C. Stair, R. C. Baetzold and G. A. Somorjai, Phys. Rev. Letters, 36, 1316 (1976).
48. J. M. Burkstrand, G. G. Kleiman, G. G. Tibbetts and J. C. Tracy, J. Vac. Sci. Technol., 13, 291 (1976).

Computation Methods of LEED Intensity Spectra*

N. Stoner, M.A. Van Hove and S.Y. Tong

Department of Physics and Surface Studies Laboratory
University of Wisconsin, Milwaukee
Milwaukee, Wisconsin   53201

A review of the dynamical methods used in low-energy electron diffraction (LEED) intensity spectra computations is given. LEED intensity spectra have been used to extract the position where a surface atom sits with respect to other atoms of a solid. To obtain a successful surface structure determination, many trial geometries must be calculated, with different registries and varying interlayer spacings. This puts a premium on computation efficiency as well as the general flexibility of the methods used. In the past five to six years, a number of computation methods have been proposed for the calculation of LEED intensity spectra. We shall give an outline of some of the most useful methods and attempt a brief comparison among the methods. Each computation scheme usually offers some unique features which make it more suitable for certain kinds of surface structures than others. Conversely, no one method can claim to be superior for the analysis of all surface structures of current interest. We shall point out the criteria used for the selection of computation methods in making surface structure determination more efficient.

---

*Work supported in part by the National Science Foundation Grant No. DMR73-02614 and by the Graduate School Research Committee, University of Wisconsin-Milwaukee.

CONTENTS

I. Introduction
   A. LEED as a Research Tool
   B. Description of the LEED System
   C. Characteristics of the LEED Spectra
   D. General Description of Calculation Schemes
   E. Description of Succeeding Sections

II. Simple Description of the LEED Process
   A. Clean Crystals and Bragg Reflections in One Dimension
   B. Electron Penetration Depth
   C. Three-Dimensional Effects
   D. Overlayer Effects

III. Exact Methods for the Calculation of LEED Intensities
   A. Definition of Exact Method
   B. The Beeby T-Matrix Method
      Introduction
      The T-Matrix Formulation
      Spherical Expansions in Momentum Components
      Lattice Sums
      Elastic Reflectivity for a Given Beam
   C. The Bloch-Wave Method
      Introduction
      Relationship Between Layer-Matrices of Pendry and Those Used Here
      Layer Reflection and Transmission Matrices and the Matrix Eigenvalue Problem
   D. Beam Intensities
   E. Layer Doubling

IV. Iterative and Perturbation Techniques

    A. Need for Fast Calculation Schemes

    B. Direct Perturbation Expansion Techniques

    C. Renormalized Forward Scattering Method

V. New Computation Methods for Dynamical LEED

    A. The L-Space Layer-Iteration Method

    B. The Combined-Space Method of Tong and Van Hove

VI. Discussion of Calculation Methods

    A. General Criteria

    B. Method-by-Method Discussion

I. INTRODUCTION

A. Low Energy Electron Diffraction as a Research Tool

Low energy electron diffraction (LEED), the elastic back scattering of low energy electrons (0-500 eV) from a crystal surface, is one of the few techniques that is currently available for resolving the surface crystallography of solids on an atomic scale. As predicted by Davisson and Germer even before their Nobel prize winning work in 1927, LEED is extremely sensitive to the state of the immediate surface region. Ironically, this fact which obstructed the growth and development of LEED after its discovery is the very reason for the resurrection of interest in LEED techniques.[1] The high sensitivity is attributed to the strong interactions between the incident electrons and the atoms of the crystal which in turn effectively confines these electrons to the first few layers of the solid. This confinement dictates that any information contained in the back scattered electrons must be information about the surface region.

LEED is strongly sensitive to any form of ordered surface geometry. Fortunately, diffraction in general tends to filter out certain forms of disorder such as small defects in crystalline structure and dilute impurities. However, other forms of disorder such as phonons, steps in the surface, and grain boundaries tend to diffuse the diffraction process. Experimentally, surface irregularities can be controlled by careful preparation of the sample whereas theoretically one can construct a perfect surface. Phonon interactions are inherent in any finite temperature situation and must be treated theoretically in a statistical manner. LEED is a valuable tool for the structural analysis of ordered geometric features of clean surfaces, reconstructed surfaces (such as semiconductors and some transition metals), physisorbed and chemisorbed overlayered surfaces, and other types (layered compounds).

B. Description of the LEED System

LEED studies are normally done on samples having two-dimensional symmetry parallel to the surface. This surface can be described by two unit cell basis vectors $\underline{a}$ and $\underline{b}$ lying in the plane of the surface. The real lattice is generated by the vectors $\underline{r} = \ell \underline{a} + m \underline{b}$ where ℓ and m are integers. A lattice of this type with a single atom per unit cell is called a Bravais lattice. Surfaces are named after the planes in the crystal to which they are parallel. The usual Miller indices $\{k\ell m\}$ offer a convenient scheme for this purpose; e.g., (001), (111), etc.

In order to establish a frame of reference, a right handed coordinate system is assumed such that x and y are parallel to the surface and $+_z$ direction is an inward pointing normal to the surface. The LEED configuration is shown schematically in Figure 1. Here, an incoming monochromatic beam of electrons of energy E and momentum $\underline{k}_o$ is incident at a certain angle on a single crystal face. A small fraction of the incident electrons will be elastically back-scattered by the crystal in specific directions that satisfy parallel momentum conservation. These directions are determined by the diffraction condition $\underline{k}'_{/\!/} = \underline{k}_{o/\!/} + \underline{g}$ where $\underline{g}$ is a reciprocal lattice vector of the surface. If the vectors $\underline{A}$ and $\underline{B}$ are the basic vectors of the surface unit cell in reciprocal space then $\underline{g} = m \underline{A} + n \underline{B}$. The backscattered electrons constituting approximately 1% of the incident flux can be post accelerated to form bright spots on a fluorescent screen or can be detected by a Faraday cup.

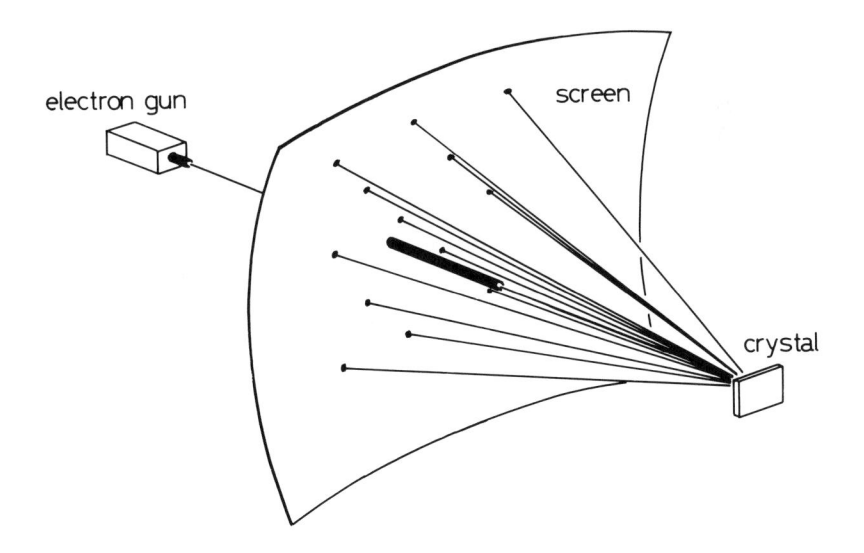

Fig. 1. Simplified sketch of an experimental LEED setup, showing incident electron beam and reflected elastic beams reaching screen.

The diffraction pattern is the Fourier transform of the two-dimensional crystal lattice and is therefore a map of the reciprocal lattice. The size, shape and orientation of the unit surface cell can be inferred directly from these patterns. The patterns also contain information about faceting and disorder. However, to obtain information about structure within the unit cell one must examine the intensities of the diffraction patterns. These intensities hold the key to the disposition of the contents of the unit surface cell: i.e. the separation between planes of atoms near the surface, the location of adsorbate atoms relative to the substrate and the reconstruction of surfaces.

Within the coherence length of the electron beam incident on the surface ($\approx 500 \text{Å}$), the beam can be described by a plane wave

$$\psi(\underline{r}) = A\, e^{i\, \underline{k}_o \cdot \underline{r}} \quad (1)$$

The electrons are backscattered at specific directions, $\underline{k}^-(\underline{g})$ given by $\underline{k}^-(\underline{g}) = [\underline{k}_{o\parallel} + \underline{g},\, -(\frac{2mE}{\hbar^2} - |\underline{k}_{o\parallel} + \underline{g}|^2)^{\frac{1}{2}}]$ where the z com-component of the momentum is obtained from energy conservation, $E = \frac{\hbar^2}{2m}[|\underline{k}_{o\parallel} + \underline{g}|^2 + (k_z^-)^2]$. The scattered wave in the region outside the solid can be written as

$$\psi_s(\underline{r}) = \sum_{\underline{g}} B_{\underline{g}}\, e^{i\, \underline{k}^-(\underline{g}) \cdot \underline{r}} \quad (2)$$

Thus, for a given $\underline{k}_{o\parallel}$ and E, $\underline{g}$ determines the diffraction pattern; one spot on the screen for each allowed value of $\underline{g}$. This fact is useful in naming the spots. The spot produced by a beam with the parallel component of its momentum given by $\underline{k}_{o\parallel} + \underline{g}$ is referred to as the (hk) spot or (hk) beam. Only a finite number of beams reach the screen because as $\underline{g}$ increases the z component of the back-scattered flux approaches zero. When

$$|\underline{k}_{o\parallel} + \underline{g}|^2 \approx \frac{2mE}{\hbar^2} \quad (3)$$

the beam is traveling nearly parallel to the surface and if $\underline{g}$ becomes larger k_z^- becomes complex and the beams are evanescent.

C. Characteristics of LEED Spectra

By varying the incident energy at a fixed angle of incidence, one can measure an intensity-voltage (IV) curve for each diffracted beam. From a purely kinematic viewpoint, one would expect intensity maxima or Bragg peaks whenever $(\underline{k}_r - \underline{k}_r') \cdot \underline{d} = 2\pi n$, where $\underline{k}_r$ is the real part of the wave vector inside the solid and $\underline{d}$ is an interlayer lattice vector. However, low energy I-V curves generally contain much more information. Figure 2 is an example of an I-V curve from 20 to 200 eV for the (001) face of nickel.[2] A number of features are evident; prominent among these are (1) the low reflectivity ($\approx 1\%$), (2) a number of sizeable peaks in addition to those identifiable as Bragg peaks. These latter peaks are usually shifted non-uniformly away from their kinematic expected energies, (3) the broad peak widths, particularly at higher energies, (4) the general reduction of intensities as the temperature increases.

An analysis of these features leads one to the identification of several important parameters that are necessary ingredients for the development of a suitable LEED theory. The mere fact that one is considering a reflected spectrum indicates that there must be a mechanism for interactions capable of reversing the electron momentum. This suggests the necessity of a well-constructed ion-core potential, which is a part of the crystal potential that describes the electron interaction. It has been found necessary to invoke multiple scattering to explain the additional non-Bragg structures in the I-V curves, at least for the surfaces that have been studied so far. The broad peaks confirm the hypothesis that the incoming electrons are confined to the surface region by a strongly absorptive medium. Since only the elastic component of the backscattered flux is measured, the inelastic processes are

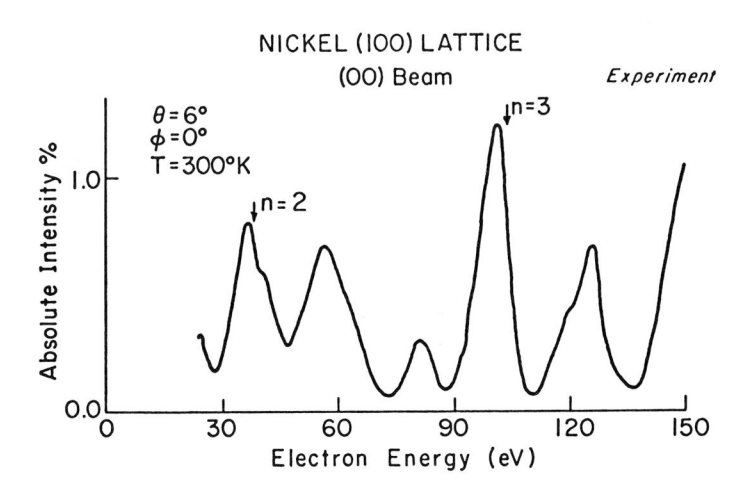

Fig. 2. The experimental I-V curve of the (00) beam reflected off Ni(001) at $\theta = 6°$, $\phi = 0°$, T = 300°K. The arrows denote Bragg energies. Data taken by J.E. Demuth, Ref. 2.

treated collectively by the introduction of an imaginary component of the crystal potential. This type of loss can be described by assigning a finite path length to the elastic electrons in the crystal. The wave function for the electron has a time variation $\psi(\underline{r},t) = \psi(\underline{r}) e^{-\frac{iEt}{\hbar}}$. If the inner potential of the crystal has an imaginary component $V_o = V_{or} + i V_{oi}$ ($V_{oi} < 0$), then the intensity of the wave function will decay away in time by the factor $e^{2V_{oi}t/\hbar}$. Thus by equating $2 V_{oi} = -\frac{\hbar}{\tau}$, electron attenuation is simulated by the introduction of V_{oi}. The diffraction must take place on the average within a time scale τ, hence from the uncertainty principle $\Delta E \Delta t \geq \hbar$, it is noted that $|V_{oi}|$ places a restriction on peak widths given by $\Delta E \geq 2|V_{oi}|$. Inelastic processes treated by this mechanism include surface and bulk plasmon excitations, auger, single particle excitations, etc., which arise from the interaction of the incident electrons with the electrons of the solid.

A very important source of electron energy loss (usually <0.1 eV) results from the electron-phonon interaction. Electrons with energy loss this small still reach the screen, however, the finite temperature lattice vibrations cause a loss in coherence in the scattering process by non-zero random momentum transfers which result in the loss in intensity of the diffracted beam. Analytically, this effect is described by modifying the elements of the scattering matrix by a Debye-Waller factor e^{-2W}.

Other parameters of importance include the shape and extent of the transition between the vacuum and the zero of potential energy inside the crystal (barrier potential). After much experience, it has been concluded that at low electron energies (≤ 50 eV) the scattering effect of the barrier potential is important while at higher energies only the refractive effect need be considered.

A model for the surface geometry is of paramount importance in LEED. The mechanisms by which the additional structure in the LEED spectra is produced are very complex. It is questionable whether or not one will ever be able to directly invert LEED data to obtain surface geometries, except possibly in systems where the scattering is dominantly kinematic. In the dynamical approach, one must postulate models for the surface geometry based on information such as atomic radius, symmetry, and at times experience and common sense. Calculations are then made for these surface models and results of the calculations are compared to experimental data. The model which exhibits the best fit is taken to represent the actual structure if the fit is good enough to satisfy "state of art" standards.

D. General Description of Calculation Schemes

There are currently available several dynamical (multiple scattering) methods to calculate LEED intensities. These methods will be reviewed in later sections. However, the intent of this section is to delineate some features common to all of the techniques. The general scheme of calculation is characterized by the following steps: 1) An accurate representation of the ion-core scattering is found by a set of phase shifts. The one electron muffin-tin potential approximation is used. 2) The ion cores are assembled into layers of mono-atomic thickness. 3) A technique is used for stacking the layers to form a crystal and 4) A formal expression is obtained for the flux in each of the backscattered directions.

The crystal potential is approximated by the muffin-tin model. It consists of the largest possible non-overlapping spheres drawn about each nucleus. The potential inside the sphere, the ion-core potential, is assumed spherically symmetric while outside it is assumed constant (this value defines the muffin-tin zero). The ion-core itself can be constructed from free atom wave functions or from wave functions more representative of the crystalline environment. At present, no general rule can be established as to which prescription is the best. For the purpose of surface structural determination by LEED, it seems most potentials used in band structure calculations are adequate. The value between vacuum level and the muffin-tin zero is often considered an adjustable parameter, since its calculation from first principles is not properly established.

For LEED purposes, it is not necessary to know the details of wave functions inside the ion-core provided one knows how plane waves are scattered by them. Scattering by the spherically symmetric ion-cores can be characterized by a set of phase shifts, one for each angular momentum quantum number, ℓ. The wave function in the region of constant potential between the muffin-tins is a superposition of plane waves. The scattering of plane waves by a single spherically symmetric potential is well known and is given in asymptotic form by

$$\psi_s(\underline{r}) = e^{i\underline{k}\cdot\underline{r}} + f(\theta)\left[\frac{e^{i|\underline{k}||\underline{r}|}}{|\underline{r}|}\right] \quad (4)$$

The scattering amplitude $f(\theta)$ can be written in terms of phase shifts as[3]

$$f(\theta) = \frac{2\pi}{|\underline{k}|}\sum_{\ell=0}^{\infty}(2\ell+1)\,e^{i\delta_\ell}\sin\delta_\ell\,P_\ell(\cos\theta) \quad (5)$$

Therefore, a knowledge of the phase shifts is an appropriate starting point from which the scattering from many ion-cores can be obtained by matching boundary conditions or solving the multiple scattering process self-consistently.

The effects of thermal motions of the ion-cores are treated by the inclusion of the anisotropic Debye-Waller factor

$$f(\theta,T) = f(\theta)\, e^{-M_p (\underline{k}-\underline{k}')^2} \qquad (6)$$

where $M_p = \frac{1}{2} \langle u_p^2 \rangle_T$, and u_p is the vibrational amplitude in the direction of the momentum transfer $\underline{k} - \underline{k}'$ (the angle θ is between $\underline{k}$ and $\underline{k}'$). In the high temperature limit ($T \gtrsim \theta_D$), u_p is related to the Debye temperature by

$$\langle u_p^2 \rangle_T = 3\hbar^2\, T/M\, k_B\, \theta_D^2(\underline{p}) \qquad (7)$$

where M is the atomic mass and k_B the Boltzmann constant. Within a crystal, u_p is layer dependent as well as direction dependent. Anisotropic thermal effects are difficult to include in a multiple scattering formalism. As a first order approximation, an isotropic model for the vibrational amplitudes is usually assumed. In this case, the temperature dependent scattering factor can be written in terms of a set of complex temperature dependent phase shifts. This is accomplished by expanding both sides of Eq. (6) in a spherical representation and, by equating each ℓ-component, one can solve for temperature dependent phase shifts.[3] The imaginary component being positive implies that flux is not conserved in the scattering process. This is the statistical procedure by which theory simulates the incoherently scattered flux that is lost by phonon scattering.

E. Description of Succeeding Sections

In the next Section, the central physics of the low-energy electron diffraction process will be cast in a form suitable for easy visualization. We shall illustrate the processes that give rise to common phenomena such as Bragg reflections, peak widths, overlayer effects, etc. Section III will deal with the exact methods of LEED. In Section IV, a number of perturbative and iterative methods are discussed. New computation schemes that are developed specifically to treat complicated systems of interest or systems where earlier methods fail are discussed in Section V. A critique and method-by-method comparison of the various schemes are given in Section VI.

II. SIMPLE DESCRIPTION OF THE LEED PROCESS

In this section, we will analyze the low energy electron diffraction process at a crystal surface in such a way that the details of the observed I-V curves are seen to follow naturally from simple effects, despite the fact that I-V curves present some unusual features, i.e., maxima shifted away from Bragg conditions, additional maxima, large peak widths and splitting of peaks.

A. Clean Crystals and Bragg Reflections in One Dimension

Let us first consider the one-dimensional case where a wave e^{ikx} (wave-number k) hits a semi-infinite row of identical, equally-spaced scatterers. Each scatterer shall have complex reflection and transmission coefficients r and t for such a wave.

In the weak-scattering limit, we have $|t| \approx 1$, with argt ≈ 0, and $|r| \approx 0$.[4] Electrons reflect off crystals composed of such atoms just as X-rays would: interference between waves reflected off successive scatterers (spacing a) produces maxima whenever the difference in "optical paths" 2ka is a multiple of 2π (the Bragg condition). We point out that in this limit the optical path difference 2ka neglects the fact that the actual path difference involves two transmissions through a scatterer, in addition to the two propagations through a distance a. This simplification is justified for weak scatterers, where t $\approx$ 1, but does not hold for strong scatterers, where each transmission through a scatterer can modify the optical path substantially. The latter is due to the attractive potential inside the scatterer which speeds up the passing electron momentarily. In wave mechanical terms, the wavelength is locally reduced as a result of a larger kinetic energy, thereby advancing the phase of the wave relative to the case of a weak potential. The optical path is therefore increased at each transmission by argt, the phase of the transmission coefficient. The Bragg condition for reflected maxima now generalizes to

$$2(ka + \text{argt}) = n2\pi \quad (n \text{ integer}) \quad (8)$$

The effect of the phase shift argt is twofold. First, all Bragg maxima are shifted downward in energy by the lowering of the average potential in the crystal. This is a shift due to the inner potential, which is a slowly varying function of the electron energy. This first effect, however, does not account for the additional irregularities in the positions of Bragg maxima. The latter effect is due to scattering resonances that occur within each atom. For this reason, argt can change substantially and without obvious regularity from one maximum to the next.

Atomic resonances alone commonly make 2 argt reach substantial values ($\approx \pi/2$), especially at low energies in strongly scattering metals (e.g. Ni and W), as can be easily imagined if one replaces each atom by a square well of sufficient depth.

Thus, the simple concept of Bragg reflection remains perfectly valid in strongly scattering LEED, as long as phase shifts argt are not neglected. Multiple scattering of electrons between atoms has no qualitative effects on these facts.

The analogy of our discussion with the case of electronic band structure is direct. Band gaps in bulk material are the result of an electron wave being reflected constructively in step with the crystal lattice, thereby being completely turned around, i.e., it is prevented from continuing forward propagation. Nearly-free electron materials exhibit the applicability of the simple Bragg conditions. In strongly scattering materials, the same reasoning applies if phase shifts argt are included. It is a familiar fact that band gaps in such materials (even in one dimension) are irregularly shifted away from the simple Bragg conditions.

Section IIC will show that maxima in I-V curves and band gaps are intimately related in the case of an ideally terminated bulk material.

B. Electron Penetration Depth

The purpose here is to emphasize that the electron penetration distance into the crystal lattice is the basic quantity that determines the width of all diffraction maxima, regardless of the magnitude of the electron absorption.

We remain with the one-dimensional clean crystal, assuming no absorption at first. When incident electrons satisfy a Bragg reflection condition, each atom reflects a fraction $|r|$ of the incident amplitude. Therefore N atoms, where $N|r| = 1$, will reflect the full incident amplitude, and the electron penetration depth will be $Na = \frac{a}{|r|}$. A typical value of $|r|$ is 0.1 and so, in the absence of absorption, Na is of the order of 10-100Å. (All penetrating amplitude is lost through diffraction and thus there is total reflection in one-dimensional diffraction when Bragg conditions are satisfied.)

The importance of the electron penetration depth Na is the following. At any Bragg condition, the crystal effectively has a finite thickness Na for diffraction purposes. But it is well known that arrays of N equally-spaced scatterers produce diffraction maxima that have a width $2\Delta k \sim \frac{\pi}{2} \frac{1}{Na}$ in reciprocal space, i.e., an energy width

$$2\Delta E \sim \frac{\pi}{2} \frac{k}{Na} = \frac{\pi}{2} k \frac{|r|}{a} \qquad (9)$$

which is just the band gap width obtained in first-order perturbation theory. In other words, we see that the width of the familiar band gap, and therefore the width of a diffraction maxima, is directly related to, and determined by, the electron penetration depth Na, as in Eq. (9).

The above argument linking the width of diffraction maxima with the electron penetration depth is commonly applied to account for the broadening of maxima as a result of electron absorption due to inelastic processes. If absorption is represented by an imaginary part V_{oi} of the crystal potential, then diffraction maxima have a width of at least

$$2\Delta E \tilde{=} 2 V_{oi} \qquad (10)$$

The positions of diffraction maxima remain unchanged as long as $V_{oi} \ll E_o$. Broader maxima will occur if $|r|$ is sufficiently large to limit penetration even more than absorption does, as can be observed (e.g. in specular reflection off a clean Ni(001) face at normal incidence near 4 eV where a wide band gap occurs). So, we can say in general that the widths of maxima are solely determined by the penetration depth. This is independent of whether the penetration is limited by elastic or by inelastic processes (there is the trivial exception of overlapping maxima). In other words, band gap widths and absorption-broadened reflection maxima are caused by the same simple diffraction mechanism.

From this discussion, it follows that the most important aspect of the electron absorption is the induced penetration distance. In contrast, the local variation of the absorption inside the volume occupied by each atom can be ignored. Therefore, it is sufficient to keep V_{oi} constant across the atomic volume; at most, it may be desirable to have V_{oi} change at the boundaries between layers of different chemical compositions.

C. Three-Dimensional Effects

The major new feature introduced by three-dimensional, with respect to one-dimensional, crystals is the increase in the number of beams in which the elastically diffracted electrons can travel. Each layer of mono-atomic thickness in the clean three-dimensional crystal surface diffracts a beam A into a beam B with a reflection coefficient $r_{gg'}$, or a transmission coefficient $t_{gg'}$. Each beam has its particular wavevector $\underline{k}(g)$; $\underline{k}(g)$ will denote the component of $k(g)$ perpendicular to the layers, and the letter a stands for the layer separation.

As in one dimension, we can generalize the familiar Bragg reflection conditions used in X-ray diffraction by including phase shifts induced by the strong potentials in the atomic layers:

$$k_\perp(g) \, a + \arg t_{gg} + k_\perp(g') \, a + \arg t_{g'g'} = n2\pi \quad \text{(n integer)} \quad (11)$$

gives conditions for maximum reflection from beam g into beam g'. This equation is rigorous for kinematic (i.e. single) scattering. In the presence of multiple scattering, it holds inasmuch as, in diffraction off a single atomic layer, the zero-angle forward-scattered beam is much stronger than all other diffracted beams. Reflection maxima are substantially shifted from the weak-scattering positions by amounts that are very much energy-dependent as a consequence of resonances. It is the systematic inclusion of the zero-angle scattering phase argt that makes Pendry's renormalized forward scattering scheme for calculating LEED intensities coverge remarkably well. This fact is further illustrated by the good convergence of a perturbation method used by Jennings, in which zero-angle forward scattering by layers is systematically included. It is important to realize that it is not sufficient to include the average energy shift due to the inner potential (or the muffin-tin constant) to obtain this good convergence. The additional energy-dependent influence of scattering resonances must be included as well (in the form of phase shifts).

The band-structure interpretations of diffraction in three dimensions can be recast in our simplified language: in the presence of multiple scattering, reflection occurs via intermediate beams between the incident and the emergent beams and a reflection maximum will appear when there is at least one generalized Bragg diffraction in the chain of scatterings (each such Bragg diffraction corresponds to a band gap).

D. Overlayer Effects

So far, we have only analyzed "clean" surfaces. The additional effects resulting from a more complicated termination of the bulk structure at the surface can be illustrated with the case of an overlayer adsorbed on a clean substrate.

We consider an overlayer of mono-atomic thickness adsorbed at a distance d from the clean substrate. The total electron reflection R can be regarded as composed of interfacing reflections from the overlayer and the substrate. In one dimension especially, this is a simple situation, the substrate reflection

R_s and the overlayer reflection r and transmission t then combine to give

$$R = r + e^{2ikd} t R_s (1-e^{2ikd} r R_s)^{-1} t. \qquad (12)$$

An analogous matrix equation holds for three dimensions (beam amplitudes being ordered in vectors). The factor $(1 - e^{2ikd} r R_s)^{-1}$ describes multiple scattering between substrate and overlayer, as proved by a geometric expansion. This factor usually plays no significant role in practice because of absorption and the not so large values of $|r|$ and $|R_s|$ (typically 0.1 and 0.5 to 0.1, respectively). Neglecting multiple scattering, Eq. (12) has interference maxima when

$$2kd + 2 \arg t + \arg R_s - \arg r = n2\pi \quad (n \text{ integer}) \qquad (13)$$

Again, we have energy dependent shifts of the maxima away from simple geometric Bragg-like conditions. These maxima are superimposed on the strong energy dependence of the substrate reflection R_s (a complex number) in such a way that only by including the relevant scattering phase shifts can one reproduce experiment faithfully. This is true with one- as well as with three-dimensional crystals. Interference also produces intensity minima in addition to maxima, i.e., total destruction can easily occur.

III. EXACT METHODS FOR THE CALCULATION OF LEED INTENSITIES

A. Definition of Exact Method

Of the full dynamical (multiple scattering) LEED methods available, the Beeby T-matrix, the Bloch wave and the layer-doubling techniques can be described as exact, in the sense that they include multiple scattering to infinite order within an infinitely thick crystal (Bloch wave method) or a crystal of arbitrary but finite thickness (Beeby and layer-doubling methods). Damping of the LEED wave functions into the crystal eliminates the necessity to consider an infinitely thick crystal, thereby making these three methods essentially equivalent in terms of exactness. In the absence of damping (such as with band structure calculations), only the Bloch wave method is applicable.

B. The Beeby T-Matrix Method

The Beeby method developed in 1968 is a self-consistent momentum space model of LEED theory.[5-10] In this formulation, a microscopic theory is developed by writing scattering and propagation processes in partial wave expansions in momentum space. Ultimately, a formal expression is obtained for the reflected intensities at the backscattered directions which satisfies the diffraction conditions. Beeby's original treatment neglected inelastic damping and thermal motions of the ion-cores, therefore, it was not capable of producing convergent results. With the incorporation of damping, Beeby's method can be used to solve self-consistently the reflectivity of a finite crystal. The Beeby formulation also provides an important framework upon which many of the faster computation methods of LEED were built.

1. <u>The T-Matrix Formulation</u>. Inside the solid, the electron wave function can be written in the standard integral equation form.

$$\psi(\mathbf{r}) = \phi(\mathbf{r}) + \int G(\mathbf{r}-\mathbf{r}') V(\mathbf{r}') \psi(\mathbf{r}') d\mathbf{r}' \qquad (14)$$

where $\phi(\mathbf{r}) = e^{i\mathbf{k}\cdot\mathbf{r}}$ represents the unscattered plane wave and the integral represents the self-consistently scattered component. The wavelets originating from all points $\mathbf{r}'$, with strength $V(\mathbf{r}')\psi(\mathbf{r}')$ propagate by $G(\mathbf{r}-\mathbf{r}')$ and combine with $\phi(\mathbf{r})$ to give $\psi(\mathbf{r})$. The potential $V(\mathbf{r}')$ can be written as a superposition of ion-core potentials by

$$V(\mathbf{r}) = \sum_{\mathbf{R}} v_{\mathbf{R}} (|\mathbf{r}-\mathbf{R}|) \qquad (15)$$

where $v_{\mathbf{R}}(|\mathbf{r}-\mathbf{R}|)$ is assumed spherically symmetric, $\mathbf{R}$ being lattice sites. The one particle Green's function is given by

$$G(\mathbf{r}-\mathbf{r}') = \frac{1}{(2\pi)^3} \int \frac{e^{i\mathbf{k}\cdot(\mathbf{r}-\mathbf{r}')} d^3k}{E - \frac{\hbar^2 k^2}{2m} + V_o} \qquad (16)$$

where V_o is the complex effective electron self-energy term.

V_{or} marks the difference between the muffin-tin zero and the vacuum zero of energy (i.e. the barrier potential), while V_{oi} represents collectively the inelastic damping due to crystal excitations with the exception of phonon losses. In general, both V_{or} and V_{oi} are energy dependent. The Green's function describes the process by which the electrons propagate, with absorption, from $\mathbf{r}'$ to $\mathbf{r}$ inside the solid.

The total scattering matrix $T(\underline{r}',\underline{r})$ of the solid is defined as

$$V(\underline{r}')\,\psi(\underline{r}') = \int T(\underline{r}'\underline{r})\,\phi(\underline{r})\,d\underline{r} \qquad (17)$$

It follows directly from Eq. (14) that

$$\psi(\underline{r}) = \phi(\underline{r}) + \int G(\underline{r}-\underline{r}_2)\,T(\underline{r}_2\underline{r}_1)\,\phi(\underline{r}_1)\,d\underline{r}_1 d\underline{r}_2 \qquad (18)$$

In operator form, Eq. (14) is

$$\psi = \phi + GV\psi$$

$$= \phi + GV\phi + GV\,GV\phi + \ldots \qquad (19)$$

and from Eq. (17)

$$V\psi = T\phi$$

therefore

$$V\psi = [V + VGV + \ldots]\phi \qquad (20)$$

and

$$T = V + VGT \qquad (21)$$

Specifically with arguments

$$T(\underline{r}_2\underline{r}_1) = V(\underline{r}_2)\,\delta_{\underline{r}_1 \underline{r}_2} + \int V(\underline{r}_2)\,G(\underline{r}_2-\underline{r})\,T(\underline{r}_1\underline{r})\,d\underline{r}. \qquad (22)$$

By a similar procedure, one can write down an expression for the t-matrix of a single ion-core at $\underline{R}$.

$$t_{\underline{R}}(\underline{r}_2-\underline{R},\underline{r}_1-\underline{R}) = v_{\underline{R}}(\underline{r}_2-\underline{R})\,\delta_{\underline{r}_1\underline{r}_2} + \int v_{\underline{R}}(\underline{r}_2-\underline{R})\,G(\underline{r}_2-\underline{r})\,t_{\underline{R}}(\underline{r}-\underline{R},\underline{r}_1-\underline{R})\,d\underline{r}. \qquad (23)$$

This expression represents all multiple scattering events by a single ion-core located at $\underline{R}$. By summing over individual ion-core potentials, one can express the total scattering matrix as a sum over lattice sites of individual ion-core t-matrices and propagators.

$$T(\underline{r}_2\underline{r}_1) = \sum_R t_R(\underline{r}_2-\underline{R}, \underline{r}_1-\underline{R}) + \sum_{R \neq R'} t_{R'}(\underline{r}_2-\underline{R}', \underline{r}_3-\underline{R}')G(\underline{r}_3-\underline{r}_4)$$

$$t_R(\underline{r}_4-\underline{R}, \underline{r}_1-\underline{R}) d\underline{r}_3 d\underline{r}_4 + \ldots \quad (24)$$

The first term in this expression represents all single ion-core scattering events from $\underline{r}_1$ to $\underline{r}_2$. The second term represents scattering from $\underline{r}_1$ to $\underline{r}_4$ by an ion-core at $\underline{R}$, followed by propagation from $\underline{r}_4$ to $\underline{r}_3$ and a scattering from $\underline{r}_3$ to $\underline{r}_2$ by a site at $\underline{R}'$ (i.e., all two ion-core scattering events). Similarly, the third term would represent all three ion-core scattering, etc. $T(\underline{r}_2\underline{r}_1)$ represents the sum of all possible intersite and intrasite scattering events that can take place for an electron going from $\underline{r}_1$ to $\underline{r}_2$ inside the solid.

2. **Spherical Expansion in Momentum Components.** The total scattering matrix $T(\underline{r}_2\underline{r}_1)$ is normally not evaluated directly, instead its Fourier transform

$$T(\underline{k}\ \underline{k}_i) = \int e^{-i\underline{k}\cdot\underline{r}_2} T(\underline{r}_2\underline{r}_1) e^{i\underline{k}_i\cdot\underline{r}_1} d\underline{r}_1\, d\underline{r}_2 \quad (25)$$

is found by expanding in momentum components. Specifically, 1) a change of variables ($\underline{\rho}_i = \underline{r}_i - \underline{R}_i$) is introduced to eliminate the site dependence from the arguments of the t factors. 2) the t factors are expanded in an angular momentum representation, assuming spherical symmetry [$(v_R(\underline{r}) = v_R(|\underline{r}-\underline{R}|)$]

$$t_{Ri}(\underline{\rho}_i\underline{\rho}_i') = \sum_L t_{Ri\ell}(\underline{\rho}_i\underline{\rho}_i')\ Y_L(\underline{\rho}_i)\ Y_L^*(\underline{\rho}_i') \quad (26)$$

3) the G factors are expanded by the use of the plane wave expansion[3]

$$e^{i\underline{k}\cdot\underline{r}} = 4\pi \sum_L i^\ell j_\ell(kr)\ Y_L^*(\underline{k})\ Y_L(\underline{r}) \quad (27)$$

Upon inspection of these results, one notes that the sums over L-space components can be represented by matrix products over scattering and propagation factors. Thus, if one defines a diagonal matrix of dimension L ($L = (\ell_{max}+1)^2$) then

$$t_{Ri}(k_o) = t_{R_i L_i L_i}(k_o)\ \delta_{L_i L_j} = t_{R_i \ell_i}(k_o) \quad (28)$$

Similarly, a square matrix of dimension L to represent the G factors is defined as

$$\underset{\approx}{G}(\underset{\sim}{R}_i - \underset{\sim}{R}_j) = G_{L_i L_j}(\underset{\sim}{R}_i - \underset{\sim}{R}_j) \tag{29}$$

The matrix elements of $t_{R_i}(k_o)$ are evaluated from a knowledge of the energy dependent phase shifts. It can be shown that[3]

$$t_{R_i}(k_o) = \frac{\hbar^2}{2m}\left(\frac{e^{2i\delta_{R\ell}} - 1}{2ik_o}\right) \tag{30}$$

The G factors in this representation have the form

$$\underset{\approx}{G}(\underset{\sim}{R}_i - \underset{\sim}{R}_j) = \sum_{L'_j} \lambda k_o i^{\ell'_j} a(L_i L_j L'_j) h^{(1)}_{\ell'_j}(k_o|\underset{\sim}{R}_i - \underset{\sim}{R}_j|) Y_{L'_j}(\underset{\sim}{R}_i - \underset{\sim}{R}_j) \tag{31}$$

where

$$\lambda = -4\pi i \left(\frac{2m}{\hbar^2}\right) F(k_o) \tag{32}$$

$$a(L_1 L_2 L_3) = \int Y^*_{\ell_1 m_1}(\underset{\sim}{k}) Y_{\ell_2 m_2}(\underset{\sim}{k}) Y^*_{\ell_3 m_3}(\underset{\sim}{k}) \tag{33}$$

and $F(k_o)$ is a complex propagator renormalization factor given by

$$F(k_o) = 2k_o / \left[2k_o + \left(\frac{2m}{\hbar^2}\right) \frac{\partial V_o}{\partial k}\bigg|_{k_o}\right] \tag{34}$$

The quantity k_o is a pole of the function

$$E - \frac{\hbar^2 k^2}{2m} + V_o = 0. \tag{35}$$

Utilizing these expressions one can write:

$$T(\underset{\sim}{k} \, \underset{\sim}{k}_i) = (4\pi)^2 \sum_{LL'} Y_L(\underset{\sim}{k}) Y^*_{L'}(\underset{\sim}{k}_i) \left[\sum_R e^{i(\underset{\sim}{k}_i - \underset{\sim}{k}) \cdot \underset{\sim}{R}} \left[\underset{\approx}{t}_R(k_o)\right]_{LL'} + \right.$$

$$\left. \sum'_{RR'} e^{i(\underset{\sim}{k}_i \cdot \underset{\sim}{R}' - \underset{\sim}{k} \cdot \underset{\sim}{r})} \left[\underset{\approx}{t}_R(k_o) \underset{\approx}{G}(R-R') \underset{\approx}{t}_{R'}(k_o)\right]_{LL'} + \cdots \right]. \tag{36}$$

Therefore, the evaluation of $T(\underset{\sim}{k}\,\underset{\sim}{k}_i)$ is reduced to a determination of the matrix elements of $\underset{\sim}{t}_{R_i}(k_o)$, $\underset{\sim}{G}(R_i-R_j)$ and successive summation over lattice sites $\underset{\sim}{R}$, $\underset{\sim}{R}'$, ...

3. **Lattice Summations.** The lattice sums of Eq. (36) are two-dimensional and in order to evaluate them the crystal is divided into planar layers parallel to the surface. If necessary, each layer is further subdivided into two-dimensional subplanes satisfying the following conditions. 1) All subplanes have the same Bravais structure. 2) Each subplane contains only one kind of atom.

From the lattice sums, one can identify the intraplanar and interplanar structural propagators,

$$\underset{\sim}{G}^{sp}(\underset{\sim}{k}_i) = \sum_{\underset{\sim}{P}\neq 0} \underset{\sim}{G}(\underset{\sim}{P})\, e^{-i\underset{\sim}{k}_i \cdot \underset{\sim}{P}} \tag{37}$$

and

$$\underset{\sim}{G}^{\gamma\beta}(\underset{\sim}{k}_i) = \sum_{\text{all } \underset{\sim}{P}} \underset{\sim}{G}(\underset{\sim}{P}+\underset{\sim}{d}_\alpha-\underset{\sim}{d}_\beta)\, e^{-i\underset{\sim}{k}_i \cdot (\underset{\sim}{P}+\underset{\sim}{d}_\alpha-\underset{\sim}{d}_\beta)} \tag{38}$$

where $\underset{\sim}{P}$ is a subplane lattice vector (R_i-R_j) and $\underset{\sim}{d}_\gamma - \underset{\sim}{d}_\beta$ is a vector connecting the origins of two subplanes. These are structural components and are not dependent on the dynamical scattering factors of the atoms. Successive scattering within a subplane are linked by $\underset{\sim}{G}^{sp}(\underset{\sim}{k}_i)$, while interplanar scattering events are linked by $\underset{\sim}{G}^{\gamma\beta}(\underset{\sim}{k}_i)$.

Within a given subplane, the scattering process can be described by a planar scattering matrix $\underset{\sim}{\tau}_\alpha$ which is a summation of all scattering events in that plane, i.e.,

$$\underset{\sim}{\tau}_\alpha(k_o) = \underset{\sim}{t}_\alpha(k_o) + \underset{\sim}{t}_\alpha(k_o)\,\underset{\sim}{G}^{sp}(\underset{\sim}{k}_i)\,\underset{\sim}{t}_\alpha(k_o) + \cdots$$

$$= \underset{\sim}{t}_\alpha(k_o)\left[\underset{\sim}{1} - \underset{\sim}{G}^{sp}(\underset{\sim}{k}_i)\,\underset{\sim}{t}_\alpha(k_o)\right]^{-1} \tag{39}$$

The sum over all scattering sequences in the crystal is done by defining a matrix $\underset{\sim}{T}^{LL'}_\alpha(k_o)$ which represents all scattering, interplanar and intraplanar that ends on the α subplane. This layer dependent total scattering matrix is

$$\underset{\sim}{T}^{LL'}_\alpha(k_o) = \underset{\sim}{\tau}^{LL'}_\alpha(k_o) + \sum_{L_1 L_2} \underset{\sim}{\tau}^{LL'}_\alpha(k_o) \sum_{\beta\neq\alpha} G^{\alpha\beta}_{L_1 L_2}(\underset{\sim}{k}_i)\, T^{L_2 L'}_\beta(k_o) \tag{40}$$

The expression for $T(\underline{k}\,\underline{k}_i)$ resulting from the direct substitution of these factors into Eq. (25) is

$$T(\underline{k}\,\underline{k}_i) = \frac{(8\pi^2)}{A}\left[\sum_{LL'} Y_L(\underline{k})\,Y^*_{L'}(\underline{k}_i)\left[\sum_\alpha e^{-i(\underline{k}_i-\underline{k})\cdot\underline{d}\alpha}\,T_\alpha^{LL'}(\underline{k}_o)\right]\right]$$

$$\sum_g \delta(\underline{k}_{i\parallel}-\underline{k}_\parallel+\underline{g}) \qquad (41)$$

4. Elastic Reflectivity for a Given Beam. The only remaining task is to obtain an expression for the reflectivity in each of the allowed backscattered direction, $\underline{g}$. From Eqs. (14) and (16), one can write the electron wave function as[3]

$$\psi(\underline{r}) = \phi(\underline{r}) + \frac{1}{(2\pi)^3}\int \frac{d\underline{k}\,e^{i\underline{k}\cdot\underline{r}}\,T(\underline{k}\,\underline{k}_i)}{E-\left(\frac{\hbar^2 k^2}{2m}\right)+V_o}$$

By writing this expression in the beam representation as

$$\psi(\underline{r}) = \phi(\underline{r}) + \sum_g C_g \qquad (42)$$

one observes

$$\sum_g C_g = \frac{8\pi}{A}\int \frac{e^{i\underline{k}\cdot\underline{r}}\,d\underline{k}}{E-\left(\frac{\hbar^2 k^2}{2m}\right)+V_o}\left[\sum_{LL'} Y_L(\underline{k})Y'_L(\underline{k}_i)\sum_\alpha e^{i(\underline{k}_i-\underline{k})\cdot\underline{d}\alpha}T_\alpha^{LL'}(\underline{k}_o)\right]$$

$$\sum_g \delta(\underline{k}_{i\parallel}-\underline{k}_\parallel+\underline{g}) \qquad (43)$$

Upon the evaluation of the integrals $dk_\parallel$ and $dk_\perp$ in the complex plane, the expression for C_g becomes

$$C_g = -\left(\frac{8\pi^2 i}{A}\right)\left(\frac{2m}{\hbar^2}\right)F(\underline{k}_o)$$

$$\frac{\left[\sum_{LL'} Y_L(\underline{k}^-(g))\,Y^*_{L'}(\underline{k}_i)\sum_\alpha e^{i(\underline{k}_i-\underline{k}^-(g))\cdot\underline{d}\alpha}T_\alpha^{LL'}(\underline{k}_o)\right]e^{i\underline{k}^-(g)\cdot\underline{r}}}{k_\perp^-(g)} \qquad (44)$$

The term C_g represents the scattered amplitude of a plane wave traveling in the direction $\underset{\sim}{g}$. The reflected intensity $R_{\underset{\sim}{g}}$ normalized with respect to the incident flux is therefore

$$R_{\underset{\sim}{g}} = \left[\frac{k_\perp^{out}(\underset{\sim}{g})}{k_\perp^{out}(\underset{\sim}{o})} \right] |C_{\underset{\sim}{g}}|^2 \qquad (45)$$

where $k_\perp^{out}(\underset{\sim}{g})$ is given by

$$k_\perp^{out}(\underset{\sim}{g}) = \text{Re}\left(\frac{2mE}{\hbar^2} - (\underset{\sim}{k}_{i/\!/} + \underset{\sim}{g})^2 \right)^{\frac{1}{2}} \qquad (46)$$

The factors $k_\perp^{out}(\underset{\sim}{g})$ and $k_\perp^{out}(\underset{\sim}{o})$ are measured outside the crystal.

C. The Bloch-Wave Method

The use of Bloch waves in the theory of solids is well documented. However, the development of the Bloch-Wave method for the treatment of LEED occured at about the same time as the T-Matrix method.[11-20] For a periodic crystal, the Bloch waves are the normal modes of the periodic potential. The normal modes in a coupled system have the property that any arbitrary motion can be written as a superposition of these modes. Further, the decomposition of complex motions into the simpler normal modes provides better understanding of the complex mechanisms that constitute the LEED process.

The Bloch-Wave method is a useful microscopic method for LEED calculations. Using this technique, one calculates the Bloch waves of an infinite crystal and by matching boundary conditions at the surface, one is able to determine the intensities of the backscattered electrons. The scheme of calculation involves a matrix eigenvalue problem of which the eigenvectors are the Bloch waves and the eigenvalues give the energy bands. This method, first introduced by McRae[11,12], has been extensively developed by Jepsen, Marcus and Jona[14,15] for extensive applications.

The infinite crystal consists of a set of identical layers of equal interlayer spacing each characterized by a single layer scattering matrix $M_{\underset{\sim}{g}',\underset{\sim}{g}}^{\pm\pm}$. In the regions between the layers, the potential is constant and the Schrodinger equation yields plane wave solutions. Assuming an origin half way between layers, the wave field between the ith and the $(i+1)^{th}$ layers is expressed by

$$\phi_i(\underline{r}) = \sum_{\underline{g}} b^+_{i\underline{g}} e^{i\underline{k}^+(\underline{g})\cdot\underline{r}} + \sum_{\underline{g}} b^-_{i\underline{g}} e^{i\underline{k}^-(\underline{g})\cdot\underline{r}} \qquad (47)$$

In principle, there is an infinite number of g beams, however, if $|\underline{k}_{0//} + \underline{g}|$ is large enough to make $k^{\pm}_{\perp}(\underline{g})$ imaginary that beam dies away exponentially at a rate that increases with $|k^{\pm}_{\perp}(\underline{g})|$. Therefore, it is sufficient to include only a finite number of beams, the propagating beams and the first few evanescent beams, to obtain a numerically accurate result.

1. **Relationship Between Layer-Matrices of Pendry and Those Used Here.** In order to give an outline of the Block-Wave method, we must first define layer-scattering matrices $M^{\pm\pm}_{\underline{g}'\underline{g}}$ in reciprocal space. Following the definition given in Eq. (39)

$$\tau_{LL'} = t_L (\underline{1} - \underline{t}\,\underline{G}^{sp})^{-1}_{LL'}$$

$$= (\underline{1} - \underline{t}\,\underline{G}^{sp})^{-1}_{LL'}\, t_{L'} \qquad (48)$$

$$= (\underline{1} - \underline{X})^{-1}_{LL'}\, t_{L'}$$

where $X_{LL'}$ is a planar matrix defined as[3]

$$X_{LL'} = 4\pi i \sum_{\underline{P}\neq 0} i^{\ell_1} a(LL'L_1)\, h^{(1)}_{\ell_1}(k_o|\underline{p}|)\, Y_{L_1}(\underline{p})\, e^{-i\underline{k}_i\cdot\underline{P}}\, e^{i\delta_\ell}\sin\delta_\ell$$

(49)

and

$$a(LL'L_1) = \int Y^*_L(\Omega)\, Y_{L'}(\Omega)\, Y^*_{L_1}(\Omega)\, d\Omega \qquad (50)$$

Pendry defines his planar matrix (see Eq. (4.47), Ref. 21) as

$$X^P_{LL'} = 4\pi i\, i^{\ell-\ell'} \sum_{\underline{P}\neq 0}\sum_{L_1} (-1)^{\ell_1} i^{\ell_1} h^{(1)}_{\ell_1}(k_o|\underline{P}|)\, a(LL'L_1)\, e^{-i\underline{k}_i\cdot\underline{P}}$$

$$Y^*_{L_1}(\underline{P})\, e^{i\delta_\ell}\sin\delta_\ell. \qquad (51)$$

It can be easily be shown that our definition given in Eq. (49) is related to Pendry's definition by

$$X_{LL'} = i^{\ell'-\ell}(-1)^{\ell_1} X^P_{LL'} e^{i\delta_{\ell'}}\sin\delta_{\ell'} \Big/ e^{i\delta_\ell}\sin\delta_\ell. \quad (52)$$

If now we define layer-scattering matrices in reciprocal space as[3]

$$M^{++}_{g'g} = -\frac{8\pi^2 i}{A}\frac{1}{k_o k^\pm_\perp(g')}\sum_{LL'} Y_L(k^\pm(g')) \tau_{LL'} Y^*_{L'}(k(g))$$

$$= \frac{8\pi^2 i}{A}\frac{1}{k_o k^\pm_\perp(g')}\sum_{LL'} Y_L(k^\pm(g'))[1-X]^{-1}_{LL'} Y_{L'}(k(g)) e^{i\delta_\ell}\sin\delta_{\ell'}.$$

(53)

This definition is identical to Eq. (4.49), Ref. 21 given by Pendry, for the case of one atom per unit cell. We note that in Pendry's definition, the quantity $|k^\pm_g|$ should be corrected to $k^\pm_g (= k_o)$. With this correction, the two definitions of $M^{++}_{g'g}$ are identical.

2. **Layer Reflection and Transmission Matrices and the Matrix Eigenvalue Problem.** The propagation factors between layers are diagonal matrices

$$P^\pm_g = e^{i\frac{1}{2}k^\pm(g)\cdot a} \quad (54)$$

where $a = (d_{i+1}) - (d_i)$ is the vector connecting the origins to the i and (i+1) layers. P^+_g represents inward propagation with wave vector $k^+(g)$ through one half an interlayer distance while P^-_g represents the corresponding outward propagation with wave vector k^-_g. It is convenient to define the following four matrices

$$T^{++}_{g'g} = P^+_{g'}(I_{g'g} + M^{++}_{g'g})P^+_g \quad (55)$$

$$T^{--}_{g'g} = P^-_{g'}(I_{g'g} + M^{--}_{g'g})P^-_g \quad (56)$$

$$R^{+-}_{g'g} = P^+_{g'} M^{+-}_{g'g} P^-_g \quad (57)$$

$$R_{g'g}^{-+} = P_{g'}^{-} \, M_{g'g}^{-+} \, P_{g}^{+} \tag{58}$$

where $I_{g'g}$ is a unit matrix. These represent transmission and reflection with propagation and are shown schematically in Figure 3. Eqs. (55)-(58) describe the fact that any beam incident on a layer can be diffracted forward or reflected backwards into a whole new set of beams. These matrix equations show the details of the processes by which the coupling of the forward and back traveling waves occurs.

Using the four matrix equations, one can express the wave amplitude couplings as

$$b_i^+ = T^{++} \, b_{i-1}^+ + R^{+-} \, b_i^- \tag{59}$$

$$b_{i-1}^- = T^{--} \, b_i^- + R^{-+} \, b_{i-1}^+ \tag{60}$$

keeping in mind that the b's are column vectors of length g and the T's and R's are gxg matrices. The normal code condition on the amplitude is

$$b_i^+(k) = e^{ik \cdot a} \, b_{i-1}^+(k) \tag{61}$$

$$b_i^-(k) = e^{ik \cdot a} \, b_{i-1}^-(k) \tag{62}$$

with $k_\parallel = k_{o\parallel}$ and $k_\perp$ is to be found. These equations must be solved to find the Bloch waves of the crystal. By the elimination of b_{i-1}^- and b_i^+ from Eqs. (59) and (60) using Eqs. (58-59), a matrix equation can be constructed

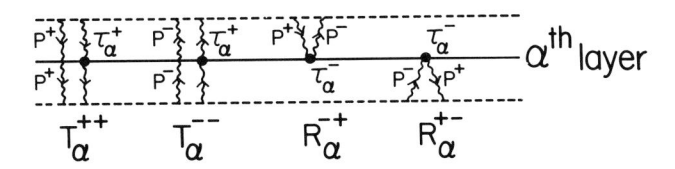

Fig. 3. Schematic diagrams of transmission and reflection matrices at the αth subplane. The broken lines are the center lines between the subplanes.

$$e^{i\underline{k}\cdot\underline{a}} \begin{bmatrix} \underline{I} & \underline{0} \\ \underline{R}^{-+} & \underline{T}^{--} \end{bmatrix} \begin{bmatrix} \underline{b}^+_{i-1} \\ \underline{b}^-_i \end{bmatrix} = \begin{bmatrix} \underline{T}^{++} & \underline{R}^{+-} \\ \underline{0} & \underline{I} \end{bmatrix} \begin{bmatrix} \underline{b}^+_{i-1} \\ \underline{b}^-_i \end{bmatrix} \quad (63)$$

Multiplication by the inverse of the matrix on the left side yields the eigenvalue problem

$$\underline{L} \begin{pmatrix} \underline{b}^+_{i-1} \\ \underline{b}^-_i \end{pmatrix} = \lambda \begin{pmatrix} \underline{b}^+_{i-1} \\ \underline{b}^-_i \end{pmatrix} \quad (64)$$

where

$$\lambda = e^{i\underline{k}\cdot\underline{a}} \quad (65)$$

and

$$\underline{L} = \begin{bmatrix} \underline{I} & \underline{0} \\ \underline{R}^{-+} & \underline{T}^{--} \end{bmatrix}^{-1} \begin{bmatrix} \underline{T}^{++} & \underline{R}^{+-} \\ \underline{0} & \underline{I} \end{bmatrix} \begin{bmatrix} \underline{T}^{++} & \underline{R}^{+-} \\ -(\underline{T}^{--})^{-1}\underline{R}^{-+}\underline{T}^{++} & (\underline{T}^{--})^{-1}(\underline{I}-\underline{T}^{-+}\underline{R}^{+-}) \end{bmatrix} \quad (66)$$

The complex matrix $\underline{L}$ has dimensions 2g by 2g, therefore there are 2g complex eigenvectors or Bloch waves. Of these, only g solutions are physically acceptable, those which are decaying exponentially into the crystal. In terms of eigenvalues, this means that the eigenvectors corresponding to eigenvalues having Im $\underline{k}>0$ are chosen.

D. Beam Intensities

It was shown earlier that the electron wave function inside the solid can be written in terms of an incident wave and a scattered wave. The total wave function just outside the solid-vacuum interface can be written as

$$\psi(\underline{r}) = \phi(\underline{r}) + \sum_g C_g\, e^{i\underline{k}^-(g)\cdot\underline{r}} \quad (67)$$

The coefficients C_g can be determined by matching $\psi(\underline{r})$ with the total wave function expressed as Bloch waves just inside the interface

$$f^{in}(\underline{r}) = \sum_{j=1}^{g} a_j \phi_o^j(\underline{r})$$

Thus, by equating $f^{in}(\underline{r})$ and $\psi(\underline{r})$ and their derivatives at the interface one obtains

$$\delta_{\underline{g}o} + \underline{C}_g = \sum_{j=1}^{g} A_j (\underline{b}^+_{og} + \underline{b}^-_{og}) \qquad (68)$$

$$k^+_\perp(\underline{o})\, \delta_{\underline{g}o} + k^-_\perp(\underline{g})\, \underline{C}_g = \sum_{j=1}^{g} A_j (k^+_\perp(\underline{g})\, \underline{b}^+_{og} + k^-_\perp(\underline{g})\, \underline{b}^-_{og}) \qquad (69)$$

where the incident beam is given as $\underline{k}_i = \underline{k}^+_o$. These expressions can be solved to obtain the $\underline{C}_g$'s. The elastic reflectivity for the gth beams then follows as

$$R_{\underline{g}} = \left(\frac{k^{out}_\perp(\underline{g})}{k^{out}_\perp(\underline{o})} \right) \underline{C}_g^2 \qquad (70)$$

E. The Layer-Doubling Method

It is possible to obtain analytically the reflection and transmission matrices for a pair of atomic layers from those of the single layers in such a way that multiple scattering between the layers is fully taken into account. Using the diffraction matrices defined in Eqs. (55)-(58), and indexing these with the letters A and B to represent two single layers (cf. top pair of diagrams in Fig. 4) or with the letter C to represent the pair A+B (second diagram from top in Fig. 4), we can write:[21,22]

$$T^{++}_C = T^{++}_B (1 - R^{+-}_A R^{-+}_B)^{-1} T^{++}_A \qquad (71)$$

$$R^{-+}_C = R^{-+}_A + T^{--}_A R^{-+}_B (1 - R^{+-}_A R^{-+}_B)^{-1} T^{++}_A \qquad (72)$$

$$R^{+-}_C = R^{+-}_B + T^{++}_B R^{+-}_A (1 - R^{-+}_B R^{+-}_A)^{-1} T^{--}_B \qquad (73)$$

$$T^{--}_C = T^{--}_A (1 - R^{-+}_B R^{+-}_A)^{-1} T^{--}_B \qquad (74)$$

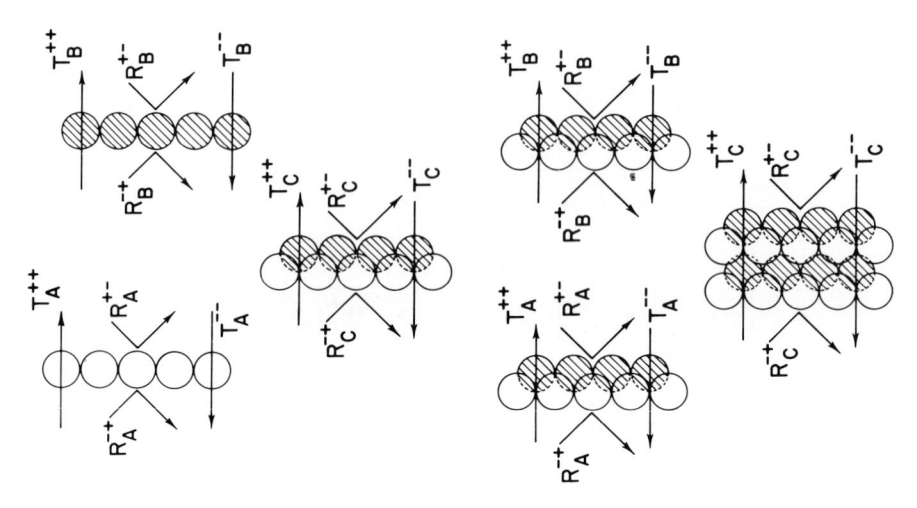

Fig. 4. Stacking planes into a crystal slab by the layer-doubling method. Parts A and B are stacked together to form slab C.

The inverted terms in brackets describe the interlayers multiple scattering, as can be readily recognized by expanding these terms into geometric series.

Because Eqs. (71)-(74) remain valid when the diffraction matrices on the right-hand side represent stacks of layers, this process of combining layers can be repeated any number of times, using the same set of equations (71)-(74), to build up the crystal stack by stack until there is convergence of the reflection amplitudes. All multiple scattering in the crystal is taken into account when doing this. The single atomic layers may be different from each other, a valuable feature of crystal surfaces whose structure can deviate from the bulk structure or involve foreign atoms.

The method is most powerful, however, in a periodic array of layers, as with the bulk that often starts at or very near the surface. Then, the layer-stacking process can proceed as illustrated in Figure 4. The pair C can be joined with another identical pair C, using the same set of equations (71)-(74), yielding a stack of 4 layers, which itself can be joined with another identical stack of 4 layers. This process doubles the crystal thickness at each step (therefore the name of "layer-doubling method" chosen by its originator Pendry).[21] The crystal thickness grows exponentially, ensuring rapid convergence of the reflection amplitudes. Three or four doublings, corresponding to the inclusion of 8 or 16 layers, are generally sufficient (absorption determines the speed of convergence).

Once the converged bulk reflectivities have been obtained using the layer-doubling method, any number of surface layers can be added, using again the same set of equations. In addition, these surface layers can be shifted to other positions at will any number of times without recomputing the bulk reflectivities.

IV. ITERATIVE AND PERTURBATION TECHNIQUES

A. Need for Fast Calculation Schemes

In sections (IIIA), (IIIC), and (3D), three exact microscopic methods of LEED were presented. All solve self-consistently the multiple scattering events of an incident electron inside a crystal and take into account all orders of intraplanar and interplanar multiple scattering events. These methods have proven very valuable for investigations of LEED intensities from clean surfaces. They converge over a wide range of parameters and by eliminating the uncertainty due to numerical accuracy, these methods have provided a valuable test of the LEED model.

Structural analysis by LEED would be limited to rather simple systems if only the exact methods were to be used. The T-matrix method requires the solution of a set of N inhomogeneous equations with complex matrices of dimension L, N being the number of subplanes included. This is a very time consuming step and overall the technique requires a large amount of computer storage. The Bloch wave method requires less computer core storage but the matrix eigenvalue problem becomes time consuming when the number of reciprocal space beams required is large. The matrix doubling method which replaces the matrix eigenvalue problems with the matrix multiplication process has been successful in reducing computer time needed for LEED calculations. It is quite apparent though that faster calculation methods must be used to investigate complex overlayer systems, reconstructed semiconductor surfaces and organic crystals.

To develop a perturbation technique, one must identify which portions of the complex LEED process can be simplified. Three possible approaches to this problem are: the adoption of a simpler model of the ion-core potential; the development of fast computation techniques to approximate the interplanar and intraplanar multiple scattering events; the development of data averaging and/or convolution techniques to process experimental data in such a way that the complicated portions of the spectra are eliminated.[23-26]

Early attempts at the formulation of approximate techniques involved the isotropic S-wave scattering model. This model reduces calculation times by taking only the $\ell = 0$ partial wave of the ion-core scattering. This model could not produce reflected intensities of sufficient strength nor could it adequately reproduce the multiple scattering structure in the experimental LEED spectra.

B. Direct Perturbation Expansion Techniques

The first-, second-, and third-order τ-matrix expansion techniques are based on the direct expansion in terms of the planar scattering matrix $\underset{\approx}{T}_\alpha(k_o)$.[3] The expansion of Eq. (40)

$$\underset{\approx}{T}_\alpha(k_o) = \underset{\approx}{\tau}_\alpha(k_o) + \sum_{\beta \neq \alpha} \underset{\approx}{\tau}_\alpha(k_o) \, \underset{\approx}{G}^{\alpha\beta}(k_i) \, \underset{\approx}{T}_\beta(k_o) \tag{75}$$

yields

$$\underset{\approx}{T}_\alpha(k_o) = \underset{\approx}{\tau}_\alpha(k_o) + \sum_{\beta \neq \alpha} \underset{\approx}{\tau}_\alpha(k_o) \, \underset{\approx}{G}^{\alpha\beta}(k_i) \underset{\approx}{\tau}_\beta(k_o) + \sum_{\beta \neq \alpha} \sum_{\gamma \neq \beta} \underset{\approx}{\tau}_\alpha(k_o) \underset{\approx}{G}^{\alpha\beta}(k_i)$$

$$\underset{\approx}{\tau}_\beta(k_o) \, \underset{\approx}{G}^{\beta\gamma}(k_i) \, \underset{\approx}{\tau}_\gamma(k_o) + \ldots \ldots \tag{76}$$

The elastic reflectivity follows from Eq. (45) as[3]

$$R_g = \frac{k_\perp^{out}(g)}{k_\perp^{out}(o)} \left| I_g^{(1)} + I_g^{(2)} + I_g^{(3)} + \ldots \right|^2 \tag{77}$$

where the factors $I^{(n)}$ represent the corresponding orders of contribution. The first order contribution is given by

$$I_g^{(1)} = -\left(\frac{8\pi^2 i}{A}\right)\left(\frac{2m}{\hbar^2}\right) F(k_o) \sum_\alpha \sum_{LL'}$$

$$\left\{ \frac{Y_L(k^-(g)) \tau_\alpha^{LL'}(k_o) Y_{L'}^*(k^+(o)) \, e^{i(k^+(o) - k^-(g)) \cdot d_\alpha}}{k_\perp^-(g)} \right\} \tag{78}$$

or in terms of the g-space representation is given by

$$I_g^{(1)} = \sum_\alpha \tau_{go}^{\alpha-}(k_o) \; e^{i(k^+(o)-k^-(g))\cdot d_\alpha} \tag{79}$$

If each layer has the same kind of atom, the α dependence can be dropped and the expression becomes

$$I_g^{(1)} = (P_o^+)^{\frac{1}{2}} (P_g^-)^{\frac{1}{2}} \tau_{go}(k_o) \left\{\frac{1}{1-P_o^+ P_g^-}\right\} \tag{80}$$

where

$$P_o^+ = e^{ik^+(o)\cdot a} \tag{81}$$

and

$$P_g^\pm = e^{ik^\pm(g)\cdot a} \tag{82}$$

Equation (80) is an algebraic function and, therefore, can be evaluated rapidly. Two second order terms can be derived following the same procedure.[3] These are as follows for the case of a periodic monatomic crystal with equal interplanar spacing

$$I_g^{(2)}(d_\alpha > d_\beta) = (P_o^+)^{\frac{1}{2}} (P_g^-)^{\frac{1}{2}} \sum_{g_1} \tau_{gg'}(k_o) \, \tau_{g_1 o}^+(k_o)$$

$$\left\{\frac{P_{g_1}^+ P_g^-}{(1-P_{g_1}^+ P_g^-)(1-P_o^+ P_g^-)}\right\} \tag{83}$$

and

$$I_g^{(2)}(d_\alpha < d_\beta) = (P_o^+)^{\frac{1}{2}} (P_g^-)^{\frac{1}{2}} \sum_{g_1} \tau_{gg_1}^+(k_o) \, \tau_{g,o}^-(k_o)$$

$$\left\{\frac{P_o^+ P_{g_1}^-}{(1-P_o^+ P_g^-)(1-P_o^+ P_{g_1}^-)}\right\} \tag{84}$$

The expansion scheme can continue to arbitrary order, however, the number of algebraic terms increase rapidly with increasing order (e.g. 8 terms for third order). For this reason, calculations have been made up to third order expansions only.

One can also expand the total scattering matrix $T_\alpha(k_o)$ and the planar scattering matrix $\tau_\alpha(k_o)$ both in terms of the ion-core t-matrix.[27-30] This is desirable because the $t_\alpha(k_o)$ matrix is a diagonal matrix and has only $\ell_{max} + 1$ distinct elements. This results in savings of both computation time and computer core storage. The transmission and reflection matrices simplify to

$$t^{\alpha\pm}_{g_1 g_2}(k_o) = -\left(\frac{8\pi^2 i}{A}\right)\left(\frac{2m}{\hbar^2}\right) F(k_o) \sum_L \left\{ \frac{Y_L(k^\pm(g_1)) t^\alpha_L(k_o) Y^*_L(k(g_2))}{k^\pm_\perp(g_1)} \right\} \tag{85}$$

$$= -\left(\frac{8\pi^2 i}{A}\right)\left(\frac{2m}{\hbar^2}\right) F(k_o) \sum_\ell \frac{(2\ell+1)}{4\pi} P_\ell(\cos\theta_\pm) \left\{ \frac{t^\alpha_\ell(k_o)}{k^\pm_\perp(g_1)} \right\}$$

The t-matrix technique was formulated and implemented to third order by Tong et al.[27,28] Explicit expressions for this formalism can be found in references 27-30. The technique is very fast and was found to have acceptable accuracy for weak scattering metals like aluminum.

Recently, Jennings[51] made a third order calculation with an additional feature. He included all orders of zero-angle scattering events (i.e., $f(\theta)$ for $\theta = 0$) in his third order scheme. Thus, the first order term becomes

$$I^{(1)}_g = (P^+_o)^{1/2}(P^-_g)^{-1/2}\left[\tau^-_{go}(k_o) + (1+\tau^+_{gg}(k_o)) P^-_g \tau^-_{go}(k_o) P^+_o (1+\tau^+_{oo}(k_o)) + \ldots\right] \tag{86}$$

$$= (P^+_o)^{1/2}(P^-_g)^{1/2}\tau^-_{go}(k_o)\left[\frac{1}{1-(1+\tau^+_{gg}(k_o))P^-_g P^+_o(1+\tau^+_{oo}(k_o))}\right]$$

In other words, instead of pure propagation factors $p^\pm_g = e^{ik^\pm(g)\cdot a}$, the repeated factors include the zero-angle scattering events $[1 + \tau^+_{gg}(k_o)]$. Jennings applied his third order calculation to Cu(001) using an APW potential by Snow and Waber.[32] Working at normal incidence, he obtained reasonable agreement with experiment.

The direct expansion techniques have the advantages that they are very fast and easily tractable. The main drawback is the rapidly increasing number of terms as one attempts to go to higher orders of the expansion.[3]

C. Renormalized Forward Scattering Method

A better perturbation expansion scheme would be one in which each additional order has exactly the same form as the previous one. This would avoid the problem of having to face diverging number of expansion terms. If each order requires an amount of computation time t, then n orders would require a time of nt. This linear scaling of time would result in large computational time savings, if t is small with respect to the times required for exact methods.

The RFS technique introduced by Pendry[33] is a perturbation scheme that utilizes this iterative principle. This method calculates the layer scattering matrices exactly and then iterates the interplanar scattering events as the electrons pass in and out of the crystal. The layer by layer iteration is carried out to successively higher orders until numerical convergence is achieved.

Matrix equations (55)-(58) describe the processes by which forward and back traveling waves couple in terms of flux exchanges as electrons are scattered by the layers. A column vector $A_\alpha^i(\underline{g})$ can be defined to represent the fraction of electron flux in each $\underline{g}$ beam propagating into the crystal. The index i is the order of the iteration (i.e. the number of times that the electron has propagated into the crystal) and α is the layer index. The incident electron beam has a well defined direction and therefore at the surface the initial inward propagation is 100% in the incident direction. Thus, we can write the starting condition at the interface

$$A_0^1(\underline{g}) = \begin{bmatrix} 1 \\ 0 \\ 0 \\ 0 \\ \vdots \end{bmatrix} \qquad (87)$$

Successive coefficients of $A^i(\underline{g})$ can be evaluated at midpoints between layers by

$$A_1^1(\underline{g}) = \sum_{\underline{g}_1} T_1^{++}(\underline{g}\underline{g}_1) \, A_0^1(\underline{g}_1) \qquad (88)$$

and in general

$$A_\alpha^1(\underline{g}) = \sum_{\underline{g}_1} T_1^{++}(\underline{g}\underline{g}_1) A_{\alpha-1}^1(\underline{g}_1) \qquad (89)$$

The iteration continues inward into the crystal calculating each new coefficient from the previous until the total electron flux $|\Sigma_{\underline{g}} A_N^1(\underline{g})|^2 < \varepsilon$ where epsilon is a cutoff parameter. Typically, epsilon is 10^{-4} and 5 to 15 layers are required before the amount of elastic flux reaching the next layer (N+1) is considered negligible.

Upon reaching the last layer N, the electrons are traced back out of the crystal and these amplitudes are stored in a column vector $B_\alpha^1(\underline{g})$. Above the deepest layer N, the coefficients $B_{N-1}^1(\underline{g})$ are formed by the reflection of the amplitudes $A_{N-1}^1(\underline{g})$ at the N layer. These coefficients are

$$B_{N-1}^1(\underline{g}) = \sum_{\underline{g}_1} R_N^{-+}(\underline{g}\underline{g}_1) A_{N-1}^1(\underline{g}_1) \qquad (90)$$

Each successive coefficient $B_{N-i}^1(\underline{g})$ has two contributions: 1) backscattering of $A_{N-i}^1(\underline{g})$ at the (N-i+1)th layer and 2) outward transmission of the amplitudes $B_{N-i+1}^1(\underline{g})$ through the (N-i)th layer. These can be written in matrix-vector form

$$B_{N-i}^1(\underline{g}) = \sum_{\underline{g}_1} R_{N-i+1}^{-+}(\underline{g}\underline{g}_1) A_{N-i}^1(\underline{g}_1) + \sum_{\underline{g}_1} T_{N-i+1}^{--}(\underline{g}\underline{g}_1) B_{N-i+1}^1(\underline{g}_1) \qquad (91)$$

Each coefficient $B_{N-i}^1(\underline{g})$ is evaluated from the previous values of $B_{N-i+1}^1(\underline{g})$ and $A_{N-i}^1(\underline{g})$ until $B_o^1(\underline{g})$ at the interface is found.

The electrons are now propagated inward then outward through the layers again. The coefficients $A_\alpha^2(\underline{g})$ and $B_\alpha^2(\underline{g})$ for this second order contribution must be found. Below the first layer,

$$A_1^2(\underline{g}) = \sum_{\underline{g}_1} R_1^{+-}(\underline{g}\underline{g}_1) B_1^1(\underline{g}_1) \qquad (92)$$

while succeeding coefficients $A_\alpha^2(\underline{g})$ has two contributions

$$A_\alpha^2(\underline{g}) = \sum_{\underline{g}_1} R_\alpha^{+-}(\underline{g}\underline{g}_1) B_\alpha^1(\underline{g}_1) + \sum_{\underline{g}_1} T_\alpha^{++}(\underline{g}\underline{g}_1) A_{\alpha-1}^2(\underline{g}_1) \qquad (93)$$

where $1 < \alpha \leq N_1$, N_1 being the deepest penetration on the second pass ($N_1 \leq N$). The coefficients $B_\alpha^2(g)$ are then formed as the electrons propagate upward once again until the surface is reached and $B_o^2(g)$ is calculated. This inward and outward propagation process is repeated, each order contributing an additional coefficient $B_o^n(g)$, until the set of coefficients numerically converge. Convergence is determined by the condition

$$\left| \sum_g B_o^{M+1}(g) \right| < \varepsilon, \tag{94}$$

where ε_1 is the convergence parameter and M is the order of iteration. The final reflectivity in each beam is given by

$$R_g = \frac{k_\perp^{out}(g)}{k_\perp^{out}(o)} \left| \sum_{i=1}^{M} B_o^i(g) \right|^2 \tag{95}$$

The iterations involve matrix-vector multiplication and therefore computation times scale as g^2. Each order of iteration involves the same number of steps. Since each iteration is very fast in comparison to exact method times the RFS technique is a useful computation technique for LEED calculations.[34]

There are, however, two circumstances under which the RFS technique fails to converge properly.[22] When there is little damping, the coefficients $A_\alpha^i(g)$ fail to converge as the electrons penetrate deeply into the crystal. Fortunately, this situation is not of primary interest to LEED for surface structure analysis since the deeply penetrating electrons cease to provide surface sensitive information. The second case is when two layers are close to each other. Backscattering between two closely spaced layers becomes large and the iteration process produces divergent terms.[3]

V. NEW COMPUTATION METHODS FOR DYNAMICAL LEED

A. L-Space Layer-Iteration Method

The iterative methods (RFS, layer-doubling) described earlier involve matrices in K-space. The dimension of the matrices depend on the number of plane wave components necessary in the expansion of a Bloch wave inside the layers. In cases of small overlayer adsorption distances (e.g. hydrogen or oxygen on open metal faces) or reconstructed structures of semiconductor surfaces, interlayer spacings can become very small. The number of plane-wave components (g beams) required increases rapidly with decreasing interlayer spacing. For a d-spacing of 0.3 Å, for example, over 1,000 beams are required. This poses a serious problem on K-space matrix methods since obviously, it is currently impossible to accommodate matrices of such a dimension. The alternative is to use the angular momentum (L-space) representation, for example, the T-matrix method (Section 3 II). In the L-space representation, the matrix dimension is independent of the number of g-beams used. This is because an explicit sum over g-beams is carried out. An analogy may be drawn with the band structure methods APW (K-space representation) and KKR (L-space representation). Thus, the L-space method begins to have an advantage in terms of computation speed when the number of beams gets big.

It is desirable to make the T-matrix method more efficient. Zimmer and Holland[35] have proposed a reverse scattering iteration scheme which is based on carrying out perturbation iterations on the exact T-matrix method. We start with the exact matrix algebraic form of Eq. (40)

$$T_i = \tau_i + \sum_{j \neq i} \tau_i G^{ij} [k^+(o)] T_j \qquad (96)$$

The incident direction is $k^+(o)$ and if we define vectors

$$T_i[k^+(o)] = T_i Y^*[k^+(o)] \qquad (97)$$

and

$$\tau_i[k^+(o)] = \tau_i Y^*[k^+(o)] \qquad (98)$$

then Eq. (96) can be rewritten as

$$T_i(k^+(o)) = \tau_i(k^+(o)) + \sum_{j \neq i} \tau_i G^{ij} [k^+(o)] T_j[k^+(o)] \qquad (99)$$

$$= \tau_i[k^+(o)] + T_i^+[k^+(o)] + T_i^-[k^+(o)]$$

In Eq. (99), the vectors $T_i^\pm[k^+(o)]$ are defined as

$$T_i^+[k^+(o)] = \sum_{j > i} \tau_i G^{ij}[k^+(o)] T_j[k^+(o)] \qquad (100)$$

and

$$\underset{\sim}{T}_i^-[\underset{\sim}{k}^+(\underset{\sim}{o})] = \sum_{j<i} \underset{\sim}{\tau}_i \underset{\sim}{G}^{ij}[\underset{\sim}{k}^+(\underset{\sim}{o})] \; \underset{\sim}{T}_j[\underset{\sim}{k}^+(\underset{\sim}{o})] \tag{101}$$

The vectors $\underset{\sim}{T}_i^+(\underset{\sim}{k}^+(\underset{\sim}{o}))$ and $\underset{\sim}{T}_i^-(\underset{\sim}{k}^+(\underset{\sim}{o}))$ denote total electron amplitudes reaching the ith layer from layers below and above respectively. At the ith layer, we put $\underset{\sim}{T}_i^{\pm(n)}(\underset{\sim}{k}^+(\underset{\sim}{o}))$ as the amplitudes of electrons that start from a layer below (above) and reach layer i with n back scatterings. Thus, it is easy to write down the iteration relations

$$\underset{\sim}{T}_i^{+(n)}[\underset{\sim}{k}^+(\underset{\sim}{o})] = \underset{\sim}{\tau}_i \sum_{j>i} \underset{\sim}{G}^{ij}[\underset{\sim}{k}^+(\underset{\sim}{o})] \left[\underset{\sim}{T}_j^{+(n)}[\underset{\sim}{k}^+(\underset{\sim}{o})] + \underset{\sim}{T}_j^{-(n-1)}[\underset{\sim}{k}^+(\underset{\sim}{o})]\right] \tag{102}$$

and

$$\underset{\sim}{T}_i^{-(n)}(\underset{\sim}{k}^+(\underset{\sim}{o})) = \underset{\sim}{\tau}_i \sum_{j<i} \underset{\sim}{G}^{ij}[\underset{\sim}{k}^+(\underset{\sim}{o})] \left[\underset{\sim}{T}_j^{-(n)}[\underset{\sim}{k}^+(\underset{\sim}{o})] + \underset{\sim}{T}_j^{+(n-1)}[\underset{\sim}{k}^+(\underset{\sim}{o})]\right] \tag{103}$$

The boundary conditions are

$$\underset{\sim}{T}_1^{-(o)} = \underset{\sim}{T}_1^{-(1)} = \ldots = \underset{\sim}{T}_1^{-(n)} = 0 \tag{104}$$

and

$$\underset{\sim}{T}_N^{+(o)} = \underset{\sim}{T}_N^{+(1)} = \ldots = \underset{\sim}{T}_N^{+(n)} = 0 \tag{105}$$

because no flux can reach from above the top layer or from below the deepest layer N. Also, we must put the starting condition as

$$\underset{\sim}{T}_j^{-(-1)} = \underset{\sim}{T}_j^{+(-1)} = \underset{\sim}{\tau}_j[\underset{\sim}{k}^+(\underset{\sim}{o})] \qquad j = 1, \ldots N \tag{106}$$

since the incident electron can propagate to any layer $j \neq i$ with no scattering and we must include the initial scattering $\underset{\sim}{\tau}_j(\underset{\sim}{k}^+(\underset{\sim}{o}))$ at the jth layer. Putting the iteration relations into the exact solution (Eq. 99), we obtain

$$\underset{\sim}{T}_i(\underset{\sim}{k}^+(\underset{\sim}{o})) = \underset{\sim}{\tau}_i(\underset{\sim}{k}^+(\underset{\sim}{o}) + \sum_{n=o}^{M} \left[\underset{\sim}{T}_i^{+(n)}[\underset{\sim}{k}^+(\underset{\sim}{o})] + \underset{\sim}{T}_i^{-(n)}[\underset{\sim}{k}^+(\underset{\sim}{o})]\right] \tag{107}$$

which is accurate up to M back scattering events and all forward scattering events for N layers of a crystal. The final reflected amplitude is

$$\underline{T}[\underline{k}^-(g), \underline{k}^+(\underline{o})] = \sum_L Y_L[\underline{k}^-(g)] \sum_i e^{i[\underline{k}^-(g)-\underline{k}^+(\underline{o})]\cdot \underline{d}i} T_i^L[\underline{k}^+(\underline{o})]$$

(108)

The matrix size of this L-space iteration scheme is $(\ell_{max} + 1)^2$, where ℓ_{max} is the largest angular momentum component used. For nearly all materials calculated, $\ell_{max} + 1 \leq 8$ in the energy range 0-200 eV. Thus, this iterative method has the advantage in speed over RFS if the number of beams gets much larger than 64. For simple crystals where the number of beams is less than 64, this method is slower, because it requires the evaluation of matrices $\underline{G}^{ij}(k^+(o))$, for $i = j$. For N layers of a solid, there are N(N-1) such matrices. The exact break-even point in terms of computation speed depends on the particular surface structure, but it is estimated that when $g \geq 100$, the L-space iteration method would have a clear advantage over RFS. Finally, since the L-space method employs the iteration scheme, it can be summed to numerical convergence like the RFS method. It is expected to have the same convergence behavior in terms of multiple scattering that RFS has.

B. The Combined-Space Method of Tong and Van Hove

In the case of chemisorption of small atoms, i.e. H, O, S, C, etc. or the rearrangement of some semiconductor and transition metal surfaces, it is generally true that a few (2 or 3) surface layers are closely spaced while deeper layers remain at the larger, bulk spacings. To treat the entire crystal with the L-space iteration method would involve evaluation of many $\underline{G}^{ij}(\underline{k}^+(\underline{o}))$ matrices. Similarly, RFS, layer-doubling, or the Bloch wave method would be inefficient because of the large number of beams associated with the small surface spacings. The simplest solution is to merge the two representations, L-space for small d-spacings (<1.0 Å) and K-space for larger, bulk-like spacings (≥1.0 Å). Such a formulation is presented in the following.

First, we define vectors for incident beams $\underline{k}^{\pm}(g)$ from above (the + sign) and below (the - sign) on the ith layer[36]

$$\underline{T}_i[\underline{k}^{\pm}(g)] = \underline{T}_i \underline{Y}^*[\underline{k}^{\pm}(g)]$$

(109)

and

$$\underline{\tau}_i[\underline{k}^{\pm}(g)] = \underline{\tau}_i \underline{Y}^*[\underline{k}^{\pm}(g)]$$

(110)

Then the interlayer matrices $\underset{\approx}{G}^{ij}$ are defined for general incident directions

$$\underset{\approx}{G}^{ij}[\underset{\sim}{k}^{\pm}(g)] = \gamma_o\, e^{-i\underset{\sim}{k}^{\pm}(g)\cdot(\underset{\sim}{d}_i-\underset{\sim}{d}_j)}\, \underset{\approx}{M}^{+(i,j)} \qquad (111)$$

$$(d_i^{\pm} > d_j^{\pm})$$

and

$$\underset{\approx}{G}^{ij}[\underset{\sim}{k}^{\pm}(g)] = \gamma_o\, e^{-i\underset{\sim}{k}^{\pm}(g)\cdot(\underset{\sim}{d}_i-\underset{\sim}{d}_j)}\, \underset{\approx}{M}^{-(i,j)} \qquad (112)$$

$$(d_i^{\pm} < d_j^{\pm})$$

where $\gamma_o = (8\pi i/A)(2m/\hbar^2)F(k_o)$ and the matrices $\underset{\approx}{M}^{\pm(i,j)}$ are defined as

$$M_{LL'}^{\pm(i,j)} = \sum_{g_1} \left(\frac{e^{i\underset{\sim}{k}^{\pm}(g_1)\cdot(\underset{\sim}{d}_i-\underset{\sim}{d}_j)}}{k_\perp^{\pm}(g_1)} \right) Y_L^*[\underset{\sim}{k}^{\pm}(g_1)]\, Y_{L'}[\underset{\sim}{k}(g_1)] \qquad (113)$$

The important thing is to note that beams g_1 in $M_{LL'}^{\pm(i,j)}$ are explicitly summed over. This is the principal logic of the combined-space method. Within closely spaced layers from i to j, the number of required beams g_1 is large (e.g. over 1,000 beams for $(d_i - d_j) \simeq 0.3$ Å). But the matrices $G^{ij}(k^{\pm}(g))$ defined in Eqs. (111) and (112) have dimension L (e.g. L = 64 for 8 phase shifts), independent of the number g_1, because g_1 is summed over in Eq. (113). In fact, the matrix $G^{ij}(k^{\pm}(g))$ can be obtained directly from a real-space summation, which can be advantageously used in extreme cases where the summation over g_1 of Eq. (113) requires an excessive number of beams. The total vectors $T_i(k^{\pm}(g))$ are first solved for the closely spaced layers (e.g. i = 1,2,3, etc.) using the L-space iteration formula given in Eqs. (102)-(107). The matrices involved have dimension L. Then, transmission and reflection matrices are formed in K-space for this group of closely spaced layers. For simplicity, let us take the case of two closely spaced layers (i and i+1) and define

$$R_g^{\pm} = e^{\pm i\underset{\sim}{k}^{\pm}(g)\cdot\underset{\sim}{a}} \qquad (114a)$$

$$P_g^{\pm} = e^{\pm i\underset{\sim}{k}^{\pm}(g)\cdot\underset{\sim}{a}/2} \qquad (114b)$$

where $\underset{\sim}{a}$ is the vector connecting origins of the two layers. Thus, the transmission and reflection matrices for the two-layer composite are

$$\{T_{i,i+1}^{++}\}_{\underset{\sim}{g}'\underset{\sim}{g}} = \gamma_0 \sum_L \frac{Y_L[\underset{\sim}{k}^+(\underset{\sim}{g}')]}{k_\perp^+(\underset{\sim}{g}')} \left[P_{\underset{\sim}{g}'}^+, R_{\underset{\sim}{g}'}^+, T_L^i[\underset{\sim}{k}^+(\underset{\sim}{g})] P_{\underset{\sim}{g}}^+ + \right.$$
$$\left. P_{\underset{\sim}{g}'}^+ T_L^{i+1}[\underset{\sim}{k}^+(\underset{\sim}{g})] R_{\underset{\sim}{g}}^+ P_{\underset{\sim}{g}}^+ \right] \quad (115)$$

$$\{T_{i,i+1}^{-+}\}_{\underset{\sim}{g}'\underset{\sim}{g}} = \gamma_0 \sum_L \frac{Y_L[\underset{\sim}{k}^-\underset{\sim}{g}')]}{k_\perp^+(\underset{\sim}{g}')} \left[P_{\underset{\sim}{g}'}^-, T_L^i[\underset{\sim}{k}^+(\underset{\sim}{g})] R_{\underset{\sim}{g}}^+ P_{\underset{\sim}{g}}^+ + \right.$$
$$\left. P_{\underset{\sim}{g}'}^-, R_{\underset{\sim}{g}'}^-, T_L^{i+1}[\underset{\sim}{k}^+(\underset{\sim}{g})] R_{\underset{\sim}{g}}^+ P_{\underset{\sim}{g}}^+ \right] \quad (116)$$

for beams $\underset{\sim}{k}^+(\underset{\sim}{g})$ incident from above, and

$$\{T_{i,i+1}^{--}\}_{\underset{\sim}{g}'\underset{\sim}{g}} = \gamma_0 \sum_L \frac{Y_L[\underset{\sim}{k}^-(\underset{\sim}{g}')]}{k_\perp^-(\underset{\sim}{g}')} \left[P_{\underset{\sim}{g}'}^-, T_L^i[\underset{\sim}{k}^-(\underset{\sim}{g})] R_{\underset{\sim}{g}}^- P_{\underset{\sim}{g}}^- + \right. \quad (117)$$
$$\left. P_{\underset{\sim}{g}'}^-, R_{\underset{\sim}{g}'}^-, T_L^{i+1}[\underset{\sim}{k}^-(\underset{\sim}{g})] P_{\underset{\sim}{g}}^- \right]$$

$$\{T_{i,i+1}^{+-}\}_{\underset{\sim}{g}'\underset{\sim}{g}} = \gamma_0 \sum_L \frac{Y_L[\underset{\sim}{k}^-(\underset{\sim}{g}')]}{k_\perp^-(\underset{\sim}{g}')} \left[P_{\underset{\sim}{g}'}^+, R_{\underset{\sim}{g}'}^+, T_L^i[\underset{\sim}{k}^-(\underset{\sim}{g})] R_{\underset{\sim}{g}}^- P_{\underset{\sim}{g}}^- + \right.$$
$$\left. P_{\underset{\sim}{g}'}^+ T_L^{i+1}[\underset{\sim}{k}^-(\underset{\sim}{g})] P_{\underset{\sim}{g}}^- \right] \quad (118)$$

for beams $\underset{\sim}{k}^-(\underset{\sim}{g})$ incident from below. There, $\underset{\sim}{g}$ is the beam number corresponding to the larger d-spacings of the deeper layers of the solid. The transmission and reflection matrices (Eqs. (115)-(118)) defined for the closely spaced layers can then be used to solve for the reflectivity of the crystal using any one of the established k-space schemes (e.g. RFS, layer-doubling, Bloch wave method, etc.). Although Eqs. (115)-(118) are written explicitly for the case of two layers (i and i+1), its extension to three or more closely spaced layers is straightforward (involving essentially kinematic relations between the set of $T_L^i(\underset{\sim}{k}^\pm(\underset{\sim}{g}))$ vectors). The dimension of the matrices defined in Eqs. (115)-(118) is given by $\underset{\sim}{g}$, a number substantially smaller than $\underset{\sim}{g}_1$, the latter being the beam set required for the closely spaced surface layers. This is the computation advantage of the combined-space method.

VI. DISCUSSION OF CALCULATION METHODS

In the preceding Sections, we have indicated that a number of methods exist for calculating LEED intensities. These methods have widely varying computational characteristics. We now wish to compare these characteristics as they affect actual calculations and especially the choice of a method of calculation.

A. General Criteria

The various calculational methods are quite dependent in their performance on the surface crystallography. Surface structures presumably occur in infinite variety; some of the structures currently under investigation are: clean, unreconstructed, but sometimes relaxed surfaces of metals, ionic, semiconductor and layer compounds; atomic and molecular physisorption or chemisorption on unreconstructed metal surfaces in one or more overlayers; reconstructed surfaces of clean metals and semiconductors. Common to all ordered surfaces that can be studied by elastic LEED however is the concept of single atomic layers parallel to the surface. A most important parameter to be considered in the selection of a computational method is the interlayer spacings of such layers.

On the one hand, it is most convenient to expand the electron wavefunction between layers in terms of plane waves, because this choice leads to diagonal interlayer propagation matrices. On the other hand, the use of plane waves requires good convergence of the plane-wave expansion: such convergence is difficult to achieve when the overlayer spacing is small (then strongly decaying plane waves can reach across and transport flux from one layer to the next). In practice, the plane-wave expansion is no longer advisable for spacings smaller than 1.0 Å, depending on the desired accuracy. For smaller spacings, the expansion in angular momentum space is to be used, the price being interlayer propagation matrices $G_{LL'}^{ij}$. The difference in behavior between angular and linear momentum space expansions is due to the fact that individual scattering atoms cannot get closer together than their bond length, while planes of scattering atoms can get arbitrarily close together on an open face.

Another consideration is the selection of a calculational method concerns the question of differing bulk and surface unit cells (two-dimensional). Overlayers of submonolayer coverage often arrange in an ordered array that has a larger unit cell than the substrate has. Similarly, reconstruction of surfaces often increases the unit cell size. The larger unit cell first appears to the observer in the form of additional diffracted beams. The bulk-like substrate also sees a shower of additional beams impinging on it, naturally described by additional plane waves. These additional plane waves cannot be diffracted by the substrate into the original set of "clean surface plane waves". More specifically, one can say that a surface that has a unit cell N times larger in area than the substrate (where N is an integer) will produce N sets of plane waves (one of them is the substrate set) that will not be diffracted into each other by the substrate. It follows that the plane-wave reflection matrices of each substrate layer and of the entire substance will have a block-diagonal form, each block corresponding to one of the N beam sets. Block-diagonal matrices present clear computational advantages in their calculation, storage and use. Bloch waves also fall into N independent simpler sets due to the block-diagonalization.

An angular momentum space treatment of the problem of differing unit cells requires the use of more subplanes (N times more subplanes in the substrate than in the case of non-differing unit cells, where N is defined as above) and many more interplanar propagation matrices. No block-diagonalization or similar simplification occurs.

Symmetry also has an influence on the choice of a calculational method. When there is symmetry between some beam intensities (because of a privileged primary incidence direction, such as normal incidence, in addition to surface-structural symmetry), this symmetry can be exploited by standard group-theoretical methods, generally considerably reducing computation times and storage. However, the use of symmetry has a tendency to make computer programs less flexible, as each type of symmetry (e.g. n-fold rotation axes, mirror planes and combinations thereof) leads to different expressions. This problem is particularly severe in angular momentum space, while it is quite manageable in linear momentum space (the difference in manageability stems from the condition $|m| \leq \ell$ in angular momentum space, which is inherently inconvenient to program, especially when symmetries add their own varying conditions). A drawback of the use of symmetry in the linear momentum space representation is that atomic layers that have different registries (lateral positions) relative to the symmetry axis or plane, will have different symmetry-reduced diffraction matrices $M_{g'g}^{\pm\pm}$. But the gain in computing efficiency obtained from symmetrization heavily outweighs such complications. For example, a 4-fold symmetry axis allows one to cut the size of the layer and crystal diffraction matrices by a factor of 4, reducing storage by a factor of 16, and reducing computation time by a factor between 16 and 64, depending on the method.

Finally, we must take into account that the techniques discussed here are normally used for the determination of unknown positions of surface atoms. The multiple-scattering process is too complicated to lend itself to direct extraction of atomic positions from experimental data (direct methods are under study for performing this inversion: constant-momentum-transfer averaging,[23,24] Fourier-transform deconvolution)[25,26]. One must therefore resort to a trial-and-error approach, i.e., making intensity calculations for a number of trial surface structures and selecting that structure which produces the best fit with experiments. It is clear that under these conditions computational efficiency is of great concern. One will therefore select methods that involve the smallest matrices, the fewest matrix products, inversions or diagonalizations, the most repeated use of partial results, while providing the greatest possible flexibility.

B. Method-by-Method Discussion

From the foregoing arguments, it is clear that the Beeby T-matrix method, which works entirely in angular momentum space, is not suitable for a structural search. This method involves the inversion of a matrix of dimension $(\ell_{max}+1)^2 N$ (N is the total number of subplanes used). This dimension is of the order of $(4+1)^2 \times 5 - 125$ for the simpler applications, and gets rapidly worse for other applications. The inversion must be repeated for <u>each</u> new trial structure. The Beeby method will, however, deal with small interlayer spacings and in principle with any complexity of the surface structure.

An ingredient for most of the methods is the single-layer diffraction matrix $M_{gg'}^{\pm\pm}$, which is obtained (as a special case of T-matrix method for N = 1) by inversion of a matrix of dimension $(\ell_{max}+1)^2$ for Bravais lattices. The Bloch wave method is appropriate for surfaces of simple materials as opposed to, for example, layer compounds[37], since it is based on periodicity perpendicular to the surface. It can accommodate arbitrarily small absorption, unlike most other methods, which is valuable at very small energies. A relatively slow matrix diagonalization is involved in the eigenvector and eigenvalue problem, it benefits however from block-diagonalization due to enlarged surface unit cells. A wavefield matching at the surface has to be repeated for each new trial structure, whereas the substrate Bloch waves are reusable.

The layer-doubling method involves two inversions of g-dimensional matrices (g is the number of required plane waves) for each iteration, 3 or 4 iterations being normally sufficient. Except for very small layer spacings, this method has proved quite convenient for simple substrates, yielding a substrate reflection matrix (in block-diagonalized form when extra beams are present) that can be repeatedly used for any number of overlayer structures. Such structures are obtained by successive stacking of additional layers on the substrate, each stacking involving one inversion of a g-dimensional matrix (no block-diagonalization occurs in the overlayer). A by-product of the layer-doubling method is the reflection matrix off the back side of the finite-layer crystal, which can have a different structure than the front side: for example, the front and back sides often have different registries with respect to a symmetry axis or plane, which is an important consideration when symmetrization is employed. A disadvantage of layer-doubling is that relatively much working space is needed for the various matrix manipulations. Also, the method is not well suited for substrates that have a long repeat distance perpendicular to the surface (e.g. layer compounds), a problem shared with Bloch wave method.

Computation times for the above exact methods scale as the third power of matrix dimensions, as a result of the matrix multiplications, inversions and diagonalizations. Approximate, perturbative methods are more efficient: the physical reason is that only significant scattering events are included rather than <u>all</u> of them, the computational reason lies in the avoidance of most of the above-mentioned matrix operations.

The kinematic (single-scattering) limit is most attractive for both physical and computational reasons, but it can be directly applied only to the restricted class of very weak scatterers (such as xenon).

Second- and third-order perturbation methods apply to more materials than the kinematic method (for instance, the third-order t-matrix method is quite successful with aluminum). The t-matrix perturbation method avoids the matrix inversion that corresponds to intra-layer multiple scattering. While it is very efficient for low orders of perturbation, higher orders become increasingly complicated and unwieldy. The τ-matrix perturbation method is more accurate, at the price of a matrix inversion for the intralayer multiple scattering. Higher orders again become unwieldy. The modification by Jennings[31] would improve the convergence of third-order methods.

The last category of calculational methods to be considered contains the iterative techniques: they represent the good compromise between accuracy and efficiency. These methods (the renormalized forward scattering theory) include all intralayer multiple scattering and allow the iteration to proceed to convergence. These methods involve only the products of matrices with vectors (apart from the matrix inversion corresponding to intralayer multiple scattering), so that computation times scale as the square of the number of beams or spherical harmonics.

The renormalized forward scattering (RFS) theory is very flexible with respect to crystal structure, including the case of varying unit cells, since it is based on the plane wave expansion. The computation times of the iterative part of the method are small compared to the time required for the generation of the single-layer diffraction matrices, resulting in very good efficiency when repeating the perturbation calculation for different surface structures.

However the RFS method converges poorly, if at all, when small layer spacings are involved: minimum spacings are of the order of 1.0 A. It is then necessary to invoke the reverse scattering perturbation (RSP) theory, which operates the iterations in angular momentum space. Here again matrices multiply into vectors, yielding times proportional to the square of the matrix dimensions. And the perturbation can be carried iteratively to convergence. The method is flexible with respect to crystal structure, but it does not, however, take advantage of beam sets due to differing unit cells and it requires many interlayer propagation matrices. Furthermore, it is not easy to incorporate symmetry in this method.

The method which is most flexible in terms of use, and most efficient in terms of computation speed is to combine L-space and K-space representations in one program and select the representation according to the surface structure and interlayer spacings. The general procedure is to solve for total layer-dependent scattering vectors in L-space for those layers that are closely spaced. The resulting layer vectors are merged to form transmission and reflection matrices in K-space for the stack of layers. The final reflected intensities are then calculated for the entire crystal by iterating in K-space the scattering matrices of the composite and those of the remaining layers of the solid.

Thus, in a short span of five years or so, we have seen the formulation, development and application of seven to eight useful computation schemes of LEED intensities. This is an active and vital area of research with an important mission: the determination of where surface atoms sit with respect to other atoms of a solid. The answer to this question will be useful to many important areas of current research in surface characterization.

REFERENCES

1. T.N. Rhodin and S.Y. Tong, Physics Today, Vol. 28, No. 10, 23 (1975).
2. J.E. Demuth, Thesis, Cornell University, Ithaca, N.Y., 1972.
3. S.Y. Tong, *Progress in Surface Science*, S.G. Davison, ed., Pergamon, New York, Vol. 7, No 1 (1975).
4. M.A. Van Hove, Surface Science, 49, 181 (1975).
5. J.L. Beeby, J. Phys. C 1, 82 (1968).
6. C.B. Duke and C.W. Tucker, Jr., Surf. Sci. 15, 231 (1969).
7. S.Y. Tong and L.L. Kesmodel, Phys. Rev. B8, 515 (1973).
8. G.E. Laramore, Phys. Rev. B8, 515 (1973).
9. S.Y. Tong and T.N. Rhodin, Phys. Rev. Lett., 26, 711 (1971).
10. G.E. Laramore and C.B. Duke, Phys. Rev. B5, 267 (1972).
11. E.G. McRae, J. Chem. Phys., 45, 3258 (1966).
12. E.G. McRae, Surf. Sci., 11, 479 (1968).
13. P.J. Jennings and E.G. McRae, Surf. Sci., 23, 363 (1970).
14. D.W. Jepsen, P.M. Marcus, and F. Jona, Phys. Rev. Lett., 26, 1365 (1971).
15. D.W. Jepsen, P.M. Marcus, and F. Jona, Phys. Rev. B5, 3933 (1972).
16. J.B. Pendry, J. Phys., C4, 2501 (1971).
17. J.B. Pendry, J. Phys., C4, 2514 (1971).
18. G. Capart, Surf. Sci., 26, 429 (1971).
19. K. Kambe, Z. Naturforsch, 22a, 332 (1967).
20. K. Kambe, Z. Baturforsch, 23a, 1280 (1968).
21. J.B. Pendry, *Low Energy Electron Diffraction Theory*, Academic Press, London (1974).
22. M.A. Van Hove and S.Y. Tong, J. Vac. Sci. Technol., 12, 230 (1975).
23. J.C. Buchholz, M.G. Lagally, and M.B. Webb, Surf. Sci., 41, 248 (1974).
24. M.B. Webb and M.G. Lagally, Solid State Physics, 28, 301 (1973).
25. T.A. Clarke, R. Mason, and M. Tescari, Proc. Roy. Soc. (London), A331, 321 (1972).
26. U. Landman and D. Adams, Phys. Rev. Lett. 33, 585 (1974).
27. S.Y. Tong, T.N. Rhodin, and R.H. Tait, Phys. Rev. B8, 421 (1973).
28. S.Y. Tong, T.N. Rhodin, and R.H. Tait, Phys. Rev., B8, 430 (1973).
29. R.H. Tait, S.Y. Tong,, and T.N. Rhodin, Phys. Rev. Lett., 28, 553 (1972).
30. S.Y. Tong, T.N. Rhodin, and R.H. Tait, Surf. Sci., 34, 457 (1973).
31. P.J. Jennings, Surf. Sci., 41, 67 (1974).
32. E.C. Snow and J.T. Waber, Phys. Rev., 157, 570 (1967).
33. J.B. Pendry, Phys. Rev. Lett., 27, 856 (1971).
34. S.Y. Tong, Solid State Comm., 16, 91 (1975).
35. R.S. Zimmer and B.W. Holland, J. Phys., C8, 2395 (1975).

36. S.Y. Tong and M.A. Van Hove, to be published.
37. B.J. Mrstik, R. Kaplan, T.L. Reinecke, S.Y. Tong, and M. Van Hove, Bull. Am. Phys. Soc., March 1975.

Discussion

On the Paper by P.J. Estrup

S.Y. Tong (*University of Wisconsin-Milwaukee*): Since the binding energies of most adsorbates are a few eV, did you see any evidence of electron beam desorption?

P. Estrup (*Brown University*): Yes, for some adsorbates the electron beam will cause desorption, dissocation, etc. at a significant rate. For example, in the case of CO on Ni (110) the cross-section for dissociation is about 2×10^{-18} cm^2 (T.N. Taylor, P.J. Estrup, J. Vac. Sci Technol. 10, 26 (1973)) and noticeable changes occur if the LEED measurements last longer than a minute or so. Cross-sections as large as 10^{-16} cm^2 and as small as $< 10^{-26}$ cm^2 have been reported (T.E. Madey, J.T. Yates, Jr. J. Vac. Sci. Technol. 8, 525 (1971), but processes for which it is less than 10^{-20} cm^2 can usually be ignored in LEED experiments.

I Nicolau (*Stauffer Chemical Co.*): In the desorption spectrum of CO on Ni(100) (desorption rate versus temperature) are several forms of adsorbed CO observed?

P. Estrup: The binding energy of CO on Ni(100) was obtained by isosteric heat measurements (39) rather than by thermal desorption experiments (which are difficult at the low temperatures used). The energy does not vary with coverage initially but drops with the "compression" of the overlayer begins - that is, when the repulsion between neighboring CO molecules becomes significant. The adsorbate appears to be homogeneous at all values of the coverage.

On the Paper by D. Adams and U. Landman

L.H. Lee (*Xerox Corp.*): Dr. Adams made excellent efforts to answer some difficult questions about LEED. This paper strongly complements three other papers of this session by Dr. Estrup, Dr. Marcus and Dr. Tong to bring us to date about theoretical and experimental problems related to LEED. I am grateful that Dr. Adams could accept our invitation to present this important work to this symposium.

On the Paper by P. Marcus

P.N. Ross (*United Technologies*): In your slide showing the best match of the calculated intensities to the experimental intensities you used the (1/2 1/2) diffracted beam. Why did you pick that beam and is the agreement as good for the specular beam?

P. Marcus (*IBM Corp.*): The (1/2 1/2) beam, like all fractional order beams, necessarily involves scattering from the overlayer, whereas integral order beams are present even without the overlayer. Hence, the effects of the overlayer are generally larger on the fractional order beams, and they are more sensitive to the overlayer position. The 00 beam is not available at normal incidence, the angle for the 1/2 1/2 beam shown. In general, the agreement for the integral order beams was good, although not quite as good as for this 1/2 1/2 beam.

On the Paper by S.Y. Tong

P. Marcus (*IBM Corp.*): Dr. Tong has commented on how close the bond lengths determined by the LEED analysis due to sums of covalent radii: I think that in fact the differences from covalent radii are significant, and are interesting magnitudes for discussion by bonding theory. For example, the covalent raduis of $\underline{O}$ is 0.66Å, but is consistently larger on Ni; Fe and W. This increase may be due to the $\underline{O}$ being slightly ionic, or it might be interpreted as decreased bond order. Other examples will be given in a table in my contribution to the proceedings.

PART IV

Secondary Ion Mass Spectrometry

PART IV

Plenary Lecture: Ion Microscopy and Surface Analysis

G.H. Morrison

*Department of Chemistry
Cornell University
Ithaca, New York 14853*

Secondary ion mass spectrometry (SIMS) can be used for microspot or localized analysis, surface analysis, and depth analysis. Ion microscopy, a microanalytical method based on SIMS, is described and some of its unique features for the study of nonhomogeneous solid surfaces are discussed. The direct imaging capability of the method for the examination of the distribution and association of chemical species in the surface with high spatial resolution is emphasized.

The fundamentals and instrumental aspects of ion microscopy are reviewed. Chemical and physical effects of the sample on the brightness of the image are covered. The surface analysis capability of SIMS is discussed in terms of sample consumption by sputtering. Examples of ion micrographs are presented.

INTRODUCTION

Secondary ion mass spectrometry (SIMS) is an important member of the arsenal of new techniques for the study of solid surfaces. The various techniques discussed in this symposium and other surface tools have been reviewed and compared in the literature.[1-4] At the outset, it must be emphasized that these surface techniques are not generally competitive. Each has its own particular advantages and features, with some overlap, so that for complete characterization of a material it is often necessary to apply several of these methods.

Modern SIMS instruments can perform a variety of analytical functions with parts per million sensitivity or better. They can be used to perform microspot or localized analyses, and depth analyses. Complete mass spectra and isotopic ratios can be obtained from areas on the micron scale. The chemical composition of the outermost layers of a solid can be determined. Ion images can be obtained showing the distribution of elements and molecular species at the surface. Finally, in-depth profiles of both thin and thick films can be performed. These analytical capabilities are possible because the sputter ion source is capable of removing successive monolayers from the surface without deep penetration into the sample. Commercially available SIMS instruments have been compared recently by Evans.[5]

These various modes of SIMS analysis have been reviewed in the literature[6-9], and some are discussed by others in this symposium. The object of this paper is to present some of the unique features of ion microscopy, a microanalytical method based on SIMS, for the study of nonhomogeneous solid surfaces. The direct imaging capability of the method for the examination of the distribution of chemical species in the surface with high spatial resolution will be emphasized. The recent paper by Morrison and Slodzian[10] provides comprehensive background material for the technique of ion microscopy.

SECONDARY ION EMISSION

When a surface is bombarded with ions of several keV energy, collisions with the target atoms produce collision cascades which result in the emission of neutral atoms, positive and negative ions, electrons, and electromagnetic radiation.[6] See Figure 1. The sputtering produced by the impact of these ions originates mainly in the first few atomic layers[11-13]. A major portion of this secondary emission is neutral atoms, whereas the yield of positive and negative ions emitted is only a small fraction of these neutrals (typically 10^{-2} to 10^{-5}) and constitutes the phenomenon of secondary ion emission. The processes involved in ion bombardment have been treated in detail[8,14,15].

A variety of ionic species are produced by the ion bombardment process and include singly-charged monoatomic and polyatomic ions, multiply-charged ions, and inter-element molecular ions. The relative abundance of the different ion species depends on the nature of the sample. In addition, "hydrocarbon" ions and oxide ions may be produced from molecules chemisorbed at the surface depending upon the vacuum conditions in the vicinity of the target surface. Generally, the most intense ion species from inorganic materials are the singly-charged monoatomic ions of the elements present. Chemical analysis based on ion bombardment makes use of mass spectrometric techniques. Because the initial energies of the emitted secondary ions vary, both momentum and energy filtering are mandatory for high mass resolution.

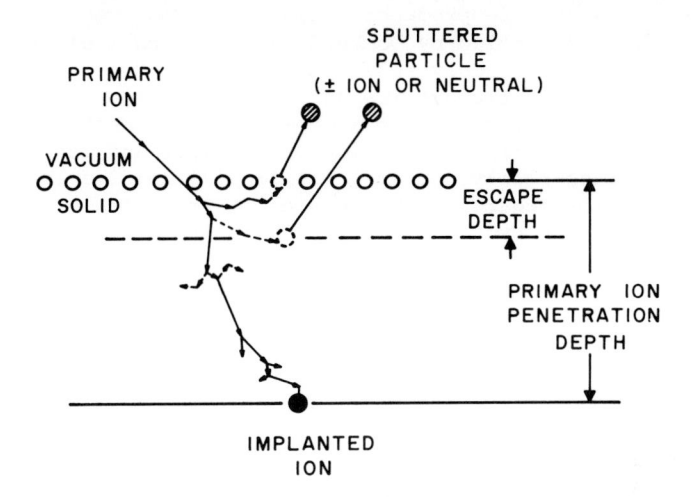

Fig. 1. Schematic of processes occuring in ion-solid interaction.

LOCALIZED VS. IMAGE ANALYSIS

In the past, emphasis in SIMS for microanalysis has been placed on the use of the method as a probe to analyze selected localized regions observed first with a light microscope, similar to the approach used with an electron probe. Such an analysis can be achieved by either using a small diameter primary ion beam to sample the local region or by bombarding a larger region and accepting for detection only those secondary ions originating from the local region of interest. When primary beam diameters of a few microns are employed, the technique is referred to as ion microprobe analysis. In practice, much larger beam sizes are used to provide adequate sampling for detection.

Since most solid materials are heterogeneous at the micron level, it is often more advantageous to examine simultaneously many of the microfeatures observed to quickly establish the compositional distribution of these features at the surface and their relationships. The ion microscope, first introduced by Castaing and Slodzian[16], is ideally suited to such studies. Because of its speed and large field of view (∼300 μm diameter), it provides an excellent preparatory step for the more judicious selection of regions for localized or probe analysis. By adjusting instrumental parameters and simultaneously monitoring the visual image, much more can be learned about a given surface than by repeated localized analyses. If quantitative analysis is ultimately required, calibration standards can be employed in the analysis of selected localized micro regions of the image.

The ion microscope produces a spatially resolved mass analysis of the surface of a solid by bombarding with a beam of primary ions and analyzing the secondary ions sputtered from the sample. The ion optics of the instrument are unique and provide a direct imaging capability by retaining a one-to-one correspondence between the point of origin of an ion on a sample and its location in the final mass-resolved beam. The final magnified image contains information regarding the spatial distribution of any selected element from hydrogen to uranium in the sample within an area of diameter less than half a millimeter with a spatial resolution of less than a micron. Since ion sputtering is used to remove material from the surface of the sample to generate data, microcompositional information can be obtained on a three-dimensional basis. Thus, in addition to lateral images of the surface described in this paper, concentration profiles can be obtained over depths varying from tens of angstroms to several microns.

INSTRUMENTATION

The ion microscope has been described in detail by Morrison and Slodzian[10] so that only a brief summary is presented here. A schematic representation of the CAMECA IMS-300 ion microscope is shown in Figure 2.

A Freon cooled cold cathode duoplasmatron gun using either noble gases or reactive gases is used to produce either positive or negative ions. A double condenser electrostatic lens system allows focusing of the primary ion beam from about 10 μm to over 300 μm and directed at approximately 45° incident to the sample.

The secondary ions emitted from the sample which is held at ± 4.5 kV from ground, are accelerated in a uniform electrostatic field by the immersion lens which forms a global ionic image containing in superposition images for each isotope present in the sample.

A mass spectrometer is used to separate the ions according to their m/e ratio to produce an image corresponding to a given element. The trajectory of ions leaving the surface normally, cross over the optical axis at point C where the entrance diaphragm to the spectrometer is located. For a monokinetic ion beam the first deflection produces an image C' of the crossover C, at which position a mass-resolving aperture or selection slit is located. The momentum filtering is achieved by the first deflection through the prism and by the selection slit at C'. By adjusting the entrance diaphragm at C, the selection slit at C', and the upper limit of the energy variation on the electrostatic mirror, a mass resolution (M/ΔM) of 1000 can be obtained.

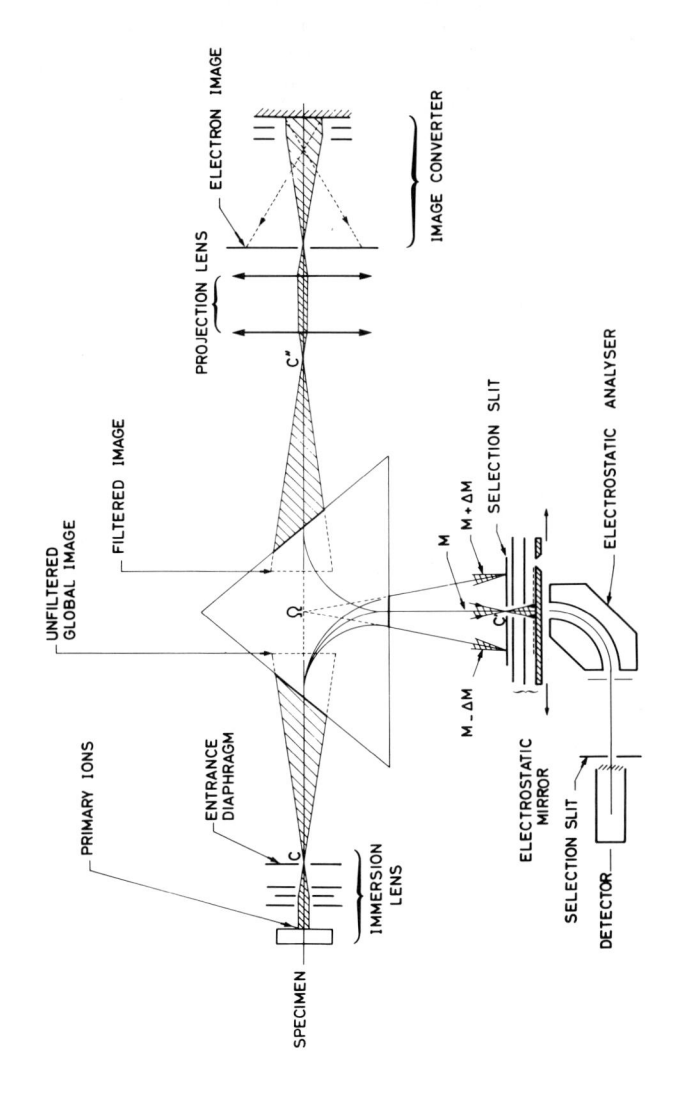

Fig. 2. Schematic representation of ion microscope(10).

After the first deflection, the mass-resolved beam enters the electrostatic mirror. Ions which enter along the axis of the mirror are decelerated, stopped, and then reversed. Ions whose kinetic energies are too high at the entrance to the mirror hit the bottom repelling electrode during the deceleration and are neutralized and eliminated from the beam. The reflected mass and energy filtered ions again enter the magnetic prism and follow a circular trajectory symmetrical to the previous one, with the mean axis of the filtered beam finally being in the same direction as the incident beam.

The filtered ion image is projected by a projection lens system onto the cathode of an electron image converter. The secondary electrons emitted from the cathode form an equivalent electron image which can strike a fluorescent screen and be converted to a photon image for visual observation or be directly recorded on photographic film. To isolate the emission coming from a small selected region of the sample, the microscope can be used as a microprobe by introducing an appropriately sized field of view aperture. The particles which form the image of this restricted region of the sample strike a light scintillator-photomultiplier detector combination where current output is proportional to the amount of a particular ion impinging on the converter cathode. The specifications of the instrument for microscopy are summarized in Table 1.

To minimize mass interference of the desired ions by various other types of ions produced during bombardment, an electrostatic analyzer attachment to the CAMECA instrument has recently been developed. The repelling electrode of the electrostatic mirror is grounded and shifted to provide an aperture which allows the mass-filtered beam after the first deflection to enter the electrostatic analyzer for energy filtering. The final beam is counted using an electron multiplier detector. Using this system, mass resolution on the order of 4000 can be achieved; however, the instrument cannot be used simultaneously in the microscopy mode.

Table 1. Geometric Specifications of Ion Microscope

Resolving power:	< 1 µm
Degree of magnification:	1000X - visual 110X - photographic
Field of view:	up to 300 µm
Information bits:	5×10^4/image
Information region:	20 Å - several µm

BRIGHTNESS OF IMAGE

The brightness of features in the enlarged image is of primary importance since the contrast obtained on the distribution image is related to local variations in concentration. The quantity to which the microscopist has direct access is the illumination i of the final enlarged image. Illumination is defined as the number of M^+ ions arriving per unit area and unit time at the detected image.

In principle, the total number of M^+ ions collected from a sample area a_o during time t could be computed knowing the angular and energy distribution patterns and integrating the various contributions over the solid angle and the energy bandpass of the spectrometer. It is more convenient and precise to measure directly the practical ion yield τ, which is defined as the ratio of the number of M^+ ions collected to the number of M atoms removed from the target. This yield also takes into account the ionization efficiency as well as the transmission efficiency of the instrument.

The illumination of the image is related to concentration by the equation

$$i = \tau C_m S i_p G^{-2} \quad (1)$$

where τ is the practical ion yield, C_m is the atomic concentration of element M (corrected for its isotopic abundance), S is the total sputtering yield (number of atoms of any kind removed per incoming particle), i_p is the number of primary particles arriving per unit time and unit area on the target surface, and G is the degree of magnification. The product of S and i_p is the sputtering rate.

Intensity information can be obtained from ion micrographs by measuring the optical density of the features of interest. A rapid scan computerized densitometer system is being developed in our laboratory for point-by-point quantitation of ion micrographic images. Present software routines allow reorientation and cross correlation between similar images.

SECONDARY ION CURRENT AND LOCALIZED ANALYSIS

If an aperture is used to limit an area A of the magnified image in front of the photomultiplier counting system as is done in localized or probe analysis, the measured secondary ion current I is related to the concentration by the equation

$$I = \tau C_m S i_p a_o \quad (2)$$

where i_p is now defined as the primary beam current density, and

a_o is the area on the target surface ($A = G^2 a_o$). The aperture for area A can be shaped to meet a specific sampling requirement for a given analytical problem.

CHEMICAL AND PHYSICAL EFFECTS OF SAMPLE

The physical and chemical nature of the sample strongly influence the illumination or secondary ion current. The practical ion yield depends on the ionization processes involved which are governed by the chemistry of the sample. This includes both the original chemical form of the element in the sample and that induced by the bombarding and vacuum conditions.[8] Practical ion yields may vary over a wide range. Under argon bombardment and clean vacuum conditions, they reflect the original chemistry of the sample. Depending upon the element, the practical ion yield of M^+ ions produced from an oxide of a metallic element may be as much as a factor of 10^3 higher than that from the corresponding pure metal. Thus, these chemical effects may be used to localize oxide inclusions by microscopy.

These effects can also be used to enhance the practical ion yields and thereby increase sensitivity of detection. Bombarding with reactive gases such as oxygen produces a chemically modified thin superficial layer (~ 100 Å) in the target. Flooding the surface of the sample with oxygen during bombardment will further enhance the practical ion yields for some elements.[17-19]

The sputtering yield is also influenced by chemical effects as well as lattice effects.[8] Thus, metals may result in higher sputtering rates than the corresponding oxides. Lattice effects are determined by the orientation of the primary beam with respect to the target lattice in the crystals. Once again, oxygen flooding of the sample can minimize this effect.[18]

In general, positive ion images are more representative of a sample, but for certain applications negative secondary ion images are useful. For example, many nonmetallic elements such as O, S, As, Te, and halogens are more sensitive using the negative ions produced by Ar^+ bombardment.[20] In addition, the negative ion molecular species produced and their relative intensities are different from those observed with positive ions so that in some cases spectral interferences can be minimized.

SURFACE ANALYSIS CAPABILITY

In general, a surface is an undefined depth corresponding to the "outer portion" of a sample and is used in general discussions of the outside regions of the sample. More precisely, one must differentiate between a physical surface and an experimental surface. A physical surface is that atomic layer of a sample which, if the sample were placed in a vacuum, is the layer in contact with the vacuum, i.e., the outermost atomic layer. An experimental surface is that portion of the sample which is sampled by the "surface technique" in use, which in turn is the volume of sample required for detection or the volume due to the escape depth for the detected radiation or particles, whichever is larger.

The escape depth of atoms producing the secondary ions used to generate an analytical signal in SIMS is on the order of 10-20 Å, thereby permitting in principle the localization of the analysis to a shallow depth. However, to obtain a significant signal for a given species present at a certain level of concentration it may be necessary to sputter deeper into the sample to accumulate a sufficient number of ions. There is a close relationship between the practical ion yield and the smallest volume of sample required to measure an element of concentration C_m with a given precision. The sputtering of a volume v produces a number of M^+ ions equal to $C_m \cdot \rho \cdot v \cdot \tau$, where ρ is the atomic density of the sample. Statistical fluctuations imply that at least $10^4/p^2$ ions have to be counted to obtain a precision of $\pm$ p/100. Thus, the volume of sample that must be sputtered to achieve this number of ions is determined by the equation

$$v = \frac{1}{C_m \rho \tau} \times \frac{10^4}{p^2} \qquad (3)$$

To optimize the analysis of a surface, therefore, it is desirable to sample a large area, thereby minimizing the depth of erosion. Figure 3 plots the calculated relationship between the detection limit and depth of sputtering for different values of practical ion yield, assuming an atomic density of 6 x 10^{10} atoms/μm³ for the sample, a precision of ± 10%, and a sampling area achieved with a 250 μm diameter field, which is the maximum used in the ion microscope. It is obvious that the applicability of SIMS to surface analysis depends greatly on the level of concentration at which the species of interest is present, as well as its chemical form as reflected by the practical ion yield. It should be noted that the sputtering rate is unimportant in determining the amount of material that must be consumed.

Experimental procedures used with SIMS for surface analysis fall into two categories: low density sputtering and high density sputtering. The static procedure pioneered by Benninghoven[21,22] uses primary beam densities in the range of 10^{-7} to 10^{-9} A/cm², so that the material consumption rate can be less than 10^{-4} monolayers per second. The dynamic procedure, which is employed in the ion microscope[16] uses primary beam densities in the range of 10^{-3} to 10^{-5} A/cm², so that the erosion rate of the sample corresponds to an average material consumption of about one mono-

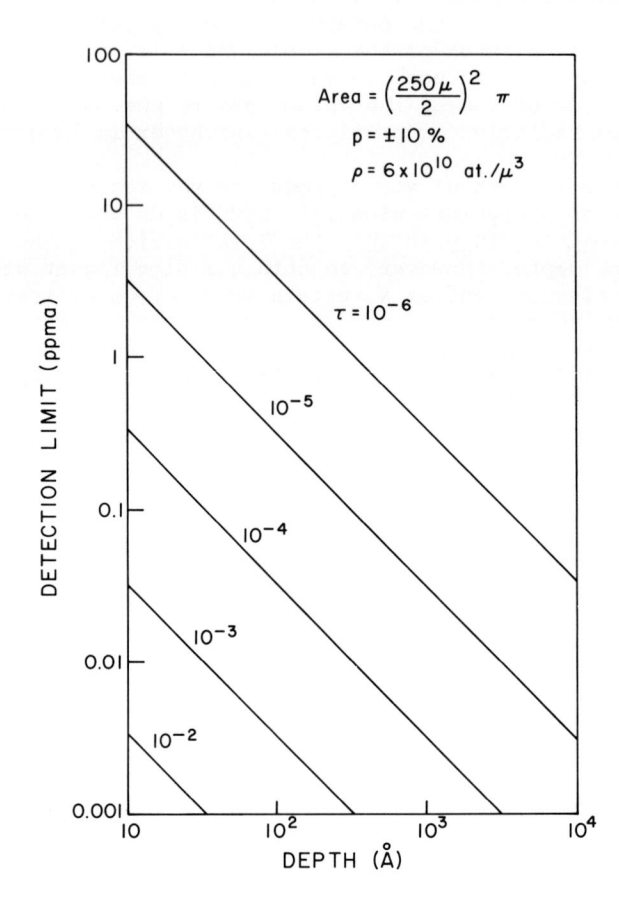

Fig. 3. Detection limit versus depth of erosion as a function of practical ion yield.

layer per second. High density sputtering is also employed with other commercial ion microprobes. The removal of a certain amount of material, however, depends on the total primary ion dose impinging on the surface and not on the manner in which this dose is distributed as a function of time.

Blaise[23] has shown that the sensitivity in surface analysis is not related to which of the two experimental procedures is employed but depends on the way in which the secondary ions are collected and transmitted by the instrument. In the static approach Benninghoven[24] uses a 2800 μm diameter beam where his instrument efficiency factor is 10^{-5}. In the ion microscope with an instrument efficiency factor of 10^{-1}, the diameter of the sampling beam is limited to 250-300 µm. Using Equation 3 to relate the depths that must be sputtered, and assuming that the minimum detectable number of ions is the same with both instruments, the minimum thickness to be removed with the ion microscope is fifty times greater than with Benninghoven's instrument. Consequently, the ion microscope appears to be more sensitive for in-depth analysis than the static SIMS instrument. With ultra high vacuum equipment, the ion microscope might be used advantageously in the static regime.

A number of large beam SIMS instruments for surface analysis are commercially available.[5] However, they provide minimal lateral resolution and they employ a quadrupole mass spectrometer with poor mass resolution. As will be shown in the next section, the high lateral resolution of the ion microscope makes it an invaluable tool for the study of nonhomogeneous surfaces.

APPLICATIONS

In the study of surface composition by SIMS, it is important that good vacuum conditions prevail during the analysis. Otherwise, chemically active components of the residual gas surrounding the sample can act upon the surface composition. Alternatively, this sensitivity of secondary ion emission of the surface to the surrounding gaseous phase can be used to advantage for investigating gas-solid interface reactions. Thus, the analytical applications of SIMS for surface analysis have involved the identification and distribution of species in the surface, as well as the study of dynamic surface processes. Some of the surface phenomena studied include catalysis, corrosion, adsorption, and diffusion.

The quantitation of secondary ion information from the outermost region or reaction zone of a surface presents some difficulty at this time. Fortunately, surface processes can often be studied with qualitative information. The types of species and the variation of secondary ion signals with time, temperature, and gas pressure provide sufficient information to elucidate many processes occuring at surfaces. Comprehensive reviews of applications in these areas have been provided by Fogel[25,26], Benninghoven[24], Blaise[23], and Bernheim and Slodzian[27].

The particular aspect of surface analysis emphasized here is the investigation of nonhomogeneous surfaces. As mentioned earlier, the ion microscope with its large field of view and high spatial resolution provides a convenient means for quickly establishing the topological distribution and association of chemical elements of importance to a given problem. The speed of obtaining ion microscopic data cannot be over-emphasized. Images can be obtained in the order of seconds, thus allowing rapid sequence photography of selected areas. Images of positive and negative species, including atomic and molecular ions, provide a means for establishing the chemical associations of the species of interest.

The application of ion microscopy to the study of elemental distribution in both fundamental and applied studies in metallurgy, geology, solid state physics, and biology has recently been reviewed by Morrison and Slodzian[10], so that only a few additional examples are presented now.

The first example is that of corrosion, an important practical problem in metallurgy. Recently, our laboratory has been involved in an examination of steel strands in the cables suspending a 525-ton feed platform in the world's largest radio-radar telescope at Arecibo, Puerto Rico. Ion microscopic examination of one of the broken strands from this relatively high stress structure revealed the presence of Na, K, Mg, Ca, Cl, etc. in the corrosion layer, suggesting the possibility of chemical action by sea salt, as might be expected in this tropical environment.

On the left hand side of Figure 4 is an ion micrograph of $^{56}Fe^+$ at the surface of the steel strand near the break point. The corroded horizontal surface region is indicated by the light area resulting from the enhancement of the iron emission due to the chemical effect produced by oxidation. This has been confirmed by a comparable distribution in an ion micrograph of FeO^+. The bulk of the sample is located in the lower half of the micrograph. Ion images of $^{23}Na^+$ and $^{24}Mg^+$, shown in the center and right images in Figure 4, have the same distribution as the corrosion pattern. Comparable figures were obtained with $^{39}K^+$, $^{35}Cl^-$, and $^{40}Ca^+$.

The second example of the use of ion microscopy in surface analysis involves contamination of semiconductor devices. Native oxide formation is used in device fabrication for preparing masks for diffusion and ion implantation, as a protective coating for the annealing step, and as the oxide layer in a Metal Oxide Semiconductor (MOS) device. Recently, interest has been shown in a technique for forming oxide layers on device surfaces by anodic growth from solution. Ion microscopy studies in our laboratory have shown that one serious problem with this approach is the introduction of impurities. On the left hand side of Figure 5 is an ion micrograph of $^{115}In^+$ from an InP substrate. The lighter region illustrates chemical enhancement due to the oxidized surface layer which is approximately 500 Å thick. The adjacent dark region has been masked during the anodizing step. The center image of Figure 5 is that of $^{23}Na^+$, a nonhomogeneously distributed impurity introduced into the oxide layer during the anodizing process. Sodium is especially critical in this application because of its high mobility. The $^{28}Si^+$ image on the right indicates that Si im-

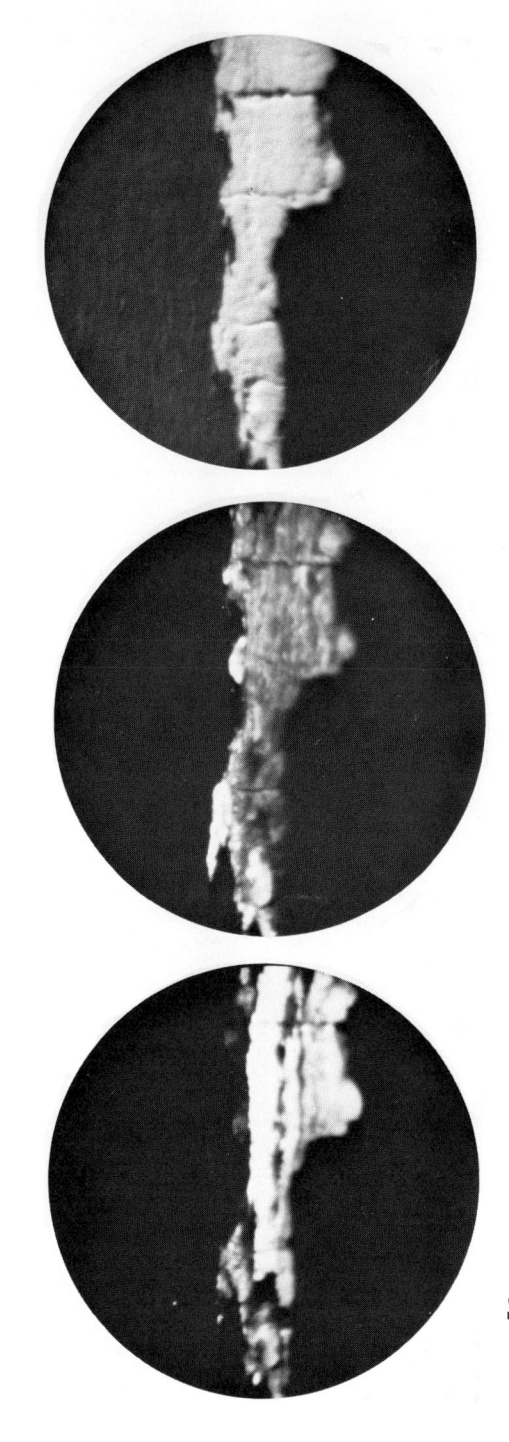

Fig. 4. $^{56}Fe^+$ ion micrograph (left) indicating corrosion of surface of steel strand. $^{23}Na^+$ (center) and $^{24}Mg^+$ (right) ion micrographs show comparable distributions in the corrosion layer.

Primary beam: O_2^+

Viewing field: 200 µm

Exposure times: Fe(0.5 sec), Na(1 sec), Mg(50 sec).

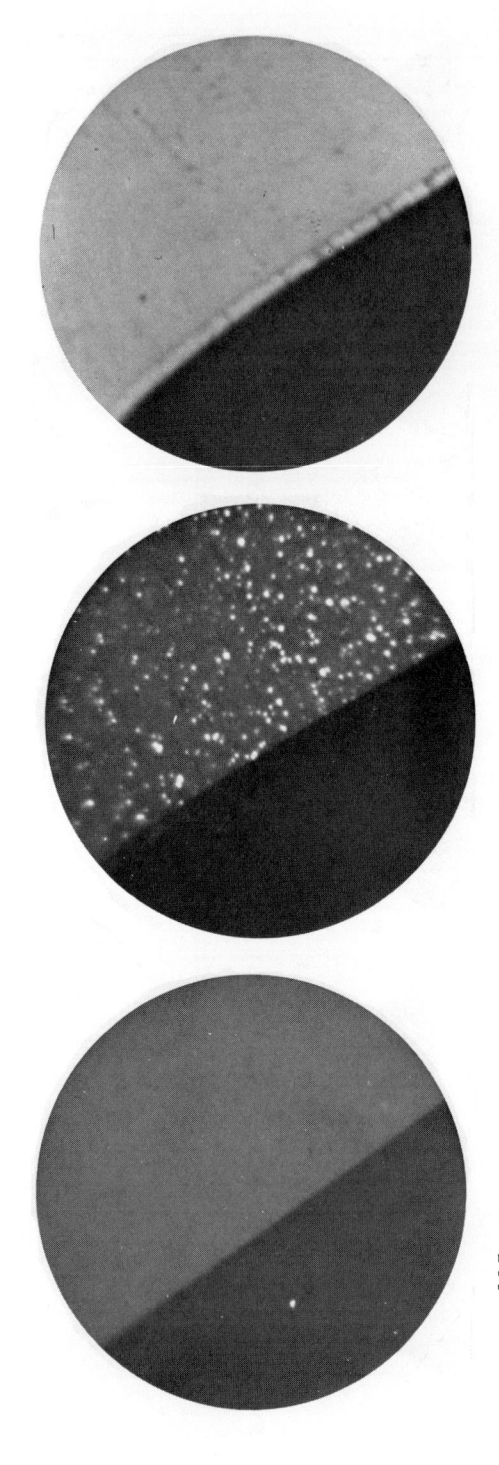

Fig. 5. $^{115}In^+$ ion micrograph (left) of anodized layer on InP.
$^{23}Na^+$ (center) and ^{28}Si (right) ion micrographs show contamination during anodic growth process.

Primary beam: O_2^+

Viewing field: 200 μm

Exposure times: In(0.5 sec), Na(10 sec), Si(100 sec).

purity is also introduced. Comparable images for K, Ca, F, and Cl further support the contamination of the oxide layer.

ACKNOWLEDGEMENTS

The author thanks G. Scilla for performing the ion micrographic studies. Research was supported by the National Science Foundation under Grant No. MPS 71-03280.

REFERENCES

1. C.A. Evans, Jr., Anal. Chem. 47(9), 818A (1975).
2. J.W. Coburn and E. Kay, CRC Crit. Rev. Solid State Sci., 4, 561 (1974).
3. H.W. Werner, "Characterization of Ceramics by Means of Modern Thin Film and Surface Analytical Techniques", *Science of Ceramics*, in press.
4. P.K. Kane and G.B. Larrabee, eds., *Characterization of Solid Surfaces*, Plenum Press, N.Y., 1974.
5. C.A. Evans, Jr., Anal. Chem., 47(9), 855A (1975).
6. J.A. McHugh, "Secondary Ion Mass Spectrometry", in *Methods and Phenomena Methods of Surface Analysis*, S.P. Wolsky and A.W. Czanderna, Eds., Elsevier, Amsterdam, 1975.
7. C.A. Evans, Jr., Anal. Chem. 44(13), 67A (1972).
8. G. Slodzian, Surf. Sci., 48, 161 (1975).
9. R.E. Honig, in *Advances in Mass Spectrometry*, Vol. 6, A.R. West, ed., Applied Science Publ., Berking, Essex, England, 1974, p. 337.
10. G.H. Morrison and G. Slodzian, Anal. Chem., 47(11), 932A (1975).
11. H. Liebl, J. Appl. Phys., 38, 5277 (1957).
12. D.D. Odintsov, Sov. Phys. Solid State, 5(4), 813 (1963).
13. D.G. Harrison, Jr., W.L. Moore, H.T. Holcombe, Rad. Effects. 17, 167 (1973).
14. R. Castaing and J.F. Hennequin, in *Advances in Mass Spectrometry*, Vol. 5, A. Quale, ed., Instit. of Petr., London, England, 1971, p. 419.
15. C.A. Andersen and J.R. Hinthorne, Science, 175, 853 (1972).
16. R. Castaing and G. Slodzian, J. Microsc., 1, 395 (1962).
17. V. Leroy, J.-P. Servais, and L. Habraken, Centre Recherche Metallurgique, Liege, Belgium, No. 35, p. 69, June 1973.
18. M. Bernheim and G. Slodzian, Int. J. Mass Spectrom. Ion Opt., in press.
19. G. Slodzian and J.F. Hennequin, C.R. Acad. Sci., Paris, B263, 1246 (1966).
20. C.A. Evans, Jr., in *Advances in Mass Spectrometry*, Vol. 5, A. Quale, ed., Institute of Petr., London, England, 1971, p. 436.
21. A. Benninghoven, Z. Physik, 230, 403 (1970).

22. A. Benninghoven, Sur. Sci., 28, 541 (1971).
23. G. Blaise, Bull. Soc. Chim. Belg., 84(6), 617 (1975).
24. A. Benninghoven, Surf. Sci., 35, 427 (1973).
25. Ya M. Fogel, Sov. Phys. Usp., 10, 17 (1967).
26. Ya M. Fogel, Int. J. Mass Spectrom. Ion Phys., 9, 109 (1972).
27. M. Bernheim and G. Slodzian, Surface Sci., 40, 169 (1973).

Surface Characterization by Ion Microprobe Analyzer

Ian M. Stewart

Walter C. McCrone Associates, Inc.
Chicago, Illinois   60616

The ion microprobe mass analyzer (IMMA) is a comparatively recent tool which permits mass spectroscopic analysis on a microscale. IMMA analyzes secondary ions induced by bombarding the target with a high energy primary ion beam. The method is sensitive to all elements in the periodic table and in most cases this sensitivity is better than 1 ppm. The secondary ions which are analyzed are generated in the sample surface, hence the method is a surface characterization method. By successively sputtering material from the surface it is possible to monitor the thickness of surface layers with resolutions of a few tens of angstroms and at the same time to perform depth profile analyses for specific elements. The primary ion beam may be focused to a probe approximately 1 μm in diameter and thus particulate contamination of the sample surface may also be analyzed. Organic materials, too, may be identified in this method provided a suitable standard of the suspect material is available. There are, however, some limitations in the general application of the ion probe to unknown organic materials principally due to the high energy of the primary ion beam. Some examples of the application of the ion microanalyzer to surface characterization problems will be described.

The ion microprobe analyzer[1,2,3] is basically a mass spectrometer, with a spatial resolution in the x-y plane of the order of 1 to 2 micrometers, and the potential for depth resolution, in the Z direction, of atomic monolayers.

The potential for mass spectrometry with spatial resolution of the order of a few micrometers was first realized by Castaing,[4] who developed an ion microanalyzer which now tends to be differentiated from the ion microprobe analyzer under the name of ion microscope. As this design is already covered by another paper in these proceedings, this paper will restrict itself to the ion microprobe analyzer. The basic difference between the two instruments should, however, be noted. In the ion microscope, a large area of the sample is flooded with a high energy ion beam, and regions of interest are selected by an aperturing and imaging system. In the ion microprobe analyzer, the region of interest is selected by focusing the impinging ion beam to a small probe whose position on the surface may be adjusted electrostatically.

Figure 1 shows a schematic of the ion microprobe analyzer; Figure 2 shows a typical instrument. A duoplasmatron gun provides a high energy ion beam which is mass sorted by passing it through a primary magnet. The object of this mass sorting is to remove any unwanted species which may be present in the original gas stream. After passing from the primary magnet, the ion beam enters the ion optical column which has provision for focusing the beam to a small probe, typically 1 to 2 micrometers across, together with capabilities for deflecting the probe in the x and y direction. When this highly focused high energy ion beam strikes the sample, secondary ions are generated from the chemical species present in the sample surface. These secondary ions are collected by a collector electrode placed very close to the sample surface, which passes them, in turn, into the mass spectrometer. The mass spectrometer consists of an electrostatic segment followed by an electromagnetic segment and has a mass resolution, somewhat poor for a conventional mass spectrometer, of around 300. The detector is an ion converter, scintillation type and the collected ions may be processed in several ways. Firstly, the ions may be merely counted to give a digital output, for any individual species. Secondly, the magnet of the mass spectrometer may be swept through the entire mass range to give an analog output of ion count versus mass number--a mass spectrum. Finally, the signal from the mass spectrometer may be used to modulate the brightness of a cathode ray tube which can then be scanned in synchronism with a scan performed on the surface of the sample. In this way the distribution of mass on the sample surface may be displayed.

Let us, then, consider the application of the instrument to some practical problems:

The Determination of Composition of Particulate Matter on the Surface of the Sample

Surface contamination either in the form of corrosion products or as settled debris is almost always on the surface of any real life material, and can frequently inhibit the correct application of this material. Surface contamination can lead to increased abrasion of mating components, can inhibit the deposi-

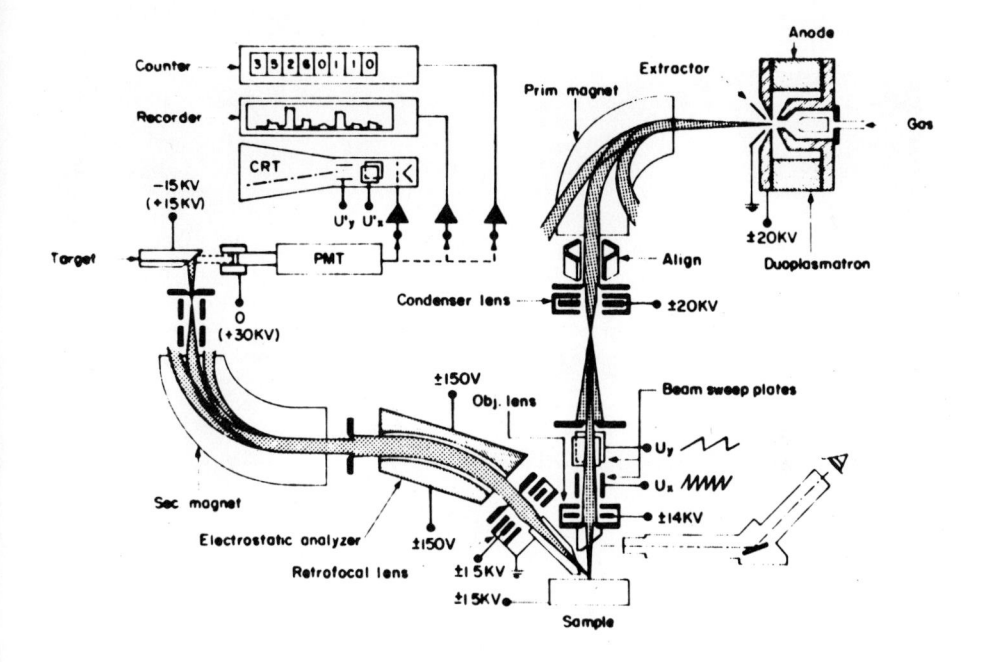

Fig. 1. Schematic of the Applied Research Laboratories ion microprobe analyzer.

tion of electro-deposits or vacuum evaporated films or may promote electrical resistivity and contact problems, for example. Knowledge of the composition of these contaminants may very well give an indication of their source, and hence enable their elimination. To determine what these contaminants are, one can typically align them, using a light microscope, under the incident ion beam and determine the composition of the particles by examining the entire mass spectrum obtained from the secondary ions. Such an analysis is shown in Figure 3. In general, however, such an approach would probably be better served by the electron microprobe which paradoxically, because of its lower sensitivity, may remove much of the source of confusion which arises in the ion probe due to the high signals given by certain elements which are readily ionized but which may actually be present only in trace amounts. The technique, however, is of value in the examination of extremely small particles scattered at random on the surface. Here, however, the information obtained can not be verified by the light microscope, thus it is not always possible to determine whether one is dealing with a uniform surface film or a sparsely distributed minute particulate contaminant.

Fig. 2.   A.R.L. ion microprobe analyzer.

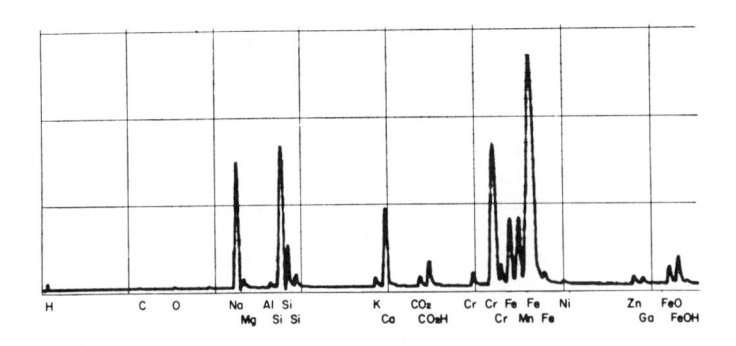

Fig. 3.   Mass spectrum of a small particle of stainless steel.

The Determination of Trace Amounts of Elements Distributed on the Surface of a Sample

Here again the approach that may be used is very similar to that used for the determination of particle contamination on the surface, assuming that it is not clear what contaminants are to be searched for. If, on the other hand, we need to obtain knowledge on the presence or absence of any specific elements this may be accomplished by monitoring only the signal for that mass number.

Elemental Distribution in the x-y Plane

This may be applied both to the particulate problem and to the problem of films on the surface. By monitoring the distribution in the x-y plane, it is possible to show how a particular element is distributed on the sample surface. This may give information on the distribution of small particles on the surface or of discontinuities in surface films. An interesting phenomenon occurs, however, when this method is applied to the analysis of particles: not only do the particles show as regions of high intensity but surrounding them we frequently see a halo of a similar element. This is due to recondensation of some of the sputtered secondary ions which are not immediately accepted by the collector electrode. These are subsequently resputtered as the primary beam traverses the sample surface giving rise to a spurious signal; spurious in the sense that this material was not originally in that position.

Depth Profiling

Because the ion microprobe technique is a destructive one in that successive layers are stripped from the specimen surface during the sputtering process, it is quite obvious that it is possible to perform an analysis at depth within the sample by sputtering material away and monitoring the signal of any desired element. If information is available on the sputtering rate of the material in question, it is possible to correlate discontinuities in depth with a definite depth interval. Figure 4 shows an example of this technique applied to the problem of can processing. Canned material typically is contained in a steel or aluminum can which on the inner surface has a protective lacquer coating. During processing the heat treatments to which the can is subjected causes sodium from the foodstuff to migrate from the lacquer contact surface into the bulk of the lacquer. The three graphs shown in the figure represent an unprocessed can which has not been packed, a can which has been packed but not processed and a fully processed can. It can be seen that although sodium is present in the original lacquer the level in-

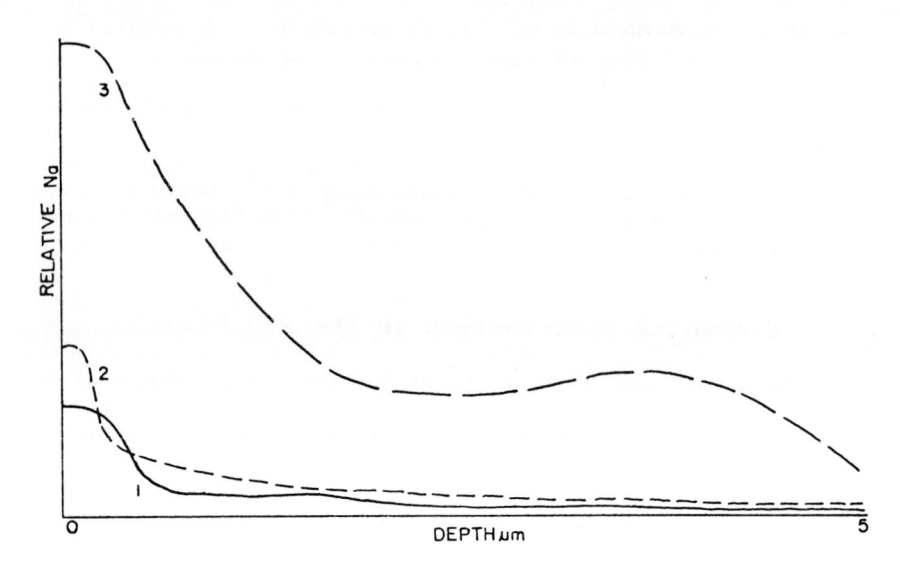

Fig. 4. Variation of Na with depth in lacquer coatings of three cans. (1) Unused can--neither packed nor processed; (2) Can packed but not processed; (3) Can packed and processed.

creases at the surface which has been in contact with the foodstuff though no processing has as yet taken place. Following processing the diffusion profile into the substance of the material can be clearly seen. This depth profiling technique has been equally successfully applied to studies of diffusion in silicon and other solid state materials. In the practical application of the method, the beam may either be scanned in a raster over a wide area on the sample or be defocussed to cover a wide area in order that sufficient depth resolution may be obtained. Operating in this mode, depth resolutions equivalent to atomic monolayers may be achievable in suitable material.

Although several applications have been discussed, the discriminating reader will note that little has been said of quantitation in the ion microprobe. At the present time, the mechanisms affecting ion production under the impingment of a high energy primary ion beam are not fully understood. There are several very complex reactions which take place at the sample surface and although a model of local temperature equilibrium does show some promise in absolute quantitation of ion probe analysis in many systems, this approach has given results which are still far from satisfactory. For many practical applications, however, absolute quantitation is not necessary and comparison with standards identical to, or near to, the suspected material is frequently a sufficient criterion for an identification to be made. It is on this basis that many industrial problems have been solved and continue to be solved on the ion probe.

REFERENCES

1. W.C. McCrone and J.G. Delly, Ed., *The Particle Atlas*, Edition II, Vol. II, pp. 168-178, Ann Arbor Science Publishers Inc., 1973.
2. J.L. McCall and W. Mueller, Ed., *Microstructural Analysis - Tools and Techniques*, Proceedings of an IMS-ASM Symposium, Plenum Press, New York, 1973.
3. J.I. Goldstein and H. Yakowitz, Ed., *Practical Scanning Electron Microscopy*, Chapter on Electron and Ion Microprobe Analysis, Plenum Press, New York, 1975.
4. R. Castaing, Thesis, University of Paris, Paris, France, 1951; Publ. ONERA, No. 55.

Study of Adhesive Bonding and Bond Failure Surface Using ISS-SIMS

W.L. Baun

Mechanics and Surface Interactions Branch (MBM)
Air Force Materials Laboratory
Wright-Patterson Air Force Base, Ohio 45433

Frequently it is not easy using visual or even microscopic examination of an adhesive joint to determine after physical testing whether an apparent adhesive failure occurred at the original interface due to improper wetting or at some new interface leaving behind a thin layer of adhesive. Elemental analysis techniques such as ion scattering spectrometry (ISS) and secondary ion mass spectrometry (SIMS) combined with Scanning Electron Microscopy (SEM) are capable of determining the locus of failure in an adhesive joint. The use of ISS and SIMS in combination is shown for investigating adhesive bonding phenomena. The operating parameters as well as advantages and disadvantages of each are summarized. ISS-SIMS data are shown for two adherend surfaces which broke in a lap shear test by apparent cohesive failure in both adhesive and adherend. Data are also shown for failure surfaces from peel test and thin adherend double cantilever beam (wedge test) specimens.

INTRODUCTION

The strength of an adhesive joint is assessed by means of physical tests such as single lap shear, double lap shear, peel, etc., in which an increasing load is placed on the joint until failure occurs. Following failure, visual (or sometimes microscopic) examination of the surfaces is made to determine the mode of failure. If adhesive remains on each adherend and the joint appears to have failed in the adhesive itself, the failure is termed "cohesive" failure. If the failure appears to have occurred at the interface between the adhesive and adherend, the failure is termed "adhesive".

Bikerman[1] says that "rupture so rarely proceeds exactly between the adhesive and an adherend that these events (that is "failure in adhesion") need not be treated in any theory of adhesive joints". He points out that apparent failures in adhesion are quite common but they take place so near the interface that the adhesive remaining on the adherend after the rupture is not visible. Good[2] has analyzed adhesive joint failure and reports that interfacial separation is highly improbable when true wetting of the surface has taken place.

Frequently it is not simple using visual or even microscopic examination to determine after testing whether an apparent adhesive failure occurred at the interface due to improper wetting or at some new interface leaving behind a thin layer of adhesive. Scanning electron micrographs are not definitive for very thin layers of adhesive. Often surface features of the original adherend are closely reproduced by a surface covered by a thin film of adhesive. Even when a failure mode appears certain where it appears that the break is primarily at the interface with a little cohesive failure in the adhesive, there is still a possibility for misinterpretation. The area of the surface which looks like the original surface may actually be covered by a thin layer of adhesive.

There is a resolution limitation of about 100Å for most scanning electron microscope which makes very thin films difficult to detect, especially when the adhesive is a pure polymer containing no fillers of higher atomic number than the polymer to increase contrast. Optical and staining methods have been reported[3] to determine the presence of adhesive films. However, if one looks at the elemental distribution on an aluminum adherend one would see, if the failure is mostly adhesive, primarily aluminum with traces of primer and/or adhesive. On the adhesive side, elements in the adhesive would be observed. The appearance of aluminum on this surface would indicate some cohesive failure in the oxide. On the adherend side, if only elements of the adhesive or primer are observed, then the failure is cohesive with a thin layer of adhesive or primer covering the alloy surface. In this paper ISS/SIMS is applied to determine elemental analysis of the interfacial surfaces and the information used to determine locus of failure.

EXPERIMENTAL

Figure 1 shows a simplified diagram of the combined ISS-SIMS instrument. The same primary ion beam is used to probe the surface in both techniques. The scattered ion energy spectrum is produced by the electrostatic spectrometer and is representative of the elements occupying the first atomic layer at the surface.[4] The primary ion beam also sputters away surface atoms which are ionized and detected by the quadrupole mass analyzer.[5] The surface finish of the specimen can have a significant effect on ISS intensities.

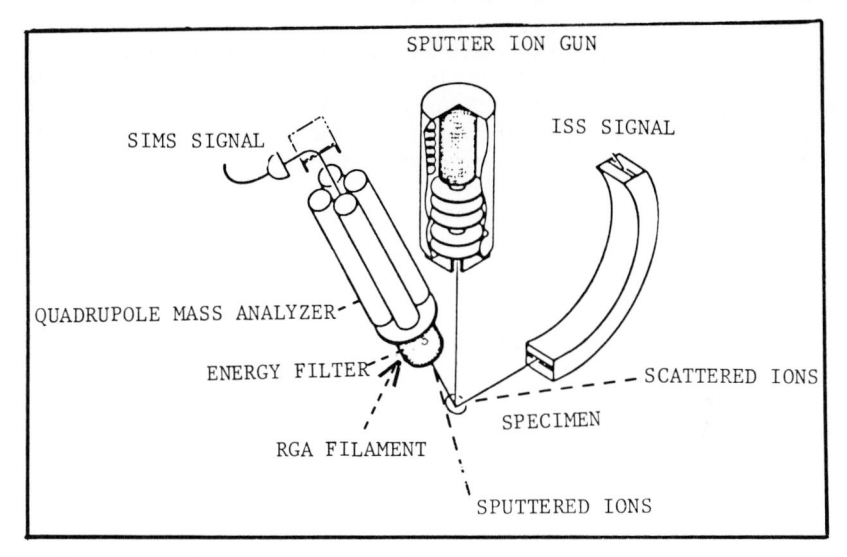

Fig. 1. Essential Components in Ultra High Vacuum For Combination ISS-SIMS Characterization of Surfaces.

Both absolute secondary ion yield and the ratio of one ion to another in SIMS can also be influenced by surface roughness. A further complicating factor results from possible mixing of surface atoms due to "knock on" effects of energy transfer by the primary beam. The system used here in the commercial ion scattering equipment Model 520 manufactured by 3M company (3M Co., St. Paul, Minnesota) to which a modified UTI (UTHE Technology International, Sunnyvale, California) model 100C quadrupole mass filter has been added to allow positive SIMS. A simple three element energy analyzer was added to the mass filter to obtain + SIMS spectra.

RESULTS AND DISCUSSION

Chemical and Thermal Treatments

Many chemical etching and oxidizing treatments are used on metals and alloys to enhance adhesive bonding of the surface. Numerous thermal pretreatments following fabrication improve strength, ductility, toughness or other property. Each of these chemical or thermal treatments affect the composition of the surface either by introducing impurities or by increasing or decreasing the concentration of alloying elements at the surface.

An example of changing surface chemistry is seen in Figure 2 where ISS and + SIMS data are shown for the aluminum alloy 7050 (nominal Zn 6.2%, Cu 2.3%, Mg 2.2% and Zr 0.12%) treated using an acid (HNO$_3$/HF) etch at the bottom and for the same surface anodized in a nonaqueous Li$_2$SO$_4$-K$_2$SO$_4$ salt mixture at the top of the figure. The etch leaves the surface fluorine rich. The fluorine remains on the surface during anodic oxide growth to produce not aluminum oxide as expected but rather an apparent aluminum oxyfluoride of unknown structure. Forest Products Laboratory (FPL) etch applied to this same alloy results in the surface becoming very zirconium rich even though the bulk concentration of zirconium is only 0.12%.

Lithium appears in the SIMS spectrum from the anodized specimen and comes from the anodization salt mixture. It also is seen on the etched sample due to contamination while washing the sample in H$_2$O following anodization. Lithium is highly mobile and rapidly covers the surface even when not originally present. Fluorine appears stronger in the anodized specimen than in the etched sample. This effect probably represents a change in relative fluorine population in the new lattice compared to oxygen and aluminum, or it may reflect oxygen removal due to decomposition of the hydroxide or H$_2$O in the hot salt mixture.

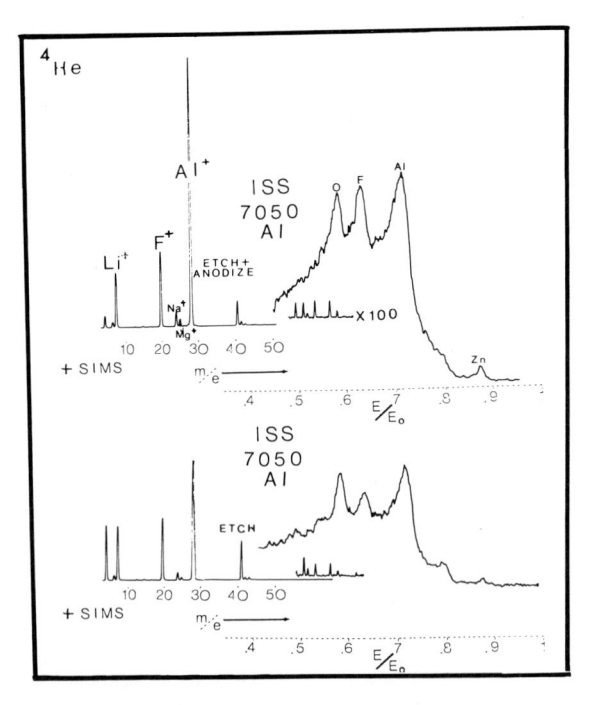

Fig. 2. ISS and + SIMS Data for 7050 Aluminum Alloy Etched With Dilute HNO$_3$/HF (Bottom) and Then Anodized (Top).

Often various processing steps on Al alloys result in the formation of "smut", a loosely adhering powdery dark-colored material on the surface which makes it unsuitable for bonding. One such smut was formed on 2024 aluminum alloy (nominal Cu 4.5%, Mn 0.6%, Mg 1.5%) with a standard chromate stripping solution. Figure 3 shows ISS and SIMS data from the smutted surface compared to an area which had been above the solution and which had not become smutted. The ISS spectrum of the smut shows it to be virtually all copper. Electrolytic deposition of copper and other elements occurs in many processing steps and is often difficult to detect by visual inspection.

Many other changes and abnormalities in surface composition have been observed. Even "as received" materials show surprising surface composition as seen in Figure 4 where SIMS and ISS data on 2024 aluminum alloy (degreased) indicate high magnesium concentration at the surface. Conventional alkaline cleaning treatments do not etch the surface appreciably, leaving the surface magnesium

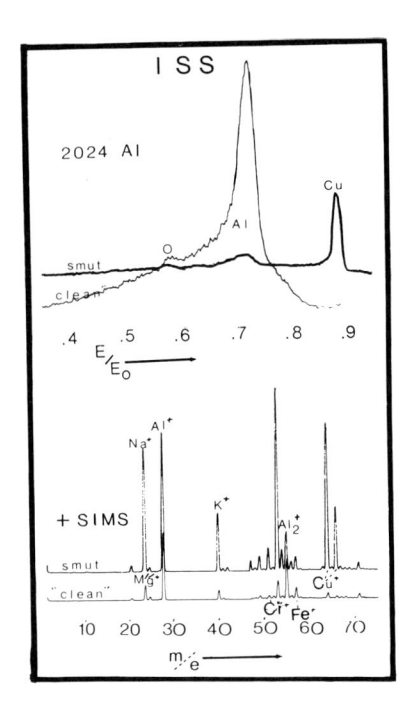

Fig. 3. ISS and + SIMS Spectra from "Smut" Formed on 2024 Aluminum Alloy Along with Spectra Obtained From the Clean Alloy.

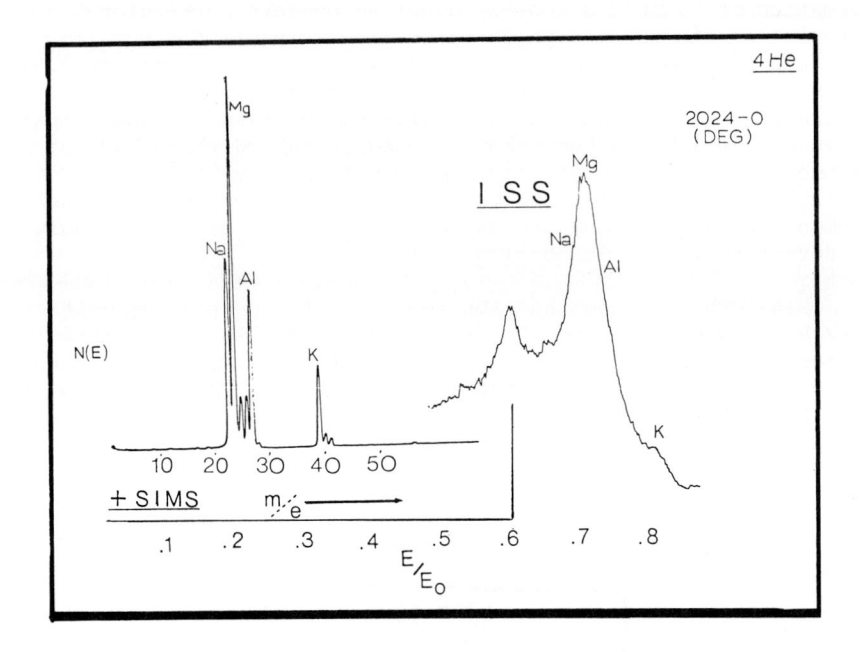

Fig. 4. ISS and + SIMS Data From 2024 Aluminum Alloy. No
treatment Except for Degreasing.

rich. Such a surface when adhesively bonded may exhibit long
time durability anomalies when compared with bonded structures
in which formation of aluminum oxide has been insured. A comprehensive report showing changes in surface composition with various
chemical treatments for 2024 and 7075 aluminum alloys has been prepared.[6] Data from AES, ISS and SIMS have been included along with
surface preparation information.

Analysis of Failure Surfaces

The majority of failure surfaces examined which visually appeared to be adhesive or interfacial failure, have been of the
type where cohesive failure in both materials appears to have
occurred in a weak boundary layer containing elements of adhesive
and adherend. It is necessary to be able to determine by ISS, SIMS,
SEM, Auger electron spectroscopy or some other elemental analysis
method whether the surface contains both adhesive and adherend,
since a test only for the adhesive on the adherend might cause an
incorrect deduction of failure mode and location. For instance,

a test for adhesive such as differential scanning calorimetry as performed by Bair and coworkers[7] to determine the amount of branched polyethylene adhering to a copper oxide surface does not in itself prove that the mode of failure was only cohesive in the polymer.

Some failures which could be classed as pure adhesive or cohesive have been examined by ISS-SIMS. Adherend surfaces which were obviously not wet by the adhesive showed no trace of the adhesive on the adherend. Often these surfaces were "dirty" and showed a thin layer of contaminating elements on the adherend. This kind of failure probably should not be considered "adhesive" if proper bonding between the two surfaces never occurred.

Other types of failure surfaces (in addition to lap shear specimens) have been analyzed by instrumental methods. A method which can assist in evaluation of initial bond and durability is the peel test (illustrated being performed under H_2O in Figure 5). In one such test using a glass fiber reinforced adhesive and FPL etched 2024 adherend, the failure mode was cohesive and peel strengths were high when the test took place in air. Both adherends appeared to retain an equal thickness of adhesive. Scanning electron micrographs at two magnifications are seen of this cohesive failure surface in Figure 6. The usefulness of SEM or other microscopic technique is shown here where although cohesive failure in the adhesive has occurred, it appears likely that supporting fiber-adhesive failure was adhesive since the fibers appear very clean and the adhesive is smooth at the points where the fibers have pulled out. Of course, as has been emphasized, the appearance of adhesive failure does not necessarily mean that the failure was truly adhesive. A micro-analysis method such as the scanning Auger microprobe (SAM) or scanning ion scattering with imaging would be necessary to conclusively prove the mode of failure.

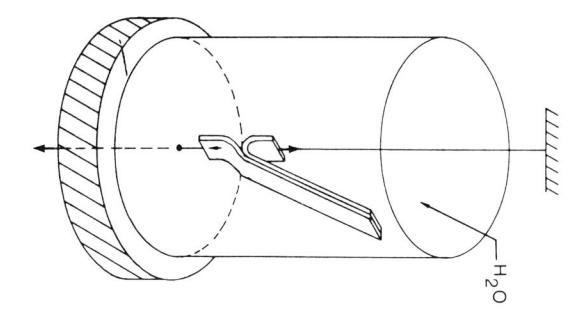

Fig. 5. Peel Test Configuration for Determination of Peel Strength Under Water.

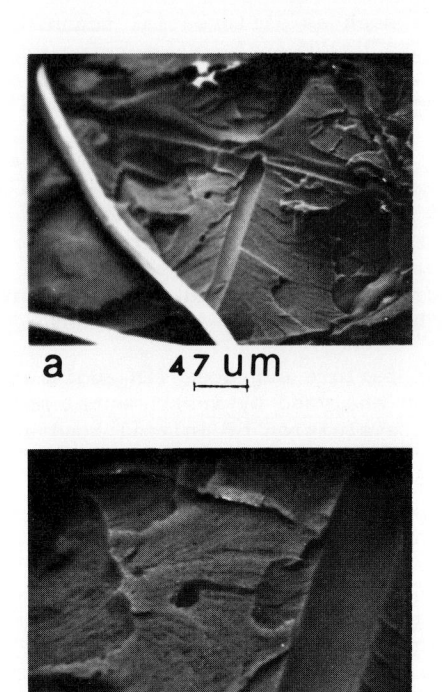

Fig. 6. Cohesive Failure Surface of Glass Fiber Supported Adhesive.

When the sample was tested in H_2O in the configuration shown in Figure 5, failure appeared adhesive and peel strength was very low. Scanning electron micrographs of the adherend showed that although the adherend surface is fairly clean, some clumps of adhesive remain. A similar distribution of torn areas exposing the fiber reinforcement was seen on the adhesive surface. Ion scattering spectra from the adherend side of this failure are shown in Figure 7. Spectrum 1 was recorded just after placing the beam on the sample and shows carbon, oxygen, fluorine, aluminum and strontium on the surface. Spectrum 2 was taken ten minutes later and shows a sharpening of the elements present in the adherend. The dashed curve shows scattering using neon recorded to identify strontium. The resolution in the region of strontium using helium is poor, allowing only an approximate identification. Strontium chromate corrosion inhibitor is used in the primer. The appearance of strontium on the adherend along with elements of the adherend indicates mixed failure at the adherend-primer interface

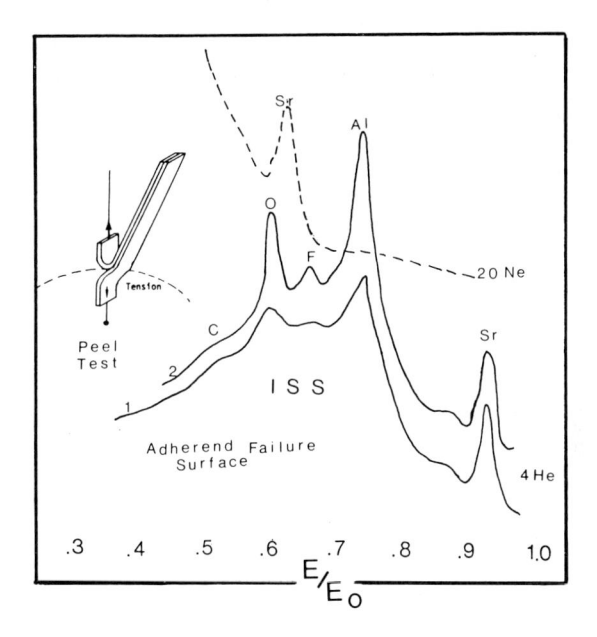

Fig. 7. ISS Data From 2024 Aluminum Alloy from Peel Test
Spectrum 1: Sputtering Begun
Spectrum 2: Ten Minutes of Sputtering

and in the primer or at the primer-adhesive interface. SIMS data in Figure 8 show elements of both the primer and adherend in agreement with ISS. In addition to the elements shown earlier in the ISS data, the mass spectra separates chromium and copper. Note the very low concentrations of both sodium and potassium in this failure surface. ISS and SIMS data from the adhesive side of the joint are shown in Figure 9. Scattering curves from both helium isotopes at 2500V are included to show the enhanced response of carbon to 3He compared to 4He. Note that aluminum is seen on the adhesive side.

Failure surfaces from thin adherend double cantilever beam (wedge test) have proved to be interesting illustrations of the ISS-SIMS technique. The double cantilever beam (DCB) method provides information about adherend surface preparation. This configuration is sensitive to different surface preparation treatments and can discriminate between bonding processes that give good and poor service performance.[8] The DCB spectrum consists of two thin adherends with a wedge driven into the bondline as depicted in the

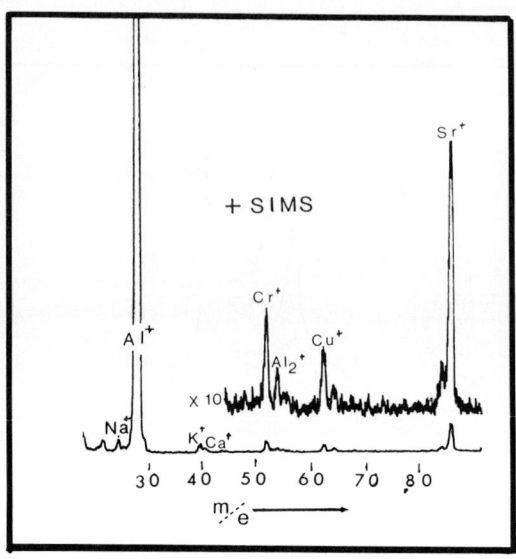

Fig. 8. SIMS Data From Peel Test Specimen Corresponding to ISS Data in Figure 7.

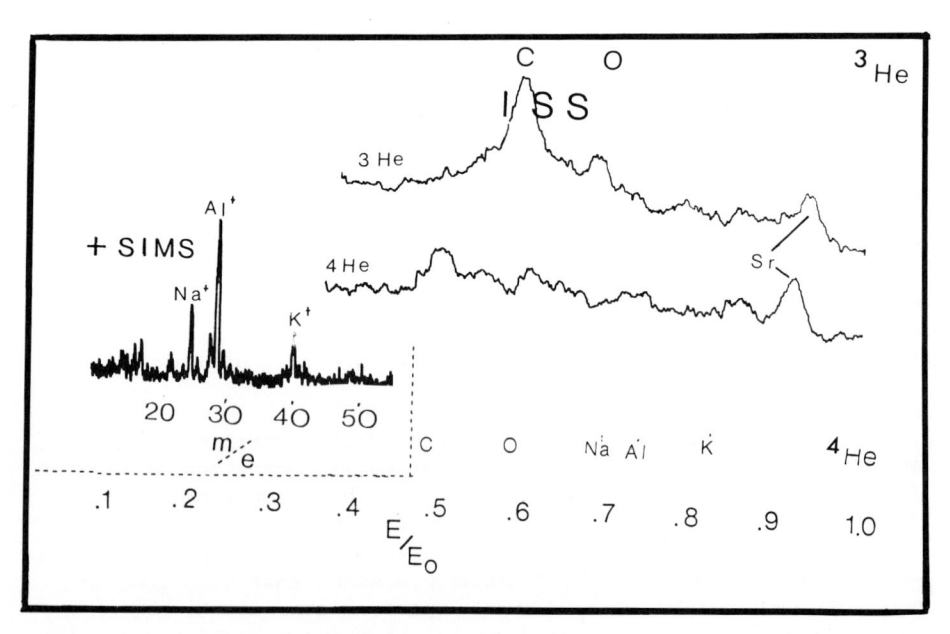

Fig. 9. ISS and SIMS Data From Adhesive Side of Peel Test Specimen.

drawing labeled DCB in Figure 10. The position of the crack leading edge is determined microscopically and then the DCB is subjected to various external stimuli such as changes in temperature and relative humidity. The propagation of the crack tip is followed with time. Sometimes when the wedge is driven into the bondline separation of the specimen occurs over a portion of the bondline as in the pictures shown for two typical DCB's in Figure 10. Here the wedge was driven in as shown first causing the cohesive failure in the adhesive at the left, then apparent adhesive failure between a and c during testing at 160°F and 95% R.H., followed at the right again by cohesive failure when the specimen is opened following the test. The areas abc ("clean adherend") and def ("clean adhesive") are matching regions from DCB's which failed. The adherend shows no indication of adhesive either visually or in the SEM, although there are slight reflectivity differences

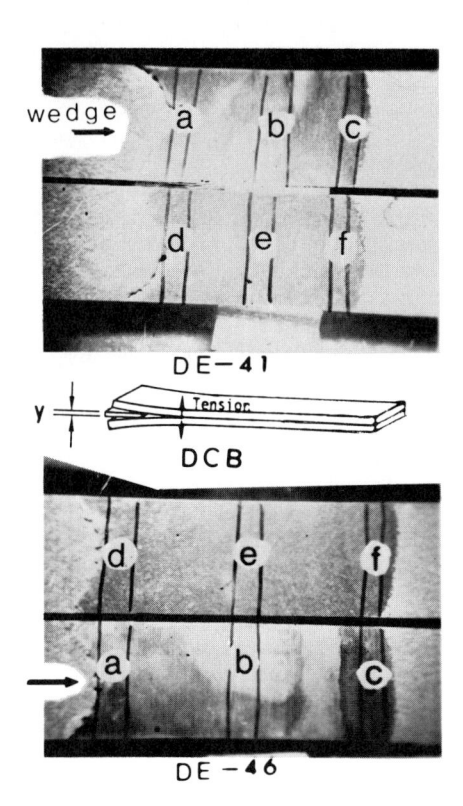

Fig. 10. Double Cantilever Beam (DCB) Test Specimens and Drawing of DCB Configuration.

across the surface. Perhaps staining techniques such as described by Brett[3] would be effective in outlining areas which contained thin films of adhesive when they exist. However, with extremely thin films it is doubtful if staining would provide any information concerning the locus of failure where a primer is used and therefore several interfaces exist. ISS and SIMS data for the DE-46 "clean" adherend at positions a, b, and c are shown in Figure 11. These spectra indicate substantially adhesive failure at the oxide-primer interface at A and B. The increase in chromium content at C may indicate some cohesive failure in the oxide since the original adherend was prepared in a standard FPL (Forest Products Laboratories, Madison, Wisconsin) etch which contains chromate ion. The matching "clean" adhesive side gave the ISS-SIMS data shown in Figure 12. Spectra at D and E appear to be contaminated primer or a mixed primer-adhesive zone, while F looks like it comes from a contaminated film of oxide which has adhered to the adhesive. The appearance of Al even at position D seems to indicate some cohesive failure in the oxide or in the oxide-metal interface all the way across the "clean" region of the specimen. ISS-SIMS data from the DE-41 "clean" adhesive side in Figure 13 indicates even more convincingly that the failure was similar to DE-46 with failure near the end of the crack occurring in the oxide or at the oxide-metal interface. Analysis of several such fractures after exposure to high humidity and elevated temperature indicates early crack propagation at the primer-oxide interface with large amounts of impurity ions (Na+, K+) present in this region. With continuing time the crack appears to progress either cohesively in the oxide or adhesively at the oxide-metal interface. An important factor is that all surface treatments do not behave in this way, and therefore the DCB is a useful indicator of the durability of a particular surface treatment for aluminum.

CONCLUSIONS

The combination ISS/SIMS method used along with other techniques such as SEM provides a powerful tool for elemental analysis of surface composition. These results, as well as earlier work in this laboratory[6], indicate that the surface composition can be significantly different from the bulk due to contamination, selective chemical etching and segregation. These same techniques also provide an analysis of the mode of failure in adhesive joints. Many failures classified as "adhesive" on the basis of visual inspection are frequently mixed mode failures or failures at a new interface containing elements of both adhesives and adherend.

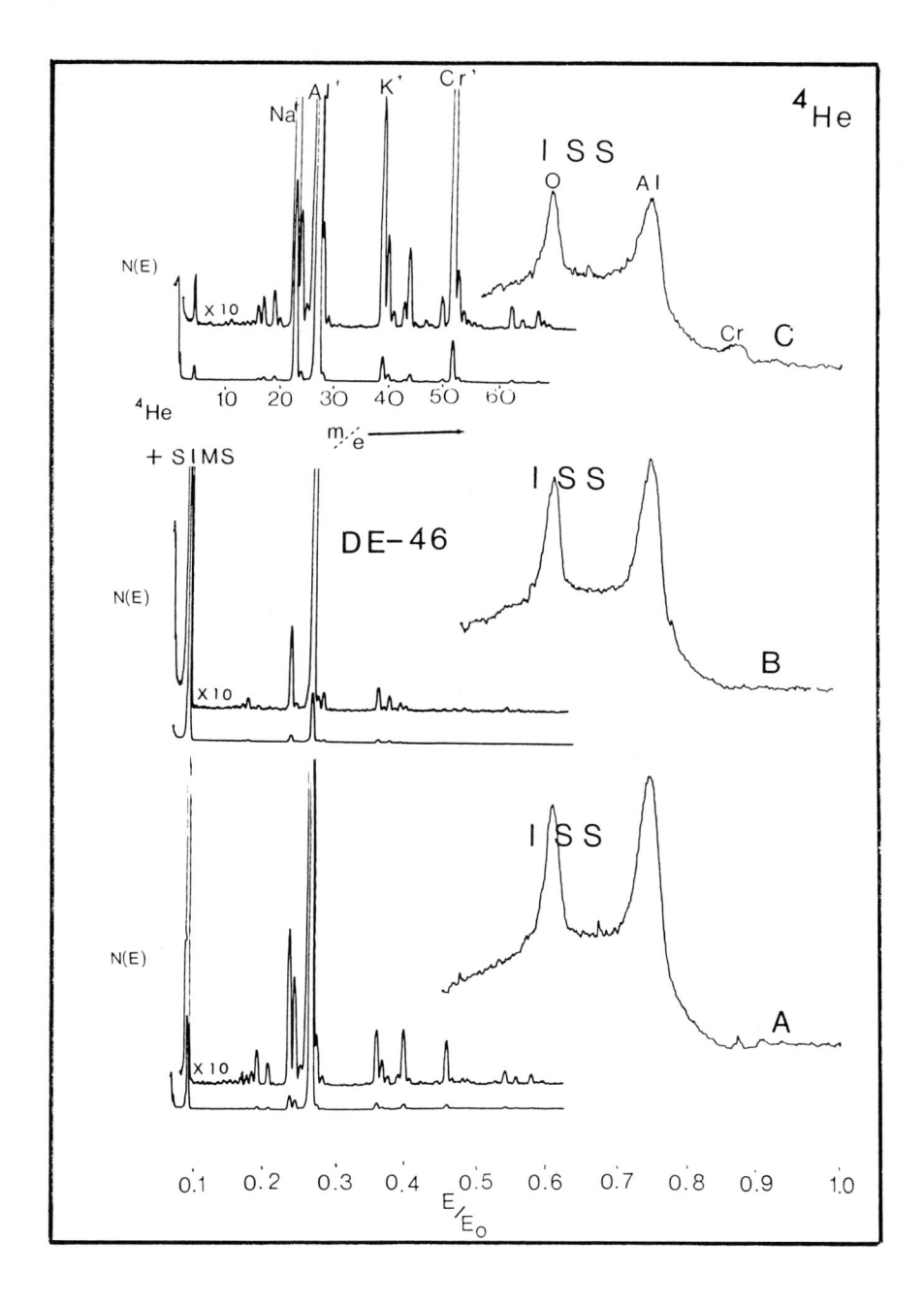

Fig. 11. ISS and SIMS Data From Adherend Side of DCB Specimen De-46.

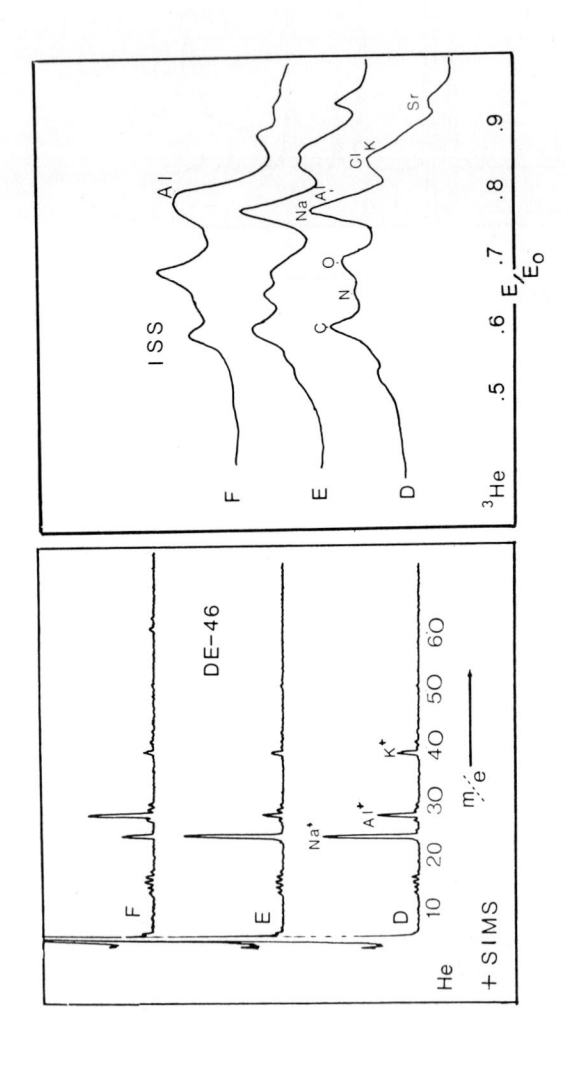

Fig. 12. ISS and SIMS Data From Adhesive Side of DCB Test of Specimen.

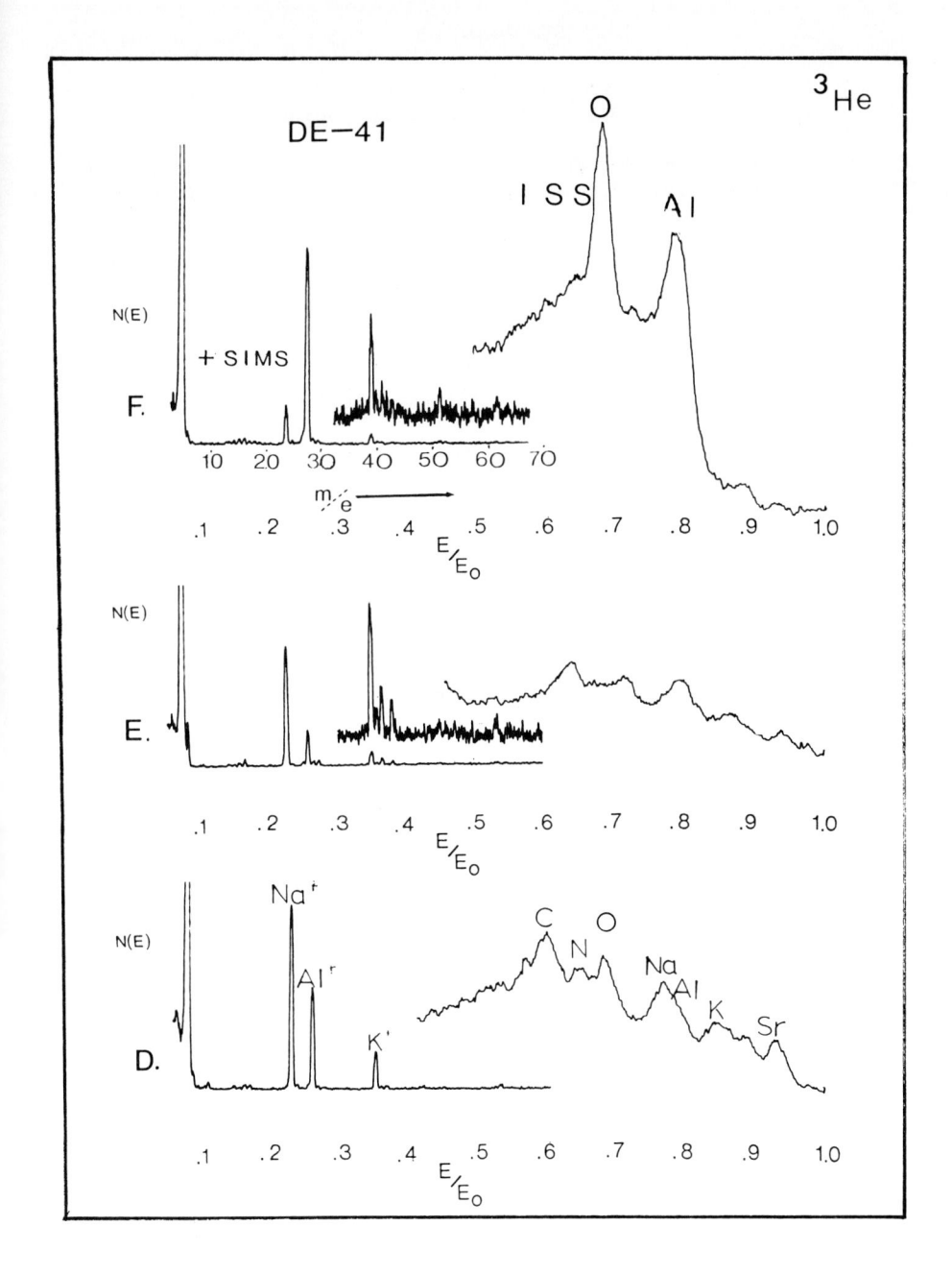

Fig. 13. ISS and SIMS Data From Adhesive Side of DCB Specimen DE-41.

REFERENCES

1. J. Bikerman in "*Recent Advances in Adhesion*", p. 351-356, Lieng-Huang Lee, Ed., Gordon and Breach Science Publishers, New York, 1973.
2. R.J. Good in "*Recent Advances in Adhesion*", p. 357-377, Lieng-Huang Lee, Ed., Gordon and Breach Science Publishers, New York, 1973.
3. C.L. Brett, J. Appl. Poly. Sci. **18**, 315 (1974).
4. R.F. Goff, J. Vac. Sci. Technol. **10**, 355 (1973).
5. A. Benninghoven, Surf. Sci. **28**, 541 (1971).
6. N.T. McDevitt, W.L. Baun, and J.S. Solomon, AFML TR 75-122, October 1975, Available from National Technical Information Service (NTIS).
7. H.E. Bair, S. Matsuoka, R.G. Vadimsky, and T.T. Wang, J. Adhesion **3**, 89 (1971).
8. J.A. Marceau and W. Scardino, AFML TR 75-3, February 1975.

Discussion

On the Paper by G.H. Morrison

D.L. Allara (*Bell Laboratories*):

1. Have you had any experience with sputtering-depth profiling analyses (using SIMS, etc.) on organic materials?

2. If so, what special problems are threre?

G.H. Morrison (*Cornell University*): SIMS has been used to determine elemental distributions in biological samples with considerable success; however, I cannot recall offhand of depth profiling studies. The problems associated with the analysis of insulating materials include surface charging which can be minimized by using conducting metal films on grids deposited on the sample surface. Also, organic fragment ions will occur in the mass spectrum which may interfere in the determination of certain elements.

B. Siskind (*Argonne National Laboratory*): What effect does preferential sputtering of one of the components of multicomponent samples have upon the quantitative aspects of the technique? I am thinking specifically of the transition metal oxides, the surfaces of which are reduced to a lower oxidation state by inert gas ion bombardment. The composition of the surface layer is altered by the bombardment.

G.H. Morrison: Preferential sputtering of one of the components of multicomponent samples adversely affects the quantitative ability of the technique. Sputtering yields are influenced by both chemical and lattice effects determined by the sample. One way to overcome these effects is to bombard the sample with a reactive gas rather than an inert gas. Flooding the surface of the sample with oxygen during bombardment also helps. These methods produce a target with a chemically modified thin superficial layer which results in more uniform sputtering and higher practical ion yield.

E. Kay (*IBM Corp.*): You showed where the secondary ion yield comes from during the analysis. Where does the total sputtering yield including neutrals (which often represent the majority species) come from? Is it an independent measurement?

G.H. Morrison: The total sputtering yield for a given element is determined prior to or after an analysis by measuring the amount of material removed for a given set of bombardment conditions. Using a Talleystep or an interference microscope, one measures the size of the crater produced to determine the amount of sample removed.

D. M. Hercules (*University of Georgia*): I have always wondered why in SIMS an electron gun source is not used to ionize the neutrals sputtered from the surface. This would certainly improve the ion yield. Do you know why this is not done?

G.H. Morrison: The use of an electron gun source in SIMS to ionize the sputtered neutrals has been tried by a number of investigators; however, the practical effect of enhancing the ion yield has not been impressive.

L.H. Lee (*Xerox Corp.*): We would like to thank Dr. James Roth for the presentation of this paper. We are happy to hear that Prof. Morrison has recovered from his recent discomfort. This paper has demonstrated a new and powerful tool for the probe of surfaces. Professor Morrison's excellent work should be congratulated. (The Editor also appreciates Dr. Morrison in answering the above questions after the Symposium).

On the Paper by I.M. Stewart

Participant: With an intense ion beam isn't the rate of erosion very rapid and if so doesn't this destroy depth resolution?

I.M. Stewart (*W.C. McCrone Associates*): The answer to both questions is yes, if the focussed beam is held stationary at one spot. Generally, however, those problems in which depth resolution is required involve comparatively large areas and are not of the locallized spot variety. In these cases, as described in the paper, the energy can be dissipated over a larger area either by raster scanning or by defocussing the beam, resulting in a lower energy input per unit area, thus lower erosion rates and hence better depth resolution.

On the Paper by W.L. Baun

W. Newby (*Aluminum Company of Canada*): The structure of aluminum can be very complex. Many alloying elements are used, Fe, Mn, Mg, Si, etc. and these occur both as solid solution and as particles which are themselves solid solutions. These particles are differently affected by various surface treatments. In addition manufacturing techniques determine size and distribution of these particles.

PART V

Photoelectron and Electron Tunneling Spectroscopy

Introductory Remarks:

Ruth Rogan Benerito

USDA Southern
Regional Research Center
1100 R.D. Lee Blvd.
New Orleans, Louisiana 70179

Recently, both chemists and physicists are contributing to our understanding of heterogeneous chemical reactions. These efforts to increase our understanding of solid surfaces of polymers, heterogeneous catalysis and of what is happening on natural fibers and living membranes are due, in part, to the availability of new research techniques and instruments. Electron emission spectroscopy for chemical analyses (ESCA) of surfaces has been applied successfully for the characterization of many surfaces. The techniques of ESCA are complementary to those of other sophisticated methods, such as those involved in Mössbauer and Auger Spectroscopy that were discussed in our earlier sessions of this symposium. It is very probable that the basic data supplied by ESCA spectra will contribute significantly to a better insight and initiate ideas about chemical changes at interfaces. Such fundamental knowledge should ultimately make possible the syntheses of tailor made catalysts, the production of clean solid surfaces, and the coating of natural fibers, synthetic polymers and metals with specific surface films required in a variety of industrial and biomedical uses.

It is timely that our symposium today should include presentations of current work in the application of ESCA towards the advancement of our understanding of heterogeneous reactions on solids.

(Editorial Note: Several papers presented by Andrade, Millard and Soignet to this Session on ESCA of polymers will appear in Part One of the second volume of the Proceedings. Instead, the paper on photoemission by Rhodin and Brucker originally presented to Session One is included in this Part.)

Plenary Lecture: Surface Characterization Using Electron Spectroscopy (ESCA)

David M. Hercules

*Department of Chemistry
University of Georgia
Athens, Georgia 30602*

    Electron spectroscopy (ESCA) and its use for surface characterization are reviewed. The fundamentals and instrumentation are briefly discussed, particularly those aspects relating to surface analysis. Three examples are given in the application of ESCA to diverse areas. First, a study has been reported of adhesion of polymer layers to a metal. When the polymer is removed, very thin (ca. 20-80A) layers of polymer remain which can be detected by ESCA. Second, ESCA has been used to study fluoridation of dental enamel. Products of fluoridation differ when NaF and SnF_2 are used as fluoridating agents. Of particular interest is the high concentration of Sn in the outer layer (<0.2μ) when SnF_2 is used. Third, cobalt molybdate and nickel tungstate hydrodesulfurization catalysts have been studied in a comparative fashion. Changes on the catalyst surface differ for the two during hydrogenation. Sulfiding of the catalyst surface produces an intermediate species between WO_3 and WS_2. The nature of the oxide and sulfide species for nickel has been established.

INTRODUCTION

    X-ray photoelectron spectroscopy (ESCA) and Auger spectroscopy currently are making a major impact on chemistry, particularly on the analytical chemistry of surfaces. These techniques offer unique advantages for the examination of surfaces and are only beginning to be exploited to study the fundamental chemistry occuring at interfaces. Therefore, it is very appropriate that electron spectroscopy be included in a symposium on the advances in the characterization of metal and polymer surfaces. For the interested reader, a number of review articles are available including a survey of recent literature and the proceedings of the most recent international conference on electron spectroscopy.[1-4]

Generally, ESCA and Auger spectroscopy are considered as separate types of spectroscopy; however I will treat them here as one. The reason they have developed separately comes primarily from development of instrumentation, rather than from fundamental differences. ESCA electrons come from primary photoionization events and Auger electrons come from secondary electron emission, but the information contained in both is very similar. I would like to divide the present paper into three sections. First, I would like to give an introduction to the general concepts of electron spectroscopy and instrumentation; second, to look at the appropriateness of these techniques for surface analysis; and third, to take three examples from our laboratory where ESCA has been applied to different surface problems. The third section will include studies on bond failure of polymers bound to metals, the fluoridation of dental enamel and the surface chemistry of hydrodesulfurization catalysts.

PRIMARY PROCESSES AND INSTRUMENTATION

The fundamental physical processes appropriate to ESCA are summarized in Table 1. There are both primary processes and secondary processes. The primary process, or atomic ionization event, can be brought about either by photoionization or by electron ionization. Part A shows photoionization where an atom A interacts with a photon (x-ray) to produce a discrete energy electron and an excited ion. This discrete energy electron is the photoelectron which is measured by ESCA. The kinetic energy of the photoelectron is essentially the difference between the energy of the incident photon and the binding energy of the electron in the atom. Thus, we can see ESCA is capable of measuring the binding energies of atomic electrons. In B, one has electron ionization where an atom A interacts with an electron e_o to produce two electrons which however do not have discrete energy. However, the excited ion produced is identical to that produced in A.

In one secondary process, the excited ion A^{+*} can relax to produce an x-ray photon as shown in Table 1, part C. This is the fundamental process for x-ray fluorescence. A process competitive with C is D. Here, the atom relaxes to form a doubly charged ion by the emission of a second electron which is known as an Auger electron. Note that the energy of the Auger electron does not depend on photon kinetic energy. It is important to note that either photoionization or electron ionization can produce Auger electrons whereas ESCA electrons can be produced only by photoionization.

Figure 1 shows the relationship between ESCA, x-ray fluorescence and Auger electron emission. On the left hand side of

Table 1

Fundamental Physical Processes

Primary Processes:

    A. Photoionization

$$A + h\nu \longrightarrow A^{+*} + e^- \quad \text{(discrete energy - ESCA)}$$

$$E(\text{kinetic}) = E(\text{photon}) - E(\text{binding})$$

    B. Electron Ionization

$$A + e_o^- \longrightarrow A^{+*} + 2e^- \quad \text{(not discrete energy)}$$

Secondary Processes:

    C. Photon Emission

$$A^{+*} \longrightarrow A^+ + h\nu' \quad \text{(x-ray)}$$

    D. Auger Electron Emission

$$A^{+*} \longrightarrow A^{++} + e^- \quad \text{(discrete energy - AES)}$$

E(Auger) does not depend on photon energy

A = atom A^{+*} = excited ion

A^+, A^{++} = ground-state ions

$h\nu$ = x-ray photon

Figure 1, the x-ray photon ejects a 1s electron to produce the ESCA electron. The middle diagram shows x-ray emission arising by a higher energy electron (2p) dropping to fill the vacancy in the 1s shell. The diagram on the right shows Auger electron emission; a 2s electron drops to fill the 1s hole at the same time ejecting a 2p electron. Again, note that x-ray fluorescence and Auger electron emission are secondary processes whereas ESCA involves a primary process.

    The characteristics of ESCA that make it a valuable technique for the examination of surfaces are summarized in Table 2. First, signals can be obtained from all elements except hydrogen and the intensities from all elements are approximately the same order of

402   David M. Hercules

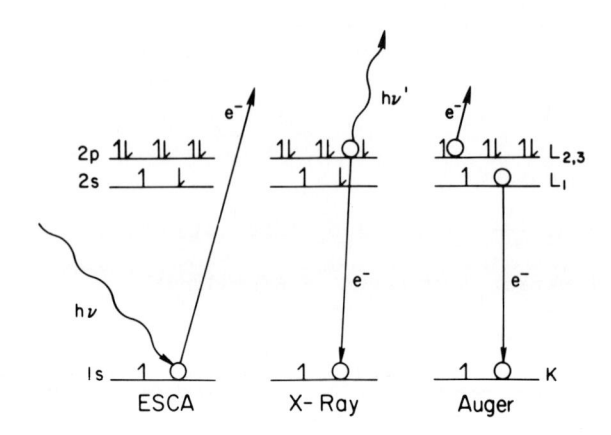

Figure 1. Diagram of ESCA, X-ray and Auger Processes.

Table 2

Experimental Characteristics of ESCA and AES

1. Signals from All Elements Except H

  intensity: same order of magnitude for all elements

2. Lines from Adjacent Elements Widely Separated

  interference: none from adjacent elements
  example: 1s electrons for C, N and Si:
   C(1s)   285 eV   N(1s)   400 eV
   Si(1s)   1840 eV
  Auger lines for C, N and Si:
   Carbon   260 eV   Nitrogen   370 eV
   Silicon 1575 eV
  resolution: 0.1 eV

3. Chemical Shifts are Observed

  correlate with oxidation state
  correlate with atomic charge
  related to fundamental groups in organic molecules
   N(1s) of amines   399 eV   N(1s) of nitro   405 eV

4. Surface Techniques

  sampling depth is ca. 20Å
  signal is from first few atomic layers

5. Techniques are Quantitative

  relative concentrations of elements

magnitude. This means that ESCA can investigate virtually all elements of the periodic table, at approximately the same level of sensitivity.

Second, lines from adjacent elements are widely separated. Therefore, one does not see systematic interferences observed in some forms of spectroscopy. For example, the 1s photoelectrons for carbon, nitrogen and silicon are at 285, 400, and 1840 eV, respectively. Similarly, the KLL Auger lines for these elements are at 260, 370 and 1575 eV, respectively. Since it is possible to resolve electron binding energies to ±0.1 eV, the separation between adjacent elements is quite adequate.

Third, chemical shifts are observed. The chemical shifts in ESCA and Auger lines can be correlated with oxidation state, or more precisely with atomic charge. Therefore, these techniques can be used to obtain information about the chemical binding state of an element. Chemical shifts are also related to functional groups in organic molecules. For example, the nitrogen 1s electrons of amines occur at 399 eV whereas the N 1s electrons of nitro groups occur at 405 eV.

Fourth, ESCA and Auger spectroscopy are surface techniques. The sampling depth of both techniques is approximately 20 Å, the exact number being dependent upon the nature of the material examined. This means that the signal is derived from the first few atomic layers. Experimental techniques are available which can enhance signals from the first atomic layer to an even greater extent.

Fifth, the techniques are quantitative. ESCA and Auger spectroscopy are capable of giving relative concentrations of elements or relative concentrations of different oxidation states of the same elements.

Figure 2 shows a block diagram of an electron spectrometer. In concept, it is relatively simple, consisting of an ionizing source, a sample, an electron energy analyzer, a detector, and some form of readout device. In Figure 2, it is shown that either an electron gun or an x-ray source can be used for ionization. The electron is ejected from the sample and passes into the electron energy analyzer where the kinetic energy of the photoelectron is sorted and detected. A variety of readout devices is available. In practice, most ESCA instruments are computer controlled because of the complexity of the scanning system and the necessity for multiple scans in order to enhance sensitivity.

Electron spectrometers operate under a vacuum of at least 10^{-6} torr. The reason for this is that at higher pressures, the photoelectrons would be scattered going through the analyzer to the de-

404   David M. Hercules

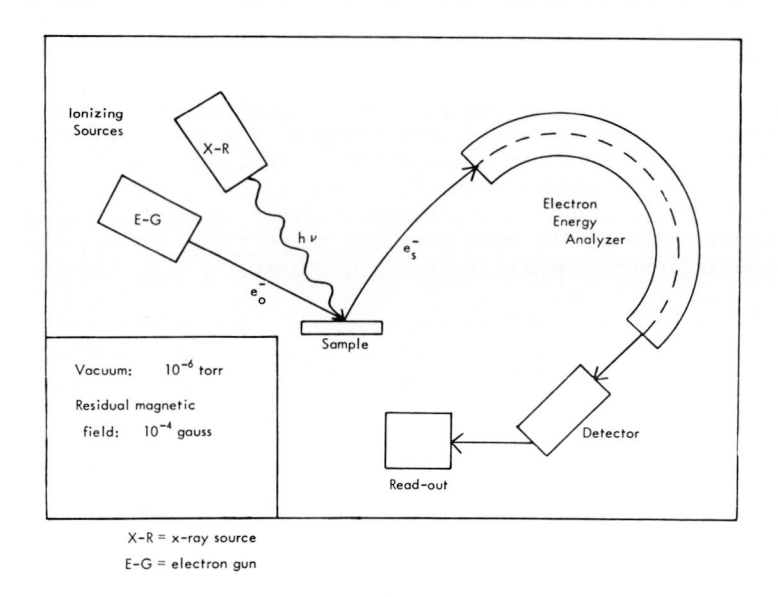

Fig. 2. Block Diagram of Electron Spectrometer.

tector. Although this necessity for high vacuum is a problem for some applications, it is generally not a difficulty for the examination of solid surfaces. The residual magnetic field in the region of the sample and electron energy analyzer must be below 10^{-4} Gauss. Reduction of the earth's magnetic field (ca. 0.5 Gauss) to this level is usually accomplished in electrostatic analyzers by magnetic shielding material. The reason for reduction of the magnetic field is that one measures low energy electrons (ca. 1 KeV) and perturbations by the earth's magnetic field become significant.

Often in the examination of surfaces, it is desirable to chemically treat a surface so it can be examined in the spectrometer and not be exposed to the atmosphere. This is usually accomplished with a reaction chamber attached to the electron spectrometer as shown in Figure 3. The spectrometer chamber is labeled SV, the electron energy analyzer EA, and the x-ray source X. To the left of the spectrometer chamber is the reaction chamber separated by an isolation value IV. In the top figure, the sample probe is withdrawn into the reaction chamber and the isolation valve closed. Here, the sample can be treated by introducing a reactant through the gas inlet system, GI. The reaction chamber has its own vacuum system (RCV) and thus noxious or corrosive gases

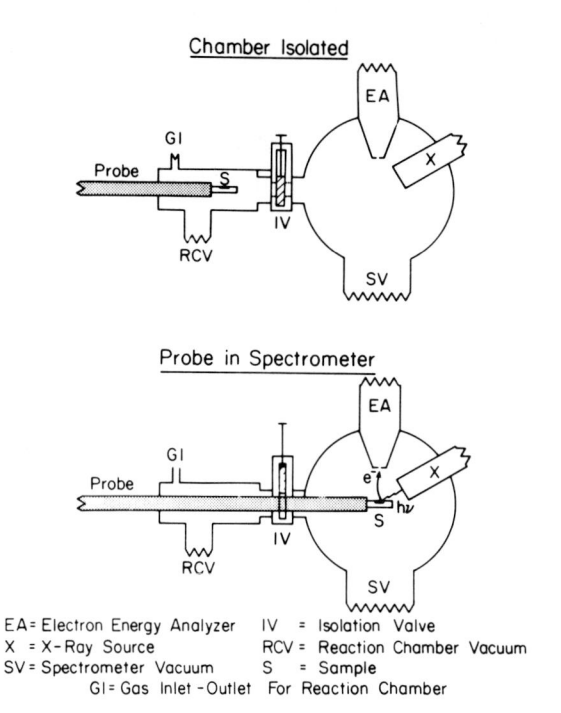

Fig. 3. Diagram of Reaction Chamber Used with an Electron Spectrometer.

can be used without causing problems to the pumping system of the spectrometer. Once the sample is treated, the isolation valve is opened and the probe inserted into the spectrometer as shown in the lower part of Figure 3. Here, the sample is being irradiated by x-rays and the electrons analyzed by the spectrometer. This type of reaction chamber is extremely valuable and is widely used by investigators working on surface characterization.

Figure 4 shows a typical ESCA spectrum both in broad scan and under high resolution. The top spectrum is a broad scan between 1500 and 300 electron volts of a sample of sodium azide, NaN_3. Note as labeled the sodium 1s, 2s and 2p lines, the nitrogen 1s lines, the sodium Auger lines and the ever present carbon 1s line. The carbon 1s line arises from contamination carbon found on surfaces that have not been scrupulously cleaned. The lower part of Figure 4 shows the region at approximately 1080 eV under high resolution; this is the region of the nitrogen 1s line. You will note that the nitrogen 1s is resolved into a doublet

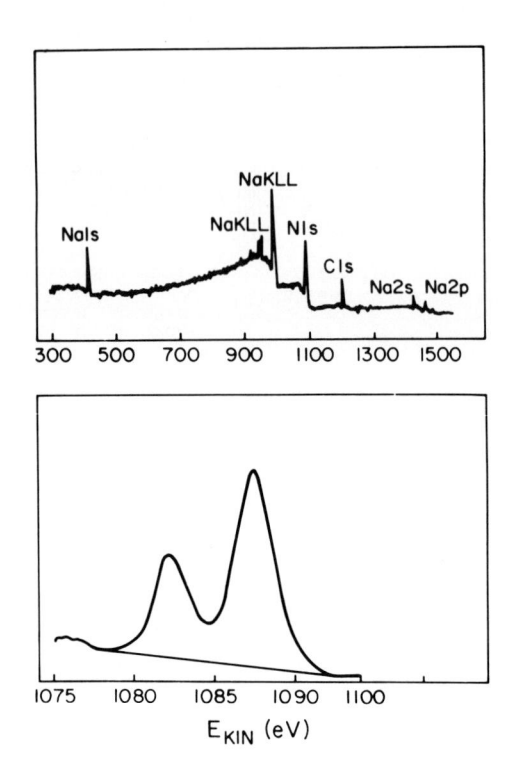

Fig. 4. ESCA Spectra of Sodium Azide (NaN$_3$).
Top: Broad-scan spectrum
Bottom: High resolution scan of N1s region

having a relative intensity of 2:1. These nitrogen lines originate from the two chemically different nitrogens in the linear azide ion; the center nitrogen being different from the two end nitrogens. Such spectra show very clearly the capabilities of ESCA for qualitative analysis over a broad range of elements as well as detailed examination of the chemical nature of the species for a particular element.

USE OF ESCA FOR SURFACE ANALYSIS

The information about surfaces which can be acquired using ESCA is shown in Table 3. First, ESCA can provide an elemental analysis of the elements on the surface for all elements except hydrogen. Second, the sensitivity of ESCA is between 0.1% and 1% of a monolayer depending upon the element; in general 1% of a monolayer for light elements, 0.1% for heavy elements. Third, ESCA is

Table 3

Information About Surfaces Derived from ESCA

1. Elemental Analysis: all elements except H

2. Sensitivity: 0.1% - 1% of a monolayer

3. Nondestructive (relatively)

4. Different Excitation Modes Give Different Information

    a. X-rays
    b. UV-photons
    c. Electrons
    d. Synchrotron

5. Surface Species Studied

    a. Detection by chemical shifts
    b. Bonding to surfaces
    c. Relative concentration

6. Depth Profiling (combined with Ar-ion etching)

---

non-destructive compared to techniques such as secondary ion mass spectrometry (SIMS). Fourth, different excitation modes are available; using different excitation modes can provide different types of information. One can use x-rays, high energy UV photons, electrons or even synchrotron radiation. The latter shows particular promise. Fifth, surface species can be studied in a number of ways; they can be detected by their chemical shifts, the nature of the bonding to the surfaces can be established, the relative concentrations of species can be measured. Sixth, depth profiling can be accomplished by combining ESCA or Auger spectroscopy with argon ion bombardment.

The model for a surface as studied by ESCA is shown in Figure 5. Notice the three layers: a contamination layer, the surface layer sampled, and the bulk of the material. The contamination layer and the surface layer sampled combine to form the ca. 20Å depth ESCA samples. The contamination layer is usually composed of species produced by the atmosphere or handling and can be removed only be chemical treatment. In order to have a surface completely uncontaminated requires ultra high vacuum techniques, 10^{-10} torr or lower. Therefore, most surfaces examined by ESCA

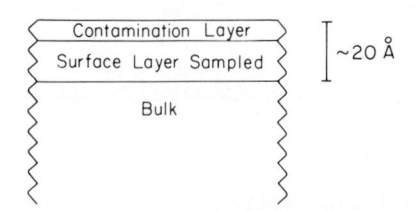

Fig. 5. Model for a Surface Studied by ESCA.

are a composite of the surface of the material and the contamination layer. If one uses oil-pumped vacuum systems the contamination layer will consist of a thin carbon film probably a monolayer or less. At times, this film can be valuable as a calibration standard.

Although one speaks of sampling the first 20Å of a surface, it must be realized that atoms closer to the surface give a higher proportion of an ESCA signal than atoms further down in the bulk. Figure 6 illustrates this. Consider atom A which lies in the second layer below the surface. The hatch marked areas indicate the path an electron emitted from A can take and not be scattered by atoms in the lattice. If, however, one considers an atom B in the fifth layer below the surface, the dotted lines show the

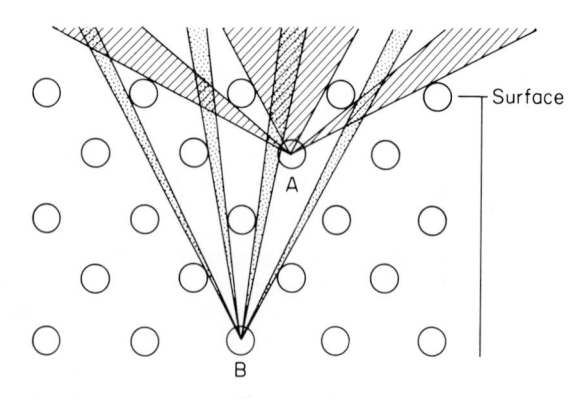

Fig. 6. Illustration of Contribution of Different Atomic Layers to ESCA Signals.
A - atom in second atomic layer; B - atom in fifth atomic layer

path an electron can take and not be scattered by atoms in the lattice. It is obvious that the integrated area in the plane of the paper for atom A is much greater than B indicating a higher probability of escape for an electron from A and therefore a higher contribution of A to the total ESCA signal. Thus, when one speaks of escape depth, one speaks only of an average escape depth and must recognize that a larger portion of the signal comes from the atoms near the surface.

The average escape depth of an electron varies with its kinetic energy. This is illustrated in Figure 7. This is a logarithmic plot of escape depth versus kinetic energy for very low kinetic energy electrons (5 eV) up to very high kinetic energy electrons (ca. 10^4 eV). Note that the escape depth reaches a minimum in the vicinity of 100 eV. Also note that the escape depth is proportional to the square root of kinetic energy from 100 eV to about 10^4 eV. Although this relationship is only approximate, it is a valuable rule of thumb. ESCA electrons generally fall in the range of 100-1500 eV; therefore one can change the effective sampling depth by measuring photoelectrons of widely different kinetic energies.

Another factor important for examination of surfaces by ESCA is sample charging. In order to retain neutrality of the surface, the electrons that are photoejected must be replaced. If sample conductivity is high, such as in metals, this is no problem. If, however, one is examining an insulator, a charge can build up on the surface. On the left hand side of Figure 8 one sees an electron escaping an uncharged surface. The only positive charge

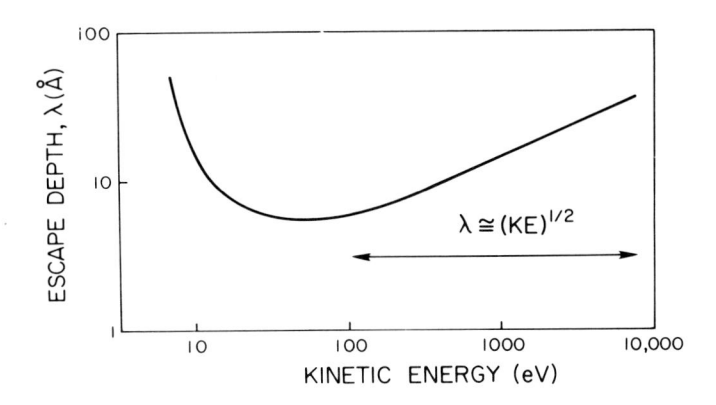

Fig. 7. Escape Depth as a Function of Electron Kinetic Energy.

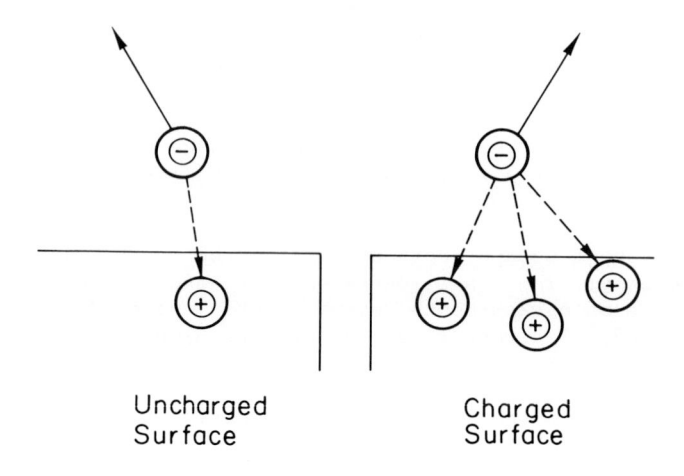

Fig. 8. Model for Effect of Surface Charging on Electron Kinetic Energy.

which the electron must overcome is that of the atom from which the electron has just left. If the surface is charged, the electron must not only overcome the attractive force of the ion it has left, but also of other ions in the vicinity, as shown on the right-hand side of Figure 8. Therefore, the kinetic energy of such an electron will be reduced. In practice, surface charge builds up rapidly to a steady state value and does not change, probably due to surface conductivity. However, this poses a problem in the accurate measurement of binding energies and therefore some sort of calibration becomes necessary. In general, deposition of materials on the surface has been used to monitor charge. Although this technique is widely accepted in electron spectroscopy, its validity has not been completely established.

It is possible to enhance the signal of the atoms from the first atomic layer by varying the angle between the surface and the spectrometer. This is shown in Figure 9. For the diagram on the left, one can see that the angle between the surface and the spectrometer is 65° whereas the angle for the sample on the right is 20°. If one considers an atom emitting at the bottom of the hatched layer in both cases it is clear that the pathlength through the material for θ = 20° is much longer than in the case of θ = 65°. Therefore, atoms further down in the surface will contribute less to the ESCA signal if low surface angles are used. This result is summarized in the equations at the bottom of Figure 9; note that the intensity of the bulk signal varies exponentially as the sine of the angle whereas the intensity of the surface layer varies reciprocally as the sine of the angle. If one considers the ratio

Bulk: $I = K \exp(-a/\sin\theta)$
Surface: $I = K/\sin\theta$

Relative Intensities

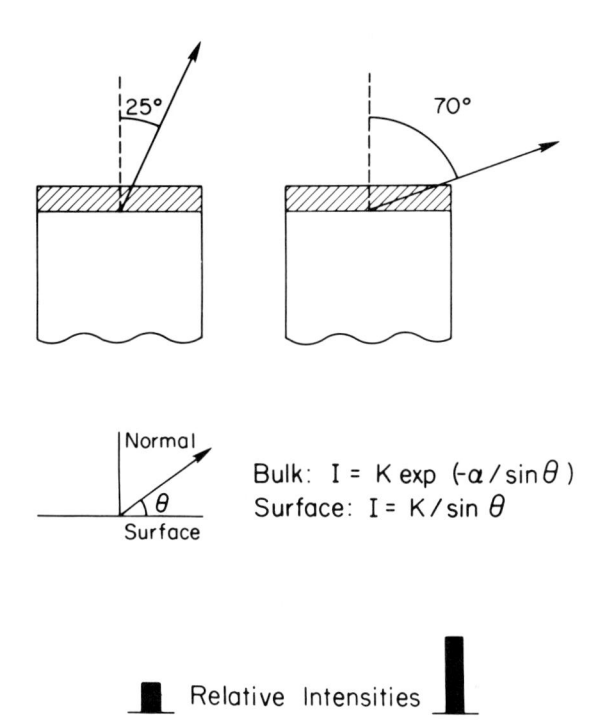

Fig. 9. Effect of Angle on ESCA Intensities Derived from Surface Layers.

of the intensity of the surface layer to the intensity of the bulk layer for the 65° and 20° cases the relative intensity ratios are shown by the black lines at the bottom of Figure 9. Simply stated the surface enhancement at 20° is about a factor of 3 over that for θ = 65°.

    Argon ion bombardment is an important technique both for cleaning surfaces for examination by ESCA and for depth profiling. Figure 10 shows a diagram of an argon ion gun. Electrons are boiled off the filament F and accelerated toward the plate P in the ionization chamber in which argon ions are produced. The argon ions are accelerated toward the grid G and focused into a beam by the deflection plates D. Ions passing through the grid are at high velocity and strike the sample. What this accomplishes is shown in Figure 11. An argon ion bombarding a surface strikes the surface and knocks loose a number of atoms at the point of impact.

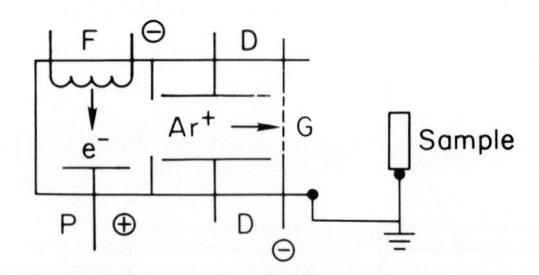

F = Filament   D = Deflection Electrodes
P = Plate   G = Grid

The reaction in the icnization chamber is:
$$e^- + Ar \longrightarrow Ar^+ + 2e^-$$

Fig. 10. Diagram of an Argon-Ion Gun.

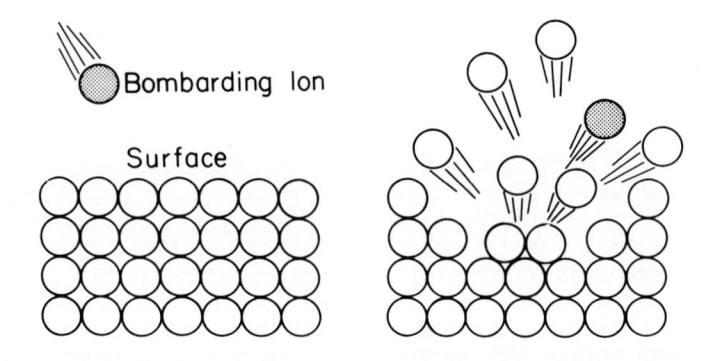

Fig. 11. Idealized Reaction of Argon Ions with Surface Atoms.

Therefore, by continuously bombarding the surface with argon ions one can cleave away a surface, hypothetically one atomic layer at a time. Although this is certainly idealized, it is possible to accomplish depth profiling using argon ions and is very valuable in establishing the distribution of species below the surface layer.

POLYMERIC BOND FAILURE ON ALUMINUM SURFACES

When one bonds a polymer to a metal such as aluminum and then removes the polymer, separation can occur at the polymer-metal interface or within either the polymer or the metal layers. As can be seen in Figure 12, if separation within one of the layers occurs (dotted lines) a thin layer of polymer will remain on the metal or vice-versa. Since it is likely that this layer is very thin[5,6] (of atomic dimensions) it is an ideal problem to be studied by ESCA. We have carried out such a study to demonstrate the utility of electron spectroscopy in determining the nature of bond failure at polymer-metal interfaces.[7] We have studied two polymers, NBS 8957 and NBS 8958. We used argon ion etching to establish the surface nature of the coating.

Initially, we examined the polymers by scanning electron microscopy to determine whether or not any gross tearing of the polymer could be observed. Typical photographs are shown in Figure 13. The photograph on the left with the stretch marks is an SEM of a piece of aluminum on which no polymer had been placed, the center is an SEM of the same piece of aluminum from which polymer had been peeled, and the right is an SEM of a bad cure in which clumps of polymer were known to be present. It is evident that no islands or clumps can be seen on the aluminum that has been stripped. Even at higher magnification (1000 x) no polymer clumps were seen. Therefore, any polymer remaining must be in particles smaller than the limits of resolution of the scanning electron microscope, or as a film.

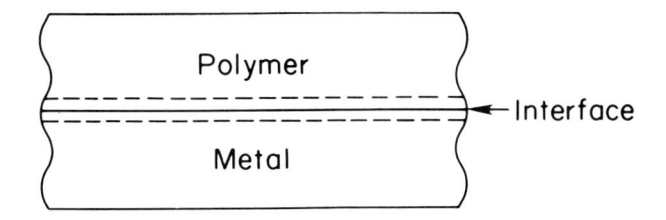

Fig. 12. Diagram of Polymer-Metal Interface.

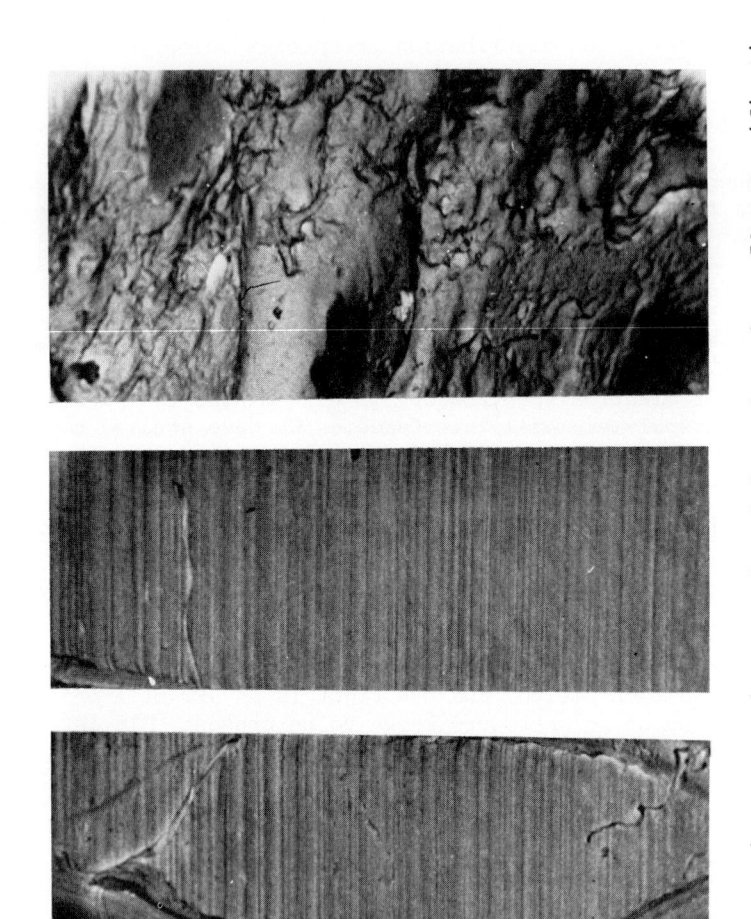

Fig. 13. Scanning Electron Microscope Pictures of Peeled Metal Surfaces 50 x Magnification; left: bare aluminum surface, center: polymer peeled from Al surface, right: polymer peeled-bad cure.

ESCA spectra of NBS 8958 revealed two peaks in the C 1s spectrum as shown at the top of Figure 14. The peak at the lower binding energy is the reference carbon while the one at the higher binding energy is the polymeric carbon. Figure 14B shows the carbon signal of the aluminum plate after the polymer had been removed. The similarity between these two spectra is striking, indicating that the same species exists on the aluminum surface after the polymer was removed as existed in the bulk polymer. Furthermore, aluminum signals were detected on the peeled surface indicating that the polymer layer remaining was quite thin.

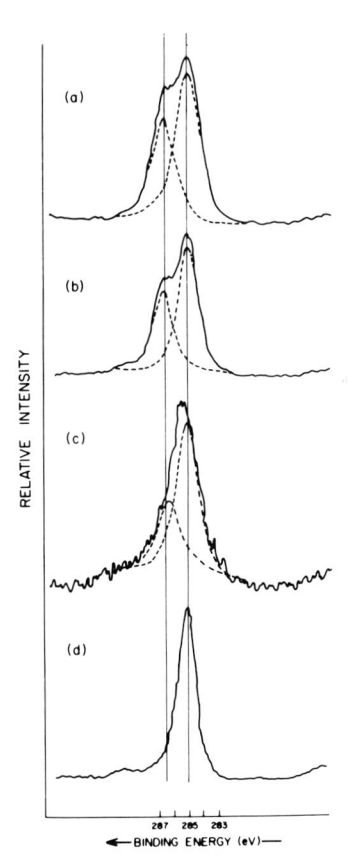

Fig. 14. C1s ESCA Spectra of Polymer 8958. a - doublet of the bulk polymer; b - doublet of the aluminum surface after stripping the polymer; c - doublet after in situ etching of the stripped aluminum surface; d - cleaned aluminum, never exposed to polymer.

Figure 14c shows the C1s spectrum of the aluminum surface after it had been bombarded with the argon ion gun for 15 minutes. Note that the polymeric carbon signal is reduced in intensity relative to the pump oil carbon after sputtering. This is further evidence that the species giving rise to the polymer carbon is on the surface. Figure 14d shows the carbon signal observed only from the clean aluminum itself indicating the presence only of pump oil carbon but not polymeric carbon.

More quantitative data obtained from the spectra in Figure 14 are summarized in Table 4. Again note that the binding energies and intensities for cases A and B are virtually identical and although the intensities are different for C again the binding energy of the polymeric species is virtually the same. This constitutes very strong evidence that cohesive failure has occurred within the polymer and that the layer on the surface is quite thin.

ESCA examination of NBS 8957 polymer indicated the presence of a silicon peak which was used to monitor its adhesion to the aluminum surface. Summary of data for 8957 are shown in Table 5. In the bulk polymer, the silicon 2s and 2p lines were observed; the aluminum line included is from aluminum oxide used as a reference. On the peeled surface note the similarities of binding energies between the three species. For C the sputtered surface, note again that the binding energies do not change significantly but that the intensity of the Si:Al lines is reduced. However, the ratio of the Si2p:Si2s line remains the same. Again, on the cleaned aluminum surface both the oxide and metal were seen but no silicon. This constitutes evidence that cohesive failure has occurred within polymer 8957 when it is bonded to an aluminum surface.

Table 4

C1s Doublet of Polymer 8958

		Compound Peak	Relative Height	Binding Energy (eV)
a.	Polymer	C1s	1.00	285.0
		C1s	0.67	286.6
b.	Peeled surface	C1s	1.00	285.0
		C1s	0.64	286.6
c.	Sputtered surface	C1s	1.00	285.0
		C1s	0.49	286.3
d.	Clean aluminum	C1s	1.00	285.0

Table 5

Reference Al2s Peak and Polymeric Silicon Peaks of Polymer 8957

		Compound Peak	Relative Height	Binding Energy (eV)
a.	Polymer			
		Al2s	----	121.0
		Si2p	----	103.0
		Si2s	----	154.2
b.	Peeled surface			
		Al2s	1.00	121.0
		Si2p	0.98	103.7
		Si2s	0.58	155.0
c.	Sputtered surface			
		Al2s	1.00	121.0
		Si2p	0.16	103.8
		Si2s	0.10	154.8
d.	Cleaned aluminum			
	Oxide	Al2s	1.00	121.0
	Metal	Al2s	0.37	118.7

As was mentioned previously, aluminum signals could be observed by ESCA on the peeled surfaces for both polymers. Thus, the thickness of the polymeric film can be approximated, if one is willing to assume that escape through the polymeric film is essentially through that of a carbon film.[8] Under the circumstances of our experiments, one could not obtain a quantitative measurement of the film thickness but one can at least produce a semi-quantitative estimate: the film must be between 20 and 80Å thick. One should note, however, this does not imply the presence of an even, thin film; once bond failure has occurred in the polymer it will not be restricted to an even tear along any one polymeric layer, thus it is not perfectly smooth but will be thicker in some places and thinner in others. ESCA approximates only the _average_ thickness.

Thus, we can conclude that when bond failure occurs between polymers and aluminum, failure is definitely cohesive rather than adhesive because some of the polymer remains on the aluminum surface. Examination of the removed polymer surface for aluminum revealed no aluminum peaks; therefore cohesive failure is entirely within the polymeric layer.

FLUORIDATION OF DENTAL ENAMEL

The topical application of fluorides to dental enamel results in the formation of protection coatings; calcium fluoride and fluoroapatite $Ca_{10}(PO_4)_6F_2$ for acid fluoride phosphate gel and $Sn_3F_3PO_4$ for stannous fluoride.[9] Generally, analyses of fluoride-enamel interactions have been performed on slabs on enamel several microns thick; thus it is impossible to distinguish the surface composition from the bulk. However, the application of ESCA combined with argon ion etching to this problem has been shown to be an effective way of studying such surface layers.[10,11] The rate of etching used was approximately 6 Å per minute and allowed us to establish depth-concentration profiles for fluoride treated enamels.

The technique used was to saw a slab of human enamel about 4 mm square and 1 mm thick and treat it either with a stannous fluoride or acid fluoride phosphate gel. ESCA spectra of the Cls, P2p, Fls and Ca2p lines were then monitored alternately with etching by the argon ion gun. This was performed in an alternate fashion in order to establish a depth profile.

The surface of untreated dental enamel was covered with an inorganic film thick enough to almost obscure any ESCA signals from the substrate. The carbon in this film showed a strong, broad peak which deconvoluted into a variety of different carbon compounds ranging from hydrocarbons to highly oxidized species. The thickness of this film must be ca. 40 to 80 Å assuming it to be uniform; however, it could be considerably thicker if there are holes allowing exposure of the enamel surface. The film was removed either by brief argon ion etching or by brushing which approximated normal oral hygiene. The inorganic substrate below the carbon film showed a Ca:P ratio of 1.4:1 which agrees well with the known ratio of 1.6:1 for enamel. Almost no carbon signal was observable from the bulk enamel. Enamel which had been treated with an acid fluoride phosphate gel was alternately analyzed and argon ion etched to produce the depth profile shown in Figure 15. Note that the calcium concentration is virtually unaffected by fluoridation at any depth. Phosphorus, however, is low at the surface and diminishes further toward the interior and then rises to its bulk ratio at ca. 0.4 microns. The concentration of fluorine at the surface was found consistently at ca. 0.4 microns. The concentration of fluorine at the surface was found consistently to be in a 2:1 ratio with the calcium clearly identifying calcium fluoride as the major product formed during acid fluoride phosphate treatment. Below the surface the fluorine concentration steadily decreases indicating a transition from calcium fluoride to an apatite structure, probably undergoing a transition from fluoroapatite.

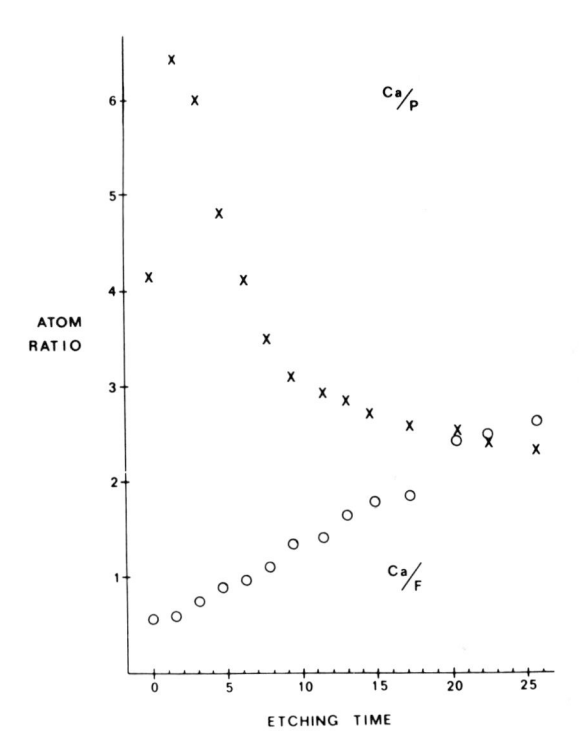

Fig. 15. Relative Concentrations of Fluorine, Calcium, and Phosphorus in Dental Enamel vs. Depth. Enamel has been treated with a commercial fluoride containing gel.

The behavior of phosphorus within the first 0.1 micron of the surface is interesting. The fact that a minimum value occurs at 500 Å and then increases reflects a steady transition from calcium fluoride to an apatite structure. The nature of the surface phosphorus could not be determined with certainty, however, it is probably a build up of non-calcium phosphate due to the presence of phosphate in the acid fluoride gel.

ESCA examination of enamel which has been treated with stannous fluoride showed the astounding result that the surface contained no fluorine and that it was primarily composed on a tin compound probably some form of tin oxide. Removal of a 0.15 micron thick layer by argon ion etching showed that the sub-surface layer contained a moderately high concentration of fluorine as well as tin.

The depth profile for SnF$_2$ treated enamel is shown in Figure 16. Note that the Ca:P ratio remains essentially constant as a function of depth whereas the Sn:F ratio decreases from ca. 12-15 at the surface to approximately 0.5 in the bulk. The elemental ratios at depths greater than 0.2 micron correspond to formation of the known fluoridation products, namely Sn$_3$F$_3$PO$_4$, CaF$_2$, hydroxy apatite, and fluoroapatite. However, the measured Sn:F ratio at the surface cannot be accounted for by any mixture of the known fluoridation products. Thus, ESCA is shown to be an ideal tool for examining depth profiles in fluoridated enamel, and even these preliminary studies have revealed interesting aspects of the first 0.2 micron in SnF$_2$ fluoridated enamel.

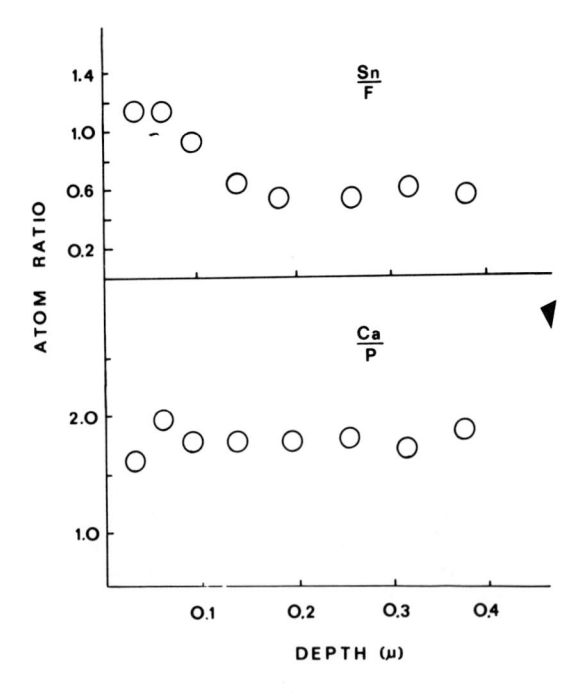

Fig. 16. Atomic Ratios as a Function of Etching Time. x=Ca/P; o = Ca/F. Etching rate ca. 300 Å/hr.

HYDRODESULFURIZATION CATALYSTS

Hydrodesulfurization catalysts (HDS) commonly consist of molybdenum or tungsten oxides supported on alumina or silica and promoted with small amounts of cobalt or nickel. These catalysts are important because of their action in hydrodesulfurization, hydrocracking and hydrogenation processes.[12] Table 6 summarizes the reaction schemes for desulfurization of organic sulfur compounds. Note the variety of sulfur compounds hydroreduced to H_2S and the corresponding hydrocarbon. There are many conflicting interpretations about desulfurization activity, although it is generally thought to derive from a reduced state of molybdenum or tungsten on the catalyst. We have carried out ESCA studies of nickel tungstate and cobalt molybdenum catalysts,[13,14] in an effort to compare their behavior, to determine whether or not the same mechanism is responsible for desulfurization in both types of catalysts. We have carried out various treatments such as reduction and sulfiding and have looked at the effect of these treatments both on the oxide as well as its promoter.

Table 6

Reaction Schemes for Desulfurization of Organic Sulfur Compounds

1. Thiols: $RCH_2CH_2SH + H_2 \longrightarrow RCH_2CH_3 + H_2S$

2. Thioethers:

    $RCH_2CH_2SCH_2CH_3R' + H_2 \longrightarrow RCH_2CH_3 + R'CH_2CH_3 + H_2S$

3. Heterocycles:

    (thiolane) $+ H_2 \longrightarrow RCH(CH_3)CH_2CH_3 + H_2S$

    (thiophene) $+ H_2 \longrightarrow CH_3CH_2CH_2CH_3 + H_2S$

    (benzothiophene) $+ H_2 \longrightarrow$ (ethylbenzene) $\longrightarrow$ (ethylcyclohexane)

In order to study HDS catalysts, a special sample probe and reaction chamber were designed as shown in Fig. 17. The probe has an inner shaft that is retractable and sealed at the end with a gold ring seal. This enables transfer of the sample from a reaction chamber into the spectrometer without exposure to air. The reaction chamber is the outside cylindrical tube shown in Figure 17 which has an inlet and outlet system for carrier gas. The entire assembly can be placed in a tube furnace used to bring the reaction chamber to the desired temperature. This technique is superior to using a reaction chamber attached to the spectrometer since several probe-chamber combinations can be used to prepare catalysts simultaneously thus speeding up the acquisition of data.

In studying HDS catalysts, a variety of chemical shifts were used to monitor the different species of tungsten, molybdenum, cobalt, and nickel on the catalyst surface. The important chemical shifts for HDS catalysts are summarized in Table 7, at the top for cobalt and molybdenum, at the bottom for nickel, tungsten and sulfur. These numbers were obtained from a large pool of data and are probably accurate to ±0.1 eV. Although chemical shifts can be measured accurately, deconvolution of spectral envelopes is necessary when several oxidation states of an element are present. Deconvolutions were carried out with a Du Pont 310 curve resolver and typical examples are shown for the molybdenum 3p and 3d spectra in Figure 18. The full width at half maximum separation of the doublets was assumed the same as for the Mo and W oxides for deconvolution. Such deconvolution allows one to establish the relative concentrations of species accurate to ±5%, when relative intensities are reported and ±8% for absolute intensity values.

The effects of calcining on both the molybdenum and tungsten spectra were very similar, namely, that the binding energies reported for the calcined catalysts do not match those of the free oxides. These measurements along with others have been shown consistent with a surface interaction complex arising from an acid-

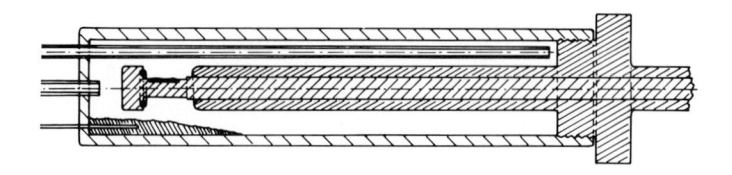

Fig. 17. Diagram of Reaction Chamber and Sample Probe

Table 7

Summary of Important Chemical Shifts for HDS Catalyst Studies

Cobalt Molybdate Catalysts

	Mo3d$_{5/2}$		Co2p$_{3/2}$
Mo (+6)	233.0	CoAl$_2$O$_4$	781.2
Mo (+5)	231.9	CoMoO$_4$	781.5
Mo (+4)	229.9	Co (fresh cat.)	782.4
MoO$_3$	232.5	Co (red cat.)	782.3
MoO$_2$	229.4		
MoS$_2$	228.9		

Nickel Tungstate Catalysts

	W4f$_{7/2}$		Ni2p$_{3/2}$	S2p
WO$_3$	35.0	Ni$_2$O$_3$	857.1, 863.0	-----
WO$_2$	32.7	NiO	854.9, 856.8 862.1	-----
W	30.7	Ni	853.1	-----
WS$_2$	31.4	WS$_2$	----	161.9
NiWO$_4$	35.2	NiWO$_4$	857.5, 863.7	-----
		NiS	854.9	162.0
		Ni$_3$S$_2$	854.1	162.1

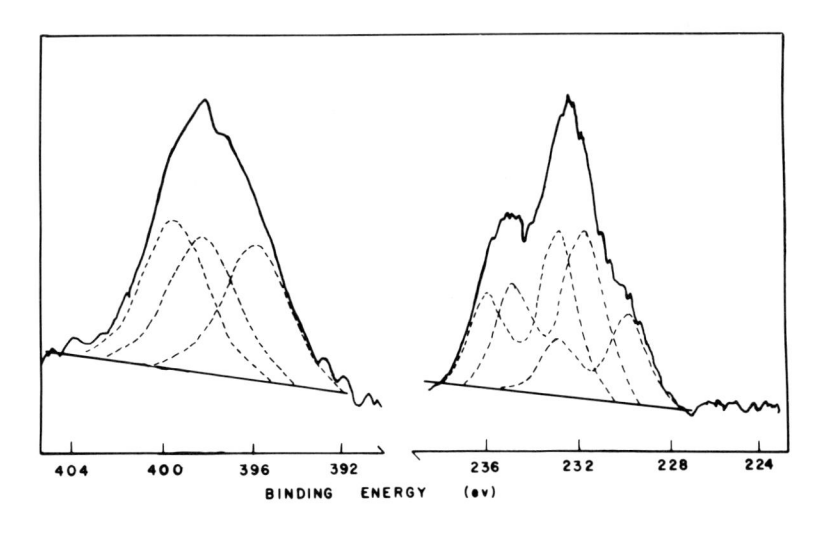

Fig. 18. Deconvolution of Mo 3p and 3d Envelopes
Left: Mo 3p envelope; Right: Mo 3d envelope

Table 8

Structures of Surface Species

Oxiding of Surface

$O_2W(OH)_2$ + Al₂O₃(OH)(OH) →Δ→ $O_2W(O)_2$/Al₂O₃ + 2H₂O

Sulfiding of Surface

$O_2W(O)_2$/Al₂O₃ + H₂S →Δ→ $O=W(=S)(O)_2$/Al₂O₃ + H₂O

$O=W(=S)(O)_2$/Al₂O₃ + H₂S →Δ/H₂→ WS₂/Al₂O₃(OH)(OH) + H₂O

---

base reaction during calcination as shown at the top of Table 8.[15] Intensity changes observed on calcination are also consistent with this interpretation.

Reduction of the catalysts in a hydrogen atmosphere showed some striking contrasts between the behavior of the cobalt molybdate and nickel tungstate catalysts. Figure 19 illustrates this. The left side of Figure 19 shows the oxidation states of molybdenum as a function of reduction time. Note that initially the molybdenum is reduced from +6 to the +5 state and then after approximately 40 minutes of reduction time molybdenum +4 is formed and after about 160 minutes all three oxidation states reach a steady state value which does not change on further reduction. During this reduction, there is essentially no change in the oxidation state of cobalt. It is interesting that the Mo(+5) curve is virtually identical with similar curves obtained using electron spin resonance.[16] These results confirm that reduction clearly influences the oxidation state of molybdenum on the surface and that Mo(+5) is an intermediate in the reduction of Mo(+6) to Mo(+4). It is interesting that no Mo(+3) could be detected although this species has been implicated as a possible active species in the catalytic behavior of HDS catalysts.

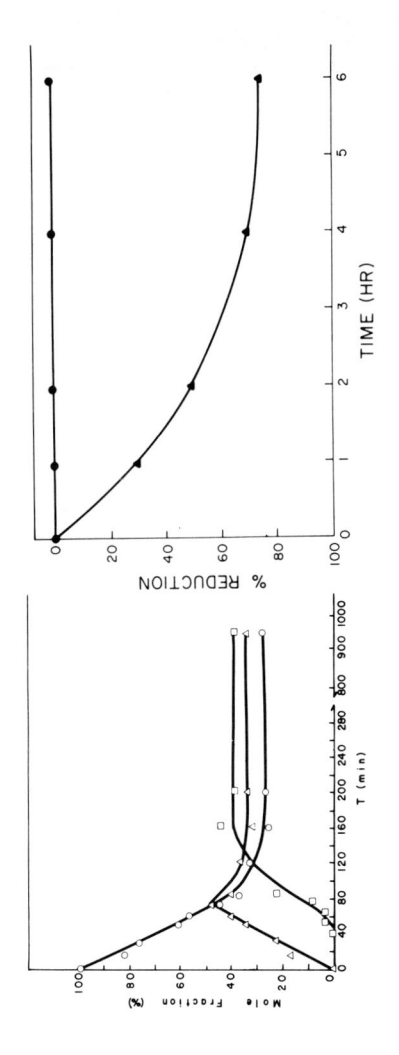

Fig. 19. Changes in Oxidation State for HDS Catalysts by Hydrogen Reduction
Left: Cobalt Molybdate Catalyst (500°C); O - Mo(+6); △ - Mo(+5); □ - Mo(+4)
Right: Nickel Tungstate Catalyst (450°C); ● - W; △ - Ni

Figure 19, right side, shows the behavior of a nickel tungstate catalyst on reduction. Note that the tungsten signal remains unchanged over 6 hours and in fact did not change even at longer reductions. Similarly, reduction of the nickel begins immediately and levels off at a steady state of about 75% reduced after 6 hours. The fact that the tungsten is not reducible and that the nickel is shows that at least the reductive behavior of nickel tungstate and cobalt molybdate catalysts is quite different.

Because of the low percentage of cobalt in the cobalt molybdate catalyst it was not possible to define the nature the the cobalt species. However, in a nickel tungstate catalyst a combination of reduction characteristics chemical shift measurements revealed that the nickel of the catalyst surface is in two forms, about 75% Ni_2O_3 and about 25% $NiAl_2O_4$.

We have also studied the effect of reduction on the sulfiding activity of cobalt molybdate catalysts. These results are summarized in Figure 20. Spectrum A shows the S2p doublet of MoS_2; spectrum B is that of a catalyst sample which has been reduced by exposure to thiophene in hydrogen. It is clear from the spectra that primarily MoS_2 is formed. Spectrum C is that of a catalyst sample which has been fired in an inert atmosphere and then exposed to thiophene. This spectrum clearly indicates that while thiophene chemisorption has occurred, little desulfurization activity has accompanied it. Spectrum D shows a catalyst sample that was first reduced in hydrogen and then exposed to thiophene. Note that the sulfur spectrum in D is much broader and has components both of the MoS_2 and adsorbed thiophene spectra in it. This indicates that desulfurization can occur to a limited extent in the absence of hydrogen but only if the surface of the catalyst is first reduced.

Figure 21 shows the behavior of both cobalt molybdate (left) and nickel tungstate (right) catalysts when sulfided in hydrogen H_2S. It is clear that the molybdenum in the cobalt molybdate catalyst undergoes a transition from molybdenum oxide to MoS_2. However, even after extended (10 hours) exposure to the H_2S/H_2 mixture, the catalyst is not completely sulfided; sulfiding is approximately 62% conversion to MoS_2. At the same time, changes in the cobalt peak indicates that the cobalt sulfides more slowly than the molybdenum and again that sulfiding is incomplete.

In contrast, in the nickel tungstate catalyst it is clear that the nickel sulfides more rapidly than the tungsten and is almost completely converted to a nickel sulfide species; even after extended exposure not all of the tungsten is sulfided. Again, it was not possible to determine the nature of the cobalt species after sulfiding, but sulfiding nickel tungstate catalysts produced NiS and Ni_3S_2 when thiophene was used as the sulfiding agent.

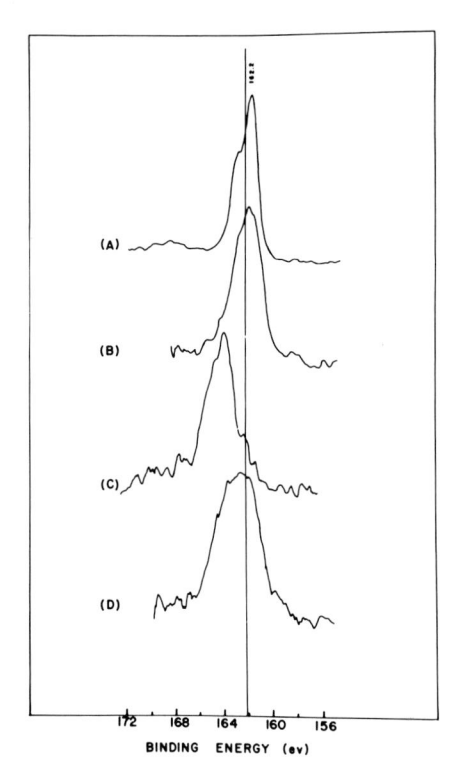

Fig. 20. S2p ESCA Spectra of Catalyst
A - Spectrum of MoS$_2$ on wire mesh; B - Spectrum of catalyst fired 4 hours at 500°C, then exposed to thiophene in hydrogen for 35 minutes at 420°C; C - Spectrum of catalyst sample fired 4 hours at 500°C, then exposed to thiophene in helium for 2 hours at 420°C; D - Spectrum of catalyst sample fired 4 hours at 500°C, reduced for 450 minutes at 500°C, and exposed to thiophene in helium for 2 hours at 420°C.

More interestingly, as the nickel tungstate catalyst is being sulfided both the WS$_2$ and sulfur signals increase, but their intensity ratio does not correspond initially to that of WS$_2$. The sulfide signal is more intense in the early states of sulfiding, the WS$_2$ signal then catches up. This implies that sulfur is being incorporated into the nickel tungstate catalyst prior to conversion of the tungsten to WS$_2$. This is consistent to the mechanism of forming an intermediate complex as shown at the bottom of Table 8. Here, replacement of one of the oxygens of the bound tungsten oxide with a sulfur would have only a minimal chemical shift since the formal oxidation state of the tungsten is unchanged.

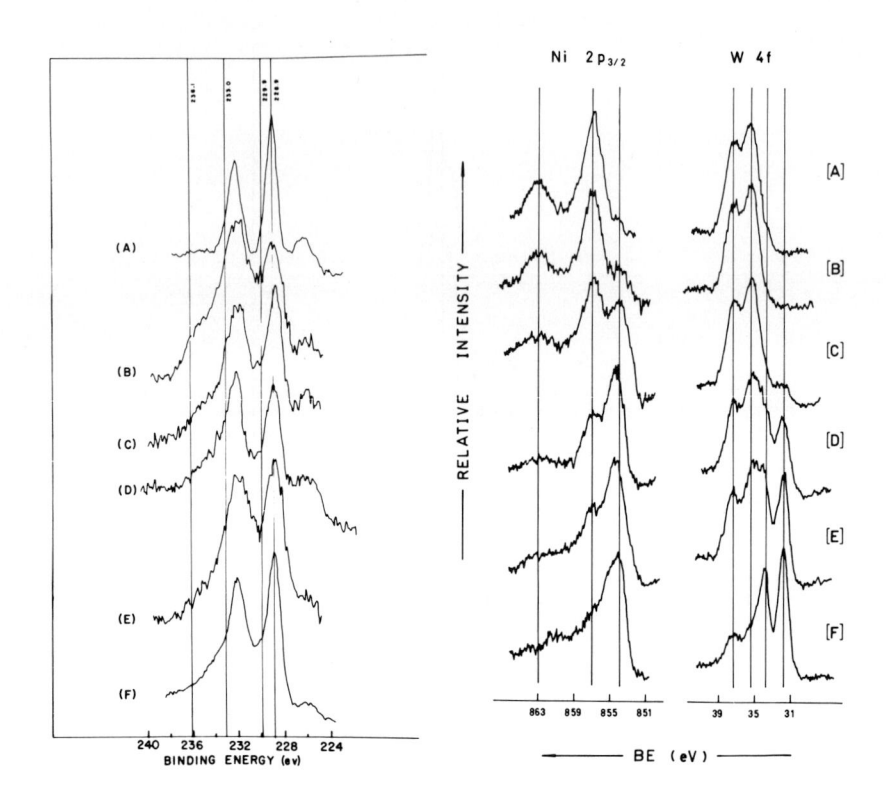

Fig. 21. Sulfiding of HDS Catalysts in H_2S/H_2
Left: Cobalt Molybdate (400°C)
   A - Spectrum of MoS_2; B - Spectrum of catalyst after 30 minutes exposure; C - Spectrum of catalyst after 2 hours exposure; D - Spectrum of catalyst after 10 hours exposure; E - Same sample as D but reduced 5 hours in hydrogen at 550°C; F - Spectrum of $Al_2(MoO_4)_3$ after 30 minutes exposure.

Right: Nickel Tungstate (310°C)
   A - as received; B - 1 minute; C - 12 minutes; D = 95 minutes; E - 240 minutes; F - 600 minutes

When the tungsten of a nickel tungstate catalyst is being converted to WS_2 and the conversion of the tungsten complex reaches 50%, the surface of the network of the catalyst changes, causing a decrease in surface charging, probably due to a decrease in the work function of the surface. Although the relationship between the surface conductivity and catalytic activity is not clear, this is an interesting effect. At the point where the conductivity changes, the total nickel signal increases without an increase in the sulfur signal; this indicates that diffusion of unsulfided nickel to the surface is occurring.

These studies show that there are both similarities and differences between cobalt molybdate and nickel tungstate catalysts but the differences are probably larger than was previously supposed. The point is, however, that ESCA represents an excellent tool for studying the nature of heterogeneous catalytic surfaces.

CONCLUSIONS

In this paper, I have tried to give examples of how ESCA may be applied to a variety of problems involving surfaces in diverse areas such as dentistry, adhesion and heterogeneous catalysis. Of necessity this type of general survey must be brief. More details can be obtained from the original papers referenced.

The important thing to remember is that ESCA is new and still an experimental technique, but one that has tremendous potential for the characterization of surfaces and surface species. The examples I have chosen here are only three of many which have already been demonstrated. In the future, we shall see improvements in both hardware and interpretations which will make the ESCA even more powerful for studies of surfaces. In fact, at the present time one can truthfully say we have not yet even scratched the surface.

REFERENCES

1. S.H. Hercules and D.M. Hercules in P.F. Kane and G.L. Larrabee, eds. "*Characterization of Solid Surfaces*", Plenum, New York, 1974.
2. K.L. Cheng and J.W. Prather, CRC Crit. Rev. Anal. Chem., 5, 37 (1975).
3. D.M. Hercules and J.C. Carver, Anal. Chem., 46, 133R (1974).
4. R. Caudano and J. Verbist, (eds.) "*Electron Spectroscopy - Progress in Research and Applications*", Elsevier, Amsterdam, 1974.
5. Y. Baer, P.F. Heden, J. Hedman, M. Klasson and C. Nordling, Solid State Comm., 8, 1479 (1970).
6. P.W. Palmberg and T.N. Rhodin, J. Appl. Phys., 39, 2425 (1968).
7. D.M. Wyatt, R.C. Gray, J.C. Carver, D.M. Hercules and L.W. Masters, Appl. Spectrosc., 28, 439 (1974).
8. R.G. Steinhardt, J. Hudis and M.L. Perlman, Phys. Rev., B5, 1016 (1972).
9. S.H.Y. Wei and W.C. Forbes, J. Dent. Res., 53, 51 (1974).
10. N.L. Craig and D.M. Hercules, Anal. Letters, 9, 831 (1975).

11. N.L. Craig and D.M. Hercules, J. Dent. Res., (in press).
12. O. Weisser and S. Landa, "*Sulfide Catalysts, Their Properties and Applications*", Pergamon, New York, 1973.
13. T.A. Patterson, J.C. Carver, D.E. Leyden and D.M. Hercules, Spectrosc. Letters, 9, 65 (1976).
14. K.T. Ng and D.M. Hercules, J. Phys. Chem., in press.
15. M. Dufaux, M. Che and C. Naccache, Compt. under Sec. C, 268, 2255 (1969).
16. K.S. Seshadri and L. Petrakis, J. Catal., 30, 195 (1973).

Photoemission Study of Chemisorption on Metals

Thor Rhodin and Charles Brucker

School of Applied and Engineering Physics
Cornell University
Ithaca, New York 14853

Recent advances in the application of photoemission spectroscopy in the vacuum ultraviolet wavelength range (UPS) have made significant contributions to the study of the interactions of gases and vapors with metal surfaces. Characterization of the surface electronic structure using UPS has proved to consist mainly of the following three general categories of information.
(1) The chemical fingerprint of the electronic structure of the clean or chemisorbed surface, (2) Information on the nature of electron excitations and charge transfer, and (3) Information on the combination of molecular orbitals and metal electrons contributing to the formation of surface chemical bonds. In addition, interpretation of UPS measurements is greatly enhanced by combined measurements with other types of measurement of electron structure together with data on surface chemical composition and surface crystallography. Useful advances have also been made in the study of kinetics of chemical rearrangements and reactions on metal surfaces. In addition to the use of fixed-energy photon sources based on the excitation emission of inert gases such as helium, argon and neon, synchrotron radiation from electron storage rings has made it possible to study the important effects of photon frequency and polarization. Typical examples of these uses for the microscopic and atomistic study of metals are discussed.

CONTENTS

INTRODUCTION

A. General Approach
B. Critical Atomistic and Microscopic Parameters

EXPERIMENTAL APPROACH

A. Photoemission Energy Distributions
   1. Fixed-Energy Photon Sources
   2. Synchrotron Radiation
   3. Energy Analyzers
B. Surface Sensitivity
   1. Bulk vs Surface Emission
   2. Relative Photoionization Cross-Sections

APPLICATIONS OF VUV-PHOTOEMISSION TO CHEMISORPTION SYSTEMS

A. Chemical Bonding on Metal Surfaces
   1. Molecular Physisorption on an Inert Substrate
   2. π-d Bonding of Chemisorbed Olefins on Transition Metals
   3. Distortion and Fragmentation During Chemisorptive Bonding
B. Synchrotron Radiation Studies Using Variable Energy and Polarization
C. Correlation of CO Chemisorption Behavior with Position in the Periodic Table
D. Use of Combined Techniques to Provide a Chemisorption Model
   1. Hydrogen and Deuterium on W(100)
   2. Oxygen on Fe(100)
E. Directional Photoemission - Application to Surface Crystallography and Bonding Symmetry
   1. Simple Electron Scattering Model
   2. Angular Resolved Photoemission Study

SUMMARY AND PROSPECTS FOR THE FUTURE

ACKNOWLEDGEMENTS

REFERENCES

INTRODUCTION

A. General Approach

The surface region of a clean metal may involve no more than the outermost two or three atomic layers. Atoms in this region, sensing the loss of translational symmetry, experience atomic potentials different from those of atoms still deeper within the bulk. Scientific interest arises from their unique properties which have no counterpart in bulk solid-state physics, properties that we are only beginning to understand. Surface scientists are vigorously exploring this exciting new field of solid surfaces with a wide and ever-increasing variety of modern surface science techniques that involve various combinations of electrons, photons, ions, electric fields, etc. The problems encountered in these experiments are sometimes rather complex and difficult to interpret in simple terms. Through the close interplay of theory with experiment, it is hoped that a truly microscopic understanding of surface chemistry of heterogeneous systems will evolve from this combined information.

B. Critical Atomistic and Microscopic Parameters

Interest in systematic studies of single crystal surfaces has expanded rapidly in recent years. This increased effort in the study of the chemistry of solid surfaces is justified by the many phenomena which occur at solid surfaces which are of technological importance. Understanding of heterogeneous catalysis, corrosion of solids, solid-state electronic devices, and many biological systems can only be as complete as our insight into the atomistic and microscopic properties from which these phenomena originate.

As with any solid or molecule, a solid surface is characterized by (1) the chemical identity of the atomic species present, (2) the structural arrangement and motion of these atoms, and (3) the distribution in space and energy of the valence electrons.[1] It is this third characteristic, the surface valence electronic structure, which largely determines the physical and chemical nature of chemical bonding at surfaces. Ultraviolet and photoemission spectroscopy (UPS) is a surface-sensitive probe of the occupied valence electronic orbitals for both the adsorbate and host solid over a wide range of electron energies. Since chemical bonding is a direct reflection of interactions among valence electrons, UPS is an ideal tool for the study of chemisorption. In this article, we discuss some recent typical applications of UPS to the study of valence electronic structure and chemical bonding for chemisorption on transition metal surfaces. While emphasizing the contributions of photoemission to our understanding of surface phenomena, we also point out, through selected examples, the importance and usefulness of combining and correlating interpretations based on photoemission spectroscopy with other measurements of critical adsorption parameters.[2] The reader is referred to other discussions[2-6] of the application of photoemission and other electron spectroscopic approaches to surface science studies for a broader review of the field.

EXPERIMENTAL APPROACH

A. Photoemission Energy Distributions

It is only within the past ten or twelve years that the fundamental processes of the photoeffect have become reasonably well understood together with the availability of reliable photoemission data. These advances are due in large part to recent technological advances in preparing clean surfaces and in attaining intense monochromatic radiation sources in the vacuum ultraviolet. The most convenient and commonly used ultraviolet sources in photoemission experiments have been hydrogen discharge lamps (4-11.6 eV) and, particularly for chemisorption studies, helium (21.2 eV and 40.8 eV) and neon (16.8 eV and 26.9 eV) resonance lamps.[7] Synchrotron radiation produced by accelerating electrons in a synchrotron has been playing an increasingly important role in surface studies[8,9]. Synchrotron radiation has many desirable features as a light source for photoemission electron spectroscopy. It is stable, intense, continuously tunable (from the ultraviolet to the x-ray region), strongly polarized and highly collimated. The photoemitted electrons can be effectively energy-analyzed in an electrostatic deflection-type analyzer with an electron multiplier for digital counting by a multichannel recorder. Satisfactory results can also be obtained using a retarding grid type of analyzer (e.g., LEED optics).

A conceptual potential energy diagram relevant to the UV-photoemission process for a metal surface with an adatom present is illustrated in Figure 1. The narrow ionization energy level of the isolated atom broadens and shifts as the surface is approached due to interactions with electrons in the metal. The observed photoemission spectrum from such a surface (Figure 1) reflects the coupling of the valence level of the adsorbate at a specific chemisorption site with the s- and d-band density of states of the metal to give a virtual adsorbate-induced level (cross-hatched area)[6] superimposed on the substrate emission. A characteristic UV-photoemission spectrum from a chemisorbed transition metal surface consists of (1) a Fermi edge cut-off (rounded by the finite resolution of the energy analyzer) at $\hbar\omega$ above the Fermi level of the emitter, (2) d-band emission peaks within a few eV of the Fermi cut-off, (3) a large increase in emission intensity at lower kinetic energies due to inelastically scattered electrons (which cannot be conveniently separated from the primary unscat-

N(E)

PHOTOEMISSION
ENERGY
DISTRIBUTION

POTENTIAL BARRIER

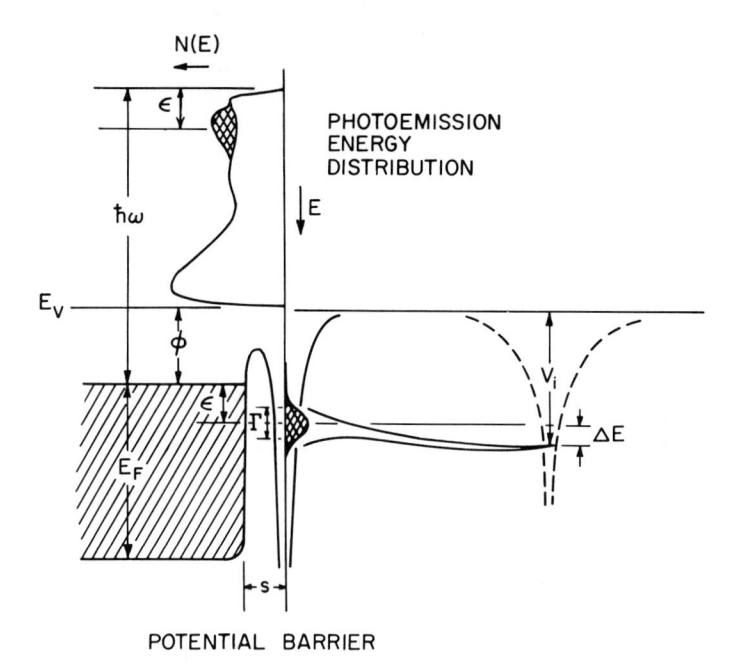

Fig. 1. Schematic potential energy diagram and UPS valence electronic energy distribution for a free-electron metal with an atom adsorbed on the surface. Natural (lifetime) broadening (Γ) and shifting (ΔE) of the adsorbate level (gas phase ionization energy V_i) due to the atom's interaction with the solid are depicted as the atom is brought from infinity (dashed potential) to a distance S from the surface (solid curve). The shaded region in the photoemission energy distribution indicates the additional emission from the virtual adsorption state (see Gadzuk, Ref. 66).

tered electrons), and (4) adsorbate-induced levels superimposed on the d-band and background emission. Spectra are generally plotted with initial-state energy referred to the Fermi cut-off, i.e., the photon energy is subtracted from the measured kinetic electron energy and the work function added:

$$E = E_{kin} + \phi - \hbar\omega.$$

An electron at E = 0 has therefore been emitted right from the Fermi level.

B. Surface Sensitivity

Any experiment designed to yield information about a surface must, to a degree, be specific to the surface. The surface specificity of photoejected electrons is enhanced by the strong electron-electron interaction which limits, in many cases, the mean free path of unscattered electrons to at most a few monolayers. From the plot of electron escape depth vs. electron kinetic energy in Figure 2 it is clear that the kinetic energy distribution of photoemitted electrons, which may extend up to about 50 eV, will reflect an initial state distribution including both bulk- and surface-like states. Bulk photoemission has been discussed in terms of a three-step process including electron excitation by photo adsorption, transport of the hot electron to the surface, and transmission through the surface[10]. While the role of the surface effect alone has been discussed theoretically[11] and experimentally[12], only very recently have more rigorous theories become available that treat both the surface and volume effects in a unified manner[13], and as such question the validity of the experimentally useful three-step model. One consequence of these considerations for chemisorption is that it often cannot be determined using photoemission alone, whether or not species are specifically adsorbed on the surface or incorporated into the outer few layers of the surface lattice.

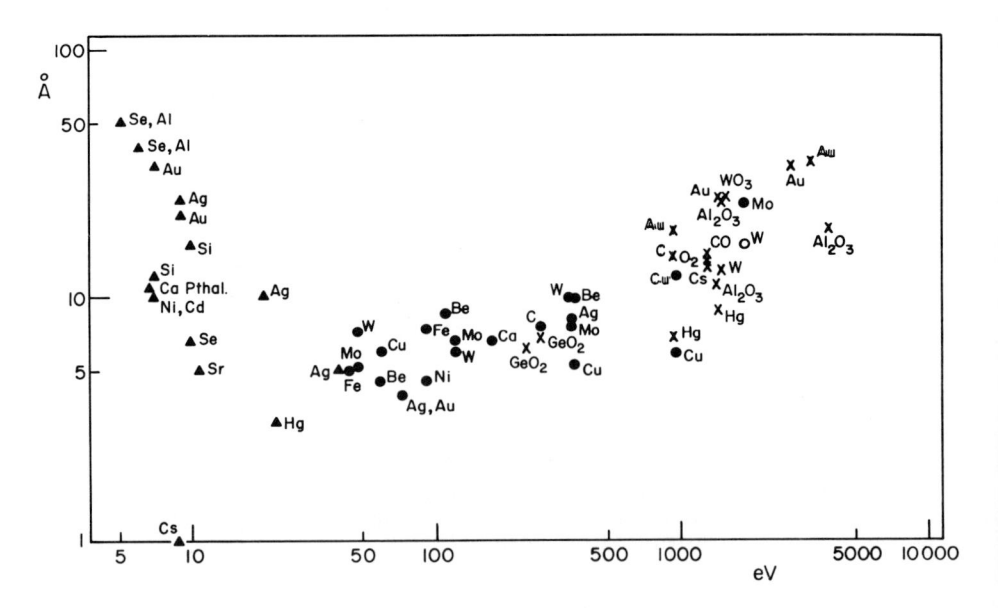

Fig. 2. Experimental values of electron inelastic mean free path lengths, $\ell(E)$, as a function of electron energy for UPS, XPS, and Auger studies (From Brundle, Ref. 67, and references cited therein).

For species chemisorbed within an escape depth of the surface, another important factor determining surface sensitivity is the relative probability of exciting electrons from adsorbate levels compared to adsorbent levels. For example, emission from the atomic 2p states of oxygen dissociatively chemisorbed on Ni dominates metallic d-band emission at ultraviolet energies[7] but the reverse is true at x-ray energies.[14] Other factors being equal, the more localized in space the initial state is, the more strongly will it couple with a higher energy final state. Thus, for chemisorption studies the choice of an appropriate photon energy is very important. As it turns out, for common residual gases or organic species adsorbed on transition metals, maximum surface sensitivity often occurs for photon energies between 15 eV and 50 eV because of the relative photoionization cross-sections.

APPLICATIONS OF VUV-PHOTOEMISSION TO CHEMISORPTION SYSTEMS

A. Chemical Bonding on Metal Surfaces

Photoemission energy distribution spectra can provide a great deal of qualitative information regarding chemisorption, some of which can be directly gathered by rather straightforward inspection of the various adsorbate-induced features. A complete interpretation of the photoemission process in terms of electronic excitations in the adsorbate-adsorbent system is much harder to achieve. In order of increasing difficulty of interpretation the more accessible information includes (1) A chemical fingerprint of the chemisorbed species, (2) Indications as to the nature of the electron binding interactions and to the magnitude and direction of charge transfer, and (3) Insights into the molecular orbitals contributing to the formation of surface bonds. Variations in the shape and position of orbital levels with kinetic energy as well as their angular- and frequency-dependency are usually the most informative features. In addition, it is often an additional objective to increase our basic understanding of the photoemission process itself.

1. <u>Molecular Physisorption on Inert Substrate</u>. (Benzene and pyridine on molybdenum disulfide) The ability to identify the "signature" or "fingerprint" of an adsorbed gas molecule on the surface is most important to the application of photoemission to the study of chemisorption and catalytic processes. Fingerprinting usually requires the comparison of the photoemission spectrum of the adsorbed species to its gaseous or condensed phase counterpart to properly identify the spectral peaks. In the simplest case, when no strong chemical bonding exists between the substrate and adsorbed molecules, little or no change in orbital ionization

energies is expected as the gas molecule is brought down to the surface. For example (Figure 3), the one-to-one correspondence with gas phase spectra in a recent UPS study[15] of condensation onto a liquid nitrogen cooled inert substrate assures that structurally and chemically the molecules are essentially unchanged upon adsorption.

The orbital ionization energies of each condensed species (Figure 3) are observed to be shifted by "relaxation" upward in energy by about 1-1.5 eV in comparison to the corresponding gas phase values[15]. This illustrates the large change in ionization energies of a non-chemical bonding nature that occur even for condensed species. It can be understood in terms of the tendency of electronic charge in surrounding molecules to screen the hole produced during photoexcitation. An additional relaxation shift of similar origin arises upon adsorption onto the surface of a metal due to the polarizability of the metal electron gas. The mobile conduction electrons lower their energy by screening the positive photohole. This additional energy is reflected in the increased kinetic energy of the outgoing photoelectron. Relaxation shifts including both effects occur for chemisorbed species on metals and are typically in the range 1-3 eV. This adds a complication to the theoretical description of the absolute ionization energies.

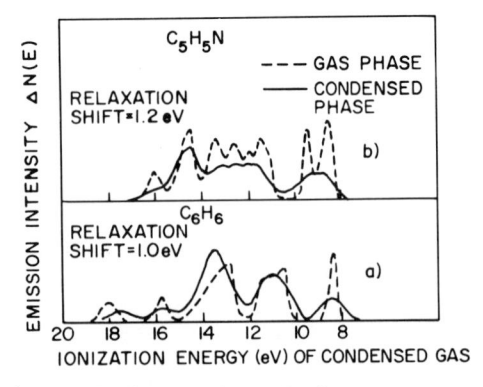

Fig. 3. Comparison of the condensed-phase spectrum with the gas-phase spectrum of (a) C_6H_6, (b) C_5H_5N. The condensed phase difference spectra were obtained by condensation onto an inert LN_2-cooled MoS_2 substrate at $\hbar\omega$ = 21.2 eV. (From Yu, et al., Ref. 15.)

It is somewhat surprising to find, at least for a molecule that "lies flat" on the surface, that all the molecular orbitals are relaxation shifted by very nearly the same amount. This is a fortunate consequence, as we shall see, in more quantitative applications of the fingerprinting technique. In principle, relaxation shifts can be different for different orbitals, however, due to the dependence of image-charge screening on the orientation and displacement of the molecule with reference to the surface.[16] Uniform relaxation becomes less plausible when the adsorbate stands end-on to the surface, e.g., as is often considered to be the case for CO-chemisorption.[17-21]

In contrast to weakly interacting physisorbed or condensed systems, chemisorption usually involves the formation of strong chemical bonds. Participation of substrate and adsorbate electrons in binding interactions is reflected by chemical binding shifts and level broadenings in the fingerprint spectrum. So-called non-bonding adsorbate orbitals may also be significantly perturbed if the bonding (and back-bonding) results in a geometrical distortion of the molecule. Although attempts at disentangling the various shifts due to a combination of relaxation, chemical binding, and molecular distortion in great detail can be complicated, much insight into this complex problem can be gained by a systematic investigation of the chemisorptive behavior of simple representative molecules on different transition metal surfaces. The following discussion is devoted to some in-depth discussions of selected photoemission studies which furnish helpful illustrations of both the complexities and the achievements resulting from some recent efforts to extend our understanding of chemisorption phenomena.

2. π-d Bonding of Chemisorbed Olefins on Transition Metals. (acetylene and ethylene on single crystal surface of nickel, iridium and iron.) In addition to measurement of other adsorption parameters defining the interactions of hydrocarbon molecules with transition metal surfaces, the wealth of accessible ionization levels at ultraviolet photon energies makes these systems well suited for photoemission studies of chemisorption. Recent publications for the unsaturated hydrocarbons acetylene (C_2H_2) and ethylene (C_2H_4) chemisorbed on Ni(111)[22,23], Ir(100)[24], and Fe(100)[25] surfaces demonstrate both the powers and limitations of applying the fingerprint technique to the interpretation of chemisorption phenomena.

Figure 4 shows UV photoemission valence orbital spectra for C_2H_2 and C_2H_4 chemisorbed on a clean Ni(111) surface compared with the corresponding gas phase spectra. Measurement of orbital ionization energies and line shapes as well as SCF-LCAO calculations of Hartree-Fock ground states for a range of possible hybridizations of the carbon atoms has been done by Demuth and Eastman.[23]

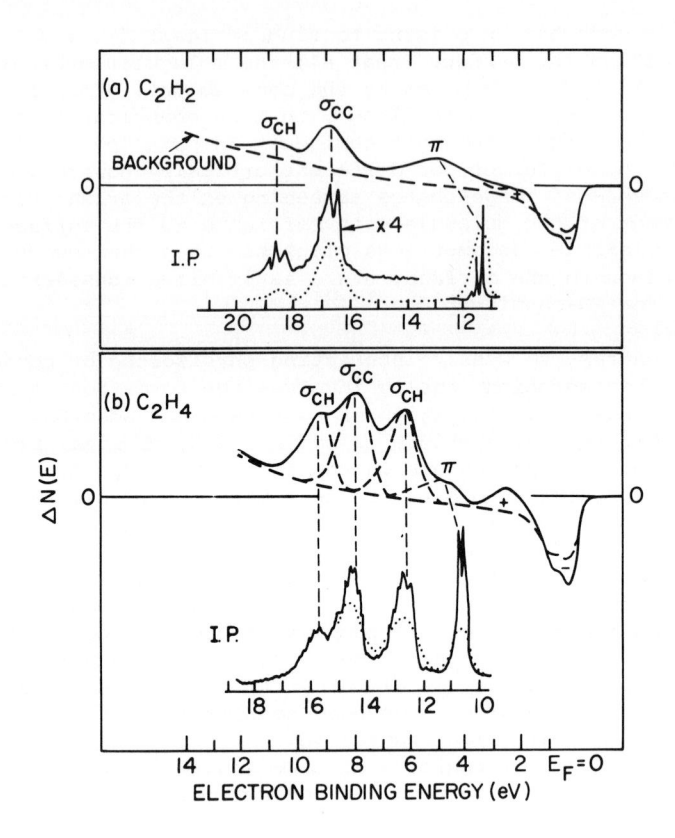

Fig. 4. UPS difference spectra for acetylene (C_2H_2) and ethylene (C_2H_4) chemisorbed on the clean Ni(111) surface. (From Demuth and Eastman, Ref. 23.)

They conclude that the chemisorptive bonding of these unsaturated hydrocarbons involves predominantly π-d bonding without any significant rehybridization relative to their gas phase counterparts. In both spectra (Figure 4) the relative positions, line shapes, and amplitudes of the non-bonding α levels are virtually unaffected by the chemisorption process, implying little or no physical distortion due to chemisorption of the surface molecule. The bonding π-levels, on the other hand, are both shifted, to greater binding energies, and broadened by chemical binding interactions with the Ni d-electrons. Estimates of π-d interaction strengths and chemisorption energies based on these π-level shifts are consistent with the energetics of observed surface dehydrogenation reactions ($C_2H_4 \rightarrow C_2H_2 + H_2$ (gas)) on Ni(111)[22]. In relating simple phenomenological chemical binding models such as Mulliken's donor-acceptor model[26] as applied by Grimley[27] to the determination of heats of adsorption from observed orbital shifts, however, Demuth and Eastman[23] point out that a significant contribution can arise between unoccupied molecular levels and substrate d-electrons. Thus, a probe of the unoccupied density of states would provide useful complementary information to UPS in this type of situation.

3. **Fragmentation and Distortion During Chemisorptive Bonding.** (acetylene and ethylene chemisorption on single crystal iridium surfaces.) Photoemission studies of C_2H_2 and C_2H_4 chemisorption on Ir(100) surfaces by Brodén and Rhodin[24] have shown that fingerprint techniques for the interpretation of valence electronic spectra can be complicated by surface bonding perturbations of the σ-orbitals not directly involved in bonding the molecule to the surface. Distinctive features associated with stretching of chemisorbed acetylene are apparent in the UPS difference spectra (Figure 5) for saturation coverages of C_2H_2 on clean (1x1) and reconstructed (1x5) surfaces of Ir(100)[24]. All three spectra have the general appearance of chemisorbed acetylene, but the relative spacing between the (non-bonding) σ-levels is both anomalously large and a sensitive function of both substrate geometry and temperature. Table 1 summarizes the significant differences between these data and both gas phase[28] and related chemisorption[22,24,29] spectra.

Table 1

Separation Between $3\sigma_g$ and $2\sigma_u$ peaks.

	eV	Ref.
C_2H_2 - Gas	2.0	28
Ir(100) (1x1) T=275K	2.8	24
+ C_2H_2 (5x1) T=132K	2.2	24
(5x1) T=300K	2.9	24
Ni(111) + C_2H_2	2.0	22
W(110) + C_2H_2	1.9	29

Brodén and Rhodin[24] have attributed the larger separations for the chemisorbed iridium surfaces to a stretching (and probably a bending as well) of the molecule on the surface. The sensitivity of the separation to substrate geometry (curves 1 and 2, Figure 5) reflects the interdependency of geometric and electronic factors in this chemisorption system. Weakening of the carbon-carbon bond and subsequent stretching is explained in terms of both π-d bonding interactions as well as back-bonding into unoccupied antibonding π-orbitals of the hydrocarbon.[24] An examination of the spatial distribution of the $3\sigma_g$ and $2\sigma_u$ molecular orbitals shows that the larger observed separation is consistent with the carbon-carbon stretching hypothesis.[24] The stretching of course perturbs the bonding $1\pi_u$ level as well, but our inability to make quantitative predictions in this regard hampers attempts at extracting π-d interaction strengths as was done for nickel.[22]

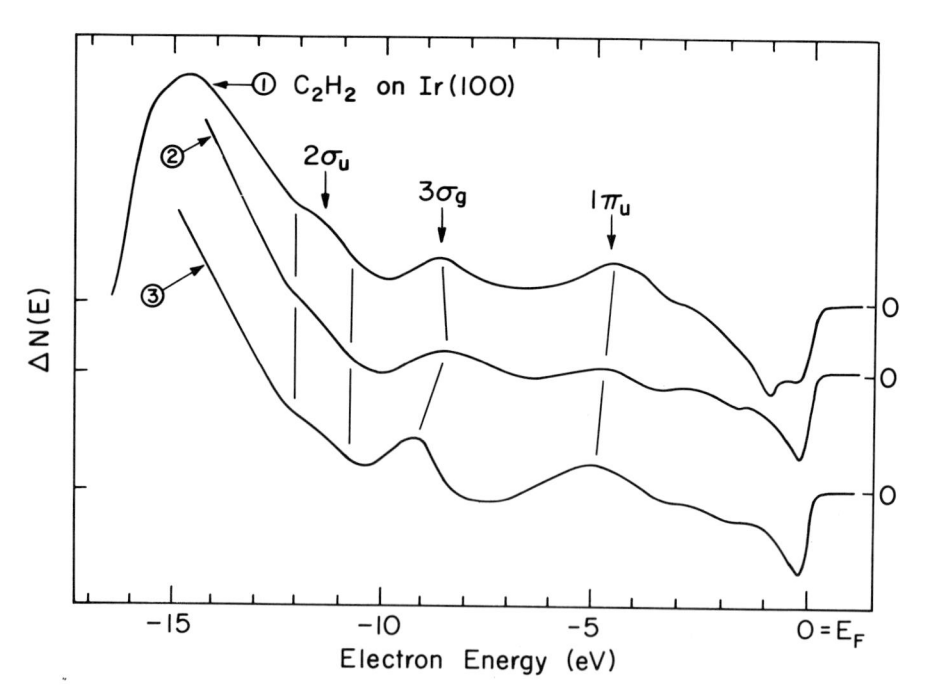

Fig. 5. UPS difference spectra for acetylene (C_2H_2) chemisorbed on clean (1x1) and reconstructed (1x5) surfaces of Ir(100) (All curves correspond to saturation coverage of acetylene.). (1) Adsorption on (1x1) surface at room temperature, (2) adsorption on (1x5) surface at room temperature, (3) adsorption on (1x5) surface at 132°K. $\hbar\omega$ = 21.2 eV. (From Brodén and Rhodin, Ref. 24.)

Further examples of molecular distortion and evidence for fragmentation is found for ethylene chemisorption on Ir(100)[24] and Fe(100)[25] surfaces. Chemisorption of C_2H_4 on the Ir(100) - (1x5) surface at 300K produces a spectrum typical of neither ethylene nor acetylene.[24] The possibility of carbon-carbon scission to produce chemisorbed methylene radicals is suggested.[24] Chemisorbed C_2H_4 on Fe(100)[25] at 93°K is apparently distorted (stretched), as evidenced by a compression of the non-bonding σ-levels. Upon warming to 123°K new features develop which bear a strong resemblance to those obtained for hydrogen adsorbed on Fe(100) with a carbon overlayer.[25] This suggests a fragmentation of the ethylene molecule into chemisorbed CH- or CH_2-fragments. Further studies are in progress to clarify these interesting implications.

B. Synchrotron Radiation Studies Using Variable Energy and Polarization (CO chemisorption on palladium, nickel, and iridium surfaces)

The interaction of CO with transition metal surfaces has traditionally been one of the more widely studied systems involving strong chemisorption. Photoelectron spectra of molecularly adsorbed CO have been reported for a wide variety of transition metal surfaces both in single crystal and polycrystalline form (see Refs. 19,21 and references cited therein). Two interesting synchrotron radiation studies of CO chemisorption from the viewpoint of clarifying the chemisorptive bonding have recently been made at the Synchrotron Radiation Center Storage Ring at the Physical Sciences Laboratory of the University of Wisconsin[8] (Figure 6).

In the first study, Gustafsson, et al.[19] made use of the variation with energy of the relative photoionization cross-sections of the CO-induced levels to remove ambiguities in molecular orbital assignments persisting from earlier investigations[30]. Figure 7 shows their photoemission spectra for CO adsorbed on an evaporated palladium film in the photon energy range 25 to 105 eV. Their measured CO gas phase spectrum is also shown (Figure 7), with the binding energy scale referenced to the Fermi energy $E_F = 0$ by subtracting the measured 6.0 eV work function for CO on Pd[18]. The Pd-CO system exhibits basically the same CO-derived energy level spectra as previously studied CO-chemisorption systems, i.e., two peaks (Figure 7) with binding energies of $\sim$ 13.5 eV (P_2) and $\sim$ 16.5 eV (P_1). Since the binding energies of these two levels are nearly identical to the binding energies of the 5σ (14.0 eV) and 1π (16.5 eV) levels of gas phase CO[28], Eastman and Cashion[30], as well as other investigators assumed that the two levels P_2 and P_1 of adsorbed molecular CO were derived from the 5σ and 1π gas phase levels, respectively.

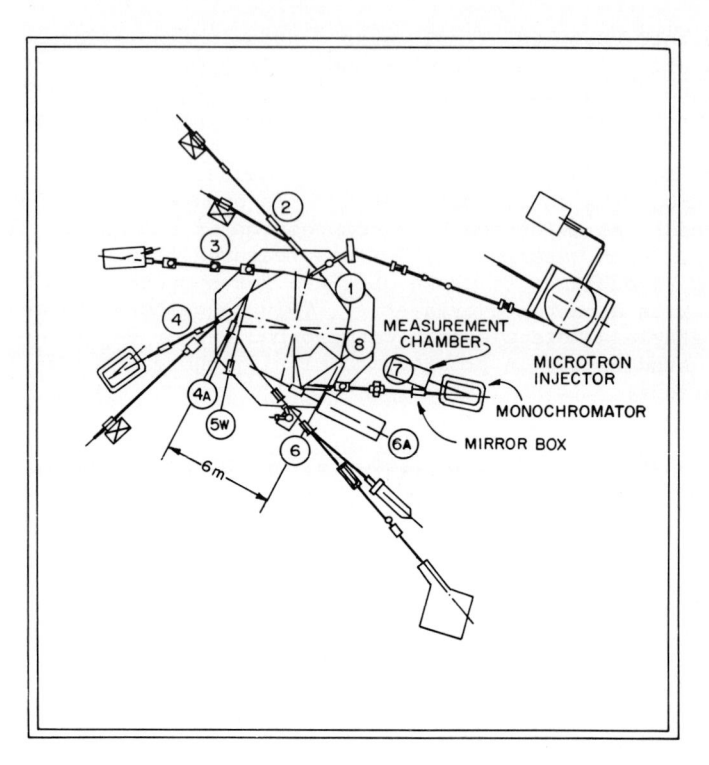

Fig. 6. Plan view of the Synchrotron Radiation Center Storage Ring at the Physical Science Laboratory of the University of Wisconsin. At Station (6), two UHV monochromators provided usable radiation over the range $4 \leq \hbar\omega \leq 120$ eV for the CO on Pd and Ni study[19]. At Station (7) measurements of CO on Ir(100) were performed up to a photon energy of 30 eV[21]. Both stations were equipped with a Physical Electronics double pass cylindrical mirror energy analyzer. (Figure courtesy E.M. Rowe)

Gustafsson, et al.[19] compared the photon energy dependence of the relative photoionization cross-sections of the levels of adsorbed CO with the gas phase levels and produced the presently accepted identification of the adsorbate energy levels for chemisorbed CO. The level at ∿ 16.5 eV (P_1) is the relaxation-shifted 4σ level while the level at ∿ 13.5 eV (P_2) is actually a double level composed of the relaxation-shifted 1π level and the 5σ level, which is shifted to greater binding energies by participation in chemical bonding with the substrate. Figure 8 shows the relative photoionization cross-sections, i.e., $\sigma(P_1)/\sigma(P_2)$, as a function of photon energy for several trial identifications of peaks P_1 and P_2 (Figure 7). Best agreement with the observed variation of

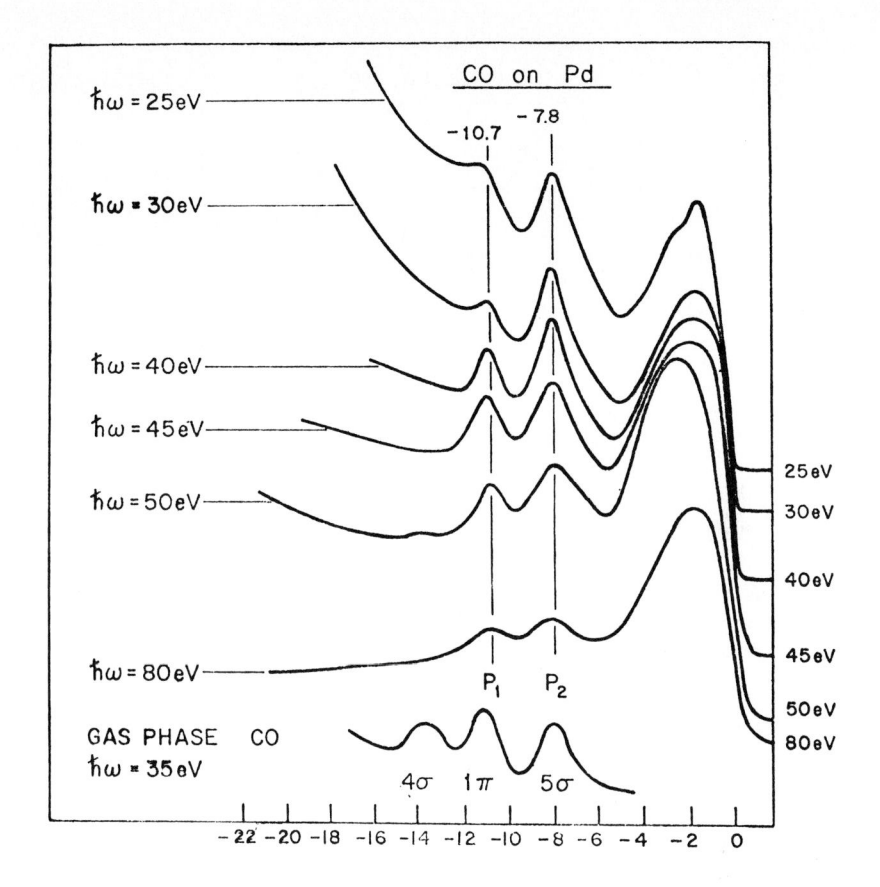

Fig. 7. Photoemission spectra for CO adsorbed on an evaporated Pd film as a function of photon energy $\hbar\omega$. The curve at the bottom is the photoelectron spectrum for gaseous CO at a photon energy of 35 eV and a resolution of 1 eV. (From Gustafsson, et al., Ref. 19.)

$\sigma(P_1)/\sigma(P_2)$ is obtained by associating P_1 with the 4σ gas phase level and P_2 with the gas phase 1π level plus the perturbed 5σ level.

Subsequent synchrotron radiation studies[19] of chemisorption on a nickel film as well as previous angular resolved measurements[18] for Co on Ni(111) indicated the possibility that the ∼ 13.5 eV level was indeed composed of two component levels. Brodén and Rhodin[20] and Brodén, et al.[21] have since definitively shown that the CO-induced 1π and 5σ levels on Ir(100) are clearly separable into two peaks in the same order as the gas phase spectrum. The

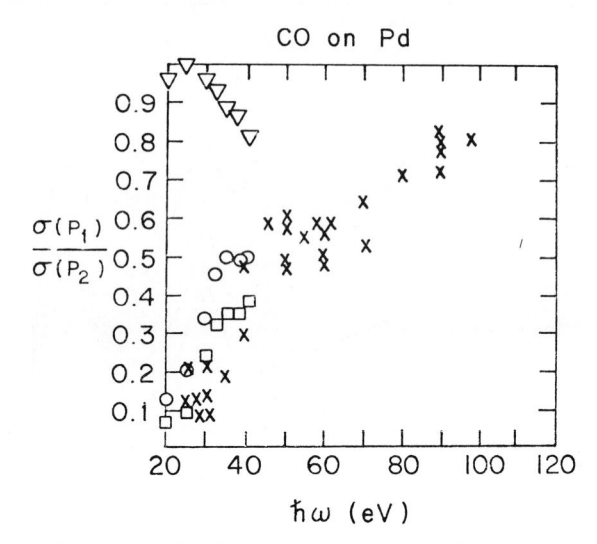

Fig. 8. Relative photoionization cross-sections for the P_1 level compared to the P_2 level (x's) (see Figure 7) as a function of photon energy for CO adsorbed on Pd. The other symbols are for relative cross-sections of gas phase levels as follows: "O" = $\sigma(4\sigma)/\sigma(1\pi)$; "☐" = $\sigma(4\sigma)/\sigma(1\pi)$; "∇" = $(\sigma(4\sigma) + \sigma(5\sigma))/\sigma(1\pi)$. (From Gustafsson, et al., Ref. 19).

resolution of this question was an important verification of the utility of photoemission spectroscopy as a technique capable of identifying the orbitals participating in the chemical bonding of an adsorbed species.

In the second study, Brodén, et al.[21] have revealed interesting dependencies of the relative photoionization cross-sections for the 4σ, 1π, and 5σ levels of chemisorbed CO on the polarization of the incident radiation (Figure 9). The use of polarized radiation to enhance and identify surface features has not as yet been widely recognized. It promises to be very useful in providing unique information on the orientation and nature of the orbitals contributing to chemisorption. The conclusion can be made from the curves in Figure 9 that the presence of an electric field component normal to the surface may be an important factor in exciting the 4σ level.[21] The 4σ level is very weakly excited for s-polarized photons using a synchrotron radiation source (curve (b)) compared to the unpolarized photons from the resonance lamp (curve (a)). Rotating and tilting the crystal so that the incident radiation has a component of the electric vector normal to the surface, therefore, enhances emission from the 4σ level (curve (c)). The peak ratio between the 1π and 5σ levels is also seen to be polarization-

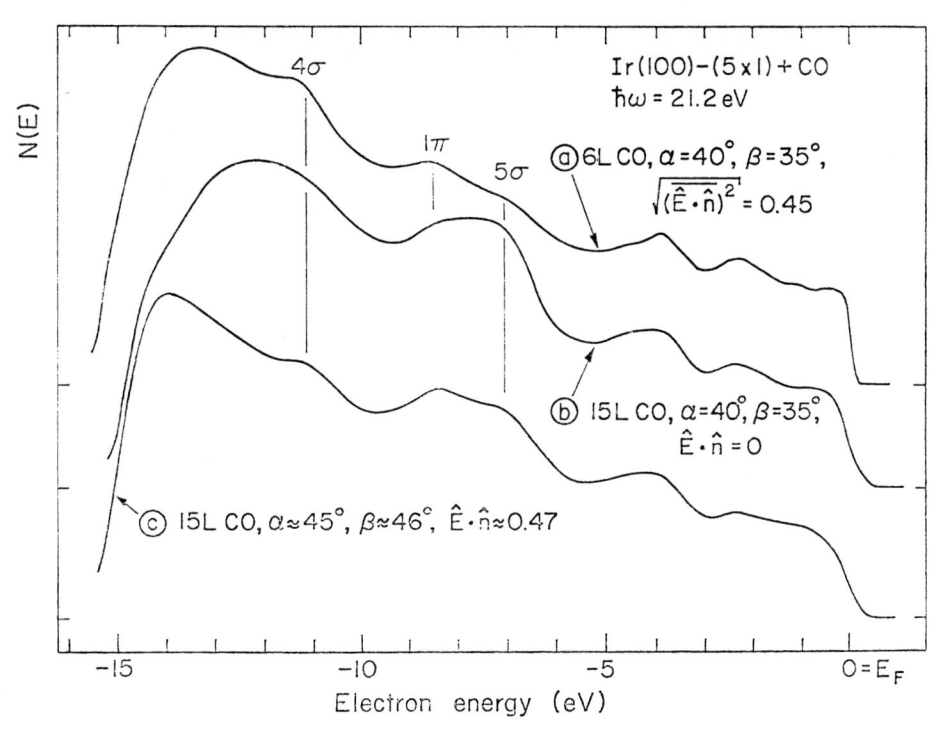

Fig. 9. Photoemission spectra showing polarization dependence of the photoionization cross-sections for the 4σ, 1π, and 5σ molecular orbitals of CO chemisorbed on Ir(100) - (5x1). $\hat{n}$ is the unit surface normal vector, $\hat{E}$ is the unit electric field vector of the incident radiation, α is the angle of incidence of the incident radiation relative to the surface normal, and β is the angle between the analyzer axis and the surface normal. (a) resonance lamp curve using non-polarized radiation; (b) synchrotron radiation curve using s-polarized radiation; (c) synchrotron radiation curve using s/p-polarized radiation. (From Brodén, et al., Ref. 21).

dependent. It is suggested from these results that the polarization effect is related to the component of the electric field vector parallel to the main axis of the chemisorbed molecule.[21] More detailed and systematic studies of other chemisorption systems from this viewpoint are needed to support this hypothesis. Another approach to the interpretation of the polarization effect on UPS spectra for chemisorbed systems is reported by Rowe.[31]

C. Correlation of CO-Chemisorption Behavior with Position in the Periodic Table

Brodén, et al.[12] have succeeded in establishing a criterion for the dissociative chemisorption behavior of carbon monoxide, nitrogen, nitric oxide and oxygen on a number of transition metal surfaces in terms of position in the Periodic Table. Use was made of a great many chemisorption studies available in the literature in which UV photoemission data was of great importance in revealing significant adsorption features. We review briefly the best documented case, i.e., that of CO-chemisorption.

We illustrate in Figure 10 the generally accepted end-on bonding configuration for CO-chemisorption on a transition metal surface in which binding interactions occur according to the donor-acceptor mechanism of the Blyholder model[32,33]. This model involves the formation of a bond between the occupied 5σ-donor level (Figure 10(a)) and unoccupied metal orbitals, implying a donation of electrons from CO to the metal. Back-bonding is then assumed to occur from occupied metal orbitals into the unoccupied 2π acceptor orbitals of CO. The 4σ and 1π orbitals (Fig. 10(b)) are thought not to participate appreciably in bond formation with the surface (Reasons for this are summarized in Ref. (21).). The net effect of donation and back-donation is thought to be a weakening of the carbon-oxygen bond, mainly because the 2π orbitals are antibonding[34,35] with respect to carbon and oxygen (See Ref. (21).). The chemisorbed CO molecule is thus stretched in the same sense that chemisorbed hydrocarbon molecules can be stretched, as previously discussed in this article.

It can be shown that an increase in the carbon-oxygen separation should in turn result in a corresponding increase in the net energetic separation of the non-bonding 4σ and 1π orbitals[21]. This separation, which can readily be deduced from the photoemission spectra for chemisorbed CO, has been plotted with reference to its position in the Periodic Table[21] (Figure 11). With one exception (Pt, 2.60 eV) the separation is observed to be larger than in the gas phase (2.75 eV), supporting the stretching prediction. Some interesting systematic trends now become immediately apparent. The separation increases as one proceeds leftwards in a given row or upwards in a Siven column. Also indicated in Figure 11 are the reported chemisorption states of CO, i.e., molecular or dissociated. This in turn shows a direct correlation between dissociation and geometric stretching. On the basis of the available evidence, Brodén, et al.,[21] have indicated a border line in Figure 11, to the right of which molecular adsorption can be expected and to the left of which dissociation of the molecule can be anticipated. For the borderline element tungsten, both types of adsorption are found depending on the crystal face.[36,37] However, in general, it seems that the primary factor determining the occurrence of molecular or dissociative adsorption is the electronic nature of the metal it-

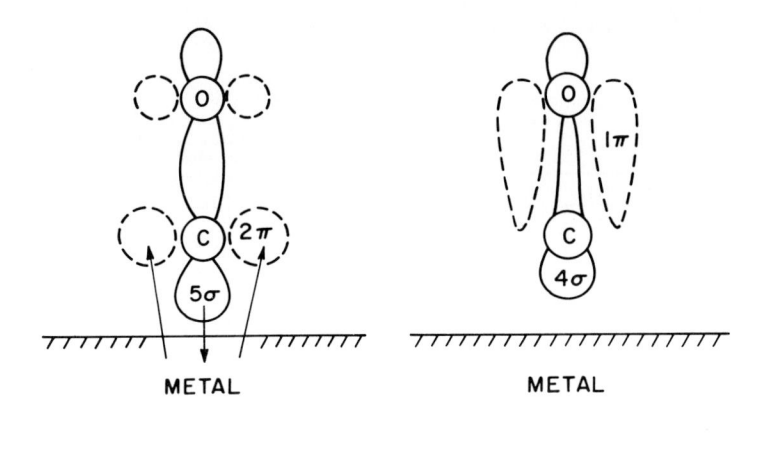

Fig. 10. (a) Blyholder Model (32,33) describing chemisorption of CO to a metal surface. The 2π and 5σ orbitals of CO are indicated by dashed and full lines, respectively. Only the orbitals participating in the surface bonding are indicated. It is assumed that the CO molecule bonds end-on with the carbon atom closest to the surface.

(b) The 4σ (solid lines) and 1π (dashed lines) orbitals of CO. These orbitals are thought not to participate in the surface bonding. (From Ref. 21).

self whereas the surface geometry is a secondary effect.[21]

Similar analyses have been performed by Brodén, et al.[21] for the N_2- and NO-chemisorption systems that lead to the following general conclusions: (1) The tendency of transition metal surfaces to dissociate diatomic chemisorbed molecules increases as one moves to the left in the Periodic Table; (2) surface crystallography and impurities influence adsorption patterns only for elements near the borderline between dissociative and associative adsorption. The primary factor determining adsorption behavior is the electronic nature of the metal surface itself.[21] A more detailed understanding of these conclusions will develop with more and improved information on the distribution and symmetry of metallic valence band orbitals.

D. Use of Combined Techniques to Provide a Chemisorption Model

1. Hydrogen and Deuterium on Tungsten. The chemisorption of

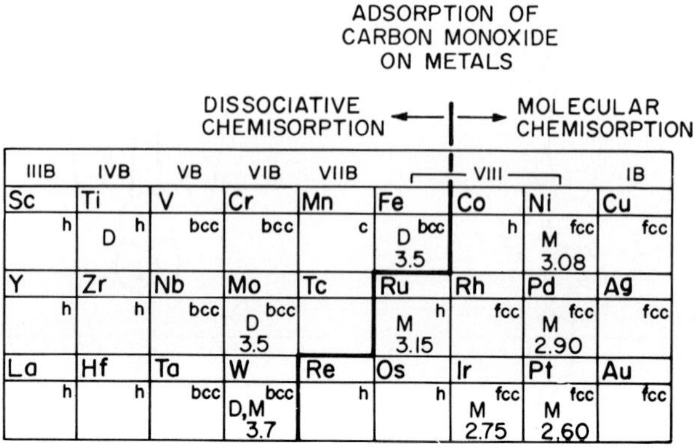

ADSORPTION OF CARBON MONOXIDE ON METALS

	DISSOCIATIVE CHEMISORPTION ←				→ MOLECULAR CHEMISORPTION		
IIIB	IVB	VB	VIB	VIIB	——— VIII ———		IB
Sc h	Ti D h	V bcc	Cr bcc	Mn c	Fe D bcc 3.5	Co h	Ni M fcc 3.08	Cu fcc
Y h	Zr h	Nb bcc	Mo D bcc 3.5	Tc	Ru M h 3.15	Rh fcc	Pd M fcc 2.90	Ag fcc
La h	Hf h	Ta bcc	W D,M bcc 3.7	Re h	Os h	Ir fcc	Pt M fcc 2.75	Au M fcc 2.60

fcc — FACE-CENTERED CUBIC
bcc — BODY-CENTERED CUBIC
c — CUBIC
h — HEXAGONAL

Fig. 11. Section of the Periodic Table showing room temperature adsorption behavior of CO. "M" denotes molecular adsorption and "D" dissociative adsorption on at least one surface, single crystal or polycrystalline. The numbers indicate the average energy separation (eV) of the nonbonding 4σ and 1π orbitals. The borderline between molecular and dissociative adsorption is indicated by the heavy line. (From Brodén, et al., Ref. 21).

hydrogen on W(100) has been investigated by a great number of experimental and theoretical techniques. This emphasis on hydrogen chemisorption probably originated from the expectation that hydrogen would be the simplest of adsorbates, and that the results from a variety of techniques would lead to a better understanding of chemisorption. In fact, even with the new data from valence band electron spectroscopies, the picture of hydrogen chemisorption remains largely incomplete.

Schmidt[38] has reviewed many techniques used to characterize this adsorption system. Briefly, the thermal desorption of a hydrogen saturated W(100) surface consists of two unequally sized peaks at ∼450K and ∼550K which have generally been designated as the β_1 and β_2 states, respectively. At low coverages all of the adsorbed hydrogen desorbs in the high temperature β_2 state with second-order kinetics, but for coverages greater than about 0.2 monolayer the desorption spectrum also shows the low temperature

β_1 state, which obeys first-order kinetics. Both states show complete isotopic mixing[39] even though the first-order kinetics often is considered to imply molecular adsorption. The LEED pattern characteristic of the β_2 regime is c(2x2), while at the higher coverages a new sequence of LEED patterns is observed. Schmidt[38] discusses the possible structures compatible with these observations. The conventional interpretation of these data would suggest two sequentially filled binding states (β_2 followed by β_1) of hydrogen on W(100).[38] The β_2 state would consist of atomic hydrogen, whereas the β_1 state would be molecular in nature.

Plummer, et al.[36,40] have also provided comprehensive overviews of hydrogen chemisorption on W(100) but with emphasis on those techniques which give direct data on energy levels of the adsorbate, i.e., UPS and vacuum tunneling field emission spectroscopy. Following the suggestion of Yates and Madey[41], Plummer and Bell[42] concluded from their tunneling measurements for hydrogen on W(100) that the β_1 and β_2 states did not fill sequentially, but that interactions between chemisorbed hydrogen species caused a reversible density-dependent conversion in the binding character, accompanied by a structural rearrangement starting at a coverage of ∿0.2 monolayer. Furthermore, no evidence for molecular hydrogen was found at any stage of the interaction.[42]

Photoemission data has been shown[36,40,43] to be consistent with a density-dependent conversion model accompanied by a structural rearrangement. Figure 12(a) shows selected difference spectra for (1) saturation coverage, denoted as $\beta_1 + \beta_2$, (2) ∿0.2 monolayer coverage, denoted as β_2, and (3) the difference between (1) and (2), denoted at β_1. The labels β_1 and β_2 do not imply that there really are two distinct states of the adsorbate on the surface, but rather that a thermal desorption spectrum would exhibit a two-peaked spectrum. An important realization to be drawn from this discussion is in fact that multiple peaks in desorption spectra can be provided by lateral interactions of the adsorbates[44,45] and are not necessarily proof of the existence of multiple binding sites[36,39-41]. The good agreement in the difference curves at different photon energies indicates that the structure in the difference curves is not induced by changes in the final state.[40]

The curves in Figure 12(b) compare the prediction of (1) a sequentially filled two state model and, (2) a simple two state density-dependent transition model, based on the difference curves in Figure 12(a). Note that the work function difference is linearly related to coverage in this system[39,46], so that Figure 12(b) is essentially a plot of peak position vs coverage. The two upper levels (at -1.2 eV and -3.5 eV, β_2 state) are thought to consist of tungsten d-orbitals influenced by the specific geometry of the site[36,40,47] and if so would be expected to be sensitive to changes in the ordering of the overlayer. The low-lying level (-5.7 eV,

[Figure showing UPS difference spectra with axes: DIFFERENCE CURVES (vertical) and ENERGY MEASURED FROM THE FERMI ENERGY (horizontal, 16 to 0). Labeled D₂ ON (100W), with dashed line 21.2 eV and solid line 16.9 eV. Three curves: #1 $\beta_1 + \beta_2$, $\Delta\phi = 0.8$ eV; #2 β_2, $\Delta\phi = 0.16$ eV; #3 β_1, #1−#2.]

(a)

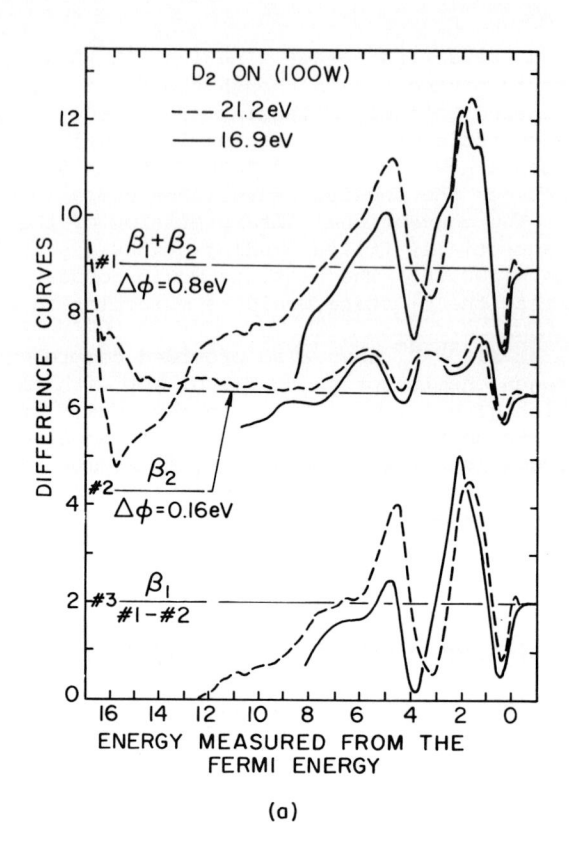

Fig. 12(a). UPS difference spectra for β_1 and β_2 states of adsorbed D_2 on W(100) at 300K. Curve 1 is for saturation coverage where both the β_1 and β_2 states will desorb. Curve 2 is for low coverage where only the β_2 state will desorb. Curve 3 is the difference between curves 1 and 2, which should give the contribution from the β_1 state.

β_2-state) could be a molecular orbital constructed from hydrogen 1s and tungsten 5d and 6s orbitals[36,40,47] and would thus be expected to be less sensitive to structural rearrangements. Figure 12(b) seems to bear this out, although as yet there is no satisfactory explanation for the linear shift of this level with coverage. The sequentially filled model predicts the peak positions vs. coverage curves shown by the dashed lines (Figure 12(b)). For coverages less than about 0.2 monolayers ($\Delta\phi = 0.16$ eV) the β_2 curve gives the peak positions whereas, for greater coverages the β_1 curve is linearly added to the β_2 curve in proportion to the coverage of β_1.[36,40] According to the two-state conversion model,

D₂ ON (100) W

○ $\hbar\omega = 21.2$ eV
× $\hbar\omega = 16.8$ eV

PEAK POSITION vs $\Delta\phi$ (eV)

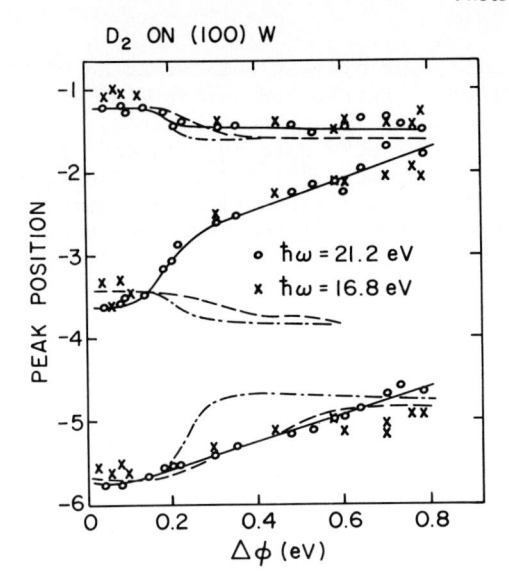

(b)

Fig. 12(b). Peak positions relative to the Fermi level for the difference curves for D₂ adsorption on W(100) at 300K, as a function of work function change. The dashed lines are the peak-position variations predicted for the sequentially filled two-state model, while the dot-dashed lines are for the two-state density-dependent transition model. The solid lines show the experimentally determined behavior for $\hbar\omega = 21.2$ eV (from Plummer, et al., Refs. 36,40).

again the β_2 state populates first up to a coverage of about 0.2 monolayers. From a coverage of ∿0.2 to ∿0.4, each adsorption event converts one β_2 state to a new state characterized by curve 1 in Figure 12(a). For coverages greater than ∿0.4, all of the β_2 state has been converted.[36,40] The dot-dashed lines in (Figure 12(b)) show the predictions of this model.

The two-state sequentially filled model fits the upper and lower peak positions well but not the middle peak, while the two-state conversion model fits only the upper peak. Neither model explains the data very satisfactorily. Two conclusions that can be drawn from Figure 12 are (1) that the inability of the sequentially filled model to fit the data rules out the existence of the low-coverage β_2 state at saturation[36,40], and (2) that even though the two-state conversion model is inadequate, the behavior of the peak positions might be explained by a modified conversion model involving more than two states as well as structural rearrangement.[40]

In conclusion, our present understanding of hydrogen chemisorption on W(100) is fairly consistent with Estrup and Anderson's[46] original model. Initially, hydrogen dissociatively adsorbs into alternate four-fold sites[48] forming islands[46] characterized by c(2x2) LEED pattern symmetry[46,48,49]. For surface coverages greater than ∼0.15-0.20, structural rearrangements are induced by the repulsive interactions of the hydrogen atoms.[50] Beyond a coverage of ∼0.26 the atoms are gradually compressed until there are two hydrogen atoms per surface tungsten atom[51], possibly located on each bridge site.[46]

2. <u>Use of UPS and LEED to Monitor Adsorption Kinetics.</u>
(Initial states of interaction of oxygen with clean single crystalline iron.) Brucker and Rhodin[52] have studied in detail the uptake of oxygen on α-Fe(100) in gradual steps using UPS in combination with LEED. Systematic evidence is produced for the occurrence of distinct stages in the room temperature oxidation process as part of a demonstration of the application of UPS and LEED to kinetic adsorption studies.

A serious experimental difficulty in surface studies of iron single crystals is in preparing and maintaining an adequately clean surface. An elaborate cleaning procedure was devised[52] to permit the single crystal iron surface study. The clean surface photoemission spectrum is shown in the bottom curve in Figure 13(a). The weak level at ∼5½ eV is due to a very small amount (estimated to be <1/20 of half a monolayer) of residually adsorbed oxygen. As the oxygen exposure is increased this 5½ eV level, which is attributed to emission from the 2p orbitals of atomic oxygen, broadened by chemical binding and multiplet splitting[52], eventually dominates the spectrum at saturation exposure (60 L) (1L = 10^{-6} Torr-sec). Also note (1) that metallic d-band emission just below E_F remains essentially unaltered for exposures <1½ L, (2) the apparent shift of the O(2p) level towards E_F for exposures between 4L and 60L, probably due to enhanced emission near 3 to 4 eV from oxidized $Fe^{2+}(3d^6)$ d-electron orbitals[52], and (3) the appearance of a new small peak at 9 to 10 eV for exposure > 4L, thought to be due to a multielectron process.[52] Although these features merit considerable interest in themselves, it is not appropriate to discuss them in greater detail here (see Ref. 52).

Changes in the work function, Φ, were followed by measuring the changes in the width of the photoemission spectra (see Figure 13(a)) and are plotted vs. exposures in Figure 13(b). Although the experimental uncertainties in this curve are rather large, it does provide an accurate indication of significant alterations in the surface dipole layer.

From the complete set of photoemission spectra for α-Fe(100) + oxygen, the integrated area under the main oxygen-derived level

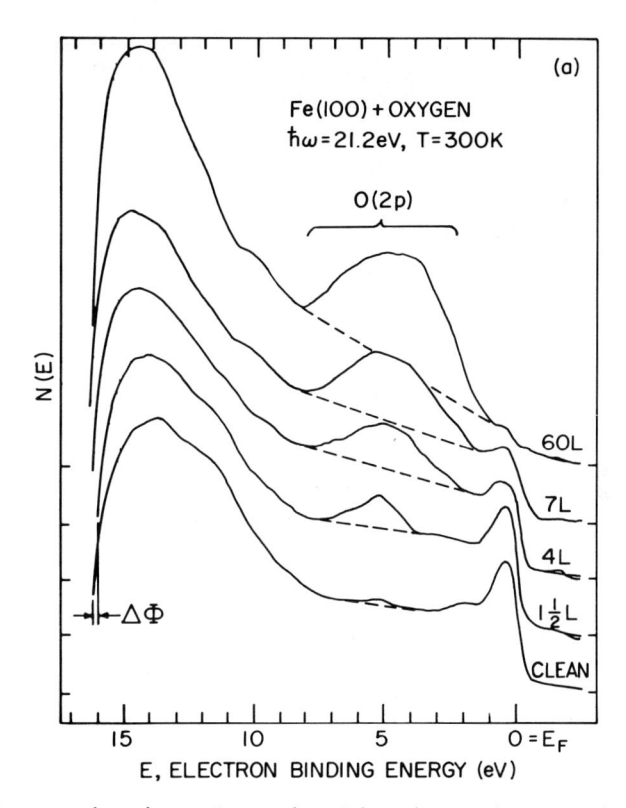

Fig. 13(a). Monitoring adsorption kinetics using UPS in combination with LEED.
(a) Photoemission spectra for oxygen-exposed Fe(100) measured with an angle-averaging retarding grid LEED optics. Exposures indicated in Langmuirs (1L = 10^{-6} Torr-sec).

(see dashed lines in Figure 13(a)) has been measured as a function of exposure (Figure 13(c)). This procedure becomes somewhat uncertain for exposures > 7L due to difficulties in accounting for background emission and contributions from oxidized iron orbitals. Nevertheless, it does indicate fundamental trends in the adsorption process. Fortunately, in the region of critical interest, 0 to 4L, these problems are minimal. The slope of the adsorption plot can be considered to be proportional to the sticking probability if the gas is adsorbed within an electron escape depth of the surface. A plot of the sticking coefficient (labelled "rate of oxygen uptake" in Figure 13(c)) is thus obtained by differentiation of the oxygen uptake curve. Also indicated in Figure 13(c) are the LEED patterns observed at various stages of the interaction. In the range 0-1½ L

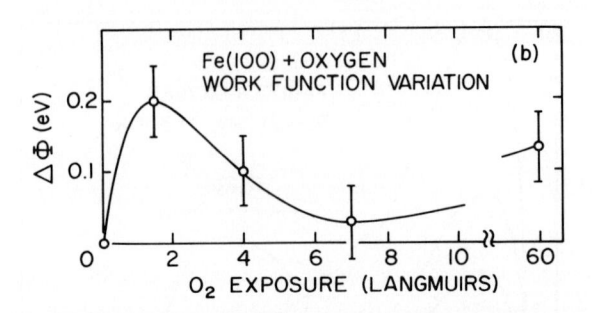

Fig. 13(b). Variation in work function, Φ, (See Figure 13(a)), as a function of exposure.

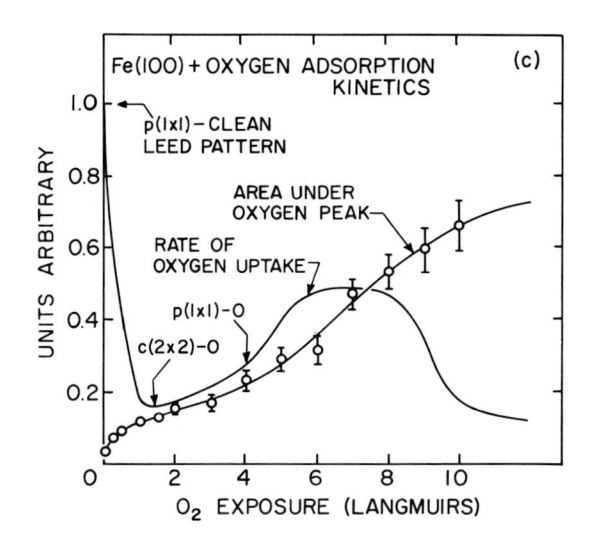

Fig. 13(c). Uptake and rate of uptake (sticking coefficient) curves vs. exposure. LEED patterns at various stages of adsorption indicated. (From Brucker and Rhodin, Ref. 52).

the LEED pattern changes from p(1x1) characteristic of clean α-Fe(100) to c(2x2)-O. The maximum intensity of the centered spots occurs between 1 and 2L. Upon additional exposure, the c(2x2) reverts to p(1x1)-O at 4L followed by a gradual fade out of all diffraction features by 7L.

These adsorption data are in excellent agreement with those obtained from LEED-Auger measurements of the room temperature adsorption of oxygen on α-Fe(100)[53], and are consistent with the following conclusions[52] (with reference to Figure 13):

(1) <u>Chemisorption Stage (0 to 1½ L)</u>. Adsorption of oxygen on α-Fe(100) corresponding to coverages less than that required to nucleate FeO (< 1½L) produces a superficially chemisorbed ordered overlayer consisting of oxygen atoms bound in alternate four-fold sites. Some controversy exists concerning the precise nature of LEED structural changes during the chemisorption stage of the interaction (See Ref. 52). The absence of change in the Fe d-orbitals supports the hypothesis that the metal substrate remains unreconstructed with little or no perturbation of the metal d-orbitals. The initial rapid increase in work function as oxygen is chemisorbed is considered to be due to charge transfer from the metal to the electronegative adatoms. The sticking coefficient falls off rapidly from its initial value of 1[53,54,55] as the availability of adsorption sites decreases. At 1½L, corresponding to saturation of the c(2x2) half monolayer, turning points are observed in the behavior of the work function, sticking coefficient, and LEED pattern.

(2) <u>Two-Dimensional Oxide Growth (1½L to 7L)</u>. Observations at higher exposures are consistent with the growth of an approximately two monolayer thick iron oxide (FeO) film. The observed photoemission features have been attributed to FeO by Eastman and Freeouf[56]. Oxygen molecules impinge on the chemisorbed layer and apparently give up their adsorption energy to the oxygen metal system[53,57] resulting in an exchange of atomic position[58] corresponding to the occurrence of some oxygen atoms beneath the surface. The work function decreases due to this rearrangement of the surface metal atoms and the electronegative adatoms. As FeO iron oxide islands nucleate and begin to grow laterally on the surface, the oxygen adsorption rate increases as the fraction of the surface occupied by the oxide increases. The occurrence of the p(1x1)-O LEED structure at 4L can be considered to mark the completion of the first oxide layer.

(3) <u>Three-Dimensional Oxide Growth (> 7L)</u>. Beyond 7L exposure, the work function begins to increase again due to three-dimensional iron-deficient oxide formation. The sticking coefficient begins to decrease again, the rate of further oxygen uptake being governed by the rate of three-dimensional oxide growth.

One advantage of using UPS (or XPS) for surface adsorption studies is that even sensitive adsorbates such as hydrocarbons are not disturbed by the photon probe used in the experimental measurement. Other probes such as high energy electron beams often alter adsorbed species.[59] In addition, desorption[59,52] and carburization[52] effects can also introduce undesirable complications when Auger spectroscopy is used. Disadvantages of UPS include the exceedingly short electron escape depth in adsorption studies involving the formation of multilayer structures and difficulties in applying quantitative analysis to the results. XPS applied to core level spectroscopy would provide useful supplementary information in these cases.

E. Directional Photoemission - Application to Surface Crystallography and Bonding Symmetry

1. <u>Simple Electron Scattering Model</u>. Recent experimental studies of angular dependence in UPS valence electron spectra[60] can provide additional information on both surface crystallography and bonding symmetry. Angular resolved photoemission measurements show promise of contributing to our basic understanding of the photoemission process, including clarification of the important distinction between the contributions from bulk and surface photoemission. In some instances photoelectrons, excited by direct optical interband transitions in the bulk, can leave the crystal unscattered. It is these electrons that carry information on the bulk band structure along a specific symmetry direction in $\underline{k}$-space. Another class of emitted electrons associated with the occurrence of surface photoemission may carry information on the surface density of initial states.

For chemisorption studies, another perhaps more important effect which may or may not be associated with the elastic scattering of the photoelectron by the surface lattice can yield valuable information on surface geometric structure and, if valence electrons are involved, on the nature of the surface bonding. Liebsch[61] has presented an exploratory theory of the angular resolved photoemission from localized adsorbate orbitals in which final state effects are discussed in detail. The symmetry of the adsorption site as well as the symmetry of the bonding orbitals is shown to dominate the angular distribution from localized adsorbate levels (see also Gadzuk[65] and Grimley[70]).

Figure 14(a) illustrates schematically the two contributions to the scattering amplitude of the adsorbate signal in this microscopic "one-step" model[61]: (1) direct emission from the orbital into a plane-wave final state, and (2) indirect emission from the orbital via back-scattering from the substrate lattice potential. When the indirect emission involves more than one scattering event, the multiple scattering analytical formation developed for LEED theory[62] may be applied. Liebsch[61] has in fact used LEED multiple-scattering theory to evaluate the indirect emission from the adsorbate for the general case of multiple elastic scattering and arbitrary atomic potentials.[61] In Figure 14(b), we show the intensity from an s orbital bound in two configurations as a function of azimuthal angle ϕ_f at constant polar angle θ_f and constant final state energy E_f. The four-fold symmetry of the substrate

Photoemission 459

(a)

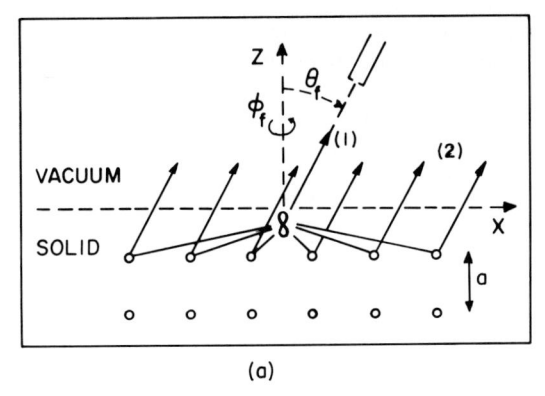

Fig. 14(a). Illustration of two processes contributing to photoemission from adsorbate orbital: (1) Direct emission into plane-wave final state, (2) indirect emission via back-scattering from the substrate. Only single scattering from the first layer is indicated.

(b)

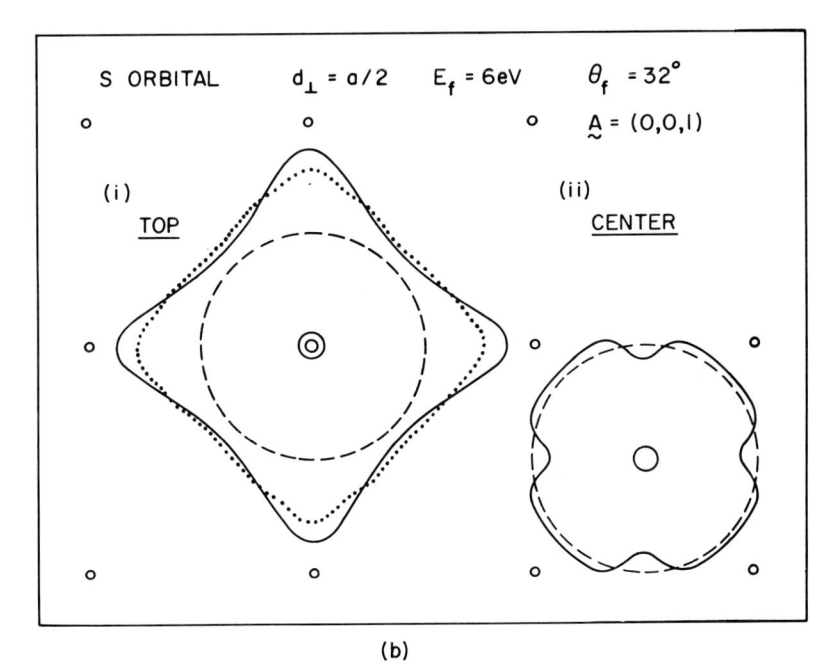

Fig. 14(b). Photoemission intensity (arbitrary units) as a function of azimuthal angle for an s orbital adsorbed in top (i) and center (ii) positions: single scattering (solid curves), multiple scattering (dotted curve, shown only in panel (i)), and no scattering (dashed curves). (From Liebsch, Ref. 61).

lattice is clearly reflected in those curves (solid and dotted) that include the effects of back-scattering from the substrate. One attractive feature of the theory is that it permits the separation of substrate and adsorbate geometries[61]: the positions at which extrema in the intensity occur are entirely determined by the substrate geometry whereas the relative intensities of these extrema are determined by the adsorption site. Also, since the adatom can act as a spherical electron wave source in contrast to the plane wave typical of LEED, single- or double-scattering approximations to the final state in photoemission from adsorbates may be more validly applied than in LEED.[61] This could be significant in facilitating the structure analysis since convergence of the iterative calculations can be readily established for the quantitative analyses of experimental data.

2. <u>Angular Resolved Photoemission Study: Hydrogen on Tungsten.</u>
Anderson and Lapeyre[63] have recently performed angle-resolved synchrotron photoemission measurements for quarter monolayer coverage of hydrogen on W(100). This coverage corresponds to full development of the c(2x2) β_2 state of adsorbed hydrogen on W(100) previously discussed. With the use of a cylindrical mirror energy analyzer modified with a movable aperture restricting the emission to a 4° cone, angle-resolved energy distribution curves were measured at azimuthal angle ϕ and polar angle θ, scanned over most of the emission hemisphere.[64]

Emission peaks within $\sim$2 eV of E_F in the UPS difference spectra for the β_2 state of H_2 on W(100) (see Figure 12(a)) have been attributed[36,40,47] to tungsten d-orbitals originating from the specific site geometry. For a given detector angle, the structure in the emission intensity as a function of final state energy is then determined by the band structure of the substrate.[38]

Anderson and Lapeyre[63] have studied the angular dependence of a hydrogen-induced doublet-structure within $\sim$2 eV of E_F over the entire emission hemisphere (Figure 15). This doublet of peaks is present for 14 eV $\leq \hbar\omega \leq$ 24 eV with maximum amplitude near 17 eV[63]. The electrons forming the doublet are emitted in narrow lobes in the four equivalent <110> azimuths. The lobes split into two sections in the angle ϕ as the polar angle moves towards the horizon. Anderson and Lapeyre propose an interpretation in which this hydrogen-induced emission is attributed to transitions between states of predominantly bulk W character, but scattered in specific new directions by an Umklapp process. This is the same Umklapp process that gives rise to the "extra" LEED centered spots indexed as ($\frac{1}{2},\frac{1}{2}$) in the well established c(2x2) LEED pattern.[63]

Their discussion makes it clear that any interpretation of photoemission spectra, especially angle-resolved spectra from chemisorbed surfaces, must account for the possibility of emission structure by the ordered overlayer.[63] Even in cases where the surface

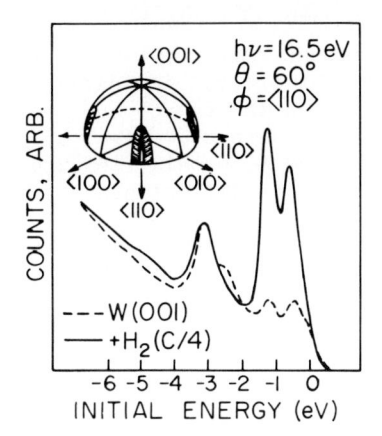

Fig. 15. Angle-resolved energy distribution curves for clean W(100) (dashed line) and for W(100) + quarter monolayer chemisorbed hydrogen (solid line). The insert is an emission hemisphere which shows (shading) the angular properties of the hydrogen-induced structure. (From Anderson and Lapeyre, Ref. 63).

periodicity is not altered, chemisorbed atoms might modify the photoemission current through their influence on the scattering amplitudes in the Umklapp processes.[63]

SUMMARY AND FUTURE PROSPECTS

In terms of special selected examples, a general overview of the present status of ultraviolet photoemission spectroscopy (UPS) as a tool for the study of chemisorptive bonding on metals is presented. UPS-spectroscopy is a simple and versatile experimental probe with minimal perturbation on the nature and configuration of the solid surface. The UV-photon beam is nondestructive to surface chemical complexes. Reasonably useful energy resolution (0.1-0.3 eV) is attainable using relatively simple energy analyzers. Primary applications to chemisorption include energy level measurements and chemical "fingerprinting" of the bonding orbitals in chemisorbed species. Photoemission is manifestly a multielectron process and phenomena such as final-state relaxation and electron shake-up complicate theoretical descriptions of absolute binding energies. Further uncertainties due to molecular distortion can arise in systems showing strong chemisorption. Despite these constraints UPS provides one of the best single techniques for the study of bonding at surfaces. When coupled with other complimentary surface-sensitive techniques, it is a promising approach to the probing of the detailed microscopic nature of heterogeneous surface systems.

There are some recent rather exciting applications which utilize the dependency of orbital ionization features on the angle of detection as well as on the energy and polarization of the incident radiation. Improved angular-resolved photoemission spectrometers are being developed to provide measurements that may allow determination of the geometrical orientation of adsorbed species and also to offer insight into the molecular orbital structure of the surface complex. Energy-dependent photoionization cross-section measurements of both adsorbed and gas phase molecules, exploiting synchrotron radiation, puts identification of surface energy levels on a firmer basis. The use of polarized radiation to enhance surface features and possibly to give indications of molecular orbital structure and orientation also promises to be very useful. At this time it is premature to claim an understanding of the microscopic electronic structure of solid surfaces with this type of technique. Close interplay, however, with theoretical methods can realistically provide a more accurate and improved understanding of electronic and geometric properties governing chemical reactions on surfaces.

Major questions have also been raised as to the connection between the interpretation of simple chemical reactions on well-defined single-crystal metals and the comparative usefulness of different approaches to the electron spectroscopy of practical catalytic materials. It has been pointed out[68] that the objective of determining electronic and atomic structure can be useful but difficult when a catalytic reaction takes place because the molecules are both larger and more complicated, co-adsorption or sequential chemisorption can occur, and order-disorder phenomena are often involved. The usefulness of photoemission studies to evaluate the role of chemical bonding and molecular properties in these systems has, as yet, not been explored.

ACKNOWLEDGEMENTS

Support from the National Science Foundation Grant DMR-71-01769-A02 and the Cornell Materials Science Center is acknowledged.

REFERENCES

1. C.B. Duke and R.L. Park, Physics Today, August, 1972.
2. For an excellent introductory survey to fundamental concepts and techniques in Surface Physics, see Physics Today, April 1975.

3. For a broad in-depth review of both theoretical and experimental approaches to surface science studies, see *Electronic Structure and Reactivity on Metal Surfaces*, NATO Advanced Study Institute, Namur, Belgium, September 1975 (to be published by Plenum Press, NATO Advanced Study Institute Series).
4. Reviews concentrating mainly on UPS include:
   (a) N.V. Smith, Crit. Rev. Solid State Sci. 2, 45 (1971),
   (b) D.E. Eastman, in *Techniques of Metals Research*, Vol. 6, Ed. E. Passaglia, Interscience, New York (1972),
   (c) W.E. Spicer, Comments in Solid State Phys. 5, 105 (1973),
   (d) B. Feuerbacher, Surface Sci. 47, 115 (1975); Surface Sci. 48, 99 (1975),
   (e) E.W. Plummer, in *Topics in Applied Physics*, Vol. 4, Ed. R. Gomer, Springer-Verlag, Berlin (1975).
5. Detailed discussions of specific aspects of both UPS and XPS can be found in:
   (a) *Electron Spectroscopy*, Ed. D.A. Shirley, North Holland, Amsterdam (1972),
   (b) J. Elec. Spectroscopy 5, 1-1136 (1974),
   (c) Proceedings of the Second International Conference on Solid Surfaces, Kyoto, Japan, March 1974, in Japan J. Appl. Phys. Suppl. 2, Part 2 (1974),
   (d) Proceedings of the IV International Conference on Vacuum Ultraviolet Radiation Physics, Hamburg, West Germany, July 1974, in *Vacuum Ultraviolet Radiation Physics*, Eds. E.E. Koch, R. Haensel, and C. Kunz, Pergaman, Braunschweig (1974).
6. Many reviews of the adsorption of gases on solids have been written and no attempt to reference all of them is made here. A modern introduction to the present state of the subject is provided by: T.N. Rhodin and D.L. Adams, Adsorption of Gases on Solids, in *Treatise on Solid State Chemistry*, Chapter 6, Vol. 6, Ed. N.B. Hannay, Plenum Press, New York (1976).
7. D.E. Eastman and J.K. Cashion, Phys. Rev. Lett. 27, 1520 (1971).
8. M.L. Perlman, E.M. Rowe, and R.E. Watson, Physics Today, July, 1974.
9. S. Doniach, I. Lindau, W.E. Spicer, and H. Winick, J. Vac. Sci. Tech. 12, 1123 (1975).
10. C.N. Berglund and W.E. Spicer, Phys. Rev. A 136, 1030, 1044 (1964).
11. (a) I. Adawi, Phys. Rev. A 134, 788 (1964).
    (b) J.G. Endriz, Phys. Rev. B 7, 3464 (1973).
    (c) P.J. Feibelman, Phys. Rev. Lett. 34, 1092 (1975).
12. (a) J.G. Endriz and W.E. Spicer, Phys. Rev. Lett. 27, 570 (1971).
    (b) J.G. Endriz and W.E. Spicer, Phys. Rev. B 4, 4159 (1971).
13. (a) G.D. Mahan, Phys. Rev. Lett. 24, 1068 (1970).
    (b) G.D. Mahan, Phys. Rev. B 2, 4334 (1970).
    (c) N.W. Ashcroft and W.L. Schaich, in *Electronic Density of States*, Ed. L.H. Bennet, Natl. Bur. Stds. Publ. No. 323 U.S. GPO, Washington, D.C. (1971).

    (d) W.L. Schaich and N.W. Ashcroft, Solid State Commun. **8**, 1959 (1970).
    (e) L. Sutton, Phys. Rev. Lett. **24**, 386 (1970).
    (f) W.L. Schaich and N.W. Ashcroft, Phys. Rev. B **3**, 2452 (1971).
    (g) P.J. Feibelman, Surf. Sci. **46**, 558 (1974).
14. G.K. Wertheim and S. Hufner, Phys. Rev. Lett. **28**, 1028 (1972).
15. K.Y. Yu, J.C. McMenamin, and W.E. Spicer, Surf. Sci. **50**, 149 (1975).
16. J.W. Gadzuk, J. Vac. Sci. Techn. **12**, 289 (1975).
17. J.C. Tracy and P.W. Palmberg, J. Chem. Phys. **51**, 4852 (1969).
18. D.E. Eastman and J.E. Demuth, Japan J. Appl. Phys. Suppl. **2**, pt. 2, 827 (1974).
19. T. Gufstafsson, E.W. Plummer, D.E. Eastman, and J.L. Freeouf, Solid State Commun. **17**, 391 (1975).
20. G. Brodén and T.N. Rhodin, Solid State Commun. **18**, 105 (1976).
21. G. Brodén, T.N. Rhodin, C.F. Brucker, R. Benbow, and Z. Hurych, to be published in Surf. Sci. (1976).
22. D.E. Eastman and J.E. Demuth, Phys. Rev. Lett. **32**, 1123 (1974).
23. J.E. Demuth and D.E. Eastman, Private communication.
24. G. Brodén and T.N. Rhodin, Chem. Phys. Lett., **40**, 247 (1976).
25. C.F. Brucker and T.N. Rhodin, to be published (1976).
26. R.S. Milliken, J. Amer. Chem. Soc. **74**, 811 (1952).
27. T.B. Grimley, in *Molecular Processes in Solid Surfaces*, Eds. E. Dranglis, R.D. Getz, and R.I. Jaffee, McGraw-Hill, New York (1969).
28. D.W. Turner, C. Baker, A.D. Baker, and C.R. Brundle, *Molecular Photoelectron Spectroscopy*, Wiley Interscience, London (1970).
29. E.W. Plummer, B.J. Waclawski, and T.V. Vorsburger, Chem. Phys. Lett. **28**, 510 (1974).
30. D.E. Eastman and J.K. Cashion, Phys. Rev. Lett. **27**, 1520 (1971).
31. J.E. Rowe, Phys. Rev. Lett. **34**, 398 (1975).
32. G. Blyholder, J. Phys. Chem. **68**, 2772 (1964).
33. G. Doyen and G. Ertl, Surf. Sci. **43**, 119 (1974).
34. W.L. Jorgensen and L. Salem, *The Organic Chemist's Book of Orbitals*, pp. 78,80, Academic Press, New York (1973).
35. G. Blyholder, J. Vac. Sci. Technol. **11**, 865 (1974).
36. E.W. Plummer, in *Topics in Applied Physics*, Vol. 4, Ed. R. Gomer, Spring-Verlag, Berlin (1975).
37. J.M. Baker and D.E. Eastman, J. Vac. Sci. Technol. **10**, 223 (1975).
38. L.D. Schmidt, in *Topics in Applied Physics*, Vol. 4, Ed. R. Gomer, Springer-Verlag, Berlin (1975).
39. T.E. Madey and J.T. Yates, Jr., *Structure Et Proprietes Des Surfaces Des Solids*, Editions Du Centre National de la Recherche Scientifique, Paris, 1970, #187.
40. E.W. Plummer, B.J. Waclawski, T.V. Vorburger, and C.E. Kuyatt, Private Communication.
41. J. T. Yates, Jr. and T.E. Madey, J. Vac. Sci. Technol. **8**, 63 (1971).

42. E.W. Plummer and A.E. Bell, J. Vac. Sci. Techn. $\underline{9}$, 583 (1972).
43. B. Feurbacher and B. Fitton, Phys. Rev. B $\underline{8}$, 4890 (1973).
44. T. Toya, J. Vac. Sci. Tech. $\underline{9}$, 890 (1972).
45. D.L. Adams, Surf. Sci. $\underline{42}$, 12 (1974).
46. P.J. Estrup and J. Anderson, J. Chem. Phys. $\underline{45}$, 2254 (1966).
47. L.W. Anders, R.S. Hansen, and L.S. Bartell, J. Chem. Phys. $\underline{59}$, 5277 (1973).
48. K. Yonehara and L.D. Schmidt, Surf. Sci. $\underline{25}$, 238 (1971).
49. D.L. Adams and L.H. Germer, Surf. Sci. $\underline{23}$, 419 (1970).
50. T.E. Einstein and J.R. Schrieffer, Phys. Rev. B $\underline{7}$, 3629 (1973).
51. T.E. Madey, Surf. Sci. $\underline{36}$, 281 (1973).
52. C.F. Brucker and T.N. Rhodin, Surf. Sci., $\underline{57}$, 523 (1976).
53. G.W. Simmons and D.W. Dwyer, Surf. Sci. $\underline{48}$, 373 (1975).
54. T. Horiguchi and S. Nakanishi, Japan. J. Appl. Phys. Suppl. $\underline{2}$, pt. 2, 89 (1974).
55. A.M. Horgan and D.A. King, Surf. Sci. $\underline{23}$, 259 (1970).
56. D.E. Eastman and J.L. Freeouf, Phys. Rev. Lett. $\underline{34}$, 395 (1975).
57. G.K. Hall and C.H.B. Mee, Surf. Sci. $\underline{28}$, 598 (1971).
58. M.A.H. Lanyon and B.M.W. Tradnell, Proc. Roy. Soc. A $\underline{277}$, 387 (1955).
59. J.P. Coad, M. Gettings, and J.C. Riviere, Faraday Disc. $\underline{60}$, (1975).
60. (a) N.V. Smith and M.M. Traum, Phys. Rev. Lett. $\underline{31}$, 1247 (1973). This article includes a comprehensive list of references to previous angular photoemission studies.
    (b) J.E. Rowe, M.M. Traum, and N.V. Smith, Phys. Rev. Lett. $\underline{33}$, 1333 (1974).
    (c) T. Gufstafsson, P.O. Nilsson, and L. Wallden, Phys. Rev. A$\underline{37}$, 121 (1974).
    (d) M.M. Traum, N.V. Smith, and F.I. Di Salvo, Phys. Rev. Let. $\underline{32}$, 1241 (1974).
    (e) M.M. Traum, J.E. Rowe, and N.V. Smith, J. Vac. Sci. Techn. $\underline{12}$, 298 (1975).
    (f) B.J. Waclawski, T.V. Vorburger, and R.J. Stein, J. Vac. Sci. Techn. $\underline{12}$, 301 (1975).
    (g) D.R. Lloyd, C. M. Quinn, and N.V. Richardson, J. Phys. C $\underline{8}$, L371 (1975).
    (h) P.M. Williams, P. Butcher, J. Wood, and K. Jacobi, Private Communication.
61. A. Liebsch, Phys. Rev. Lett. $\underline{32}$, 1203 (1974); to be published.
62. A complete and up-to-date list of articles on multiple-scattering processes in LEED can be found in Reference 6.
63. J. Anderson and G.J. Lapeyre, Phys. Rev. Lett. $\underline{36}$, 376 (1976).
64. The geometry is summarized by N.V. Smith, M.M. Traum, J.A. Knapp, J. Anderson, and G.J. Lapeyre (to be published).
65. J.W. Gadzuk, Solid State Commun. $\underline{15}$, 1011 (1974).
66. J.W. Gadzuk, Surface Sci. $\underline{43}$, 44 (1974); Surf. Sci. $\underline{6}$, 133 (1967); Phys. Rev. B $\underline{1}$, 2210 (1970).
67. C.R. Brundle, Surf. Sci. $\underline{48}$, 99 (1975).

68. T.N. Rhodin and J.F. Antonini, in *The Physical Basis for Heterogeneous Catalysis*, Eds. E. Drauglis and R.I. Jaffee, Plenum Press, New York (1975), p. 247.
69. G. Ertl, *Ibid.*, p. 189.
70. T.B. Grimley, Faraday Disc. Chem. Soc. **58**, 1 (1974).

The Study of Organic Reactions on the Surface of Magnetic Pigments by X-Ray Photoelectron Spectroscopy (ESCA)

Robert S. Haines

IBM Corporation
General Products Division
P. O. Box 1900
Boulder, Colorado 80302

    The study of organic reactions on the surface of pigments submicrometer in size requires more sophisticated methods of analysis of the XPS spectra since curve tracing, Fourier transform deconvolution, etc. are inadequate to obtain meaningful results. An APL program that uses a modified Van Cittert algorithm was developed to deconvolute the x-ray source component of the XPS spectra.

    This method is then used to deconvolute XPS spectra of the surface of submicrometer magnetic metal oxide particles and their interaction with water and two organic materials.

INTRODUCTION

    The more demanding tribology requirements, higher magnetic density and longer life required of particulate magnetic surfaces require a continued look at other methods of studying particulate particle surfaces and their interactions with solvents and polymeric binder components. Commercial infrared units have been used in this area for more than two decades, but since the magnetic pigments are opaque to infrared, solvents have been used to extract components from the interface. An attempt is made to use x-ray photoelectron spectroscopy at this interface area since it does not exhibit this opacity.

    The two magnetic pigments studied are the non-conductive ferrimagnetic gamma iron oxide and the conductive ferromagnetic chromium dioxide. These magnetic pigments are submicrometer in size and as powders show more x-ray scattering than continuous smooth films. Thus, it is essential to use proper analytical techniques to obtain meaningful spectra. Although the reactions of many solvents and a few polymers have been studied at the pigment interface, this discussion will be limited to water, cyclohexanone, and the 40/60 copolymers of hexafluoropropene with vinylidene fluoride.

EXPERIMENTAL

The x-ray photoelectron spectrometer used was a McPherson 36 with a Mg k$\alpha_{1,2}$ x-ray source. This instrument had a work function of 4.3 eV.

It is essential to deconvolute the line width contributions of the x-ray source used in the x-ray photoelectron spectrometer to obtain meaningful spectra from the broad bell shaped curves obtained from the spectrometer. See CrO_3 bell shaped curve (Fig. 1).

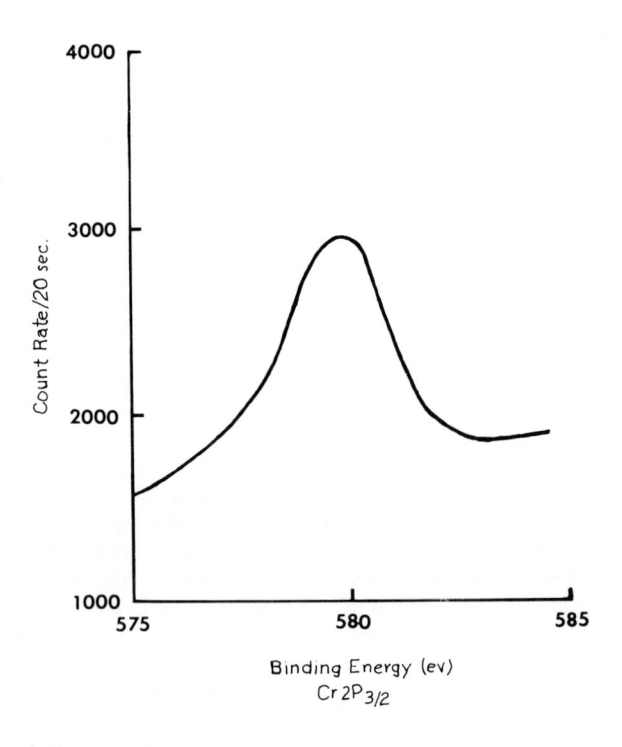

Fig. 1. CrO_3 Bell Shaped Curve

In order that these spectra be accurate, it is necessary that the best method possible be used for this deconvolution. To do this, a number of deconvolution programs were developed using various deconvolution methods on closely spaced peaks of different heights of hypothetical data to see how well the peaks were resolved. The methods used were the Fourier transform, the inverse filtering, the matrix and the Van Cittert algorithm[1-2]. The Van Cittert algorithm method was chosen as the preferred method since it gave less blow up when small errors were introduced into the program.

The Van Cittert algorithm is an iterative method in which one goes through repetitive iterations in the time domain. Usually less than ten iterations are needed to fit the curve.

$$f_i(x) = F_{i-1}(x) + f_0(x) - g(x) * f_{i-1}(x)$$

$f_i(x)$ is the ith iteration

$f_0(x)$ is the original data

$g(x)$ is the instrument response function or source.

This original algorithm has been modified with a convergency factor γ and a three point smoothing factor S which is used at the end of each iteration. Thus, the modified algorithm is:

$$f_i(x) = S(x) * \{f_{i-1}(x) + \gamma[(f_0(x) - g(x) * f_{i-1}(x)]\}$$

It is noted that in all cases, the deconvoluted spectra shows binding energy increasing from left to right along the abscissa. This is opposite to the way the data comes from the McPherson 36, but it is the only way one can operate programming languages.

The deconvoluted spectrum (Fig. 2) is convoluted again with the x-ray line width function and plotted sequentially with the original spectrum to make certain the original spectrum has not been altered in the deconvolution process. The sample spectrum shown which uses this method is that of $Cr2P_{3/2}$ of a sample of CrO_3 powder (Fig. 3). This AR grade CrO_3 powder was ground and dried by heating it over barium perchlorate desiccant for a week at 150°C before running an XPS spectrum on it. The sample was prepared by rubbing the CrO_3 powder on a rectangular piece of lead. Some graphite particles were also rubbed on the lead to give a C_{1s} reference.

The $Cr2P_{3/2}$ peak at 577.1 eV of the Cr_2O_3 impurity is brought out sharply by this deconvolution process (Fig. 2).

Standards were prepared for $Cr2P_{3/2}$ spectrum for chromium in the III and VI valence state from AR grade dried samples of Cr_2O_3

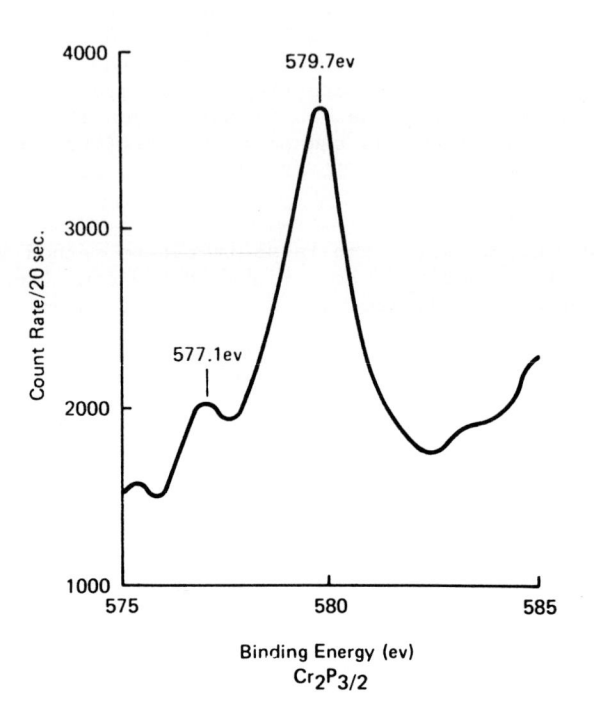

Fig. 2. Deconvoluted CrO₃ Spectrum

and CrO₃ with corresponding O_{1s} spectra since literature values were not in agreement and the spectra were not properly deconvoluted[3-4].

All powdered samples were prepared for insertion into the spectrometer by being rubbed on lead slides. Since the powders were harder than the lead, they embedded easily.

γ Fe₂O₃ and CrO₂

Commercial grades of γFe₂O₃ and CrO₂ were used. The γFe₂O₃ had $SO_4^=$ ions on it and the CrO₂ had $CrO_4^=$ ions on it as determined by wet chemical analysis.

The physical and magnetic properties determined on these two pigments are listed in Table 1.

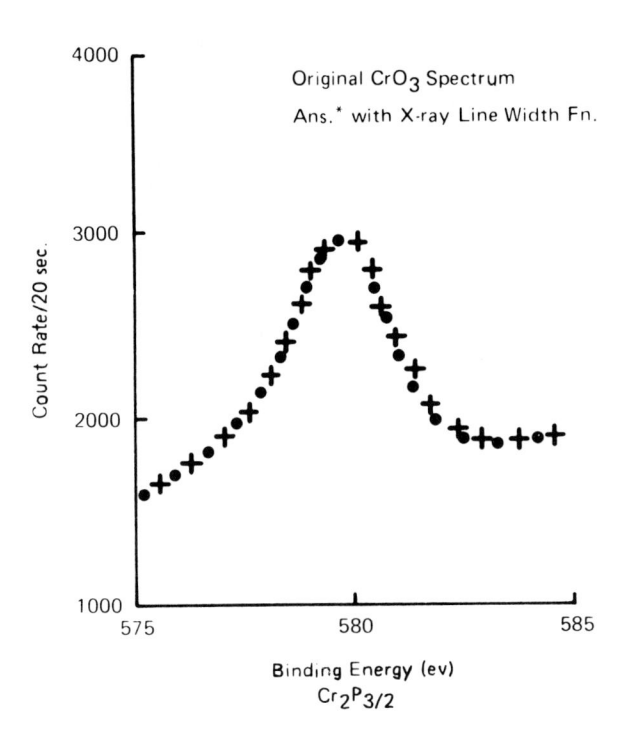

Fig. 3. Composite CrO₃ Spectrum

The deconvoluted XPS binding energy peak values obtained for these two pigments are tabulated below with the values for Cr_2O_3 and CrO_3 in Table 2.

The γFe_2O_3 shows both ferrous and ferric sulfate as an impurity. The presence of two $Fe2P_{3/2}$ peaks of equal intensity and only one O_{1s} peak needs further enlightening. However, γFe_2O_3 has a cubic super-structure with deficient oxygen sites when it is prepared by oxidizing the spinel Fe_3O_4.

The $Cr2P_{3/2}$ binding energy values for the Cr III, IV and VI oxides do not correspond to the valence values. This can be explained by the electronic relaxation effect of electronic charge flow to the core hole in the positive ion produced by photoemission since CrO_2 is a conductive oxide. The conductivity of CrO_2 is 2×10^4 mho/m at room temperature[5], whereas, that of Cr_2O_3 is 8×10^{-2} mho/m at room temperature.

Table 1

Physical and Magnetic Properties

Pigment	Surface Area*	σ_s 4000 oe[+] emu/gm	Hc[+] oersteds	Average Acicular Particle [t] Size
γFe_2O_3	20 sq. meter/gm	73	264	0.6 um x 0.1 um
CrO_2	33 sq. meter/gm	74	508	0.4 um x 0.04 um

*Determined by BET method using nitrogen adsorption.

+Determined on a vibrating sample magnetometer (VSM) with a 4000 oersted field.

[t]Determined by a Bausch and Lomb Omnicon image analyzer of transmission electron miscroscope pictures.

Table 2

Binding Energy of Elements

Pigment	Fe2P$_{3/2}$ (eV)	Cr2P$_{3/2}$ (eV)	S2P$_{3/2}$ (eV)	O$_{1s}$ (eV)	C$_{1s}$ (eV)
γFe_2O_3	710.6 711.4		168.7 169.6	530.2	284.9
CrO_2		575.6		528.0	284.9
Cr_2O_3		577.1		531.1	285
CrO_3		579.7		530.7	285

The binding energy of the O$_{1s}$ peak for the more covalent Cr_2O_3 is higher than that of the O$_{1s}$ peak in CrO_3.

This has been observed before by Allen et al.[4].

Water Adsorption

Since these two pigments cannot be degassed at elevated temperature because of their thermal instability, the oxygen O_{1s} peak shows some water after the pigments have been heated to 220°C for 16 hours; the CrO_2 more so than the γFe_2O_3. After they have been refluxed in water for 16 hours and air dried, they have very pronounced O_{1s} peaks due to water as seen in the deconvoluted O_{1s} spectra below (Fig. 4 & 5).

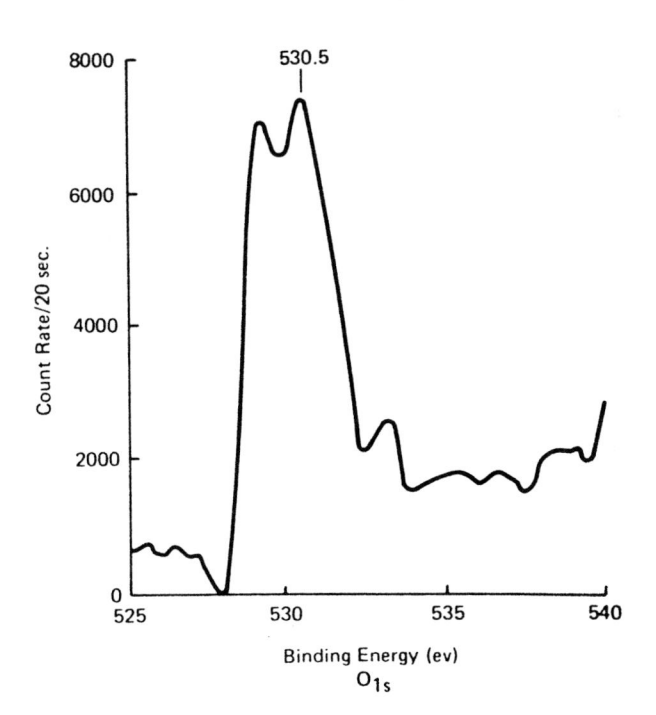

Fig. 4. Water Adsorbed on CrO_2

The O_{1s} peak (Fig. 5) has shifted from 535.1 eV for liquid water[6], to 530.5 eV for water adsorbed on CrO_2 to 531.3 eV for water adsorbed on γFe_2O_3. Thus, the water is more strongly adsorbed on the CrO_2 than on the γFe_2O_3.

This adsorption of water is accompanied by a change in saturation magnetic moment in the case of the ferromagnetic CrO_2 which is largely reversible when water is adsorbed and desorbed on the CrO_2.

Cyclohexanone Adsorption

Cyclohexanone was tumbled separately with the two magnetic pigments for five days at room temperature to assimilate conditions produced during pigment dispersion. The pigment was then precipitated with a magnet and the supernatant solvent evaporated and the residue analyzed by infrared. The small amount of residue from the γFe_2O_3 showed a dimer of (2-cyclohexanone)-1-cyclohexanol had been formed. The larger amount of residue from the CrO_2 was mainly adipic acid. XPS analysis of the precipated pigments dried for one hour at 116°C showed nothing for the γFe_2O_3, but a strong band at 287.7 eV for the -CO- binding energy of the C_{1s} spectra and a strong band at 530.6 eV for the -CO- binding energy of the O_{1s} spectra of the CrO_2. This is no doubt due to the adipic acid.

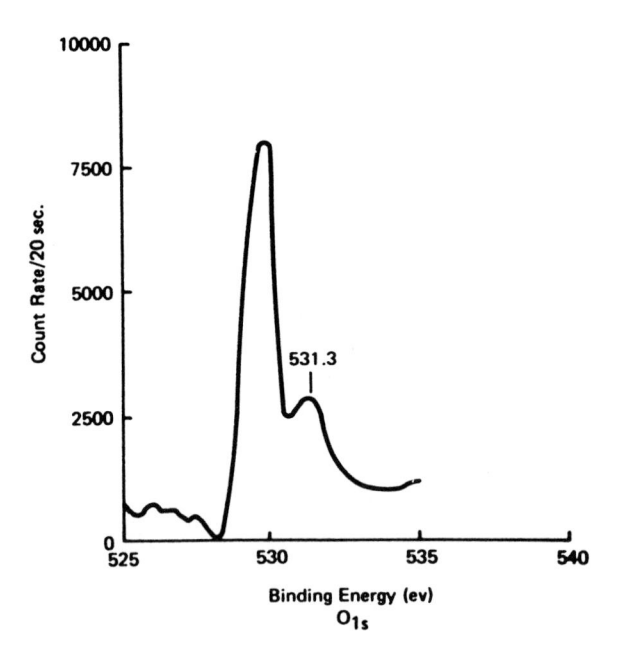

Fig. 5. Water Adsorbed on γFe_2O_3

Copolymer Adsorption
(40/60 Hexafluoropropene and Vinylidene Fluoride)

The two magnetic pigments were tumbled separately with a four percent solution of fluoro-copolymer for five days. The magnetic pigment was separated with a magnet and a portion of the supernatant copolymer deposited as a thin film on lead slides. The precipitated pigments were dried at 116°C for one hour.

When deconvoluted XPS spectra was obtained for C_{1s} and F_{1s} binding energies, no abnormalities were seen from the standard ratios of the four binding energy peaks for the C_{1s} spectra on the copolymer obtained from the supernatant solution. However, strong differences were seen in the copolymer adsorbed on the γFe_2O_3 CrO_2 pigments. Fig. 6 is the copolymer adsorbed on CrO_2.

The -CF_2- peak is no longer the dominant peak, but rather the -CH_2- peak with a slight 287.7 eV peak. This 287.7 eV peak is due to unsaturation and oxidation in the copolymer.

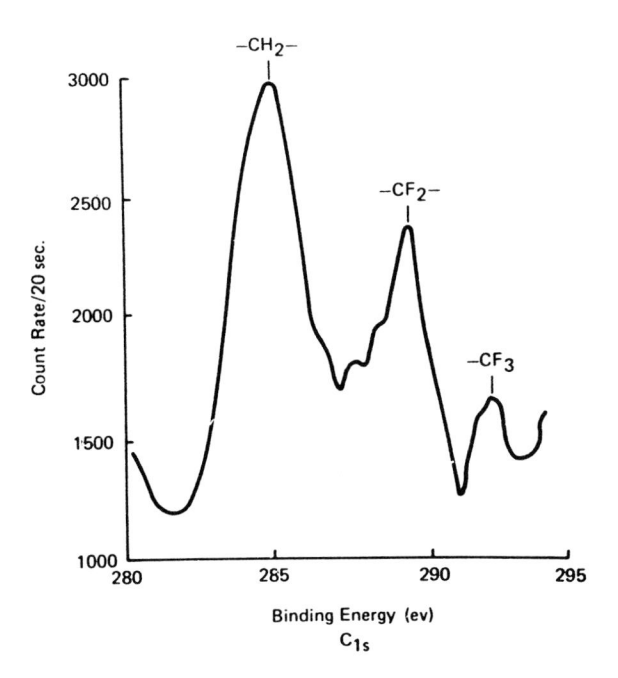

Fig. 6. Deconvoluted C_{1s} Spectrum of Copolymer Adsorbed on CrO_2

The binding energies of the fluorocarbon peaks of the fluoro copolymer on CrO_2 are approximately 0.8 eV lower than on the γFe_2O_3. This can be explained as due to the extra atomic relaxation effect of electronic charge flow from the CrO_2 neighboring atoms to the core hole in the positive ion produced by photoemission in the fluoropolymer since CrO_2 is a conductive oxide[7,8].

CONCLUSIONS

The Van Cittert algorithm deconvolution method deconvolutes XPS spectra very well so that meaningful reproducible binding energies can be obtained on small particle size pigments. The XPS spectra, besides characterizing the oxide surfaces, also has successfully interpreted the interface of the oxide surfaces with water, a solvent and a copolymer.

ACKNOWLEDGEMENTS

The author wishes to acknowledge the assistance of Roy Dent and Larry Viele in developing the Van Cittert algorithm deconvolution program; Gerald Sage for preparing samples and operating the VSM and Harry McCabe for operating the McPherson 36 spectrometer.

REFERENCES

1. P. H. Van Cittert, Z. Physik, 69, 298 (1931).
2. G. K. Wertheim, J. of Electron Spectroscopy and Related Phenomena, 6, 239-251 (1975).
3. J. C. Carver, G. K. Schweitz and T. A. Carlson, J. Chem. Phys. 57, 973 (1972).
4. G. C. Allen, M. T. Curtis, A. J. Hooper and P. M. Tucker, J. Chem. Soc. DA1973 (16); 1675 (1973).
5. G. V. Samsonov, *The Oxide Handbook*, p. 271, IFI/Plenum (1973).
6. J. C. Fuggle, L. M. Watson and D. J. Fabian, Surface Science 49, 66 (1975).
7. P. Citrin, R. W. Shaw, A. Packer and T. D. Thomas, *Electron Spectroscopy*, D. A. Shirley, Ed., p. 691, North-Holland Publishing, Amsterdam, 1972.
8. D. A. Shirley, Chem. Phys. Lett., 16, 220 (1972).

Molecular Spectroscopy by Inelastic Electron Tunneling*

Kenneth P. Roenker** and William L. Baun

*Air Force Materials Laboratory
(AFML/MBM)
Wright Patterson A. F. B., Ohio 45433*

    This paper examines the capabilities and limitations of the technique of inelastic electron tunneling (IET) as a molecular spectroscopic tool for surface studies. Following a brief review of the literature, an outline of the theory is given and the experimental technique discussed. The tunneling spectra of both clean and doped Al-Al oxide-Pb junctions are considered in detail. A comparison with existing surface analysis techniques concludes the paper.

INTRODUCTION

    When a thin oxide ($\sim$ 30 Å) is sandwiched between two metal electrodes to form a junction and a small potential difference is applied, a current due to electrons tunneling through the oxide can be detected. Electron tunneling is a quantum mechanical effect involving the penetration of the electron through a potential barrier whose height exceeds the maximum electron energy. As a result of the applied voltage, electrons near the Fermi level in one metal penetrate through the oxide into empty states above the Fermi level in the second metal. For voltages ($\lesssim$ 1 volt) small compared to the barrier height, this tunneling is essentially elastic and the resulting current varies nearly linearly with the voltage[1]. In addition to this elastic current, small sharp increases ($\lesssim$ 1%) in the conductance $\left(\frac{dI}{dV}\right)$ occur at certain characteristic voltages V. Each increase signals the opening of a new tunneling channel which parallels the elastic channel. The interaction of

---

*This paper is based on Technical Report AFML-TR-76-75 which was released for publication.
**National Research Council Research Associate

the electron with one of the vibrational modes of a molecule contained in the junction is the mechanism by which each of these inelastic tunneling channels conducts. Hence, each conductance increase at a characteristic voltage V is associated with a particular vibrational frequency ν by $eV=h\nu$.

In 1966, Jaklevic and Lambe[2] reported the first observation of the interaction of tunneling electrons with the vibrational states of molecules contained in a metal-oxide-metal junction. They observed peaks in the $\frac{d^2 I}{dV^2}$ versus voltage curve at 4.2K for Al-Al oxide-Pb junctions and for similar junctions whose oxide was exposed to the vapor of a simple organic contaminant. By comparison with the infrared spectra of the contaminant, a vibrational mode could be associated with a majority of the peaks in the tunneling spectrum[3]. Subsequent investigations revealed the remaining peaks to be associated with Raman active vibrational modes.

Since that initial study, a number of compounds, including amino acids, have been examined[4-11] and a variety of metals employed as electrodes[3,8,12,13]. The excitation of phonons within the oxide and in the adjacent metal surfaces by tunneling electrons[14-16] have been reported, as well as an electronic transition[17]. In addition, Klein and Leger[18] have observed the excitation of the vibrational-rotational levels of OH radicals by tunneling electrons.

These studies have demonstrated the applicability of inelastic electron tunneling (IET) to the study of surfaces, particularly oxides on metals. IET can be viewed as a surface tool in that, by an appropriate choice of the second metal electrode, the chemical nature of the oxide surface plus dopant is not altered appreciably. Handy[19] has shown that for Pb and similar metals with large atomic radius, negligible penetration of the oxide occurs during vapor deposition. In addition, by a suitable choice of metals, the reduction of the oxide and contaminant by the second metal electrode can be minimized. Hence, the bonding occurs principally to the original metal oxide surface and the second metal serves only to seal the junction and to act as the second electrode. A prime example of a tunnel junction exhibiting these properties is Al-Al oxide-Pb.

In contrast to the several techniques which provide information on the elemental character of the surface species, this technique offers information regarding the chemical form of the species at the oxide surface. Consider the structure of a typical tunnel junction: Al-Al oxide-contaminant-Pb. Previous studies have indicated that approximately one monolayer or less of dopant can be deposited[3,7] so that tunneling electrons excite vibrations of bonds within and to the oxide. Consequently, the IET technique provides a new means of probing the chemical nature of the oxide itself and the oxide-dopant interface.

THEORY

A detailed account of the theory can be found in the literature[3,20]. Only an outline of the theory is presented here. In Figure 1, an electron energy level diagram for an Al-Al oxide-Pb tunnel junction near T=4.2K is shown. The two Fermi levels in the metals become separated in energy by an amount eV when a voltage V is applied across the junction. Provided the oxide is thin enough, electrons near the Fermi level in one metal can tunnel horizontally (elastically) into empty states above the Fermi level in the second metal. This process accounts for the bulk of the current. However, a small number of electrons traversing the oxide do so inelastically. A tunneling electron can excite a vibrational mode ω_n of a molecule, thereby losing an energy $\hbar \omega_n$, and still reach an empty state above the Fermi level in the second metal. This inelastic tunneling channel becomes accessible when $eV \geq \hbar \omega_n + \Delta_{Pb}$, where $2\Delta_{Pb}$ is the width of the superconducting energy gap at the Pb Fermi energy.

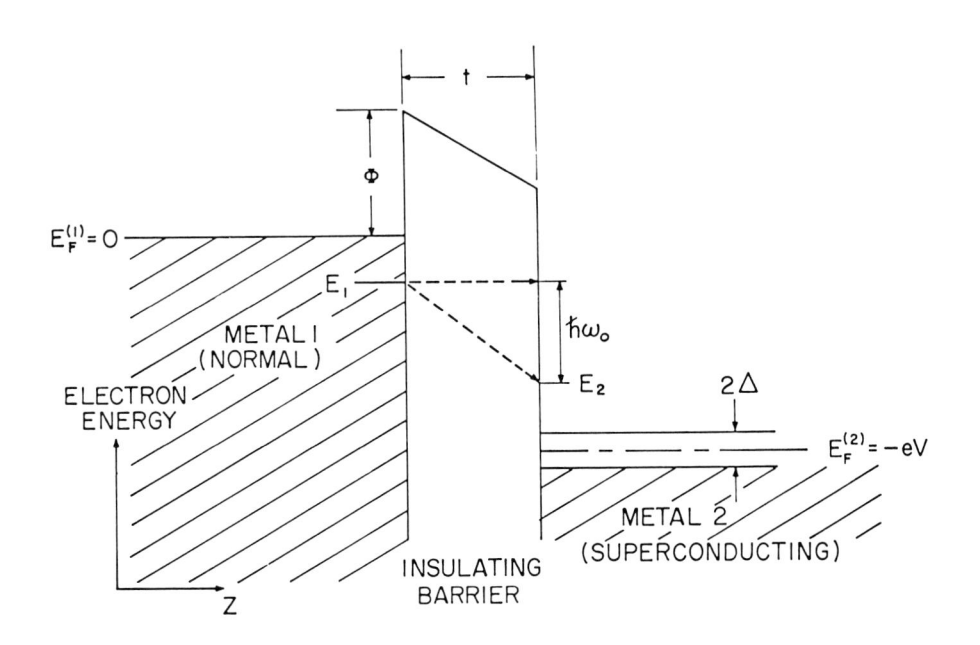

Fig. 1. Energy level diagram for tunneling electrons.

The onset of an inelastic tunneling process can be detected as a peak in the d^2I/dV^2 versus V curve. The mechanism whereby the excitation of molecular vibrational modes by tunneling electrons produces an increase in the conductance can be thought of in terms of a lowering of the potential barrier height by the dipolar field of the molecule. Lambe and Jaklevic[3], as well as Scalapino and Marcus[20], have considered in detail the situation where a molecule of dipole moment $\vec{p}$ is positioned near the metal-oxide interface. They derived an expression for the second derivative of the total inelastic current due to N molecules of a particular kind.

$$\frac{d^2I}{dV^2} = E\left(\frac{dI}{dV}\right)_0 \left[\frac{4\pi me}{\hbar\Phi}\right] \ln\left(\frac{t}{r_0}\right) \sum_m |<m|p_z|o>|^2 \delta(eV - \hbar\omega_m - \Delta_{Pb})$$

where Φ is the maximum barrier height, t is the oxide thickness and the sum is over the vibrational states of the molecule of interest. The matrix element leads directly to the infrared absorption selection rules.

The Raman-type interaction in which the tunneling electron interacts with the polarizability of a molecular vibrational mode has also been examined by Lambe and Jaklevic[3]. They found the interaction to be comparable in magnitude to the infrared type interaction and derived the following expression

$$\frac{d^2I}{dV^2} = N\left(\frac{dI}{dV}\right)_0 \left[\frac{4\pi me^3}{\hbar^2\Phi 16t^6}\right] \left[\int_{r_0}^{t} f^2(r_1) r_1 dr_1\right] \sum_m |<m|\alpha|o>|^2$$

$$\delta(eV - \hbar\omega_m - \Delta_{Pb})$$

where α is the polarizability associated with a vibrational mode and

$$f^2(x) = \frac{1}{x^2}\left[\frac{1-x^2}{(1+x^2)^2} + \frac{1}{x}\tan^{-1}\left(\frac{1}{x}\right)\right].$$

Thus, both types of modes are expected and have been observed[5,7]. However, there appears to be no direct correlation between the relative intensities of observed infrared or Raman lines and the relative intensities of the corresponding tunneling peaks. The calculated magnitude of the effect is in agreement with experiment[20]. While the initial theory[3,20] indicates that tunneling electrons can excite only longitudinal vibrations, i.e., only vibrations perpendicular to the film surface, Adler et al.[21] have recently concluded that transverse vibrations may also be excited.

For IET studies, low temperatures (T∼4K) are essential to obtain adequate resolution. When both metals are in the normal state, the peak half-width due to thermal broadening is given by 5.4kT(2meV at T=4.2K)[3]. For Al-Al oxide-Pb junctions at 4.2K, Al is normal while Pb is superconducting so that thermal smearing at the electron distribution at the Al Fermi level limits the linewidth. When the Pb metal becomes superconducting at 7.2K, the electron density of states becomes sharply peaked immediately above and below the energy gap and reduces drastically the Pb contribution to the linewidth. The presence of at least one superconducting metal in the junction at low temperatures is essential for resolution purposes.

EXPERIMENTAL

Tunnel Junction Preparation

The tunnel junctions were prepared in the following manner. On thoroughly cleaned 12x25mm glass slides, an aluminum strip (1.5mm x 25mm) was evaporated at a pressure $\sim 1 \times 10^{-6}$ Torr to a depth of ~ 500 Å. Its surface was oxidized to 30 Å using the glow discharge method[22] in an environment containing equal partial pressures ($\sim 50\mu$) of water and oxygen for ~ 45 min. After evacuation, the oxide was contaminated by one of several techniques. For high vapor pressure liquids or solids and gases, the oxide was exposed to the vapor for ~ 10 Torr-sec. The liquid or solid was heated directly below (~ 3cm) the samples in the case of low vapor pressure materials. Recently, Hansma and Coleman[10] have suggested an alternative means of contaminating the junction. After dissolving the material in a solvent, a drop is placed on the oxide and the excess is removed either by centrifuging or drying in an inert gas atmosphere. A sufficient residue remains for tunneling purposes. The evaporation of the Pb cross strip at $\sim 1 \times 10^{-6}$ Torr, similar in thickness and width to the Al strip, completes the sample preparation process. See Figure 2.

The entire tunnel junction fabrication process was carried out in an oil diffusion pumped bell jar with an ultimate pressure of 2×10^{-7} Torr. See Figure 3. A liquid nitrogen trap was positioned at the throat of the diffusion pump and another between the rotary and diffusion pumps. Evacuation of the bell jar from ambient pressure was accomplished with LN₂ cooled sorption pumps.

Electronics

The onset of an inelastic electron tunneling channel produces a minute change in the slope of the I-V curve, a miniature step in the $\frac{dI}{dV}$ - V curve and a small peak in the $\frac{d^2I}{dV^2}$ - V curve. (See Figure 4.) By far the easiest of the three features to detect and measure is the third. The usual means of detection employs conventional harmonic techniques[23,24]. As the dc bias current through

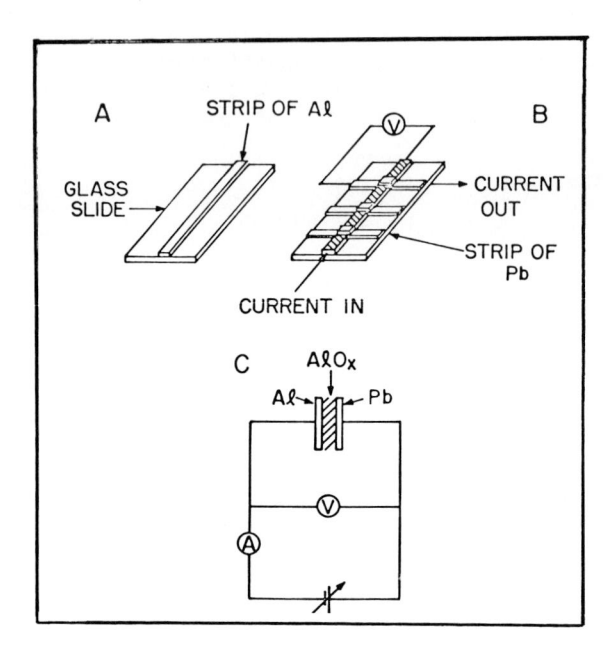

Fig. 2. Sample construction for electron tunneling.

the junction is slowly increased, a 1000 Hz ac current of constant amplitude (∼few μ amps) is added. For weakly nonlinear junctions, the voltage generated at 2000 Hz is proportional to $\frac{d^2 I}{dV^2}$. This signal is detected using a Kelvin bridge circuit and amplified with a preamp and lock-in amplifier before being displayed on the Y axis of an X-Y recorder. In Figure 5, a block diagram of the instrumentation is shown[25]. Electrical connections to the sample were made by indium soldering. The samples were immersed directly in liquid helium for all measurements.

The information contained in the tunneling spectrum is extracted by measurement of the several parameters of the inelastic peaks: position, height and width. The observed values of these parameters are influenced significantly by the instrumental parameters of ac modulation amplitude and the product of lock-in time constant and sweep rate. In Figure 6, the distortion due to the later instrumental parameter is shown. Significant distortion (∼10%) occurs when the time constant times sweep rate is approximately one-sixth

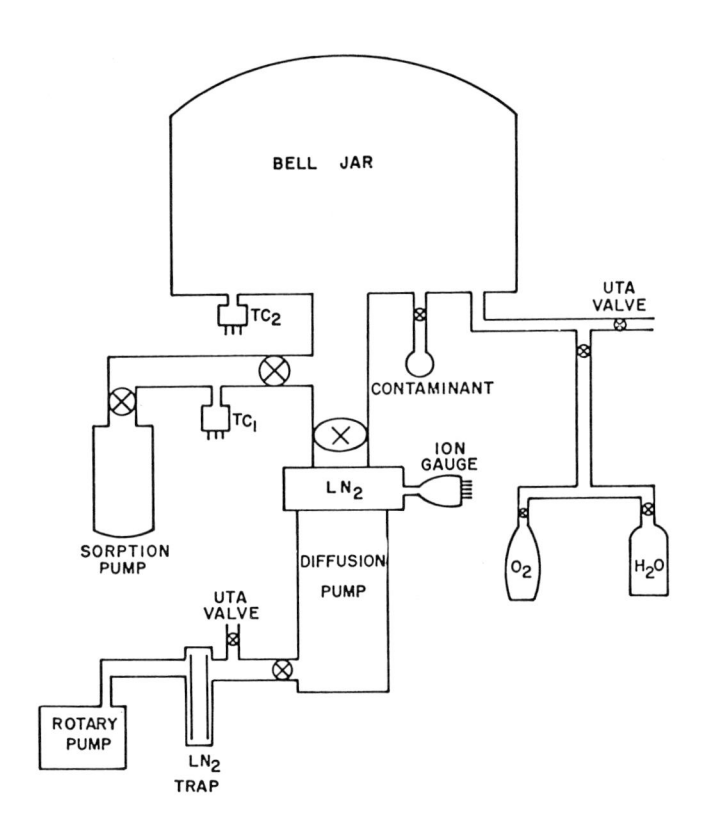

Fig. 3. Sample preparation apparatus.

of the unbroadened linewidth. Figure 7 shows the similar effects of modulation on the peak parameters. To avoid distortion the modulation amplitude should be kept less than one-fifth of the natural peak width. While, in principle, these guidelines were followed, in practice large modulation levels were sometimes required to obtain measurable signals. The ability to perform signal averaging would allow improved signal-to-noise ratios with less distortion. In Figures 6 and 7, the parameters are normalized with respect to the FWHM of the unbroadened line.

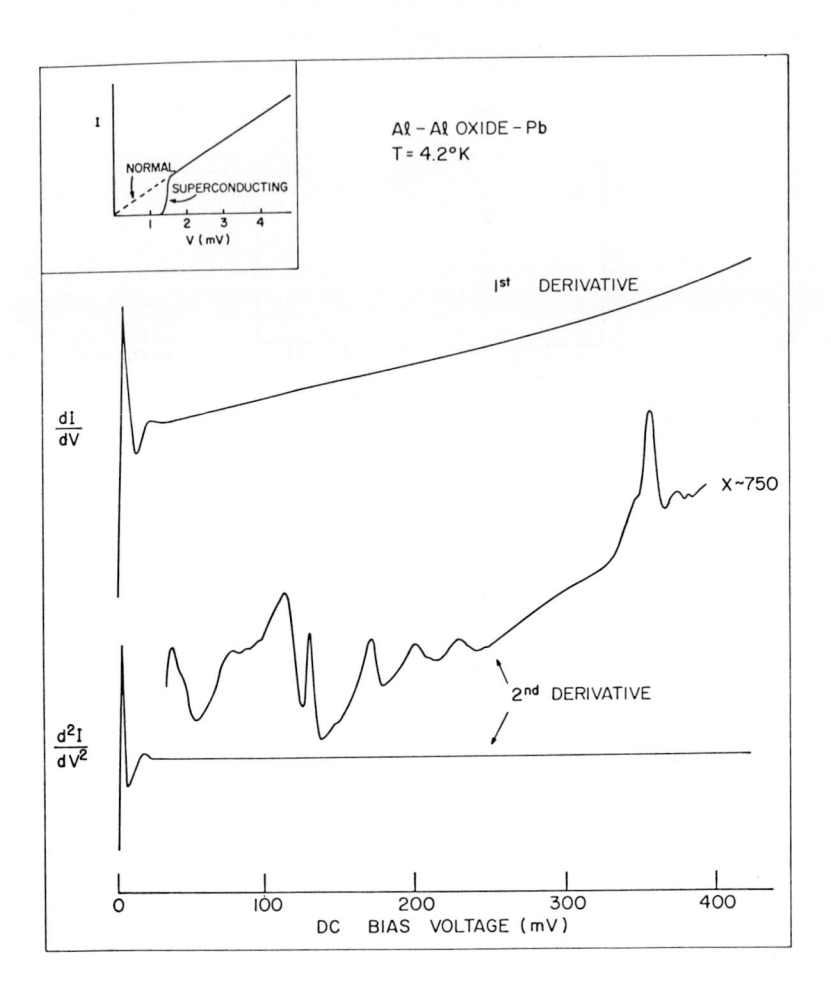

Fig. 4. I-V and derivative curves for tunnel junction.

TUNNELING SPECTRA

Clean Junction

The IET spectrum of a 'clean', i.e. undoped, Al-Al oxide-Pb junction is shown in Figure 8. In preparing a 'clean' junction, a reference with which to compare junctions containing contaminants is established. The various features observed in such a 'clean' junction are then characteristic of the metals and insulator composing the junction and of the residual impurities present in the vacuum system during preparation.

Inelastic Electron Tunneling

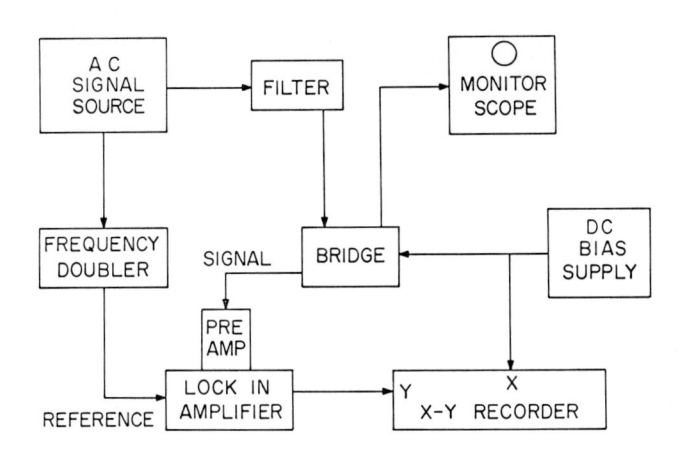

Fig. 5. Block diagram for tunneling spectrometer.

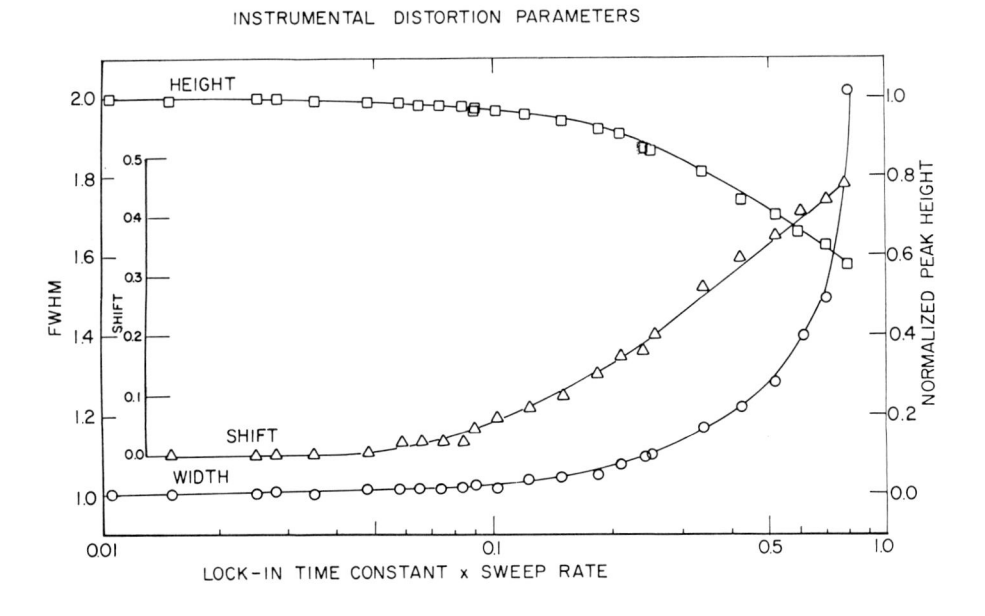

Fig. 6. Instrumental distortion of lineshape due to lock-in amplifier.

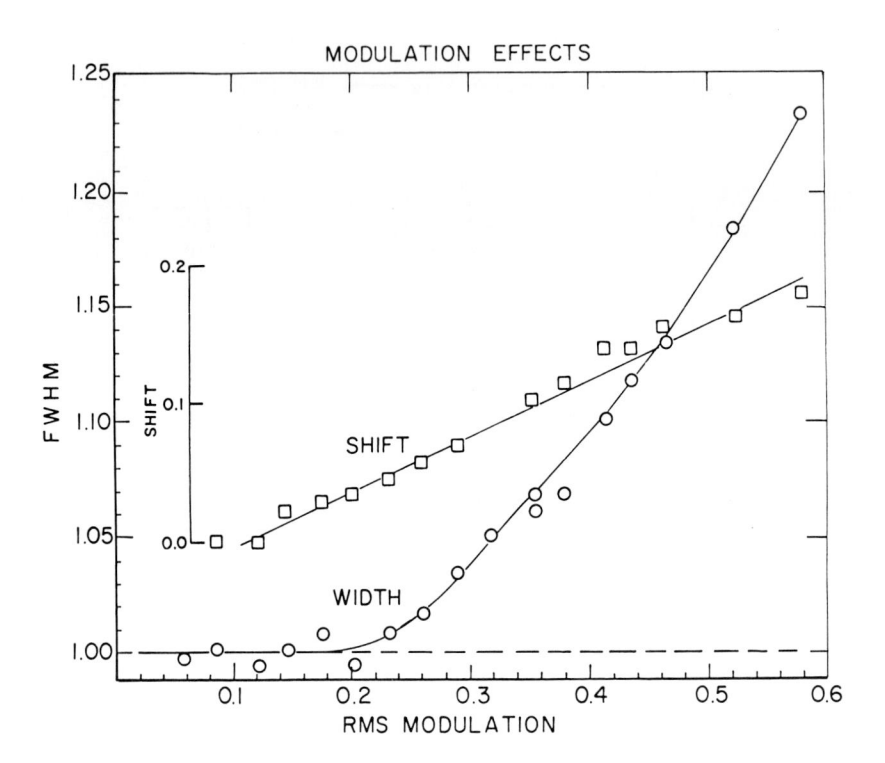

Fig. 7. Instrumental distortion due to modulation.

Although the various features of this spectrum are well agreed upon, the source of each is not. The feature at 450 meV is associated with the O-H stretching mode[26] of hydroxyl groups incorporated into the oxide during its formation. The residual water present in the vacuum system during oxide formation is sufficient alone to produce an observable 450 meV peak. In fact, no sample was prepared in which this peak was not present. This result strongly implies that the insulator formed on the Al metal is in fact an alumina-hydrate or aluminum hydroxide[12]. The isotopic replacement of H by D results in the expected $1/\sqrt{2}$ downward shift in this peak position[3,12], thus confirming this assignment. The low energy shoulder (75 meV) on the 118 meV peak is correlated with the O-H bending mode[12]. With isotopic substitution, it is shifted downward by the appropriate amount[3,12] to 52 meV (= $(1/\sqrt{2})$ x 75 meV). The overtone of the 118 meV peak appears at 234 meV.

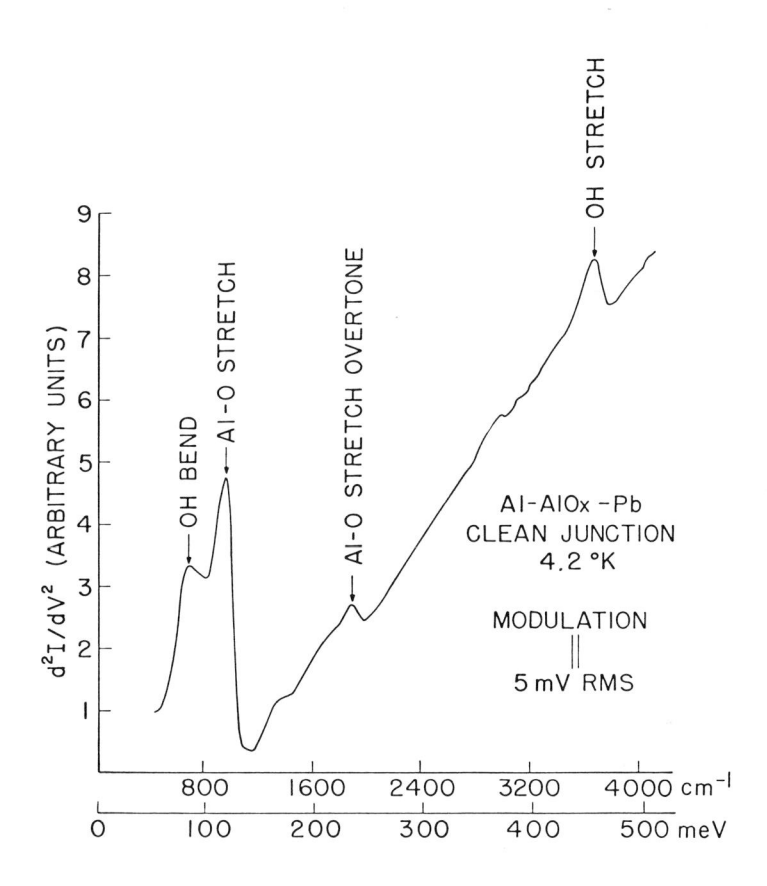

Fig. 8. Tunneling spectrum of a 'clean' junction.

The strongest feature in the spectrum of a 'clean' junction appears in the vicinity of 118 meV. The origin of this peak is currently in dispute. Geiger et al.,[12] and Skarlatos et al.,[11] attribute it to the bending motion of OH in Al-O-H based on the IR reflectance data of Dorsey[27,28,29]. He observed only a single peak (118 meV) in the range of 5 to 500 meV for barrier-type anodic oxides on aluminum and assigned it to the Al-OH bend[27]. The absence of an O-H stretch was noted and attributed to strong hydrogen-bonding broadening. From similar IR reflectance studies, Vedder and Vermilyea[30] and Takamura et al.[31] conclude that the single peak at 118 meV is due to the Al-O stretching mode. They interpret the lack of an O-H stretch as evidence that no OH is present. By anodizing in dimethylformamide containing 10 wt. % ammonium adipate, i.e., containing no hydroxyl groups, Takamura et al.[31] obtained a film producing exactly the same IR spectrum and thus substantiated their assignment.

The results of the isotopic substitution D for H are inconclusive regarding the 118 meV peak. Dorsey[29] and Takamura et al.,[31] observed almost no shift (< 1% in the IR spectrum upon deuteration). The latter interprets this fact as additional evidence that no hydroxyl (or OD) is present. However, hydrogen is much more reactive than deuterium and readily replaces it when possible making the actuality of the deuteration questionable. Geiger et al.[12] and Lambe and Jaklevic[3] have performed the same isotopic substitution to determine its effects on the 'clean' IET spectrum. In spite of the fact that D_2O was directly added (and not H_2O) during the oxide formation, the IET spectrum changed only slightly. The main features of the 'clean' spectrum remained: the O-H stretch at 450 meV, the O-H bend at 75 meV and the 118 meV peak. In addition to these, several smaller peaks due to the isotopic substitution were added: at 330 meV (O-D stretch), at 80 meV and at 53 meV (O-D bend). They interpreted the 80 meV peak as the Al-OD bend, the downshifted counterpart of the Al-OH bend (118 meV/$\sqrt{2}$ = 83 meV), and not the sharpened O-H bend.

The reconciliation of these various results is not obvious. This author believes the 118 meV peak to be associated with the OH bend of Al-O-H primarily because of the presence of OH in all tunneling junctions. Even in junctions where D_2O is added to the glow discharge after the Al evaporation, the residual water in the vacuum system causes the junction to contain predominately OH. This suggests that an aluminum-hydroxide layer is formed almost immediately after the Al evaporation and that it severely restricts the formation of Al-OD. This interpretation is not inconsistent with the available IR and IET results. If the bending of Al-O-H occurs between Al and OH, isotopic replacement results in a change of mass from 17 to 18 and causes a downshift by a factor $\sqrt{\frac{17}{18}}$ = 0.97 (not $\frac{1}{\sqrt{2}}$). This shift is the approximate size observed (∼1%) in the IR spectrum and of the order of the experimental uncertainty (∼1.3%)[29,31]. The strong absorption of OH by evaporated Al in a vacuum system suggests that the absence of any change in the IR spectrum when the metal is anodized in a solution not containing hydroxyl groups is due to the presence of OH on the surface prior to anodization. The IET results can be similarly understood. As mentioned above, only a small downshift (0.97 and not $\frac{1}{\sqrt{2}}$) is to be expected in the 118 meV peak. In addition, its intensity is expected to be small because of the limited replacement of D for H possible and so probably unobservable in the vicinity of the large 118 meV peak. The 80 meV peak in the deuterated 'clean' spectrum is most likely the O-H bend, though its narrower width is difficult to explain.

In the very low energy regime ($\tilde{<}$ 20 meV), a series of nearly equally spaced peaks (4.0, 7.7, 11.2, 14.4, 17.9 meV) are seen. These peaks are relatively intense ($\sim$ factor of 10) by comparison with the structure described earlier. Their separation is in agreement with the frequency of the transverse acoustical phonon of lead (4.1 ± 0.8 meV) and prompted Rowell, et al.[32] to conclude that tunneling electrons are exciting multiple phonons in the lead surface. No repeating structure due to the longitudinal acoustical phonon (8 meV) was apparent. For both metals in the normal state, Rowell, et al.[1] observed only a single phonon peak of each type. At slightly higher energies (< 40 meV), the phonons associated with aluminum occur. In our tunneling junctions, only the longitudinal acoustic phonon at 36 meV was visible. Both the transverse (22 meV) and longitudinal modes have been observed by Klein and Leger[14] in Al-Al junctions, while Adler[15] detected only the latter.

Doped Junctions

If, after an oxide is grown on the aluminum, the material of interest is allowed to adsorb on the oxide before the Pb is applied, a doped tunnel junction is formed. Typically, these junctions possess considerably more structure in the tunneling spectrum than 'clean' junctions. Via the intensity, width and position of the additional peaks, the spectrum yields information regarding the bonding of the absorbent to the oxide.

An example is seen in Figure 9, where cyanoacetic acid was added to the junction by exposing the oxide to the vapor. The tunneling spectrum is seen at the top; the Raman data is displayed as a bar graph underneath and the infrared data[33] appears below. The Al-OH bend and O-H stretch of the oxide are readily seen. Near 360 meV the unresolved symmetric and antisymmetric CH_2 stretches are observed. Note the sensitivity to these modes in comparison with the IR where the O-H stretch all but obscures them. At 284 meV the C ≡ N stretch is visible. The C = O stretch is downshifted to 208 meV in the IET spectrum from 214 meV in the IR and Raman spectra indicating a weakening of the bond. By contrast, the C-O stretch near 148 meV in the IR spectrum is shifted up to 160 meV in the IET spectrum - a strengthening of the bond. The OH bends of the acid are visible near 178 meV and the CH_2 scissors at 172 meV. At 113 meV, the C-C stretch is partially observable on the side of the 118 meV Al-OH bend. The out-of-plane O-H bends of the acid are seen near 100 meV, while the C-C-O and C-C≡N bends appear at 55 meV. With reduced modulation, improved resolution is obtained and a more detailed comparison with IR spectrum becomes possible. The above results do suggest a model for the attachment of the acid to the surface[11]. The C-O upward and C=O downward shifts indicate an equalizing of these bonds probably by the partial bonding of the two O atoms to the surface. The small intensity of the O-H stretch in comparison to the CH_2 stretch supports this model. In addition, the absence of a significant shift in the CH_2 stretch indicates a lack of hydrogen bonding to the surface.

In Figure 10 the IET, Raman and IR spectra of polyphenyl ether are displayed. The Al-OH bend and its overtone, O-H bend, and the O-H stretch characteristic of the 'clean' junction are readily observable. The intensity and width of the O-H stretch indicate considerable water in the oxide. An intense ring vibration is visible at 133 meV and weaker ones near 80 meV. Near 170 meV the C-O stretching mode appears. The source of the peak at 200 meV is uncertain, possibly O-H or C-H bend associated with the ether. More interesting, however, are the C-H stretching modes near 360 meV. Three peaks are easily recognized, each associated with a particular kind of hydrogen. They are significantly downshifted in the IET spectrum supporting the hypothesis that these large planar molecules are lying flat on the surface held there by hydrogen bonding.

CONCLUSIONS

There exist a number of techniques for determining the elemental composition of the outermost layers. Among these are Auger electron spectroscopy (AES), ion scattering spectroscopy (ISS), secondary ion mass spectroscopy (SIMS) and photoelectron spectroscopy (PES). Of these only AES and PES are sensitive to chemical effects. Inelastic electron tunneling (IET), however, is most sensitive to the outermost or bonding electrons, and so provides information comparable to infrared and Raman data about the chemical form of the surface species. In surface studies, both Raman and infrared spectroscopy are possible alternatives but tedious in practice. In Figure 11, an IET spectrum and an IR spectrum of C_6H_5OH on alumina are compared. The former provides increased resolution, range and sensitivity in addition to the ability to see both Raman and infrared active modes. Hence, IET may be viewed as complementary to the elemental surface analysis techniques in the information it provides.

Like other surface analysis techniques, IET possesses both advantages and limitations. As mentioned above, the primary advantage of this technique is that it provides information about bonding electrons through the excitation of these bonds. Both Raman and infrared active modes are observable from the far infrared to beyond 8000 cm^{-1}. This information pertains to the structure of the insulator separating the metal electrodes of the junction and the impurities trapped within. Some information can be also extracted concerning phonons in the adjacent metal surfaces. Due

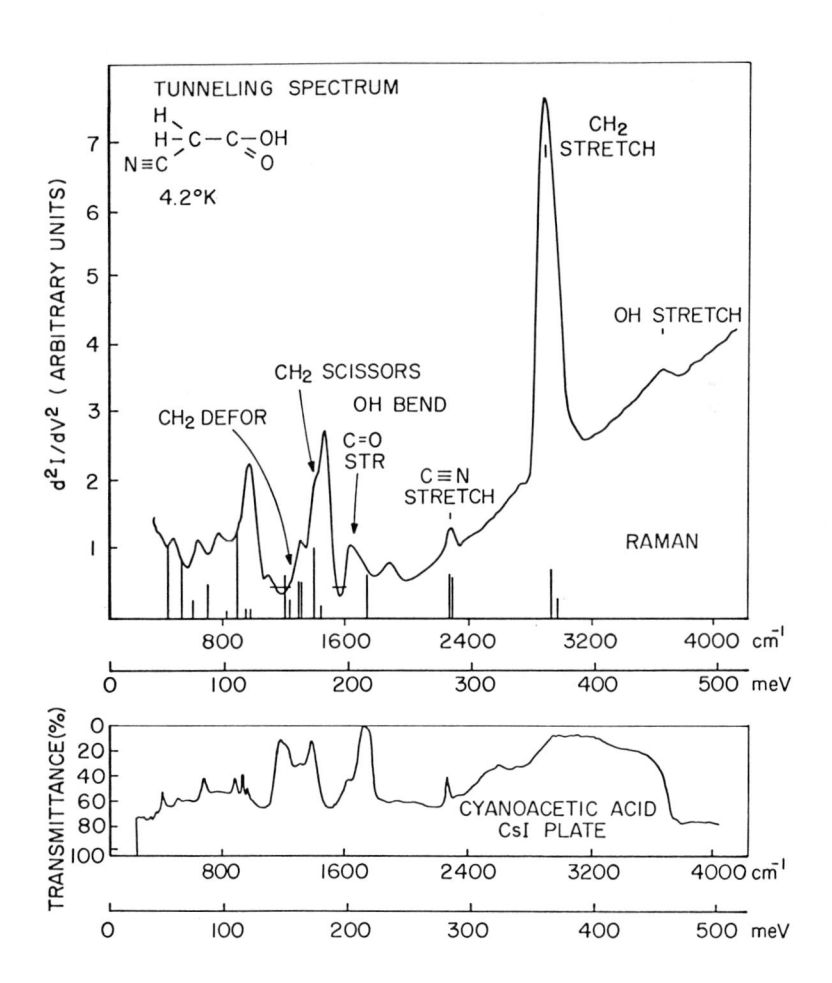

Fig. 9. Tunneling, I.R. and Raman spectra of cyanoacetic acid.

to the structure of the junction, contamination of the sample after preparation such as exposure to the atmostphere, is not a problem. In addition, the IET technique is nondestructive in its analysis so that samples may be examined repeatedly. One very beneficial property of this technique is that it allows the examination of organic materials on the insulator surface.

Perhaps the principal advantage of this technique is its extreme sensitivity to materials trapped in the insulator or at its

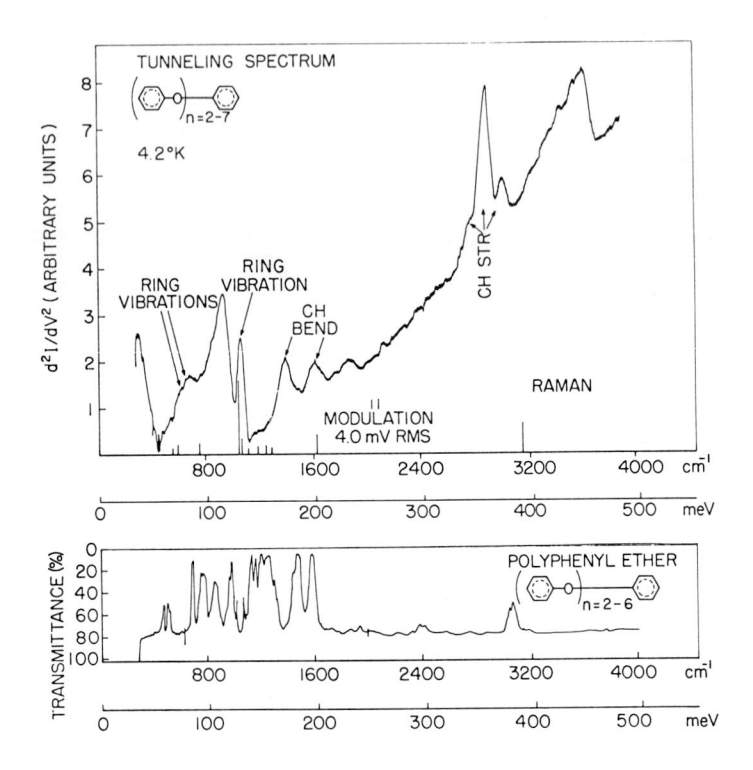

Fig. 10. Tunneling, I.R. and Raman spectra of polyphenyl ether.

surface. Lambe and Jaklevic[3] have estimated that 1/100 of a monolayer or a total of 10^{10} molecules/mm^2 are detectable. Skarlatos et al.[11] have reported the detection of organic molecules in water in the parts per million range by IET. Hence, IET exceeds both IR and Raman spectroscopy of surfaces in sensitivity. In addition, IET possesses slightly better resolution approaching that of IR and Raman of bulk materials.

There exist several limitations on the technique but none are unduly restrictive. First, samples must be prepared in an ultra-high vacuum system or a clean (well-trapped) high vacuum system. The presence of even trace amounts of back streamed oil will negate sample preparation by contamination. Second, low temperatures (T=4K) are required to obtain adequate resolution. Hence, liquid helium is essential. The presence of at least one metal that is superconducting (e.g., Pb) at this temperature is beneficial. Third, the IET technique is only capable of examining metal-

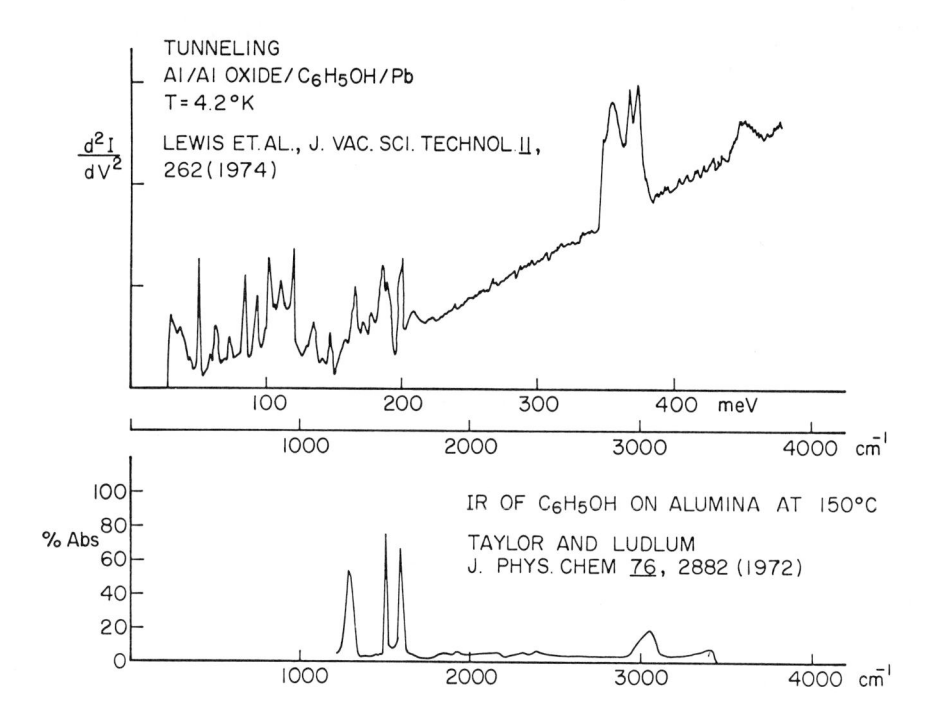

Fig. 11. Comparison of tunneling and ATR spectra.

insulator-semiconductor junctions. Although only the former have been discussed here and only with the oxide as the insulator, tunneling through nitride films has been reported[34]. Junctions employing semiconductors have also been examined[35-38].

In summary, IET is a productive surface analysis technique, which offers both new and valuable information, and provides excellent sensitivity to molecular species.

REFERENCES

1. J. M. Rowell, W. L. McMillan and W. L. Feldman, Phys. Rev. 180, 658 (1969).
2. R. C. Jaklevic and J. Lambe, Phys. Rev. Lett., 17, 1139 (1966).
3. J. Lambe and R. C. Jaklevic, Phys. Rev. 165, 821 (1968).

4. B. F. Lewis, M. Mosesman and W. H. Weinberg, Surface Sci., **41**, 142 (1974).
5. B. F. Lewis, W. M. Bowser, J. L. Horn, T. Luu and W. H. Weinberg, J. Vac. Sci. Technol. **11**, 262 (1974).
6. M. G. Simonsen and R. V. Coleman, Nature **244**, 218 (1973).
7. M. G. Simonsen and R. V. Coleman, Phys. Rev. B **8**, 5875 (1973).
8. I. K. Yanson, N. I. Bogatina, B. I. Verkin and O. I. Shklyarevskii, Soviet Phys. JETP **35**, 540 (1972).
9. J. Klein, A. Leger, M. Belin, D. Defourneau and M. J. L. Sangster, Phys. Rev. B **7**, 2336 (1973).
10. P. K. Hansma and R. V. Coleman, Science **184**, 1369 (1974).
11. Y. Skarlatos, R. C. Barker, G. L. Haller and A. Yelon, Surface Sci. **43**, 353 (1974).
12. A. L. Geiger, B. S. Chandrasekhar and J. G. Adler, Phys. Rev. **188**, 1130 (1969).
13. R. C. Jaklevic and J. Lambe, Phys. Rev. B **2**, 808 (1970).
14. J. Klein and A. Leger, Phys. Lett. **28A**, 134 (1968).
15. J. G. Adler, Phys. Lett. **29A**, 675 (1969).
16. J. G. Adler, Solid State Comm. **7**, 1635 (1969).
17. A. Leger, J. Klein, M. Belin and D. Defourneau, Solid State Comm. **11**, 1331 (1972).
18. J. Klein and A. Leger, Phys. Lett. **30A**, 96 (1969).
19. R. M. Handy, Phys. Rev. **126**, 1968 (1962).
20. D. J. Scalapino and S. M. Marcus, Phys. Rev. Lett. **18**, 459 (1967).
21. J. G. Adler, H. J. Kreuzer and W. J. Wattamaniuk, Phys. Rev. Lett. **27**, 185 (1971).
22. J. L. Miles and P. H. Smith, J. Electrochem. Soc. **110**, 1240 (1963).
23. J. G. Adler and J. E. Jackson, Rev. Sci. Instr. **37**, 1049 (1966).
24. J. G. Adler, T. T. Chen and J. Strauss, Rev. Sci. Instr. **42**, 362 (1971).
25. For a detailed description of the circuitry see T. P. Graham and R. G. Keil, Technical Report AFML-TR-74-32, Feb. 1974.
26. J. B. Peri and R. B. Hannan, J. Phys. Chem. **64**, 1526 (1960).
27. G. A. Dorsey, J. Electrochem. Soc. **113**, 169 (1966).
28. Ibid., 172 (1966).
29. Ibid., 284 (1966).
30. W. Vedder and D. A. Vermilyea, Trans. Faraday Soc. **65**, 561 (1969).
31. T. Takamura, H. Kihara-Morishita and U. Moriyama, Thin Solid Films **6**, R17 (1970).
32. J. M. Rowell, A. G. Chynoweth and J. C. Phillips, Phys. Rev. Lett. **9**, 59 (1962).
33. D. Sinha and J. E. Katon, Appl. Spect. **26**, 599 (1972).
34. Y. Uemura, K. Tanaka and M. Iwata, Thin Solid Films **20**, 11 (1974).
35. P. Guetin and G. Schreder, Solid State Comm. **8**, 291 (1970).
36. L. B. Schein and W. D. Compton, Phys. Rev. B **4**, 1128 (1971).
37. W. A. Thompson, Phys. Rev. Lett. **20**, 1085 (1968).
38. I. Giaever and H. R. Zeller, Phys. Rev. Lett. **21**, 1385 (1968).

Discussion

On the Paper by D.M. Hercules

L.H. Lee (*Xerox Corp.*): First, I would like to thank Prof. Hercules for presenting this interesting paper on ESCA. From this paper, we learn basic principles related to ESCA and some new applications of ESCA to surface analyses. Secondly, I would like to take this opportunity to express our sincere appreciation to the Division of Analytical Chemistry for co-sponsoring this Symposium. The developments of new surface techniques have made analytical chemistry once more an exciting field to be involved. We believe that this Symposium is timely in presenting the state-of-the-art of surface analyses to both analytical and polymer chemists.

On the Paper by T. Rhodin and C. Brucker

D.T. Clark (*Durham University, England*):

1. Relevant to your last comment, I would like to mention the great potential of employing diffraction data for photo-emitted core electrons from single crystals as originally described by Siegbahn. This should be capable of development as a complement to LEED.

2. I would like to know how you see the whole field developing, keeping in mind some of the problems associated with UPS studies of adsorbed species. It seems to me that some of these problems will make it difficult to develop the technique for the study of large molecules of most interest to chemists. The problems I have in mind are: the spectra involve convolution of many factors which are difficult to separate. (The range of escape depths spanned in going across the valence bands for low energy photon sources, final state effects, differential changes in relaxation energies, changes in lattice sublimation energy terms for molecule and ion (usually neglected in such treatments, changes in correlation energies which could cause reordering of levels, etc.).

3. As a consequence of (2), the development of cheap polarized sources such as that provided by Siegbahn's scheme for successive reflection from gold mirrors could well add a new dimension to the whole area by allowing close consideration of the asymmetry parameter for angular resolved studies. Do you feel however, that this will provide sufficient extra data to raise the interpretation of the photoemission data above that of qualitative discussions of the interaction involved?

T.N. Rhodin (*Cornell University*):

1. In principle, analysis of any elastically scattered photoelectrons through a surface can provide information on surface crystallography. A detailed analysis of how this could be done for valence level excited photoelectrons was published by Liebsch.

2. Although a detailed understanding of many factors involved in UPS-photoemission spectra is not likely to be developed in the near future, interpretation of photoemission data continues to provide new and significant information on the electron structure of clean and chemisorbed surfaces. In conjunction with other measurements of atomistic and microscopic parameters it provides critical clarification of hitherto inaccessible concepts of chemical bonding at surfaces.

3. The use of planar polarized photon excitation sources show great promise in exploring the atomic and electronic sturcture of solid surfaces. Studies on the vacuum ultraviolet frequency range are particularly useful because they reflect valence-level excitations related to chemical interactions. Sources of polarized photons in the X-ray region should also be useful. At present their application to surface studies have not been as extensively demonstrated to my knowledge.

P.N. Ross (*United Technologies Corp.*): I would like to offer a comment on an important problem that surface chemists and physicists have neglected. Prof. Rhodin properly emphasized the importance of understanding what changes take place in all the bonds in the adsorbate when gases chemisorb on metal surfaces. Similarly, it would be equally important to observe the metal-metal and metal-substrate bonding when small metal clusters are formed on substrate surfaces, such as we have in supported catalysts. One suspects that in the case of strong metal-substrate bonding, the orbitals in the metal cluster are significantly different than in the bulk metal.

On the Paper by R.S. Haines

G.P. Ceasar (*Xerox Corp.*): Do you try using the algorithm used by McPherson in their spectrometer and how does this compare with the present results?

R.S. Haines (*IBM Corp.*): The McPherson algorithm, which is basically a Fourier transform method, did not work since it blew up when small errors were introduced into the program.

On the Paper by K.P. Roenker and W.L. Baun

L.H. Lee (*Xerox Corp.*): I sincerely appreciate the contribution by Drs. Roenker and Baun after the symposium. This paper clearly demonstrates a new tool by inelastic electron tunnelling to be as important as IR or Raman spectroscopy for the molecular structure of metal and oxide surfaces. Because of the importance of the subject matter, I chose to include this paper in this volume.

About Authors

Lieng-Huang Lee is a Senior Scientist of Xerox Corporation. Dr. Lee is the Editor of <u>Recent Advances in Adhesion</u>, <u>Advances in Polymer Friction and Wear</u>, and <u>Adhesion Science and Technology</u>. He was the Chairman (1975-76) of the Division of Organic Coatings and Plastics Chemistry of the American Chemical Society.

Erwin W. Müller is the Evan-Pugh Research Professor of Physics at the Pennsylvania State University. Dr. Müller is the recipient of the Davisson-Germer Prize of the American Physical Society, Centenary Lectureship Medal of the Chemical Society (London), the first Welch Award of the American Vacuum Society, and membership in the National Academy of Engineering and the National Academy of Science.

Gary W. Simmons is Associate Professor of Chemistry and member of the Center for Surface and Coatings Research, Lehigh University.

Henry Leidheiser, Jr. is Professor of Chemistry and Director, Center for Surface and Coatings Research, Lehigh University. Dr. Leidheiser is the author of "Corrosion of Copper and Tin and Their Alloys".

Mary L. Good is the Boyd Professor (Chemistry) at the University of New Orleans and a Board Director of the American Chemical Society. Dr. Good is the recipient of ACS Garvan Award.

Robert L. Park is Professor of Physics and Director of the Center of Materials Research of the University of Maryland. Dr. Park was the supervisor of the Surface Physics Division of Sandia Laboratories. He is a senior member of the American Vacuum Society and a Fellow of the American Physical Society.

Ian M. Stewart is Manager of the Electron Optics Group at Walter C. McCrone Associates. Mr. Stewart was in charge of the Physical Metallurgy/Physical Analysis Section of the English Electric Co.

J. T. Grant is a Senior Research Scientist with Universal Energy Systems, Inc. Dr. Grant (Ph.D., University of New South Wales) has been actively engaged in the uses and development of Auger electron spectroscopy over the last seven years.

About Authors

Carl E. Locke is Assistant Professor in Chemical Engineering, University of Oklahoma. Dr. Locke has published articles in corrosion and polymer science.

Peder J. Estrup is Professor of Chemistry and Professor of Physics at Brown University. Dr. Estrup has published articles on LEED, electron spectroscopy and related techniques. He is a Fellow of the American Physical Society.

David L. Adams is Associate Professor of the University of Aarhus, Denmark. Dr. Adams was associated with Xerox Research Laboratory. He has published papers on LEED and other related techniques.

Paul M. Marcus is Manager of the Electronic Structure Theory Group at IBM Research Center. Dr. Marcus' publications include superconductivity, lattice dynamics, band theory and LEED.

S. Y. Tong is Assistant Professor in Physics, University of Wisconsin, Milwaukee. Dr. Tong is also a Research Consultant for Naval Research Laboratory, Washington, D. C.

George H. Morrison is Professor of Chemistry and Director of the Materials Science Center Analytical Facility at Cornell University. Dr. Morrison received the American Chemical Society Award in Analytical Chemistry in 1971. He was Guggenheim Fellow at the University of Paris (1974-75).

William L. Baun is a Research Chemist, Mechanics and Surface Interactions Branch, Air Force Materials Laboratory. Dr. Baun has published over 100 papers and received twelve local and national awards for scientific achievements.

K. P. Roenker did post-doctorate work as a National Research Council Research Associate at the Air Force Materials Laboratory. Dr. Roenker is now teaching at the Physics Department of Thomas More College in Kentucky.

David M. Hercules is Professor of Chemistry and Head of the Analytical Chemistry Division at University of Georgia. Dr. Hercules was a Guggenheim Fellow and has published over ninety articles.

Thor N. Rhodin is Professor of Applied Physics at Cornell University. Dr. Rhodin's research interests have been electronic structure and electron excitation at surfaces and the microscopic nature of chemical bonding on metals. He is one of the leaders in the field of surface science.

Robert S. Haines is a Senior Chemist of IBM General Products Division. Mr. Haines holds over 20 patents and has received two IBM Awards.

Author Index

A

Adams, D.L., 15, 208, 209, 211, 268, 297, 344, 465
Adams, I., 96
Adawi, I., 463
Adler, I., 152
Adler, J.G., 494
Afzal, M., 155, 179
Aida, K., 97
Alexander, L.E., 96
Allara, D.L., 185
Allen, A.D., 97
Allen, G.C., 476
Amelio, G.F., 152
Anders, L.W., 465
Andersen, C.A., 365
Anderson, J., 209, 465
Anderson, J.L., 17
Anderson, J.R., 209
Anderson, P.W., 209
Andersson, S., 124, 209, 268, 297, 298
Ando, K.J., 95
Andrews, R.D., 16
Antonini, J.F., 466
Appleton, B.R., 208
Arakawa, E.T., 124
Ardenne, M.V., 15
Aren, J.J., 14
Arlinghaus, F.J., 269
Arnold, D., 95
Arnott, D.R., 152
Ashcroft, N.W., 463, 464
Auger, P., 14, 151

B

Badgley, R.E., 123
Baer, Y., 429
Baetzold, R.C., 298
Bair, H.E., 390
Baker, A.D., 464
Baker, B.G., 179
Baker, C., 464
Baker, J.M., 464
Bancroft, G.M., 96
Barber, M., 96
Barker, R.C., 494
Barofsky, D.F., 47
Baron, K., 210
Barrett, J.H., 208
Barrington, A.E., 15
Bartell, L.S., 465
Bartunik, H.D., 96
Baturforsch, Z., 344
Bauer, W., 152
Baun, W.L., 375, 390, 477
Bean, C.P., 96
Beck, W., 96
Becker, G.E., 16, 209
Beckey, H.D., 46
Beeby, J.L., 344
Belin, M., 494
Bell, A.E., 465
Benerito, R.R., 397
Benninghoven, A., 15, 365, 366, 390
Berger, A.S., 46
Berglund, C.N., 463
Bergmark, T., 15
Berndt, W., 298
Bernheim, M., 365, 366
Bhasin, M.M., 152
Bikerman, J., 390
Birkes, L.S., 15
Bishop, H.E., 152
Blaise, G., 366
Blakely, J.M., 152, 209
Block, J.H., 48
Blomquist, J., 96
Blyholder, G., 464
Bockris, J.O'.M., 179
Bogatina, N.I., 494
Bohn, G.K., 151
Bommel, H., 96
Bonchev, Z., 63, 95
Bonzel, H.P., 208, 210
Bottomly, F., 97
Boudart, M., 97

Author Index

Bovey, F.A., 16
Bowker, M., 209
Bowkett, K.M., 17
Bowser, W.M., 494
Boyen, G., 464
Boyes, E.D., 47
Bozzoni, C.B., 124
Bradshaw, A.M., 124
Bragg, W.H., 14
Bragg, W.L., 14, 268
Brainard, W.A., 16
Brandon, D.G., 14, 47
Brauer, W., 124
Braundmeier, A.J., 124
Brenner, S.S., 46, 47, 48
Brett, C.L., 390
Briner, J.S., 17
Broden, G., 464
Brongersma, H.H., 208
Brown, N.M., 16
Brucker, C.F., 16, 431, 464, 465
Brundle, C.R., 464, 465
Buchanan, D.N.E., 96
Buchholz, J.C., 208, 269, 297, 344
Buck, T.M., 208
Buckley, D.H., 16, 152
Buerger, M.J., 268
Bunshah, R.F., 152
Burkstrand, J.M., 269, 298
Burton, J.W., 95
Butcher, P., 465
Butler, K.D., 96
Buttner, P., 298

C

Caal, J., 97
Camp, W.J., 123, 125
Capart, G., 344
Cares, W.R., 95
Carlson, T.A., 476
Carver, J.C., 429, 430, 476
Cashion, J.K., 16, 463, 464
Castaing, R., 15, 132, 365, 373
Caudano, R., 429
Chambers, R.S., 47
Chandra, S., 96
Chandrasekhar, B.S., 494
Chang, C.C., 14, 125, 152
Channing, D.A., 63
Chavka, N.C., 152
Che, M., 430
Chen, M.H., 152
Chen, T.T., 494
Cheng, K.L., 429
Chynoweth, A.G., 494
Citrin, P.H., 96, 476
Clabes, J., 209
Clark, D.T., 16

Clarke, T.A., 268, 344
Clausen, C.A., III, 14, 65, 96
Coad, J.P., 465
Coburn, J.W., 16, 152, 365
Cochran, W., 15
Codling, K., 123
Cohen, R.L., 63
Coleman, R.V., 494
Collins, R.L., 95
Compton, W.D., 494
Constabaris, G., 95, 96
Cooley, J.W., 269
Cowley, J.M., 208
Craig, N.L., 429, 430
Craseman, B., 152
Creutz, C., 96
Curtis, M.T., 476
Czanderna, A.W., 365
Czyzewski, J.J., 208

D

Dalins, I., 152
Dalla-Betta, R.A., 96
Dambe, K., 344
Daniels, R.D., 179
Dash, J.G., 208
Davis, L.E., 152
Davisson, C.J., 15
Defourneau, D., 494
Dekker, A.J., 124
Delgass, W.N., 96, 97
Delly, J.G., 132, 373
Demuth, J.E., 209, 269, 296, 297, 298, 344, 464
Den Boer, M., 105
Di Salvo, F.I., 465
Doniach, S., 463
Dooley, G.J., 152, 207
Dorsey, G.A., 494
Dove, D.B., 152
Dranglis, E., 464
Dufaux, M., 430
Duke, C.B., 14, 124, 207, 269, 344, 462
Duncumb, P., 132
Dwyer, D.W., 465

E

Eastman, D.E., 16, 463, 464, 465
Echenique, P., 297
Ehrlich, G., 47
Einstein, T.E., 465
Ekelund, S., 208
Ellis, W.P., 209
Endriz, J.G., 463
Ertl, G.L., 210, 464, 466
Estrup, P.J., 15, 187, 207, 208, 209, 465
Evans, C.A., Jr., 14, 15, 152, 365
Ewing, G.W., 14

Author Index

F

Fabian, D.J., 476
Fadley, C.S., 96
Fahlman, A., 15, 16
Fain, S.C., 209
Farmer, J.B., 14
Farnsworth, H.E., 15, 124
Farrell, H.H., 207
Fedak, D.G., 209
Feibelman, P.J., 463, 464
Feldman, W.L., 493
Felter, T.E., 208, 209
Ferrell, R.A., 124
Ferroni, E., 210
Feuerbacher, B., 17, 463, 465
Fitton, B., 465
Fogel, M. YA, 366
Forbes, W.C., 429
Forstmann, F., 298
Freeouf, J.L., 464, 465
Friedrich, W., 14
Froitzheim, H., 208
Fuggle, J.C., 476
Fukuda, Y., 105

G

Gadzuk, J.W., 125, 208, 464, 465
Gager, H.M., 95, 96
Gallon, T.E., 152
Gandhi, H.S., 97
Garten, R.L., 97
Geiger, A.L., 494
Gen, M.Y., 96
Gerlach, R.L., 124
Germer, L.H., 15, 209, 465
Gersten, J.I., 208
Gettings, M., 465
Getz, R.D., 464
Giaever, I., 494
Gjostein, N.A., 152, 209
Gland, J.L., 210
Godwin, R.P., 95
Goff, R.F., 390
Goldanskii, V.I., 63, 96
Goldstein, J.I., 132, 373
Gomer, R., 46
Good, M.L., 14, 65, 96, 103
Good, R.J., 390
Goodman, S.R., 47
Graham, M.J., 63
Graham, T.P., 494
Grant, J.T., 15, 133, 152, 208
Gray, R.C., 429
Greenaway, F., 14
Grieser, R.H., 16
Grimley, T.B., 464, 466
Gronlund, F., 209

Guetin, P., 494
Gufstafsson, T., 464, 465
Guggenheim, H.J., 96
Guinier, A., 209, 268

H

Haas, T.W., 152, 207
Hachenberg, O., 124
Haensel, R., 463
Hagstrum, H.D., 16, 209, 269
Haines, R.S., 467
Hains, R.O., 97
Hakraken, L., 365
Hall, G.K., 465
Hall, T.M., 46
Haller, G.L., 494
Hamilton, J.C., 268
Hammarquist, H., 124
Hamrin, K., 15
Handy, R.M., 494
Haneman, D., 152
Hannan, R.B., 494
Hannay, N.B., 463
Hansen, R.S., 465
Hansma, P.K., 494
Haque, C.A., 152
Hari, H., 97
Harker, D., 268
Harrick, N.J., 16
Harris, L.A., 14, 125, 151
Harrison, D.G., Jr., 365
Harrower, G.A., 151
Hart, T.R., 16
Hartman, C.D., 15
Hauptman, H. 268
Haworth, L.J., 151
Hayward, D.O., 209
Heckingbotton, R., 14
Heden, P.F., 429
Hedman, J., 15, 429
Heiland, W., 208
Heilig, J.A., 152
Henderson, J.E., 123
Hendra, P.J., 16
Hennequin, J.F., 152, 365
Henrich, V.E., 124
Henzler, M., 209
Herber, R.H., 63
Hercules, D.M., 16, 399, 429, 430
Hercules, S.H., 16, 429
Herzog, R.F.K., 15
Hieland, W., 297
Hightower, J.W., 95
Hinthorne, J.R., 365
Hoar, T.P., 179
Hobert, H., 95
Hobson, J.P., 14

Author Index

Hobson, M.C., 95, 96
Hochman, R.F., 47
Hojlund Nielsen, P.E., 208
Holcombe, H.T. 365
Holland, B.W., 152, 344
Holloway, P.H., 152, 210
Holsboer, F., 96
Hondros, E.D., 152
Honig, R.E., 365
Hooker, M.P., 152, 207
Hooper, A.J., 476
Horgan, A.M., 152, 465
Horiguchi, T., 465
Horn, J.L., 494
Houston, J.E., 16, 124, 125, 152, 208
Howorth, L.J., 124
Hren, J.J., 14
Hrynkiewic, A.Z., 95
Huang, C.H., 209
Hubbard, A.T., 210
Huber, W.K., 15
Hudis, J., 429
Hudson, J.B., 210
Hufner, S., 464
Hughes, T.R., 96
Huheey, J.E., 96
Hutchins, B.N., 297
Hutchins, G.A., 15, 132

I

Ibach, H., 124, 208
Iberl, F., 208
Ignatiev, A., 209, 297, 298
Inghram, M.G., 46
Isett, L.C., 152
Iwata, M., 494

J

Jackson, A.G., 152, 207
Jackson, J.E., 494
Jacobi, K., 465
Jacobs, I.S., 96
Jaffee, R.I., 464
Jaklevic, R.C., 493, 494
Janow, R., 208
Jenkins, L.H., 208
Jennings, P.J., 344
Jepsen, D.W., 207, 209, 268, 296, 297, 298, 344
Johnasson, G., 15
Jolly, W.L., 96
Jona, F., 207, 209, 268, 269, 296, 297, 298, 344
Jones, R.O., 297
Jordanov, A., 63, 95
Jorgensen, W.L., 464
Joshi, A., 152
Jostell, V., 209

K

Kaiser, J.F., 269
Kambe, K., 344
Kaminska, T.J., 208
Kane, P.F., 14, 15, 16, 125, 132, 152, 365
Kanski, J., 124
Kaplan, R., 345
Kapur, S., 48
Karle, J., 268
Karlsson, S.E., 15
Kaster, J.S., 268
Katon, J.E., 494
Kay, E., 16, 152, 365
Keil, R.G., 494
Kellerman, E., 63
Kennett, H.M., 16
Kesmodel, L.L., 297, 298, 344
Kihara-Morishita, D.A., 494
Kim, K.S., 96, 97
King, D.A., 209, 465
Kirchhoff, G., 14
Kirschner, J., 152
Kjems, J.K., 208
Klasson, M., 429
Kleiman, G.G., 269, 298
Klein, J., 494
Klimisch, R.L., 97
Klug, H.P., 96
Knapp, J.A., 465
Knipping, P., 14
Koch, E.E., 463
Koch, J., 210
Koliwad, K.M., 152
Korduk, S.L., 95
Koutsoukos, E.P., 179
Kowalczyk, S.P., 152
Kowe, E.M., 463
Krakowski, R.A., 63, 95
Kreuzer, H.J., 494
Krishnaswamy, S.V., 14, 21, 46, 47, 48
Kundig, W., 95, 96
Kunz, C., 463
Kuo, F.F., 269
Kushner, R.A., 152
Kuyatt, C.E., 464

L

Laegreid, N., 152
Lagally, M.G., 207, 208, 209, 268, 269, 997, 344
Laing, K.R., 97
Lambe, J., 493, 494
Lambrecht, V.G., Jr., 96
Landa, S., 430
Lander, J.J., 14, 124, 151, 208
Landman, U., 15, 211, 268, 297, 344
Langreth, D.C., 125
Lanyon, M.A.H., 465
Lapeyre, G.J., 465

Laramore, G.E., 123, 124, 125, 269, 297, 344
Larrabee, G.B., 14, 15, 16, 125, 132, 152, 365
Larsson, R., 96
Lee, A.E., 16
Lee, L.H., 1, 16
Lefelhocz, J.F., 95
Leger, A., 494
Legg, K.O., 297, 298
Leidheiser, H., Jr., 14, 49, 63
Lenwald, S., 208
Leroy, V., 365
Leubner, R., 97
Lewis, B.F., 494
Lewis, P.A.W., 269
Ley, L., 152
Leyden, D.E., 430
Leygraf, C., 208
Liebl, H.J., 15, 365
Liebsch, A., 465
Lindau, I., 463
Lindberg, B., 15
Lindgren, I., 15
Lingquist, R.H., 95, 96
Lippincott, E.R., 96
Lisichenko, V.I., 95
Little, L.H., 124
Lloyd, D.R., 465
Locke, C.E., 155
Low, J.R., 47
Lumsden, J.B., 179
Lunsford, J., 97
Luu, T., 494

M

MacDonald, N.C., 16, 152
Mackie, W., 124
Mackintosh, W.D., 16
Madey, T.E., 208, 464, 465
Maglietta, M., 210
Mahan, G.D., 463
Marceau, J.A., 390
Marcus, H.L., 152
Marcus, P.M., 15, 208, 209, 268, 271, 296, 297, 298, 344
Marcus, S.M., 494
Mark, H.F., 16
Markarov, E.F., 96
Marton, L., 132
Mason, R., 268, 344
Masters, L.W., 429
Masud, N., 208
Matsuoka, S., 390
Matthew, J.A.D., 152
May, J.W., 208
Mc Kinney, J.T., 46
Mc Lane, S.B., 46
McCall, J.L., 373

McCaughan, D.V., 152
McCrone, W.C., 132, 373
McDavid, J.M., 209
McDevitt, N.T., 390
McDonald, L., 152
McDonnell, L., 268
McFeely, F.R., 152
McGuire, E.J., 124, 152
McHugh, J.A., 365
McKinney, J.T., 46, 47, 48
McLane, S.B., 14, 46, 47, 48
McMenamin, J.C., 464
McMillan, W.L., 493
McRae, E.G., 207, 296, 344
Mee, C.H.B., 465
Menzel, D., 208
Meyer, F., 152
Miles, J.L., 494
Miller, J.W., 208
Miller, R.B., 63, 95
Milliken, R.S., 464
Minkova, A., 95
Mitchell, K.A.R., 268
Mohan, G.D., 125
Momorjai, G.A., 298
Monch, W., 209
Moon, A.R., 208
Moore, A.J.W., 46
Moore, W.L., 365
Morabito, J.M., 152
Morgan, A.E., 209
Moriyama, U., 494
Morrison, G.H., 351, 365
Moseley, H.G.J., 14
Mosesman, M., 494
Mössbauer, R.L., 14
Mrha, J., 96
Mrstik, B.J., 345
Mueller, W., 373
Müller, E.W., 14, 21, 46, 47, 48, 208
Murata, Y., 124
Murphy, V.T., 152
Musket, R.G., 152

N

Naccache, C., 430
Nakanishi, S., 465
Natta, M., 125
Ng, K.T., 430
Ngoc, T.C., 268
Nilsson, P.O., 124, 465
Ninkova, A., 63
Nixon, W.C., 15
Nobe, K., 179
Noggle, T.S., 208
Noonan, J.R., 208
Nordberg, R., 15
Nordling, C., 15, 16, 429

Novaco, A.D., 208
Nyberg, C., 124

O

Oatley, C.W., 15
Odintsov, D.D., 365
Ohtani, S., 124
Okamoto, G., 179
Olson, D.H., 97
Onoda, G.Y., 152
Orlov, O.L., 95
Ozaki, A., 97

P

Packer, A., 476
Pallack, H., 63
Palmberg, P.W., 125, 151, 152, 209, 429, 464
Panitz, J.A., 14, 46, 47, 48
Pantano, C.G., 152
Park, R.L., 15, 16, 105, 123, 124, 125, 208
Pascoe, R.F., 179
Passell, L., 208
Patil, H.R., 209
Patterson, A.L., 268
Patterson, M.S., 269
Patterson, T.A., 430
Pease, R.F.W., 15
Peavey, J.H., 155
Pendry, J.B., 207, 208, 268, 296, 297, 298, 344
Penn, D., 124
Pepper, S.V., 152
Peri, J.B., 494
Peria, W.T., 151
Perlman, M.L., 429, 463
Petrakis, L., 430
Phillips, J.C., 494
Piper, T.C., 124
Piron, D.L., 179
Plummer, E.W., 463, 464, 465
Pocker, D.J., 152
Pollack, R.A., 152
Poschenrieder, W.P., 15, 47
Powell, C.J., 124, 152, 269
Prados, R.A., 96
Pratesi, F., 210
Prather, J.W., 429
Press, R., 15
Propst, F.M., 124

Q

Quinn, C.M., 465
Quinn, J.J., 124
Quinn, T.F.J., 16
Quinto, D.T., 152

R

Ralph, B., 47
Ranganathan, S., 14
Rao, B., 179
Rawlingson, W.F., 15
Reddy, A.K.N., 179
Regan, B.G., 46
Reinecke, T.L., 345
Reinsaln, V.P., 97
Revie, R.W., 179
Rhodin, T.N., 15, 16, 208, 268, 296, 297, 344, 429, 431, 464, 465, 466
Riach, G.E., 152
Rich, M., 95
Richardson, N.V., 465
Richardson, O.W., 124
Rincon, O., 155, 179
Ritchie, R.H., 124
Riviere, J.C., 152, 465
Robbins, M., 96
Robertson, W.D., 152
Robinson, H., 15, 125
Roenker, K.P., 477
Röntgen, W.C., 14
Rosseland, S., 151
Rovida, G., 210
Rowe, J.E., 464, 465
Rowell, J.M., 493, 494
Ruttenberg, F.E., 152

S

Sakurai, T., 46
Salem, L., 464
Samsonov, G.V., 476
Sangster, M.J.L., 494
Sato, N., 179
Scalapino, D.J., 494
Scardino, W., 390
Schaich, W.L., 463, 464
Scheibner, E.J., 151
Schein, L.B., 494
Schmidt, L.D., 209, 464, 465
Schmidt, W.R., 48
Schow, O.E., III, 208
Schreder, G., 494
Schreiner, D.G., 16
Schrieffer, J.R., 465
Schweitz, G.K., 476
Schwoebel, R.L., 209
Seah, M.P., 152
Seidman, D.S., 46
Selhoffer, A., 15
Senoff, C.V., 97
Seo, M., 179
Servais, J.P., 365

Author Index

Seshadri, K.S., 430
Sevier, K.D., 124
Shaw, R.W., 476
Shelef, M., 97
Shelton, J.C., 209
Sherry, H.S., 97
Shih, H.D., 209, 296, 297, 298
Shimizu, H., 152
Shirley, D.A., 152, 463, 476
Shklyarevskii, O.I., 494
Shreir, L.L., 179
Sickafus, E.N., 208
Siegbahn, K., 15, 16, 152
Simmons, G.W., 14, 49, 63, 465
Simonsen, M.G., 494
Singh-Bopoari, S.P., 209
Sinha, D., 494
Skarlatos, Y., 494
Sloane, R.H., 15
Slodzian, G., 15, 365
Smith, D.A., 17
Smith, D.P., 16
Smith, J.V., 97
Smith, N.V., 463, 465
Smith, P.H., 494
Smoilovskii, A.N., 95
Snow, E.C., 269, 344
Solomon, J.S., 390
Somorjai, G.A., 17, 152, 207, 208, 209, 210, 297
Southon, M.J., 46, 47
Spicer, W.E., 463, 464
Spijderman, J.J., 63
Spink, J.A., 46
Springer, R.W., 152
Staehle, R.W., 179
Staib, P., 152
Stair, P.C., 208, 298
Stein, R.J., 465
Steinhardt, R.G. 429
Stern, E.A., 124
Stevens, J.G., 96
Stevens, V.E., 96
Stewart, I.M., 127, 367
Stoner, N., 15, 297, 299
Stratton, P.M., 16
Strauss, J., 494
Strayer, R.W., 124
Strozier, J.A., 207, 297
Strozier, J.S., Jr., 297
Sunderland, R.J., 152
Sunjic, M., 125
Sutton, L., 464
Suzdalev, I.P., 96
Svartholm, N., 15
Swanson, K.R., 63
Swanson, L.W., 124
Swartz, W.E., 96
Szalkowski, F.J., 152

T

Taglauer, E., 208, 297
Tait, R.H., 344
Takamura, T., 494
Takasu, Y., 152
Tanaka, K., 494
Tarng, M.L., 152
Taub, H., 208
Taylor, K.C., 97
Taylor, N.J., 125, 152
Terada, K., 124
Terrel, J.H., 63
Tescari, M., 268, 344
Thapliyal, H.V., 297
Tharp, L.N., 151
Thomas, J.M., 96
Thomas, R.E., 124
Thomas, T.D., 476
Thompson, J.J., 14
Thompson, W.A., 494
Thomson, G.P., 15
Thornton, P.R., 16
Tibbetts, G.G., 298
Toenniew, J.P., 268
Tong, S.Y., 15, 208, 268, 296, 297, 298, 299, 344, 345
Tosatti, E., 209
Tousimis, A.J., 132
Toya, T., 465
Tracy, J.C., 151, 152, 209, 298, 464
Tradnell, B.M.W., 465
Traum, M.M., 465
Trout, B.L., 179
Tsang, T., 152
Tsong, T.T., 46, 47, 48
Tucker, C.W., 268, 269, 344
Tucker, P.M., 476
Turner, D.W., 464
Turner, P.J., 46, 47
Tzoar, N., 208

U

Uemura, Y., 494
Uhlig, H.H., 179
Unertl, W.N., 297

V

Vadimsky, R.G., 390
Vainshtein, B.K., 16
Van Cittert, P.H., 476
Van Hove, M.A., 15, 297, 298, 299, 344, 345
Van Wierigen, J.S., 95
Vanderhoff, J.W., 16
Vanselow, A., 48
Vedder, W., 494
Verbist, J., 429
Verkin, B.I., 494

Vermilyea, D.A., 494
Viaris, De Lesegno, P., 152
Viehobock, F.P., 15
Vijh, A., 179
Von Koch, C.V., 124
Von Laue, M., 14
Vorburger, T.V., 464, 465
Vrakking, J.J., 152

W

Waber, J.T., 344
Waclawski, B.J., 464, 465
Wagner, A., 46
Wakoh, S., 269
Waldrop, J.R., 16, 152
Walko, R.J., 46
Wallden, L., 465
Walmsely, D.G., 16
Wang, G.C., 208, 297
Wang, T.T., 390
Warren, B.E., 209
Watson, L.M., 476
Watson, R.E., 463
Wattamaniuk, W.J., 494
Watts, J.C., 96
Watts, P.H., 96
Waugh, A.R., 46, 47
Webb, M.B., 207, 268, 269, 297
Weber, R.E., 151, 152
Wehner, G.K., 152
Wei, S.H.Y., 429
Weinberg, W.H., 494

Weisser, O., 430
Welch, P.D., 269
Werner, H.W., 15, 365
Wertheim, G.K., 17, 63, 96, 464, 476
Wickman, H.H., 63
Williams, M.W., 124
Williams, P.M., 465
Winick, H., 463
Winograd, N., 96, 97
Wolff, P.A., 124
Wood, A., 209
Wood, J., 465
Woodruff, D.P., 152, 268
Worsburger, T.V., 464
Wyatt, D.M., 429

Y

Yakowitz, H., 132, 373
Yang, M.G., 152
Yanson, I.K., 494
Yates, J.T., Jr., 208, 464
Yellin, E., 152
Yelon, A., 494
Yin, L.I., 152
Yonehara, K., 465
Yu, K.Y., 464

Z

Zanazzi, E., 297, 298
Zehner, D.M., 208
Zeller, H.R., 494
Zimmer, R.S., 344

Subject Index

A

Absorption
  coefficient of electron, 134
  electron, 310
Accelerating voltage, 129, 190
Acetylene, 439
  on nickel, 291
Activated charcoal, 86
Adatom, 191, 457
  interaction, 205
  surface force, 205
Adhesion, 147
  of polymer, 399
Adhesive
  bonding, 375
  for glass fiber, 381
  joint, 376
Adipic acid, 474
Adsorbate, 203, 222, 304, 347
  -absorbent, 437
  coverage, 227
  localized level, 458
  orbital, 439
Adsorbed gas, 113
Adsorbed layer, 203
Adsorption
  heat of, 441
  isostere, 204
  kinetics, 454
  rate, 203
Alkali metal
  promoted ruthenium, 86
Alloy, 201
Aluminum, 105
  A1 (001), 274
  adherend, 376
  alloy, 137, 378, 379
  filter, 113
  surface, 214, 415
  with polymer, 413
Aluminum hydroxide, 486
Aluminum oxide, 380, 477
Aluminum oxyfluoride, 378
Ammonia
  on ruthenium, 81

Ammonium adipate, 488
Angle
  of emission, 141
  of reflection, 191
Angular momentum, 334
  quantum number, 307
Anodic film, 61
Anodic polarization, 156, 159
Anodization, 488
  salt mixture, 378
Appearance potential spectroscopy, 113
Application of ion microscopy, 361
Argon, 431
  bombardment, 358
  ion, 155, 411
  ion gun, 418
Atomic density, 359
Atomic monolayer, 372
Atomic number, 135
Atomic resonance, 310
Atom-probe, 21
  time-of-flight, 32
Attractive force, 410
Auger
  analysis, 182
  electron, 53, 133, 192
  emission, 135, 401
  energy, 141
  peak overlap, 141
  process, 109, 135
  spectroscopy, 3, 11, 120, 133, 155, 188, 402
  spectrum, 134, 137, 159
  transition, 119
Auto exhaust, 90
  control catalyst, 73, 87
Auto-ionization, 135
Azimuthal angle, 191

B

Back-bonding, 442
Background
  current, 143
  subtraction, 143, 144
Band-gap, 310
Barium ruthenate, 90

Subject Index

Barrier
  height, 480
  potential, 306
Beam
  double cantilever, 383
  intensity, 324
  reflectivity, 333
Beeby T-Matrix, 313
Benzene, 437
Berylium hydride, 41
Binding energy
  of core electron, 123
  of electron, 72, 135, 415, 443, 471
  of ruthenium-oxygen system, 85
Bloch-wave, 272, 313, 320
Block-diagonalization, 341, 342
Blyholder model, 448
Boltzmann constant, 203, 308
Bond failure, 400
Bonding symmetry, 458
Born approximation, 213
Box-car window, 238
Bragg
  diffraction, 312
  Law, 129
  peak, 304
  reflection, 309
  spectrum, 272
Bravais lattice, 302
Bremsstrahung emission, 109
Bulk photoemission, 436
Butene dehydrogenation, 71

C

Calamel electrode, 176
Calcium fluoride, 418
Calorimetry differential scanning, 381
Carbon, 403
  Auger signal, 139
  -carbon bond, 442
  with Ni (111), 203
Carbon monoxide, 21
  chemisorption, 439
  on clean metal, 141
  on ruthenium, 81
Carburization, 457
Cascade process, 119
Catalyst, 147, 433
  nickel tungstate, 426
Cathode ray tube, 368
Characteristic loss of energy, 114
Charge transfer, 431
Chemical composition of nickel, 158
Chemical shift, 72, 402
Chemisorption
  of oxygen, 156
  of small atoms, 336

  on metal, 431
Chemisorptive bonding, 441
Chloride
  -free solution, 161
  ion, 155
Chlorine in hydrocarbon, 130
Chromate stripping solution, 379
Chromium, 157
  content, 386
  dioxide, 467
  oxide, 155
Clean crystal, 309
Clean metal, 106, 283
Clean surface, 201, 312
Cloud chamber, 133
Cobalt, 60, 424
  Co-doped specimen, 57
  CO on, 141
  on Ni (100), 205
Cobalt molybdate, 399, 423
Cobalt oxide, 61
Coherence zone, 199
Cohesive failure, 375, 380
Coincidence lattice, 199
Collector electrode, 368
Combined-space method, 336
Computation method of LEED, 299
Conductance, 477
Conduction band, 113
Constant momentum transfer averaging, 341
Contact potential, 106
Contaminant, 127, 478
Contamination
  layer, 407
  surface, 368
Convergence factor, 333, 469
Convolution theorem, 228
Copolymer adsorption, 475
Copper, 379
  hydride, 41
  sulfate, 214
Core electron, 110
Core hole, 113
Core level
  of solid, 110
  spectroscopy, 123
Corrosion, 49, 147, 362, 433
  film, 56
  inhibitor, 382
Coster-Kronig transition, 135
Covalent radii, 348
Crack propagation, 386
Cross-section, 192
Cross-transition, 135
Crystal
  lattice, 310
  orientation, 141
  potential, 306

Subject Index 511

Crystallite size, 74
Cyclohexane adsorption, 474
Cylindrical mirror analyzer, 134

D

de Broglie
 hypothesis, 6
 relationship, 190
Debye temperature, 202
Debye-Walker effect, 199, 306, 308
Decay time, 114
Deconvolution, 422
 Fourier transform, 467
Densitometer, 357
Density of state, 110, 181, 188
Dental enamel fluoridization, 399
Depth profile, 133, 134, 379, 407, 411
Depth resolution, 367
Desorption, 203, 457
 of electron beam, 347
 rate, 203
Deuterium, 488
 oxide, 488
Diffracted beam intensity, 191
 of crystal, 129
 specular, 261
Diffracted maxima, 311
Diffracted pattern, 190, 304
Dipole moment, 480
Dirac delta-function, 228
Dislocation, 22
Disordered solid, 201
Distribution
 elemental, 379
Donor-acceptor model, 440
Doped junction, 489
D-orbital tungsten, 460
Duoplasmatron gun, 354, 368
Dynamical LEED, 333

E

Elastic current, 477
Elastic electron, 306
Elastic reflectivity, 319, 325
Elastic scattering, 274
Electric field gradient, 67
Electrochemical treatment, 159
Electrode potential, 155, 165
Electromagnetic radiation, 352
Electron
 amplitude, 335
 backscattered, 114, 141
 beam, 128, 303
 bombardment, 127
 -electron interaction, 436
 energy analyzer, 403, 404
 excitation, 431
 flux, 332
 gas, 275
 inelastic scattering, 273
 ionization, 401
 inner-shell, 110
 low-energy, 105
 K-conversion, 56
 L-conversion, 56
 momentum, 305
 phonon interaction, 306
 straggling, 123
 tunneling, 477
 wave, 107
 wave function, 339
Electron density
 incident, 226
 self-convolution, 215
Electron diffraction
 RHEED, 189
Electron gun, 106, 159
 grazing incidence, 134
Electron microprobe, 7, 12, 127
 analyzer, 129
 scanning, 128
Electron multiplier, 159
 detector, 23
Electron spectrometer, 74
 du Pont, 650B, 403
Electron spectroscopy for chemical analysis, 8, 12, 65, 71, 399
 of ruthenium, 71
Electronic band structure, 310
Electronic device, 433
Electrostatic
 defeflection analyzer, 29
 lens, 354
 mirror, 354
Element trace amount, 379
Elemental
 abundance, 181
 distribution, 134
 sensitivity factor, 121
Ellipsometry, 156
Embrittlement of steel, 147
Emission
 d-band, 434
 Mössbauer spectroscopy, 57
 spectra, 114, 119
Energy
 cylindrical mirror, analyzer, 460
 distribution of electron, 119, 137, 434
Erosion, 360
Error indicator, 250
Escape depth, 409, 436
 elastic, 109
 of atom, 359
 of electron, 56, 71, 141
Ethylene, 439

Subject Index

Evaporation pulse, 23
Evaporation voltage, 31
Ewald sphere construction, 191
Excitation
    of phonon, 478
    state spectrum, 106

F

Failure
    cohesive, 416
    locus of, 376
    surface, 375
Faraday cup, 303
Fermi energy, 107, 110, 443
Fermi level, 116
Ferric sulfate, 471
Ferrite catalyst, 71
Ferromagnet, 68
Ferromagnetic material, 201
Ferrous sulfate, 471
Field adsorption
    of helium, 37
    of neon, 37
Field desorption, 41
    microscope, 25, 40
    of hydrogen, 40
Field emission
    microscopy, 2
    retarding potential, 106
Field ion
    microscope, 21
    microscopy, 2, 189
Flame spraying of coating, 129
Fluorescent
    screen, 303, 356
    signal, 127
    yield, 136
Fluorine, 378
    in hydrocarbon, 130
Fluoroapatite, 418
Forward scattering, 331
Fourier transformation, 215
    deconvolution, 341
    of diffraction pattern, 304
    scattering matrix, 316
Frank-Hertz experiment, 110
Frequency factor, 203
Friction and wear, 147
Friedel's law, 219

G

Gamma radiation, 50
Gamma-ray resonant, 55
Gas adsorption, 147
Gaseous adsorbate, 28
Germanium lithium detector, 73
Gibbs oscillation, 226, 235
Gold, 129
    epitaxial film, 107
    reconstruction of, 201
Grain boundary, 302
Graphite, 189
    noble gas on, 205
    particle, 469
Graphite-like overlayer, 141
Green's function, 314

H

Hartree-Fock ground state, 439
Heat of adsorption, 204
Helium, 135, 431
    ion image, 34
    gas image, 27
Hematite, 53
Heterocycle, 421
Hexafluoropropene and vinylidene fluoride copolymer, 475
High energy electron diffraction, 6
Hydrocarbon antibonding orbital, 442
Hydrocracking process, 421
Hydrodesulfurization catalyst, 399
Hydrogen, 135
    deaeration, 159
    embrittlement, 50
    on W (100), 205
Hydrogenation process, 421
Hydrogen disulfide
    on rhodium, 42
    on ruthenium, 81
Hyperfine interaction, 51

I

Image
    analysis, 353
    illumination, 357
    voltage, 23
Incident beam, 106
Incident electron current, 113
Incident flux, 320
Incontel X-750, 155
Inelastic damping, 314
Inelastic electron
    scattering, 105, 109
    tunneling, 497
Inert gas
    implantation, 84
    sputtering, 136
Inner potential, 226, 312
Inner shell, 114
    vacancy, 135
Insulator, 491
    -semiconductor junction, 493
Intensity
    profile, 129, 191
    voltage curve, 304
Interaction energy, 135

Subject Index 513

Interatomic distance, 129
Interband transition, 115
Interface
 oxide-dopant, 478
 solid-vacuum, 105, 324
Interfacial failure, 380
Interference, 313
Interlayer spacing, 192, 201, 235, 283, 329
Ion
 beam, monokinetic, 354
 bombardment, 352
 converter, 368
 implantation, 362
 mobilization, 141
Ion-core
 potential, 305, 314
 scattering, 275
 thermal motion, 308
Ion microprobe, 8, 12, 353, 367
 mass analyzer, 367
Ion microscope, 8, 353, 354, 362, 368
 geometric specifications, 356
Ion microscopy, 351
Ion scattering
 equipment, 377
 spectrometry, 375
Ion sputtering, 354
Ionic character, 285
Ionic conductivity, 178
Ionic diffusion, 53
Ionization
 efficiency, 357
 energy, 434
Iridium, 439
 crystal, 107
 helide, 38
 neide, 38
 tip, 27
Iron, 157, 439
 nuclear level, 50
 single crystal, 454
Iron oxide, 57, 457, 467
Isomer shift, 67
Isotopic
 abundance, 68
 ratio, 351
 replacement, 486
Iterative and perturbation, 327

J

Jellium, 275

K

Kaiser window, 235
Kinematic diffraction theory, 202
Kinetic energy of electron, 190, 409, 435
KLL transition, 138
KVV transition, 138

L

Lacquer coating, 379
Lap shear, 381
Lateral distribution of element, 134
Lateral periodicity, 201
Lattice
 effect, 67, 358
 parameter, 191
 phonon, 114
 sublimation, 318
 summation, 318
 vector, 215, 318
Layer
 doubling method, 325, 326
 matrice of Pendry, 321
 reflection, 322
 -stacking process, 326
Lead junction, 478
LEED
 analysis of, 211
 intensity measurement, 271
 intensity spectrum, 299
 measurement, 347
 of surface layer, 187
 pattern, 455
 spectra, 304
Legendre polynomial, 236
Light microscope, 369
Linear absorption coefficient, 56
Lithium, 378
 silicon diode, 129
Localized analysis, 357
Loss spectrum, 114
Low energy electron diffraction, 6, 13,
Luggen capillary probe, 161

M

Madelung potential energy, 72
Magnesium, 379
 Auger feature, 138
Magnetic
 hyperfine interaction, 49
 moment, 473
 pigment, 467
Magnetite, 53
Mass
 analyzer, 376
 filter, 377
 interference, 351
 spectra, 351
 spectrometer, 2, 23, 354, 368
Matrix
 eigen value, 322
 inversion, 238
McPherson algorithm, 497
Metal
 coating, 147
 electrode, 477

electron gas, 438
helide, 21, 39
hydride, 21
-metal bonding, 496
neide, 21, 40
passive-, 156
-substrate bonding, 496
superconducting, 481
surface bonding, 437
Metal oxide
metal junction, 478
of Mg, AL or Si, 141
Metallic impurities, 42
Miller indice, 219
Modulation
amplitude, 148
wave form, 140
Modulated technique, 143
tailored (TMT), 146
Molecular
adsorption, 451
orbital, 431
physisorption, 437
spectroscopy, 477
Molybdenum, 149, 202, 424
disulfide, 426, 437
isotope, 27
Momentum conservation, 303
Mössbauer
effect, 3
spectroscopy, 3, 49, 65
Muffin-tin
potential, 275, 307
zero, 307
Multiple
ionization, 135
phonon, 489
scattering, 211, 227, 307

N

Neon, 27, 431
Neutral atom, 352
Neutron diffraction, 189
Nickel, 155, 437, 439, 442
carbonyl, 206
chemisorption, 445
CO on, 151
LEED, 201
oxide, 156
sulfide, 426
tungstate, 399, 421, 423
Niobium, CO on, 141
Nitrogen, 203, 403
adsorption, 21
on tungsten, 206
Noble gas, 37
ion, 189
Nucleation center, 197

O

Olefin
chemisorbed, 439
Optical
path, 309
transform, 235
Orbital
ionization energy, 437, 438, 439
Order
-disorder transformation, 206
of iteration, 333
Organic coatings, 130
Organic surfur compound, 421
Oscilloscope, 134
Overlayer
compression, 347
effect, 312
reflection, 313
structure, 277, 285
transmission, 313
Oxidation, 53
of surface, 147
Oxide
growth, 457
inclusion, 358
metal interface, 386
Oxygen
adsorption, 198, 457
flooding, 358
on molybdenum, 193, 205

P

Palladium, 443
catalyst, 74
CO on, 151
hydride, 41
neide, 40
Particulate matter, 368
Passivation potential, 156
Passive
film by AES, 155
metal, 155
surface, 163
Passivity, 50, 155
Pathlength, 410
Patterson function, 215, 218
degradation of, 222
Peak
interference, 147
to peak height, 143
Peel test, 381
Penetration depth, 311
of electron, 127, 202, 310
Periodic
crystal, 320
potential, 320
table, 448
Perturbation
expansion technique, 328

method, 214
scheme, 331
theory, 311
Phase shift, 308
Phonon, 302
scattering, 308
Phosphorous, 418
Photoelectric
detector, 113
effect, 6, 133
Photoelectron, 133, 403, 496
spectroscopy, 8, 13, 399
Photoemission, 188, 431, 496
Photoexcitation, 133
Photoionization, 400, 401, 437
cross-section, 443, 444
Photon
analysis of, 105
emission, 401
energy, 437
excitation, 496
short wave length, 106
s-polarized, 446
Pi-d bonding, 439, 440
Planar matrix, 321
Plane wave, 307
expansion, 316
solution, 320
Plasmon
decay, 119
excitation, 109, 119
satellite, 123
threshold, 114
Plastinum, 129
CO on, 141
helide, 44
reconstruction of, 201
Polar angle, 191
Polarizability, 480
Polarization, 443
curve, 155
function of, 61
radiation, 446
Polybutadiene, 60
Polyethylene, 381
Polymer
film, 127
-metal interface, 413
technology, 1, 9
Polymeric silicon, 417
Polyphenylether, 490
Potential modulation, 107
Practical ion yield, 357
Primary
beam current, 141, 357
electron, 190
ion, 354
ion beam, 367
magnet, 368

Pyridine, 437

Q

Quadrupole
moment, 67
splitting, 49, 67
Quantitative analysis of AES, 141

R

Radial Schrödinger equation, 236
Radiation quantum, 133
Radiative emission, 113
Radio-radar telescope, 361
Rationless transition, 157
Reciprocal lattice vector, 303
Recoil
energy, 80
free fraction, 53, 69, 71
Reference spectra
Auger, 140
Reflection
coefficient, 309
Mössbauer spectroscopy, 55
Rehybridization, 440
Relative abundance, 352
Relaxation energy, 135, 495
Renormalized forward scattering, 343
Repelling electrode, 356
Retarding potential
curve, 107
energy analyzer, 145
threshold, 107
Reverse scattering perturbation, 343
Rhenium helide, 44
Rhodium
carbide, 42
nitride, 41
tip, 37
Ruthenium
catalyst, 65
oxide, 80
trichloride, 75

S

Saturation voltage, 451
Scaling constant, 250
Scanning
Auger spectroscopy, 139
electron microscopy, 7, 11, 413
ion microscopy, 375, 382
Scattered electron quasielastically, 114
Scattering
amplitude, 307
factor, 227, 236
matrix, 318
peak (elastic), 116
scheme, 312
Schrödinger equation, 320

Subject Index

Scintillator
 photomultiplier, 356
Secondary
 electron, 134, 192
 electron yield, 110
 emission coefficient, 106
 ion, 354, 367, 368
 ion current, 357, 358
 ion emission, 352
 ion mass spectrometry, 8, 12, 351, 375
Self-convolution
 of electron density, 110
Selvedge, 201
Semiconductor, 147, 201
 metal oxide, 362
Sensitivity
 factor, 181
 of ESCA, 407
 of IMMA, 367
Sigma-orbital, 441
Signal length
 AES, 139
Silica gel, 75
Silicon, 201
 Si (III), 107, 372, 403
 surface, 105
Silver, 143
Single cyystal, 187, 495
Single scattering, 211
Snow's potential, 244
Sodium, 362, 379
 azide, 405
 chloride, 155
Solid
 solution, 203
 state detector, 129
 surface, 105
Sorption, 203
Spatial resolution, 127, 367
Spectral envelop, 422
Spectrometer
 McPherson-36, 468
Spectrum
 subtraction technique, 149
Specular beam, 348
Speed
 of convergence, 327
Spherical
 expansion, 316
 potential, 246, 273
Spin polarization, 51
Sputtering
 gun, 192
 low density, 359
 rate, 357
 yield, 357
Stannous fluoride, 418, 419
Sticking
 coefficient, 455

 probability, 203, 455
Strontium chromate, 382
Structural
 imperfections, 199
 propagator, 318
 transformation, 201
Substrate
 -subtraction procedure, 261
Subsurface layer, 129
Sulfiding activity, 426
Sulfur, 427
 in hydrocarbon, 130
 in nickel, 277
Sulfuric acid, 155
Surface
 adsorbed, 302
 adsorption, 187, 201
 analysis, 36, 406
 carbide on Mo, 141
 characterization, 127
 chromatography, 431
 clean, 302
 composition, 121, 182
 conductivity, 410
 content, 177
 coverage, 203
 crystallography, 302, 458
 geometry, 188, 306
 oxidation, 424
 periodicity, 196, 199
 phase transition, 187, 201
 reaction, 187, 201
 reconstruction, 187, 201
 roughness, 140
 science, 1
 sulfidation, 424
Synchrotron, 431
 photoemission, 460
 radiation, 443, 445
 storage ring, 106

T

Tantalum
 hydride, 41
 tip, 37
Temperature
 on diffracted intensity, 202
Thermal
 desorption, 450
 treatment, 377
 vibration, 199
Thermionic
 emission, 129
 emitter, 106
Thioether, 421
Thiol, 421
Thiophene
 chemisorption, 426

Subject Index

Threshold
  energy
  potential, 113
Tin, 57
  bromide, 57
  oxide, 57
Total electron
  reflection, 312
Transform
  deconvolution method, 214
Transition
  metal, 135, 336
  non-radiative, 133
Translation vector, 191
Transmission
  coefficient, 309
  efficiency, 357
  matrice, 322
  Mössbauer spectroscopy, 52
Transpassive
  potential, 61
Trapezoidal rule, 237
True secondary peak, 114
Tungsten
  carbide, 42
  deuterium on, 449
  filament, 129
  helide, 38
  hydride, 41
  hydrogen on, 460, 449
  oxide, 41, 427
  platinum alloy, 43
  tip, 37
Tunnel junction
  preparation, 481
Tunneling spectrometer, 485

U

Ultrahigh vacuum, 136, 187
Ultraviolet
  photoemission
  spectroscopy, 8, 13, 433
  photon, 407

Umklapp process, 460
Unit cell, 221, 303
Unwanted effects
  of Auger electron, 141

V

Vacuum ultra-violet, 496
Valence
  band, 113
  effect, 67
  level excitation, 496
Van Cittert
  algorithm, 467
Vibrational amplitude, 188, 202
Vidicon tube, 193
Von Laue condition, 221

W

Water adsorption, 473
Wave
  field, 341
  form, 148
  length, 309
  vector, 311
Work function, 106, 188, 307, 435, 454

X

X-ray, 405
  absorption, 117
  crystallography, 211
  diffraction, 2, 312
  ejection, 128
  fluorescence, 127, 400
  Kγ, 53
  soft, 113, 117

Z

Zeeman splitting, 51
Zeolite, 73, 91
Zirconium, 378
  hydride, 41